Advances in Geographical and Environmental Sciences

Series Editor

R. B. Singh, University of Delhi, Delhi, India

Advances in Geographical and Environmental Sciences synthesizes series diagnostigation and prognostication of earth environment, incorporating challenging interactive areas within ecological envelope of geosphere, biosphere, hydrosphere, atmosphere and cryosphere. It deals with land use land cover change (LUCC), urbanization, energy flux, land-ocean fluxes, climate, food security, ecohydrology, biodiversity, natural hazards and disasters, human health and their mutual interaction and feedback mechanism in order to contribute towards sustainable future. The geosciences methods range from traditional field techniques and conventional data collection, use of remote sensing and geographical information system, computer aided technique to advance geostatistical and dynamic modeling.

The series integrate past, present and future of geospheric attributes incorporating biophysical and human dimensions in spatio-temporal perspectives. The geosciences, encompassing land-ocean-atmosphere interaction is considered as a vital component in the context of environmental issues, especially in observation and prediction of air and water pollution, global warming and urban heat islands. It is important to communicate the advances in geosciences to increase resilience of society through capacity building for mitigating the impact of natural hazards and disasters. Sustainability of human society depends strongly on the earth environment, and thus the development of geosciences is critical for a better understanding of our living environment, and its sustainable development.

Geoscience also has the responsibility to not confine itself to addressing current problems but it is also developing a framework to address future issues. In order to build a 'Future Earth Model' for understanding and predicting the functioning of the whole climatic system, collaboration of experts in the traditional earth disciplines as well as in ecology, information technology, instrumentation and complex system is essential, through initiatives from human geoscientists. Thus human geosceince is emerging as key policy science for contributing towards sustainability/survivality science together with future earth initiative.

Advances in Geographical and Environmental Sciences series publishes books that contain novel approaches in tackling issues of human geoscience in its broadest sense—books in the series should focus on true progress in a particular area or region. The series includes monographs and edited volumes without any limitations in the page numbers.

More information about this series at http://www.springer.com/series/13113

R. B. Singh · Dongying Wei · Subhash Anand
Editors

Global Geographical Heritage, Geoparks and Geotourism

Geoconservation and Development

Springer

Editors
R. B. Singh
Department of Geography
University of Delhi
Delhi, India

Dongying Wei
Beijing Normal University
Beijing, China

Subhash Anand
Department of Geography
University of Delhi
Delhi, India

ISSN 2198-3542 ISSN 2198-3550 (electronic)
Advances in Geographical and Environmental Sciences
ISBN 978-981-15-4955-7 ISBN 978-981-15-4956-4 (eBook)
https://doi.org/10.1007/978-981-15-4956-4

This Springer imprint is published by the registered company Springer Nature Singapore Pte Ltd.
The registered company address is: 152 Beach Road, #21-01/04 Gateway East, Singapore 189721, Singapore

Preface

Our pious planet Earth is blessed with distinguished geographical as well as geological diversity and heritage that helps us in understanding its evolutionary trend and history. This has led to the laying of much emphasis upon the preservation and conservation of the rich terrestrial heritage. Scientific and proper exploration and preservation of the rock and fossils is the key for understanding the evolutionary processes and the environments leading to the present situation of the planet (Rana 2020). According to Yang (2015) the concepts of national park, protected area system, world heritage and biosphere reserves have been constructed in the United States. Much of the research and training available on interpretation in parks comes from western tradition and literature directly or indirectly (Wilson 2013). Presently, at global level, the concept of national park is recognized from the IUCN (Peng 2018). Subsequently, it is anticipated that many more geoheritage sites will be identified and notified, thus affording these sites the necessary protection during the course of development. Human beings have achieved ecological dominace through series of long and unprecedented steps (Singh and Singh 2014). The natural calamities and anthropogenic activities have led to rapid destruction of various geosites. Consequently, new emerging concepts such as geosites, geodiversity, geoheritage, geomonuments, geoparks and geotourism, geoconservation have evolved and being patronized by the conservationists. Geoparks are well-defined areas that contain one or more geoheritage sites selected on the basis of scientific importance, rarity, scenic quality, or relation to geological history, events and processes. Geotourism is a new concept (Farsani et al. 2012). The different geoheritage sites encompassed within the various geoparks possess educational, scientific, aesthetic and cultural values.

According to UNESCO (2016), global geoparks are single, unified geographical areas where sites and landscapes of international geological significance are managed with a holistic concept of protection, educational, and sustainable development. In 2020, there are 161 UNESCO global geoparks in 44 countries in addition to a number of national-level geoparks. But the number of geoparks are increasing

(Kohmoto 2014). China leads the world in the number of geoparks (Li et al. 2018). But in South Asia there is no any single geopark conforming to the UNESCO definition has yet been created and developed, although different fragmented efforts are on to construct geoparks there.

Despite their unique geological and scientific importance, the distribution pattern of many significant geosites within the country remains poorly documented in the literature. Land use also put threat to the geographical diversity (Singh and Anand 2013). The holistic approach toward such sites will sensitize common people on the need for geoconservation of the various significant geosites and promoting those through geotourism. The geographical perspective on geoparks includes integrating landscape as its main concept requires a holistic approach and transdisciplinary research to promote multicultural participatory bridges (Wei 2013). The introduction of geoparks can be an initiative toward conservation of our prosperous geological and geographical heritage, thereby leading to economic rejuvenation and sustainable development of local communities. Geoparks use innovative strategies to promote conservation of natural heritage (Anand et al. 2014). Sustainable development is buzz word and now scientific communities are talking about the sustainability and for giving it to more emphasis, International Council of Science (ICSU) and International Social Science Council (ISSC) launched the Future Earth, international research platform in June 2012 for maintaining the sustainability (Himayama 2018).

The more spectacular is a site, greater is the potential for converting it into a geopark and renowned geotourism hub. The endeavor needed in this context is to influence the policy-makers for adopting a draft legislation aimed toward protection of the various national prospective geoheritage sites and also to encourage the various tourism departments to popularize these sites through well-designated tourism circuits. Geoheritage and geoparks should be incorporated within the fields of scientific studies and education books, research papers and review and popular articles describing the significance of geoconservation in the geoheritage sites, the need and significance, critical analysis of geoparks and the promotional aspects of geotourism.

Keeping in mind these concern and understanding of the urgency, it was planned out to bring out a springer volume on *Global Geographical Heritage, Geoparks and Geotourism: Geoconservation and Development under the Advance in Geographical and Environment Sciences*. Present edited book is a compilation of the valuable contributions made by eminent scientists, research scholars, and professionals who are trying to develop alternative strategies, solutions for the sustainability of geoheritage and geoparks and others through various empirical research and experiments ranging from local to global scale. This edition would be of immense use to the policy-makers, environmentalists, conservationist, academician, and research scholars who are directly or indirectly involved.

Elizbarashvili et al. in their chapter discussed the geographical diversity and perspectives of planning of geoparks highlighting the geoparks networks and tourist attraction of the Georgia. They emphasized that all forms of topographical features are present in the nation. To make the impression on the tourist, the country has 100 of monuments but it is disappointed that the nation is not yet been presented in the world geographical parks (geoparks) networks. In the next chapter, Sadry et al. identified and assessed geodiversities around Takht-e-Soleiman world heritage site to propose the territory as the third geopark in Iran, studied about the UNESCO listed world heritage site of northwest Iran, Takht-e-Soleiman. The place is rich from biodiversity, geodiversity, and historical and cultural point of view. According to the investigation of geoparks and geotrial map, the finding of the study is that the region has the potential for further community engagement and research in future. It is important that through private and government sector investment and sound infrastructure the region can be listed as a national and later as a global geopark in UNESCO. Next chapter has been contributed by Lugeri et al. entitled "Geoheritage: A Strategic Resource for the Society in the Anthropocene", tried to give answer to the meaning and importance of geoheritage in contemporary society. They felt the requirement of the unconventional and new system of commuting, sciences, particularly earth sciences for the involvement of society in successful interaction with the environment.

Santos et al. authored a chapter on UAV's Multimedia Technology and Augmented Reality (Geointegration): New concept and new paradigm of geodiversity presentation. The study acquired the image data with Unmanned Aerial Vehicles (VANTs) and Geographical Information System (GIS). The paradigm is that even when faced with several functionalities in some cases, this technology is not always adequate considering the available representation resources, some aspects of the structure to be imaged and the conditions of scale. Mansilla et al. in the chapter entiteld "Virtual Heritage: a model of participatory knowledge construction toward Biogeocultural Heritage Conservation", highlights that by the application of the digital imaging technology, geoheritage values and sites can be conserved. There are the association of the remarkable cultural and historical values with the ancestral communities. For the effective protection of the geoheritage site in Chile they have made the proposal. This target is being accomplished through the practices of Geovisualisation proposal of new language, fundamentally reflected as virtual heritage. Nhlabathi and Maharaj, have been described the communitarian ethics, environmental conservation and development, state that sustainable usage of both natural and man-made resources of heritage, predominantly, parks, monuments, and other natural resources are the significance for its conservation and tourism development. Both authors argued for acceptance of the communitarian ethics as a frame of reference in any development project. In the principle of common good ethics of communitarian is founded. The finding of the paper state that development would meet the sustainable development requirement only if was predicated, not on unilateral securities but on the interests of the world-wide community at large.

In next chapter entitled "Ecological and Socio-Economic Vulnerability to Climate Change in Some Selected Mouzas of Gosaba Block, The Sundarbans" written by Mukherjee and Siddique attempted to highlight the ecological and socio-economic vulnerability to climate change in some selected *mouzas* of Gosaba Block, Indian Sundarbans. By conducting intensive studies on plant morphology and stomata index, vulnerability in the ecology has been identified. Data collected from various sources such as socio-demographic profile and livelihood vulnerability index has been used to analyses the vulnerability of the regions. A study on evaluation of Zonguldak hard coal basin (NW Turkey) fossils as geosite done by Cakti et al. stated that it is important to guard the regions of geoheritage and geosite along informing the people about their significance. It has been established that building a museum comprising fossil assemblage, presenting, protecting and transmitting the geosite to the coming generation are significant for the improvement of sustainable regional development and geotourism.

Zgłobicki et al. in his chapter discussed the advantage, disadvantage, and importance of gullies and badlands as the significant site of geoheritage. They also state that scientists all over the nation are attracted by the spectacular nature and beauty of the permanent gullies and badlands which are found all over the continents of the world. They highlight that the geoheritage sites of the Africa and India are rarely visited when compared to the site in Europe and America which are frequently visited. Escorihuela, in the chapter, analyzed that geoeducation and geoethics among the children for sustainable tourism and development of aliage geological park in Spain. It creates awareness that as most of the study conclude their finding through the secondary studies without basic notions about the earth sciences, it is important to conduct in the field studies to study about the geosites, geoparks, and geological parks. The finding of the study displays that children, who have been attending the summer courses had an advanced knowledge than adults, not only about the environment, but also about the implications of the human activities in the territory.

Clary contributed chapter on the "Location, Location, Location: Challenges of Effective Geoscience Education within Geotourism Opportunities at Coastal US Fossil Park Sites" described that in US with the fossil park a unique geotourism venue exists. The park is conserved for the purpose of imparting informal education of geoscience to the visitors as well as permitting the retention of a small number of personal fossils and promoting sustainable collection. The finding of the study is that even with additional interpretive challenges, the parks can extend visitor geoliteracy for optimized geotourism opportunities. Occhipinti discribed that geoparks and geosites: geological "learning objects", mentioning that, on the school of Italy there is lack of general sensitivity toward the earth science. The author highlighted the problem resulting from the lack of sensitivity in Italian context. It directly leads to a lack of attention to the territory which whereas requires to be protected and "geo-preserved".

Tripathi analyzed the geoheritage sites and scope of geotourism in land of Chhattisgarh, discussing that geoheritage sites are scientifically, culturally, and educationally important for civilization. The study found that the overexploitation of the minerals from the region for the utilization in the industries has led to the extinction of geoheritage site in the area. The author suggested that geotourism with right laws and ethics will not merely led to conservation, management, and protection of these geosites but also help in socio-economic upliftment of the local people. Dongying, contributed a chapter entitled "The exploration into evaluation index system for the protection effectiveness of natural heritage protected areas", emphasized that protected area of the natural heritage has an important role in the cultural, historical, and scientific values of the geological heritage so it is important to strengthened their protection. The paper aimed to explore the evaluation index system for the security effectiveness of natural heritage protected areas, so that the general understanding of the current situation, problems of the heritage and the rational development of nature reserves can be obtained. Wadhawan, highlighted the importance of the geoheritage and potential geotourism in geoparks with special reference to India, describing that it has a variety of domains of the geological sites ranging from Achaen/Precambrian to active Anthropocene and Neogene with various world class interesting sites of geoheritage that display unique geological features, fossils, various rock types and tectonic discontinuities, and its geological boundaries, processes and landscapes. It has been found that geotourism is generally dependent on earth's geoheritage, educative through, geo-guides, geo-interpretation and increased consciousness, locally favorable through sustainable economic viability community involvement and foster geoconservation.

Sterquel et al. in the next chapter entitled "New Routes of Geotourism for the La Campana-Peñuelas Biosphere Reserve, Chile" concluded the Mediterranean ecosystem of Central Chile is reflected as the global "*hotspot*" of biodiversity due to its rapid rate of destruction of its habitats and high level of endemism. They have felt that its conservation and protection is needed to invite, local, regional, and national tour operators to evaluate critical points and gaps in tourist services and the management of the territories.

Hose, in the chapter on modern geotourism's UK antecedents, highlighted an impression of development of geotourism from the late seventeenth century to the contemporary day, drawing an example from the United Kingdom (UK), in which the geotourism concept was first formally defined and recognized. Especially, in relation to its role in geoconservation and the provision of geo-interpretation, a twentieth paradigm. Kubalíková et al. discribed the geomorphological resources for geoeducation and geotourism, considered that main resources for geoeducational and geotourist events are closely related and support each other. They believe that increment of the reorganization of the geodiversity and geoheritage at inclusive level can be attained by the geoeducation. It can help in maintenance of the geotourism activities in rational scale and avoid the overexploitation of the geoheritage for geotourism purposes. In their chapter, Cayla and Megerle discussed in the detailed about the Dinosaur geotourism in Europe, a booming tourism niche, mentioned that very popular science topic is not only comprised of Dinosaurs, throughout the entire world but at contemporary period paleontological tourism emphasizing the paleontological sites are booming. However the finding of the study suggested that success of the tourism projects must be based on the structured network of the gathering scientists, actor networks, stakeholders of the tourism supported by the local policy-makers and local population course.

The chapter on use of geoheritage, geopark, and geotourism concepts to conserve and sustain tourism development in Zambia, Mwamulowe, and Nyambe analyzed how geopark, geoheritage, and geotourism concepts can be utilized as the tools for the conservation and sustainable development in Zambia. This chapter introduced the various concepts in heritage of the earth, namely, geology, geological heritage, geomorphology, conservation of geoheriatge site, geotourism, and geopark. It has been found that the concepts of the geoparks, geoheritage, and geotourism have not be applied till now but it will be applied in coming days for fostering conservation and sustainable development of tourism in Zambia for its socio-economic development.

In the next chapter, Tefogoum et al. in their chapter entitle "highlight of geotouristic values of a volcanic landform on the mount manengouba eastern slopes: Case of Djeu-Seh basin" discussed that along the Cameroon volcanic line mount manengouba, is one of the most significant and voluminous volcanic apparatus and it is comprised of one of the nested calderas noticeably Elegoum and Eboga. It has been found that cultural value of the site is magnificent and the site is characterized by good accessibility and quiet environment. Jangra and Kaushik estimated the carrying capacity in a high mountainous tourist area: a destination conservation strategy, emphasized that in the tourism sector entire issues are associated with the magic number of tourist that visit the certain destination. As per the calculation, it has been found that the number of PCC, ECC, and RCC of the selected spots of tourist in Kinnaur are 64835, 5928, and 9595, respectively. It has been commonly accepted that ECC type is the most acceptable type of carrying capacity and its presentation of the status of the activity of the tourism are highly underexploited in the study area. The last chapter of the book on collaborative creation of educational geo routes: a strategy for teaching and learning sciences and geography, Puchuncaví, Chile, has been contributed by Martija at al. The authors described the collaborative development of a teaching innovation called educational geo routes. Focusing on the relationship between the socio-environmental context of the students and their schools with the national curriculum.

The aforesaid chapters of the present book look into the various critical issues related to geoheritage and geoparks in the detail. The papers also suggest solutions for the sustainable development and sustainability of the geosites in the different regions of the world. Looking the situation, it is high time to conserve and manage the existing geosites and geoparks so that geotourism can speed up. This edition would be of significant use to the policy-makers, environmentalists, conservationist, academician, and research scholars who are directly or indirectly involved.

Delhi, India R. B. Singh
Beijing, China Dongying Wei
Delhi, India Subhash Anand

References

Anand S, Saluja V, and Singh R B (2014) Varkala as a national geopark of India: geographical realities, geotourism and community participation. Int J Geoheritage and Parks 6(1): 65–81

Farsani T N et al (2012) Geoparks ang geotourism : new approached to sustainability for the 21st century, Brown Walker Press, Florida

Himiyama Y (2018) (ed) Exploring sustainable land use in Monsoon Asia, Springer, Singapore, v

Kohomoto D (2014) Characterstics of map inforamtion on the website of global geoparks and Japanese geoparks. Int J Geoheritage and Parks. 6(1):10–108

Li Y, Cen S, Li Y (2018) The evaluating and development of the geoheritage landscapes in dalaian jinshitan national marine park. Int J Geoheritage and Parks. 6 (1):60–73

Peng F (2018) The practice and exploration on the establishment of national park system in China. Int J Geoheritage and Parks. 6(1):1–16

Rana Y S (2020) Geo-heritage preservatin must for promoting geotourism in Himalaya. Punjab, News Express, 25 January

Singh R B, Anand S (2013) Geodiversity, geographical heritage and geoparks in India. Int J Geoheritage. 1(1):10–26

Singh R B, Singh S (2014) Human-induced biomes and livelihood security. In: Singh R B and Hietala R (eds) Livelihood Security in Northwestern Himalaya, Springer Tokyo, p 53.

UNESCO (2016) UNESCO Global Geoparks, Celebrating Earth Heritage, Sustaining Local Communities, Paris, accessed http://unesdoc.unesco.org/images/0024/002436/243650.pdf, pp. 1–20

Wei D (2013) Chief editor words. Int J Geoheritage. 1(1):1

Wilson, J (2013) Knowing our visitors: considering the differences. Int J Geoheritage. 1(1):2–9

Yang R (2015) Guard Chinese national parks against deterioration. Env Prot. 43(14):34–37 http://www.unesco.org/new/en/natural-sciences/environment/earth-sciences/unesco-global-geoparks/

Contents

Part I
Geographical Diversity and Geoheritage

Chapter 1
Geographical Diversity of Georgia and Perspectives of Planning of Geoparks (Geo Heritages)

Nodar Elizbarashvili, Zurab Laoshvili, Giorgi Dvalashvili, and Rusudan Elizbarashvili

Abstract Georgia is in the middle of the area (70, 000 km^2) almost 25 in the Europe. The formation of the country's territory counts for one billion years. All forms of relief are presented in Georgia, which is known worldwide. There are hundreds of natural monuments on the territory of Georgia that make a special impression on the visitors. It is also important that such monuments of nature are concentrated on a small area that increases their tourism and recreational purpose. With the diversity of nature, Georgia can be presented in the world geographical parks (geoparks) networks, and will further increase its tourist attraction. Georgia has not been presented in the world geographical parks (geoparks) networks, what is an interesting as Caucasus Office of Nature Protection World Fund, as state and local tourist organizations. The relevant nomination can be submitted by Elizbarashvili (2018): **1. Javakheti—volcanoes and Vardzia canyon** is located in southern Georgia (Akhalkalaki administrative district). It is made of lavish streams flowing from the slopes of the Mordian direction. There are several volcanic cones on this ridge that create effective peaks. The western boundary of the Plateau passes through the Vardzia canyon (the administrative district of Aspinza), with an average depth of 500 m, and a length of 7 km. It is in the middle of the medieval city of Vardzia (13-storey cave building) and monastery complex, which is nominated for a UNESCO cultural heritage monument. Some volcanic lakes are presented under geopark, which present Ramsar Convention site and are in composition of protected territories of Javakheti. **2. Imereti—The dinosaur footprint and the Imereti Cave Complex** (including the Kutaisi city and Tskaltubo administrative districts) in central Georgia—There are over 200 unique footprints of dinosaurs, which are found in 1933 on the mercury limestone. The footprint is located at 500 m above the sea level, with subtropical humid conditions and the western exposition winds (slopes from the Black Sea). Nearby, there are several karst caves, geological, paleontological, zoological, and botanical monuments. The Tskaltubo Cave is one of the most remarkable places in Europe. **3. Khevi—Glacier (Kazbegi), Gudauri volcanic plateau and Dariali Gorge**—The glacier is located on the central Caucasus in northern Georgia (Stepantsminda administrative district).

N. Elizbarashvili (✉) · Z. Laoshvili · G. Dvalashvili · R. Elizbarashvili
Tbilisi State University, Tbilisi, Georgia
e-mail: nelizbarashvili@yahoo.com

R. B. Singh et al. (eds.), *Global Geographical Heritage, Geoparks and Geotourism*, Advances in Geographical and Environmental Sciences,
https://doi.org/10.1007/978-981-15-4956-4_1

Its height is 5033 m. The local population call it "Khevi bride". The glacier is a "dormant" volcano that lasted 6,000 years ago. Here the lavish deer runs in three—south, east, and north-east. One of the mythological and one historical (Gergeti Trinity) monument is the monument on the edge of the glacier. Mythology is connected to Amiran (the analog of Prometheus) attached to the Caucasus. According to the tradition, God Amiran right here climbed the cliff, and "on top of the snow-ice gear and the glacier of the glacier". At the foot of the glacier, Dariali canyonic gorge is represented in the rocks formed in the form of volcanoes developed over 300–400 million years ago, central Caucasus. Its length is 11 km, and the relative height is more than a thousand meters. There is also a volcanic plate of Gudauri, whose vertical slopes are formed by lava cooling (suspension).

Keywords Geographical · Geopark · Natural · Diversity · Volcanic · Georgia · Caucasus

1.1 Introduction

Natural diversity of Georgia is distinguished in the world, especially in subtropical and moderate midfield. Here are presented more than seven dozens type of landscape, beginning with humid subtropical or semi-arid bright woods, ending with wetland and nival landscapes. Such diversity is connected to several factors, from which are important: geographical location (tropical and on boundary of moderate midfield), effect of unfrozen black sea, height distribution of geographical factors and characteristics (from sea level till 5200 m) and thousand-year-old history of using of trust territory (Elizbarashvili and the other 2018a).

Georgia is also distinguished in the world with "initial" in other words with high comparative lot of natural landscape, which includes 17% of country's area. We can find natural environment as an undamaged, as on protected and wetland territories, as in conditions of moderate and mountainous relief.

According to the covered territory (up to 70,000 m^2) and population (3.7 million), Georgia takes 25th, almost the middle place in Europe. Here is represented all well-known forms of relief, several types of climate (humid subtropical, dry subtropical, arid, semi-arid, and semi-humid), several dozens of plants and grounds. A huge ecological resource is forests which cover 2/5 of territory of the country. More than 15,000 plants grow in Georgia of which more than 4000 are pomaceous, 75 covered in bracken, and 2600 aquatic plant. 6.0% (i.e., almost 900 species) of plants are endemic and relict. Animal world is featured with the same diversity of which more than 100 species are mammals and more than 300 are birds. Pursuant to the number of life species, endemic regions and biodiversity, Georgia is included within five of a European Countries which distinctly confirms its diversity and irresponsibility of natural environment (Geography of Georgia 2000).

Almost the whole altitude spectrum of life world of Europe is seen in Georgia within miniature sizes. Georgia with total number of flora species takes the V place in Europe. Georgia with this index highly exceeds the UK, Ireland, Eastern Europe, and countries of Scandinavia. According to the number of mammals and bird species, Georgia takes the first place in Europe while pursuant to the life of endemic region backwards only such countries such as Spain, Greece, Italy, and Bulgaria (Biological 2000).

By the number of landscapes, Georgia holds the 12th place in the world while the first place in Europe. Georgia backwards only big countries by this index: China, USA, Russia, Australia, Mexico, India, Canada, Brazil, Argentina, Turkey, and Chile. Even though, due to proportion of territory size with landscape diversity, Georgia holds the first place in the world. It 20 times exceeds with landscape diversity the average index of the world ("Biological" 2000).

Georgia as the part of Caucasia includes (Fig. 1.1)

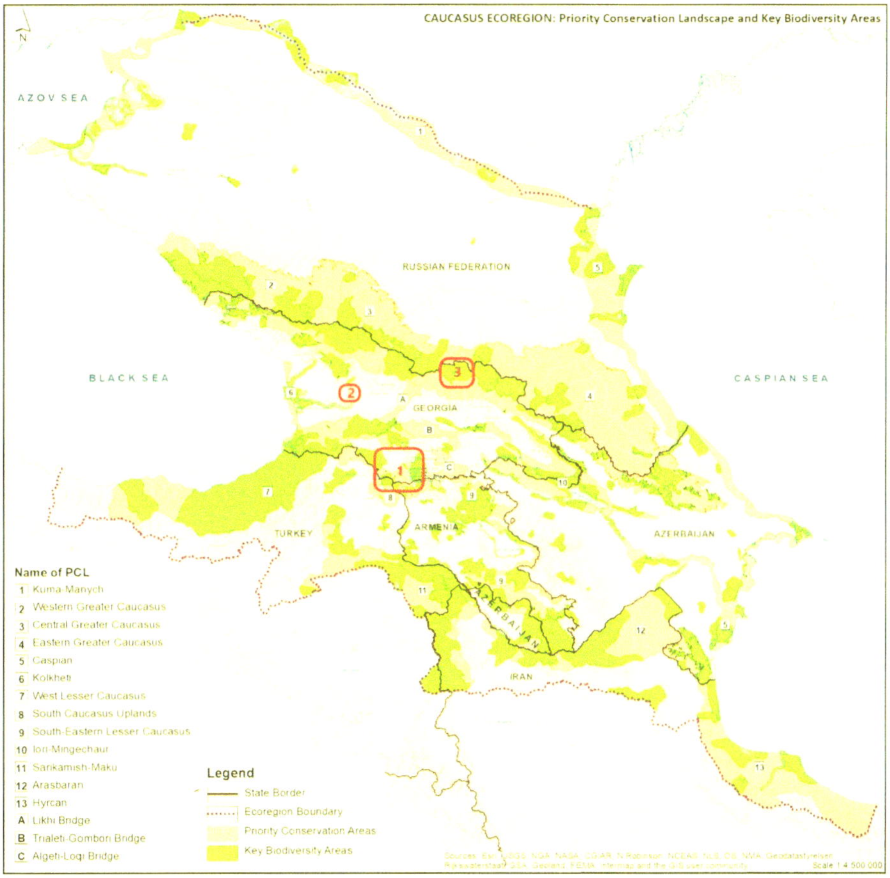

Fig. 1.1 Caucasus Ecoregion: priority conservation landscapes and key diversity areas (WWF Caucasus 2018). Perspective Geoparks Area in Georgia—Javakheti (1), Imereti (2), Khevi (3)

- In the richest biologically and being under danger 34 "hotpoints" of the world;
- In the most sensitive, vulnerable 200 ecoregions of the world, which are distinguished with the high biodiversity;
- In the location of endemic birds;
- In agro-biodiversity centers of the world;
- In the "hotpoints" of spreading of large grass eaters.

Georgia is distinguished in Europe with fertility of single components of nature (read earth, tertiary's relict Colchian forests, unimpaired forests, etc.), relict and rare species, inorganic monuments of nature (vocluses, natural bridges, location of petrous flora and fauna, etc.). According to this index and subject matter, Georgia is one of the greatest touristic-recreational centers of Europe (Elizbarashvili and authors 2018).

Georgia in the list of the world-historical heritage is presented with 2 objects including Village Chazhashi (Zemo Svaneti, unique combination of natural landscape, middle centuries architecture and traditional agriculture) and Mtskheta's (historical capital (up to fourth century) world heritage monuments (status granted in 1994). Georgian alphabet, traditional polyphony (singing), wine-jar, and Meskhian "Tenili" cheese is included in the non-material heritage of world by UNESCO. Georgia also is considered as the homeland of grapevine (316 cultural species of grapevine), location area of endemic species of cultural plants and domestic animals (5 of 14 species of world well-known wheat is Georgian, endemic Georgian bee, sheep, horse, cow is also well known). Oldest honey (5500 years old) and oldest textile (35 thousand years old) manufactured in the world is discovered in Georgia. Georgia is also considered as the homeland of first Europeans. Results of Dmanisi archeological excavations changed the geography of humans spreading from Africa to Eurasia ("Human Odyssey") and, respectively, historical–geographical development tendencies of mankind. The oldest skull of human recorded in the continent of Eurasia is 1.8 million years old and is discovered in Dmanisi. Georgian writing system and language is unique in the world. Georgian alphabet and writing system is oldest while writing system is included within 14 writing system of the world.

1.2 Results

Georgia has not been presented in the world geographical parks (geoparks) networks up to this date, what is an interesting Caucasus Office of Nature Protection World Fund, as state and local tourist organizations. The relevant nomination can be submitted by

1. **Javakheti—volcanic plateau of Javakheti, canyon of the River Mtkvari, Vardzia Monastery Complex**

Territory represented for first nomination of Geopark in Georgia is concluded with three main parts.

First or volcanic plateau of Javakheti involves several volcanic conus of Abul-Samsari Ridge (Big and Small Abuli, Chikiani, Tavkvetila, Samsari, Shavnabada,etc.) and that drift flow which creates volcanic carcass of the region. Height of Big Abuli is 3303 m above sea level. Now it represents extinct volcano which actively operated 10–12 thousand years ago. Mountain Small Abuli (2800 m) is located in South from it of which hillsides and foundation are covered with interesting denudative forms so-called "rock flows". At the south hillside of Small Abuli, 2670 m above sea level is located dry-constructed cyclopic castle which is unique in Caucasus. Chikiani Mountain located in the north part of Javakheti Ridge is constructed with volcanic glass. Word "Chika" meant glass in Old Georgian Language (both artificial and natural). Chikiani is almost wholly constructed with obsidian and the name of mountain is resulted hence. This peak reflects sun rays in the sunny weather and magnificently shines. Obsidian had a huge role in the history of human: as yet in Stone Age, the ancestor of human started creating labor and battle instruments from obsidian. Deposit of obsidian and workshops for manufacturing different instruments is left on the hillsides of Chikiani. Artificial, cultic rock columns of 3.5 m height is found near Chikiani Mountain which is dated in II millennium B.C.

Second, i.e., canyon of the river Mtkvari is located from several kilometers of Javakheti Volcanic Plateau. Its depths sometimes exceed even 500 m. The main part of canyon is steep which creates irreconcilable view. It is the biggest canyon in Georgia. It had the greatest historical meaning because roads connecting Middle East and Caucasus were passed nearby (Fig. 1.2).

Third Object directly connects to the Canyon of the river Mtkvari. Several complexes of artificial caves (Vardzia, Vani Caves) and fortress (Tmogvi, Khertvisi) remained here. They belong to eleventh–twelveth centuries monuments. Vardzia is rock-carved monastery ensemble, located 1300–1460 m above sea level. It has 13 floors and represented closely populated city. Middle ages residential and social infrastructure is founded in Vardzia which attracts thousands of visitors.

Within the First Nomination of Geopark may combine the National Park of Javakheti (trans-border territory between Georgia, Armenia, and Turkey), 6 Wildlife refuge (represented in the crater of volcanic origin lakes), several over-saturated

Fig. 1.2 Vardzia Canyon and Monastery, Volcanik plato and Volcanik conusys

territories (which represent the site of Ramsar Convection), four type landscapes (middle mountain bushes and arid sparse forests, mountain volcanic plateau with steppe plants, fields of high mountain sub-alpine and high alpine), and part of two historical–geographic province (Meskheti, Javakheti).

Meskheti and Javakheti products traditional "Tenili" cheese which is distinguished with the most complicated production techniques and is enrolled in the World Heritage list of UNESCO. Cheese material is framed for aging in the pot (of which the name is originated) and is covered with milk cream. It is produced in single families and small enterprises. Cheese is a delicatessen.

2. **Imereti—Okatse Canyon, dinosaur footprint, Thermal waters of Tskaltubo, Caves of Imereti and Katskhi Column.**

Territory represented for the Second Nomination of the Geopark in Georgia includes four main parts.

In the central part of Georgia, within the historical–geographic province of Imereti, at a several km distances is located the longest and narrowest canyon of the river Okatse and the highest waterfall—Kinchkha. River Okatse runs from Askhi limestone massive. Its length is 16 km while average width is 4 meter. Sides of canyon almost connect each other at several places and create natural rock-bridges. At the natural rock bridge ("Kvakhida") the depth of canyon reaches even 100 m. The river creates lakes which have blueish-greenish color. There are many waterfalls in gorge of which Kinchka waterfall reaches 80 m. It has the status of nature monument.

In the land of imereti, within several kilometers from each other, is situated one of the tallest Karchkhi waterfalls in Georgia and the deepest Okatse Canyon. Right tributary of river Tskhenistskali—river Okatse—flows from Askhi massif. In the area surrounding the village of zeda Gordi, it sires 16-km long canyon with the depth of 35–50 m and width mean 4 m. In some places, the walls of the canyon nearly align with each other and create natural stone bridges. One of them is "Boga" wherefrom it is possible to see even the bottom of the canyon. In the environs of the village of zemo Gordi, the river gives birth to a wonderful natural monument—2-km-long canyon, on the bottom of which natural bridges, caves, and karst sources are placed. By one of the natural bridges (called "Kvakhida", "stone bridge" in Georgian), depth of the canyon attains 100 m. River creates lakes, and beyond their transparent bluish-greenish water white bottom is reflexed, and also numerous waterfalls, the biggest being more than 80 m tall. This waterfall known under the name of Kinchkhi waterfall has been entered into the red List as a miracle of nature (Fig. 1.3).

Sataplia-Tskaltubo's limestone massive is located near the river Okatse with 92 m^2 area. Here are represented the prominent caves of Imereti and almost 200 footprints of dinosaurs on the limestone surface.

Natural monument of Prometheus (Kumistavi) karstic cave reaches with stalactites, stalagmites, helictites, petrified waterfalls, and hang rock drapes. The area of cave is 46.6 ha, length of some halls reaches 80–100 m. The total length of Sataplia Cave Complex is 900 m, depth 10 m, width 12 m. It has been forming for 30 million

Fig. 1.3 Canoyn of Okatse and waterfall of Kinchkhi

years. In 1935, Sataplia National Park was created which covered geological, paleontological, speleological, zoological, and botanical monuments. 95 % of the territory is covered with relict Colchis forest which is unique in Caucasus.

Within the villages of Imereti, one of the oldest and most beautiful is village Katskhi which has rich and interesting history. Natural limestone column is towered there with height of 40 m while top area is 150 m^2. The column mainly is towered sheer. Column is eroded and more narrow near the column while its higher part. This shape created negative hillside which is apparently unscalable. Vicinities and top of Katskhi Column is covered with broadleaf plants which creates irreconcilable landscape. During previous middle-ages, there were hermits place and church house. A small church is constructed on the upper place of the column.

Among the villages of upper Imereti, village of Katskhi is one of the oldest and most beautiful with rich and interesting history. The term "Katskhi" stems from the Svan vernacular and means "peak". This is proved also by geographic location of the village. The Pillar is approximately 40–45 m high, while the space of its square top surface is nearly 150 m^2. The Pillar stands upright. However, in its upper part, there are several small-size terraces. Near the foot, the Pillar is emaciated and narrower than its upper part, creating negative tilt and making it even more inaccessible. The environs of the Pole of Katskhi and its top are dressed in deciduous plants and lianas creating an inimitable landscape (Fig. 1.4).

Dzudzuana Cave is widely known in the the world's historical and archeological scientific circles, where was found the oldest linen thread in the world (34 thousand years ago). International expedition, which consists Georgian, American, and Jewish specialists constantly work here. The purpose of scientists is identifying of migration ways of Homo Sapiens of Paleolithic and studying of their activity. During the last 20 years, were revealed seven residential layers of upper Paleolithic on the cave's territory, where was discovered about 100 samples of tools of bone, hangers, amulets, stone.

It is well known that karst development is expressed with full force in the limestone sediments. The big part of Chiatura municipality area is built exactly with this kind of sediments. That is why there are a lot of caves and cavities of different kinds

Fig. 1.4 Katskhi pillar and Dzudzuana Cave

on the territory of this municipality. Notwithstanding, the big part of them has not been explored and reviewed, yet. Dzudzuana cave which is located in 2 km from the village of Mgvimevi to the East, in the gorge of river nekrisi, is well known in the world's dzudzuana cave happened to be home to the oldest known flax fibers in the world scientific circles. Since 1996, it is under the scrutiny of international expedition consisting of the Georgian, American, and Israeli specialists. The goal of these academics is establishment of migration ways of homo sapiens, belonging to Paleolithic period. Seven domestic layers were discovered on the territory of this cave belonging to this age whereby up to 100 items like ornamented bones, hangers, mascots, lucky pieces, stone tools were unearthed. Explorer found also remnants of stranded fibers made of wild flax fibers. Lab research has shown that due to certain technology, fiber was proccessed and dyed by a human being manually at about 34 thousand years ago.

3. **Khevi—Volcanic massive of Mkinvartsveri, Lava bed of Khorisari, Dariali gorge**

Territory represented for the Third Nomination of the Geopark in Georgia includes four main parts.

Tkarsheti lava bed created with the throwing up of Mkinvartsveri (5033 m above sea level) volcanic massive congested river Tergi. At the terraces of right bank of Tergi is bared row of lacustrine coverings where petrous plants of forest are represented. Stem of vertically standing trees witnesses the fast dynamic of accumulation of lacustrine sediments. These plants are high mountain aspen, birch, and juniper. Fossil forest was seen in 1968 due to catastrophic deluge of Tergi when water smashed down nearby hillside. Radiocarbon analyses of tree bark determined its age with 5950 + 60 age which specifies the date of volcanic actions.

In the end of the stream, in terrace precipices of Tergi river bank, there is a host of naked pool deposits collected here as a result of conjestion of river Tergi by lava

Fig. 1.5 Mkinvartsveri and Gergeti's Sameba, Lava stream of Kazbegi Region

stream. In a slew of these water pools, we come across fossilized forest. Vertically standing trunks of trees speak for swift collection of lake deposits. These trunks belong to asp-trees, birch-trees, juniper, and other plants of high mountains. Petrified Forest was revealed in 1968, during catastrophic flooding of Tergi River, when water destroyed terrace edge. Carbonaceous test of the wood determined its age at 5950 +60 years old that gives us grounds to date the time of eruption of Tkarsheti lava stream as happening approximately 6000 years ago (Fig. 1.5).

Here in 1976 with the purpose of protection flora and fauna was created Kazbegi National Park. Almost treeless park is the shelter of rare species of fauna—East-caucasian tur, chamois, gazelle, and brown bear. There are birds of Georgian red list: vulture, Caucasian heath cock, Caucasian snowcock, lammergeyer, war eagle, etc.

"Bat" volcanic cave popular within tourists is located 3 km north of Daba Stepantsminda. The ceiling of lava cave is created with quaternary andesite-basalt lava which cracked into six-sided prisms, i.e., columns within turning cold process. It resembles "lava organ" for fan-siting of columns. Cave is located in Darial Gorge. The name "Dariali" originates from Persian "Dar-i Alan" (Gate of the Alans). Roads connecting North and South Caucasus were passing through the gorge from the oldest times. Relative altitude of relief reaches 3300–3500 m for which the gorge was deemed practically impenetrable.

Popular among tourists, Volcanic Cave of Gamura is located in 3 km to the north from Stepantsminda. The ceiling of lava cave was formed by Quaternary system andesite and basalt lava which in the freezing process was broken into six-sided prisms, or pillars. "Pole separation" of the lava is known in many places. Special character of "Lava organ" developed on top of the Gamura cave is in the fan-like succession of poles giving it an original and beautiful appearance. The cave is created in the friable gravel under the lava as a result of erosion and with the help of a man—in old times flock of sheep was often kept here at night.

Gamura cave is located in Darial Gorge. The word “dariali” came to us from Persian “Dari Alani” (the gates of Alans). From the times immemorial the gorge was always a conduit of links between Western and Eastern Caucasus. In Dariali Gorge, near the village of Gveleti, in 7 km from the township of Stepantsminda, is located Gveleti waterfall. In Dariali Gorge, on the high rocky mountain on the left bank of river Tergi, northern and eastern sides of which are unapproachable, while to the west the mount neck is covered by walls, is situated a monument of Georgian architecture—Dariali fortress. Georgian historic tradition ascribed construction of this fortress to Vakhtang Gorgasali (fifth century A.d.).

The Christian Cathedral is located near Mkinvartsveri, where a beautiful view of the Dariali Gorge is found. It is visited by ten thousand visitors every year. There is also a cave of Bethlehem located at an altitude of 4100 m above sea level.

Quaternary volcanic activity is widely represented at the tableland of crater with up to twenty volcanic centers. Two volcanic centers of this tableland are well visible from the east (from Gudauri settlement)—Big and Little Nepiskalo. Besides, within axial zone of Greater Caucasus mountain range, up to the Gudauri Settlement, is also well-known volcano centers Sadzele and Sakokhe. Mleta–Gudauri lava flow runs from the volcanic center of Sadzele which is about 12 km and ends in the Village Kvesheti at the left bank of river Aragvi.

The Georgian sheep breed is produced by multi-year folk selection in the region in the nomadic cattle breeding conditions (thirteen–fourteen centuries). It freely withstands relocation at the long distance, feeding on the sparse pasturelands, has high-quality meet and white flexible elastic wool. From thread obtained from this sheep is used for knitting highest quality carpets and warm clothes.

1.3 Conclusion

Reviewed regions nowadays represent one of the most important touristic units in Georgia. Awarding geoparks status will more increase their international awareness for which is interested as the Government of Georgia also touristic organizations and local population.

References

Biological and Landscape diversity of Georgia (2000) TBilisi, WWF Caucasus office, pp. 275

Elizbarashvili N, Meladze G, Meeseni H, Khoetsiani A, Koleri T (2018a) Stable development of mountinous teritorries. Manual for higher education students. – Tbilisi., Publishing house “Dani”, 304 p. (into Georgian, English and Armenian Languages)

Elizbarashvili N, Meladze G, Elizbarashvili R (2018b) Ecological and landscape indicators of estimation of stable development. International conference: VIII International scientific-technical conference “The modern problems of water management, environmental protection, architecture and construction”materials. Tbilisi, GTU, pp 99–108

Geography of Georgia – Tbilisi (2002), 367 p
WWF Caucasus (2018) Materials of final stakeholders meeting: revision of the ecoregion conservation plan for the Caucasus. Tbilisi, 215 p

Chapter 2
Identifying and Assessing Geodiversities Around Takht-e Soleyman World Heritage Site to Propose the Territory as the Third Geopark in Iran

Bahram N. Sadry, Azar Mohammadi-Aragh, Fatemeh Fehrest, Ali Bayatani, and Ali Haji-Moradi

Abstract In the twenty-first century, humans paid special attention to applying criteria for the assessment of geodiversity and the introduction of the country's geoheritage to conserve this valuable abiotic natural heritage and to use it in a sustainable way, especially via new emergence of the geotourism industry. In Iran, scientific discussion of geotourism was not introduced in academic textbooks until the registration of the Qeshm Island (for the first time) in the Global Geopark Network (GGN) in 2006. However, in spite of being delisted in 2012, it made it to be listed again in GGN in 2017. Meanwhile, the academics, including students, have made valuable unorganized national efforts to identify more territories as potential national/global geoparks by choosing to work on the topic as a master's or Ph.D. thesis. These efforts accompanying the published textbooks have played a significant role in eliminating the merely aesthetics point of view toward geotourism and geopark own-made philosophy which was first introduced mistakenly by the Geological Survey & Mineral Explorations of Iran. Today, the number of those interested in the field is growing and thanks to the academic books published, the awareness of the issue has been increased. Fortunately, this is the first successful academic project related to a region's Geopark feasibility study in Iran completed in 2014, which had been completed partly by the financial support of Geological Survey & Mineral Explorations of Iran and by the cooperation of some geo-colleagues from this organization as a teamwork with this university project. The study area is located around the UNESCO-listed World Heritage Site, namely Takht-e Soleyman in the northwest of Iran that is rich in terms of geodiversity, biodiversity, and historical–cultural diversity. However, unfortunately, the study area is in a poor condition economically. Accordingly, the establishment of a geopark in the region can increase job creation, prevent immigration of the villagers to the cities, and flourish the economic conditions of the region. In this study, two assessment methods have been used for evaluating selected

---

B. N. Sadry (✉) · A. Mohammadi-Aragh · F. Fehrest · A. Bayatani · A. Haji-Moradi
Ecotourism Department, Faculty of Tourism Sciences, University of Science and Culture; and Consultant of the Project for the Geological Survey & Mineral Explorations of Iran (GSI), Tehran, Iran
e-mail: bahram.sadry@usc.ac.ir

R. B. Singh et al. (eds.), *Global Geographical Heritage, Geoparks and Geotourism*, Advances in Geographical and Environmental Sciences,
https://doi.org/10.1007/978-981-15-4956-4_2

geodiversities in the field, where each of them includes some parameters and sub-parameters. Based on calculated scores for each geological/geomorphological site and making use of interpolation methods in a GIS environment, the geopark territory, and the best preliminary geotrail in the study region were selected. The results of the study show that both methodologies prove high conservation importance of the geoheritage and other cultural–environmental (geographical) heritage perspective for this territory which represents the importance of sustainable application of resources by registering a geopark as a managerial strategy in this area. Consequently, related maps for future studies and filed works have been produced. According to the geopark and geotrail preliminary maps, the area has the potential for more studies, further community engagement (about 44 villages) studies, and consequently governmental and the private sectors' investments and preparing sustainable tourism infrastructures to be listed as a national geopark and later as a global geopark.

Keywords Geotourism · Geodiversity · Geotrail · Geo-assessment · Takht-e Soleyman geopark · Iran's third geopark · GIS

2.1 Introduction

Geoparks have been increasingly proved to be effective in alleviating poverty around the world (Sadry 2020a). Therefore, geopark development at the internal state scale could be an effective strategy to alleviate poverty in developing countries, raise awareness among people, and conserve geological heritage and socioeconomic improvements of villages and small towns especially in the Middle East.

There has been a growing interest in the twenty-first century to care for the abiotic nature which had been neglected previously. This interest is also growing among Iranian academics and also some governmental executives to develop geotourism, identify potential sources to establish geoparks via master's or Ph.D. theses, and more recently via the feasibility studies in governmental projects such as free zones, provincial governments, governorates, and recently in the Geological Survey & Mineral Explorations of Iran and apparently in the near future, via municipalities to establish urban geotourism. The diversity of landscapes and geological forms and processes along with the knowledge related to the story and history of the earth has given birth to abiotic nature-based tourism called "Geotourism" (Sadry 2009a, 2020a). However, protecting the geological heritage before and during the geotourism development is not only compatible with sustainability but also essential in various regions. Currently, the assessment methods have the highest priority not only for scientific goals, but also for the management and conservation of geoheritage. Therefore, it seems necessary to qualify and document the value of geoheritage and the assessment of geodiversity by making use of a standard method that is acceptable for geologists and related international or national accredited associations.

Regarding the culture and geographical features of the regions, even those of a vast country, applying a standard method would increase the effectiveness of the measures related to the geoheritage for both conservation, as the main goal of these measures, and geotourism, as a means to realize this objective, depending on a reliable assessment process (Fasoulas et al. 2012). Therefore, quantitative evaluation of geoheritage provides solid evidence for the authorities to make decisions on allocating monetary resources for geotourism development and conservation in the frame of geoparks based on the numerical data and also based on the geographical features of the region to establish a geopark within it.

Considering the absence of a domestic model of evaluating geological diversity in Iran, the present study investigates geosites of the Takht-e Soleyman region which are determined by making use of geoheritage evaluation methods used in Europe and Brazil.

In summary, this chapter is aimed at introducing a vast area located in the northwest of Iran for heritage preservation and geotourism development as a concept defined and emerged in the twenty-first century. The authors believe that this geopark is the third potential area to be listed in the national geopark committee and later as a network member of UNESCO Global Geoparks. Therefore, in this chapter, following a brief review of geotourism and geopark concepts, a historical review of their progress in Iran, the study area, and then the methodology have been introduced and subsequently, it will lead to the findings of the field assessment, conclusion, and produced maps.

2.2 A Glimpse into Geotourism and Geopark Concepts

Modern geotourism has been defined by many scholars from 1994 till now including Bahram N. Sadry who defines the term, on English summary of his work, Fundamentals of Geotourism (SAMT 2009); and later concluded and extended the concept in the recent book edited by him to vertical attractions and destinations, briefly as, geotoursim focuses not only on the earth's abiotic attributes such as any geo-objects, geosites, and cultural geosites but also on smaller geological features such as a grain of sand, small fossil, etc., to geomorphological landscapes on Earth to Mars and other planets in the universe (Sadry 2020a). Also about the geopark concept, according to Sadry (2011): "Geopark is an under protection area which is enriched with geosites and may contain historical-cultural sites and eco-sites; making use of effective management (e.g.: facilitation) and proper training of local people and emphasizing on the edutainment of geological and environmental concepts would probably attract tourists to fill their leisure times through listening to the interpretation of all attractions in a way that these measurements cause improvement in the socio-economic situation of the local community as well as all people in the country in a whole image." As stated by Turner (2006): "The Geopark concept has brought a revitalization of economics to many rural regions assisting local communities and indigenous populations." According to Sadry (2020b), in plain language, "geopark is a place, full of the geological heritages with a local community living nearby, which contributes to tourism and non-formal education of geology while the main

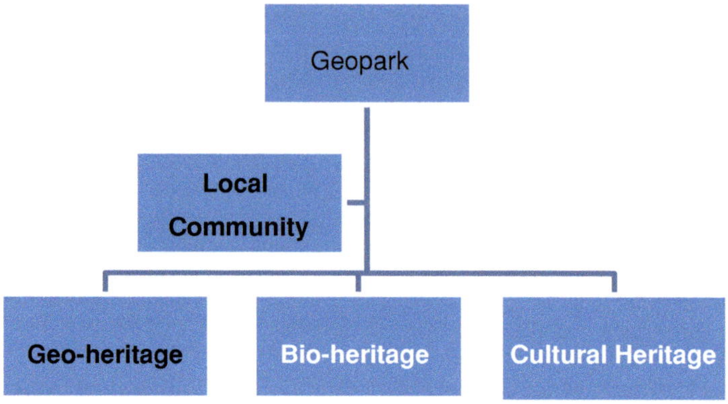

Fig. 2.1 Represents the existence of geological heritages and the local communities which are essential in establishing a geopark (Sadry 2012b). In this figure, the words in black are fundamental elements, and words in white are subsidiary and value-added factors for geopark establishment

livelihood of local people is through geology." There are two main prerequisites to nominate a territory as a geopark, the existence of geological heritage and local communities (see Fig. 2.1). Although it would be possible to register a territory as a geopark without other heritages, the existence of cultural and natural heritage would add to its richness and make it more achievable to be listed in the network and for the promotion of other types of tourism sub-segments such as agriculture tourism (Vafadari and Cooper 2020), ecotourism, and wildlife tourism (Sadry 2020a).

A geopark is established with the purpose of tourism development [with a special focus on geotourism sub-sector of tourism industry], conservation of geoheritage, and training of domestic people (UNESCO 2000). In fact, geoparks are the ideal territories to develop geotourism. Briefly, in the 1990s, UNESCO in cooperation with IUGS developed a common concept with the aim of creating parks focusing on geoheritage conservation and sustainable utilization of its resources; they described this concept as "geopark" (Patzak and Eder 1998).

In 2001, with the establishment of the European Geopark Network in four regions (Zorous and Martini 2003), this concept gained attention. Developing geoparks was first started in Europe and China exclusively (Chen et al. 2015) and thanks to the efforts made by European pioneers, the UNESCO considered the advances and the proposed Europeon model for a global geopark network. The main objective of the geoparks is the management of natural abiotic heritage as well as cultural heritage by creating a network among countries to promote high standards of conservation and to realize economic development (Fasoulas et al. 2012). There are currently 161 UNESCO Global Geoparks in 44 countries (www.unesco.org/geoparks).

2.3 Geotourism and Geopark's History in Iran

2.3.1 The History of Geotourism in Iran

Identifying Iran's geo-phenomenon (from aesthetic perspective) was first introduced by Mohammad Hossein Nabavi's crucial speech in the 18th Symposium on Earth Sciences in 2000 in Iran (Sadry 2020b), which attracted many geologists and GSI managers' attention to the Geological Survey & Mineral Explorations of Iran (GSI). Following this event, geotourism was practically started by one of the young participants of this Symposium in 2001. Alireza Amrikazemi was the young person who followed Nabavi's ideas enthusiastically in Iran. He was a new employee at GSI with a great taste for art and a background in photography who had the experience of holding photo expos previously for geology students in 1990 at Shahid Beheshti University in Tehran (Ghasemi 2010). In 2001, a project named geotourism was started by financial and immaterial support of Geological Survey & Mineral Explorations of Iran, considering the identification of geodiversity in Iran from an aesthetic perspective, despite the lack of defining any official organizational chart and the lack of sound knowledge on geotourim at the time.

Later, M. H. Nabavi (*pers. comm.*, June 03, 2014) explained, "I proposed the term "Geotourism" in Iran according to what I had heard in the sub-panels of the European academic conferences and through the inspirations, I got from western academics." He asserted, "Western people prefer to sell the photography of their country's soil to the tourists instead of their soil (as a raw mineral material) itself; this was the reason I introduced geotourism word in Iran."

However, since any practical step taken would align with the theoretical and philosophical foundation, a practical step in an isolated manner was started to promote geotourism with an aesthetic point of view toward geodiversity (called the geological phenomenon, at that time) of Iran. Nevertheless, according to Sadry (2020b), 12 years after proposing the subject, Nabavi (Nabavi and Mohammad-Hasan 2011, pp. 94–95) criticizes the merely aesthetic approach toward the geological diversity by the Geological Survey & Mineral Explorations of Iran: "unfortunately, *we simply neglected the nature of geo-phenomenon… geologists would try to introduce geo-phenomena to public people, the GSI accepted my proposal and geotourism department was established… but what I had expected did not happen. The beauty should be used as a mean to promote geology science among people… however, as we are supposed to mention the beauty of a phenomenon, geological [learning] dimension of the phenomenon should not be ignored*".

Literally, Nabavi was not acquainted with geotourism literature and definitions in 2000 and 2011; what he was trying to propose is called geo-popularization and geo-interpretation, which are connected to two specific majors, the first one in the field of Environmental Education and the second one in the field of tourism management and generally, non-formal education management. He was well aware of the necessity of these factors; however, none of these factors were practiced. Geotourism development is a multidisciplinary field; a geologist or a geomorphologist would not be able to act independently on investigations, in spite of their expertise.

Therefore, a one-dimensional point of view toward such activities would be a destructive managerial perspective or what is called "Toxic management." It would appear that a team working experience among experts of various majors in multidisciplinary fields (and even educated in several fields) is fundamental in understanding and promoting geotourism.

Nabavi (2000), the retired expert in Former Geological Survey of Iran, played a significant role by suggesting to visit the amazing geological features with a tourism approach, which led to identifying and documenting Iran's selective geodiversity sites (from aesthetic perspective) by Alireza Amrikazemi and his colleagues in the Geological Survey & Mineral Explorations of Iran. Nevertheless, visiting geodiversities would never be considered as a geotourism activity in today's world without the infrastructures, tourism industry components, especially interpretation provision as an essential tool (e.g., Hose 2006; Dowling and Newsome 2006; Sadry 2009b; Reynard and Brilha 2018; Dowling and Newsome 2018; Sadry 2020a), and geo-conservation infrastructure and community and all stakeholder engagements.

B. N. Sadry voluntarily developed the first Geotour Guide training courses as a proposed national curriculum in 2006 concurrently with the registration of Qeshm geopark by the UNESCO. The curriculum was then approved by the former Cultural Heritage, Handicrafts, and Tourism Organization of Iran in 2007. However, not being supported and actually rejected by GSI, it failed to be delivered by the curriculum designer voluntarily in Qeshm UNESCO Global geopark, which was recently established at that time. So, these official training courses were executed 7 years later, in 2014, with the support of Dr. Jahanian who was the Dean of the faculty then, for the first time to a group of 20 learners at the Tourism faculty of the University of Science and Culture by the curriculum designer, in Tehran.

Delisting Qeshm as the first Iranian geopark from UNESCO's Global Geopark Network could be attributed to the absence of these factors and essential supports which are not the subject of the present study.

As for the scientific appearance of the geotourism concept in Iran, Sadry (2006) proposed a new phrase "abiotic nature-based tourism" in the first conference of "the opportunities and threats of promoting tourism industry" held in Tabriz, Iran. Later, Sadry (2008) classified some types of geotourism (e.g., road cuttings geotours and geotours of mines) and introduced his primary classification through outlining Iran's potential for geotourism in the first Geotourism conference in GSI, (Oct. 12, 2008) in Iran. These classifications and conceptual frameworks were being proposed years after the aesthetic documentation of GSI. In the following, he consequently introduced his first definition of Geotourism in 2009 by publishing a book "*Fundamentals of Geotourism: with a special emphasis on Iran*" in the Persian language. This definition covers the cultural aspects of geosites and defines Geotourism as:

"Geotourismis knowledge-based tourism, interdisciplinary integration of the tourism industry with conservation and interpretation of abiotic nature attributes, besides considering related cultural issues, within the geosites for the general public". At the same time, the subject of mining geotourism as a sub-segment of

geotourism was first introduced and included in the categorizations with its own sub-sectors in Iran. However, it was not embraced by GSI in their efforts toward identifying the aesthetics of geodiversity.

Regarding the emergence of scientific geotourism in Iran, Yamani et al. (2012) and also Maghsoudi et al. (2013) and etc., assert that the theoretical framework of geotourism science was first introduced by Sadry (2009a) in his book "Fundamentals of Geotourism" in Iran. Consequently, Sadry (2012a) co-authored a geotourism textbook with his colleague, B. Hajalilu, for geology undergraduate student for Iran's Payam-e Noor[mega]university in 2011 in Persian.

But, despite all academic progress at national scale, Geotourism would be scientifically and practically commence in Iran, if Qeshm Island UNESCO Global Geopark will be confirmed and received the Green Card by UNESCO evaluators in a revalidation process in 2021 (for the second time after being de-listed) and also if the new and first geotourism project of Iran, namely "abandoned gold mine of Torqabeh" in Mashhad (private sector) in the northeast of Iran, introduced as the first cultural geosite, can receive governmental permissions successfully in 2021 and attract geotourists as a mining geotourism site in the near future in Iran, in this case, it could be said that geotourism has been scientifically and practically started in Iran and is consistent with international standards. The revalidation process was supposed to be done on August, 2020, but it is postponed due to the Coronavirus Pandemic situation according to the UNESCO official website. Many private sector projects in Iran have also been postponed to the post-pandemic situation.

2.3.2 The History of Iran's Geoparks

In order to encourage developing countries, proclaimed to UNESCO in 2005 by [the late] Dr. Abdholazim Haghipour, former Director, Iran Geological Survey (Turner 2008), Qeshm became the site of the first Iranian Geopark. Actually, Qeshm was listed in GGN as the only geopark in the Middle East (Qeshm geopark website[1]) in the Paris conference on March 21, 2006. Unfortunately, discussing Qeshm as a geopark and its global registration was proposed before the publication of geotourism theories and textbooks in Iran; as mentioned above, the focus was on its aesthetic dimension of geological area when it was registered as a geopark. While there were no geoparks at that time, specialized books were scarce and the knowledge of the field was quite limited; there were only a few specialists who were not in charge of geopark responsibilities; even the national geopark committee did not exist at that time. Susan Turner, then UNESCO International Advisory Group for geoparks (pers. comm., Dec. 15, 2018) conveyed that "their neglect of Western style geotourism and poor participation in the GGN may have been underlying factors." Seemingly, It was actually the lack of knowledge on true geotourism among GSI cadres and not

[1] http://qeshmgeopark.ir/en.

allowing other experts to participate in the geopark project (Sadry 2020a). Unfortunately, Qeshm was delisted from GGN after 6 years in 2012. However, the geopark's case was resubmitted to UNESCO in 2014 after various measures were taken to solve the problems and to be considered as a candidate in the network again (IRNA News Agency) and in 2017, it was listed in the UNESCO's Global Geoparks Network again. But before this, in order to establish a national geopark committee in Iran, the Geological Survey & Mineral Explorations of Iran proposed its suggestion through the Ministry of the Industry, Mine and Trade in 2010 to the Infrastructure, Industry, and Environment Division of Iran government. This was the first time that geological heritage discussion was proposed among the top-rank executives. Finally, at the beginning of 2011 the regulations were approved by the government office (Safavi and Bahare 2011). According to the regulations, GSI is supposed to identify and investigate potential areas (actually geological heritages from the aesthetic viewpoints) to introduce them as the geoparks' regions all through the country. GSI is also responsible for setting regulations to utilize and preserve the areas with other organizations' cooperation. As a result, Iran was recognized among countries that have prioritized geological heritage and geoparks at a national scale. The suggestion of establishing a national committee of Iran's geopark was approved by the government (A. Amrikazemi, *pers. comm.,* Aug. 03, 2013). Unfortunately, despite the potential in geological diversity, heritage, and unique landscape, only one of the national geoparks has been listed in UNESCO's GGN thus far. Sadry (2020a, p. 546) utters that "The issue hindering the establishment and development of geoparks is preferring personal interests to national interests by those engaged in geoparks projects [from the governmental side]. ... We need many good and enthusiastic people to foster the geopark movement and not a narrow clique." Some of the other academically studied and proposed geoparks such as Takht-e Soleyman (the subject of the present chapter), Damavand, Sabzevar, and Tabas are under considerations next to Qeshm Island UNESCO Global Geopark and Aras Inspiring Geopark (in a preliminary semantic work by Sadry 2009a), which would not be discussed here anymore. Unfortunately, the national geopark committee in Iran is at the Green Paper stage and it has not commenced its activities practically; the members were not listed among the informed experts and entities.

However, it would appear that there is a bright future ahead of geopark movement in Iran, both theoretically and practically. In the following section, the methodology of the study area (the proposed Takht-e Soleyman geopark in the northwest of Iran) will be briefly discussed.

2.4 Methodology

Due to the large scale of the territory of the study area and also the absence of a domestic model to assess the geological diversity in Iran, the present study has applied assessment methods that have been used in Europe and South America. According to Lima et al. (2010) "for large territories it is not possible to make a geosites inventory based on systematic fieldwork," therefore, in defining a geological framework in Takht-e Soleyman with an area of 5500 km^2, the main focus is on regional geological contexts and geological domains, such as geomorphology and specially "cultural geomorphology" proposed by Panizza and Piacente (Panizza et al. 2009, pp. 35–48).

On the one hand, due to the lack of a specific scientific assessment method for the inventory of geodiversity in Iran, proposed by a person, a University, or by the Geological Survey & Mineral Explorations of Iran, a methodological proposal applied in Brazil by Lima et al. (2010) "for large territories" has been used in the present research (e.g., see Table 2.2 for one of the scoring sheets). On the other hand, according to Fassoulas et al. (2012) "sustainable development, education and conservation are core issues for successful management of any protected area." Therefore, the existence of protected areas and a World Heritage Site in the study region has given us a persuasive reason to apply Fassoulas et al. (2012)'s methodology besides that of Lima et al. (2010). In addition, determining values of Fassoulas and his colleagues' methodology concerning touristic, educational, and preservative requirements of geosites also well suits the present research (see Table 3.1 for another sample of a scoring sheet). According to Fassoulas et al. (2012), quantitative evaluation is a beneficial tool that explains both management and preservation requirements of a geoheritage site, since it establishes factors of sustainable tourism development, geotourism/educational tourism activities, and geosite conservation.

In this way, highly potential geosites would be easily identified and focused for essential steps toward tourism development or environmental education. Equally, priorities for preservation are established through the conservation rate and whether the geosite is likely to be vulnerable.

The research method is based on the quantitative study of the geomorphological and geological features and mines as well as determining their relationship with the sustainable tourism development in the framework of geotourism and geoparks.

Scoring sheets proposed by Lima et al. (2010) and Fassoulas et al. (2012) were used in the field and office works of this study. On this basis, in the current research in addition to direct observations and usage of GPS for locating potential geosites on digitized topographic maps, other subjects such as satellite images and aerial photos, digitized geological and topographic maps, road maps, location of rural points, cultural and historical sites, ancient buildings, and fauna and flora zones have also been indirectly considered. After gathering and preparing mentioned layers, they were stored in a spatial databank in a GIS environment to be used in the next stages. In this respect, basic theoretical frameworks including (Dowling and Newsome 2006), classification and definition of geotourism and geopark (Sadry 2009a, 2012b, c), and Geomorphosites (Reynard et al. 2009), GGN guidelines (2010), UNESCO (2000) have all been used in the present study in 2013–2014. Quantitative studies in the region based on geodiversity evaluation parameters (Lima et al. 2010) and management of geoheritage (Fasoulas et al. 2012) have been applied for methodology and for tourism development in the region; GIS models have been used.

In this study, aforementioned two quantitative methods have been used for the assessment of sites; each of them includes some parameters and sub-parameters.

Fasoulas et al. (2012) method is based on some criteria that considers not only the geological and geographical importance of a geosite, but also its ecologic, cultural, aesthetic, and economic importance. According to this method, scientific, ecologic, cultural, aesthetic, economic, and potential uses obtained for each geosite are used to estimate touristic, educational, protection necessity values of each geosite, respectively, on a scale from 1 to 10. According to the obtained results from different criteria, three parameters are recognized that pointed out to the touristic (V_{tour}), educational (V_{edu}), and protection necessity (V_{prot}) values of each geosite. In this study, the proposed formula and scores of the aforementioned criteria are analyzed and based on the studied value; coefficient with different weights are calculated for each criterion (see below Eqs. 2.1–2.3 and Table 2.1).

Table 2.1 Estimation of touristic, educational and protection values based on Fassoulas et al. (2012)

Site:					
Explanations:					
Criteria/score	1	2.5	5	7.5	bmolij10
1-Scientific value					
1-1 Geological history	One type history	Compilation of at least two types	Compiling from most types	Local report	Indicates all geological history of the region
1-2 Representativeness	No	Low	Intermediate	High	Very high
1-3 Geological diversity	Less than 5%	25%	50%	75%	More than 75%
1-4 Rarity	More than 7	More than 5 and less than 7	More than 3 and less than 4	More than 1 and less than 2	Unique
1-5 Degradation condition	Relatively destroyed	Strongly degradation	Moderately degradation	Weak degradation	Intact
1-Ecologic value					
2-1 Ecologic effect	No	Low	Intermediate	High	Very high
2-2 Protection condition	without protection	Limited	In some parts	In most parts	Completely
2-Cultural value					
3-1 Customs	No	Low	Intermediate	High	Very high
3-2 Historical	No	Low	Intermediate	High	Very high
3-3 Art and culture	No	Low	Intermediate	High	Very high
3-Aesthetics					
4-1 Observation site	No	1	2	3	More than 4
4-2 Different landscapes	No	Low	Intermediate	High	Very high

(continued)

Table 2.1 (continued)

Site:					
Explanations:					
Criteria/score	1	2.5	5	7.5	bmolij10
4-Economic value					
5-1 Visitors	Less than 5000	More than 5000	More than 20000	More than 50000	More than 75000
5-2 Attraction	No	Local	Regional	National	International
5-3 Official protection	International	National	Regional	Local	No
5-Potential uses					
6-1 Usage severity	Very sever	Sever	Intermediate	Weak	Unused
6-2 Effects	Very high	High	Intermediate	Low	No
6-3 Frangibility	No	Low	Intermediate	High	Very high
6-4 Accessibility	Near the walking path	Near the gravel or forest road	Near the local paved road	Near the regional road	Near the highway or town
6-5 Acceptable changes	No	Low	Intermediate	High	Very high

$$\text{Educational value } (V_{edu}) = 0.4 \text{ scientific} + 0.2 \text{ cultural} + 0.2 \text{ aesthetic} + 0.2 \text{ ecologic} \tag{2.1}$$

$$\text{Touristic value } (V_{tour}) = 0.4 \text{ aesthetic } + 0.2 \text{ cultural} + 0.2 \text{ potential uses} + 0.2 \text{ economic} \tag{2.2}$$

$$\text{Protection value } \left(V_{prot}\right) = \{\text{scientific} + F_{ecol} + (11 - \text{integrity})\}/3 \tag{2.3}$$

F_{ecol} = Ecologic score/protection necessity score

In the Lima et al. (2010) method, numerical indicators related to the value of each geosite are estimated among 1–100. The "Lima" method is suitable for large territories; consequently, this method was also used for establishing geoparks to assure the obtained result. For this inventory, describing the general and geological characteristics is enough. However, regarding the fact that in the next steps, these geosites will be used as a part of geo-conservation strategy, it is proposed that the mentioned characteristics would be developed and geosites' potential use (in addition to the scientific use), degradation risk of the geosites, and their educational and touristic use would be evaluated.

In order to save time and cost, it seems necessary to carry out complete evaluation during the general and geological evaluations. For evaluating the educational and touristic potential use of a geosite, both criteria must be evaluated (e.g., see

Table 2.2 Criteria used for evaluating educational and touristic potential uses of a geosite and related scores for final evaluation (Lima et al. 2010)

Criteria	Educational use score	Touristic use score
Representativeness	5	–
Quality of outcrop	10	5
Diversity in the state	5	–
Educational potential	30	–
Logistics	15	10
Number of inhabitants in distance 25 km	10	5
Accessibility	10	10
Vulnerability due to human activities	5	15
Association with other values (ecologic or cultural)	5	10
Structural greatness	5	15
Recreational potential	–	20
Social situation	–	5
Proximity to the recreational facilities	–	5
Total score	100	100

Table 2.2). It would be possible to rank them based on the final scores and to compare the results for accuracy. These are the same potential usability criteria for both types: for example, quality of outcrop, logistics, and accessibility. For a detailed evaluation, the deterioration risk of each geosite must be evaluated which is due to its potential use. Not all geosites considered in the inventory would need emergency management actions and therefore, it seems necessary to prioritize the existing monetary resources for the sites and to identify the high priority actions. After the evaluation of educational, touristic, and protection necessity values, the total score is estimated for each geosite and in order to estimate values between the studied geosites, the Inverse Distance Weighting (IDW) method is used in a GIS environment. In order to determine geotrails in the study area, the obtained raster file was used and high-value pixels became linear. Then, the resultant geotrail was corrected based on the type and characteristics of the existing roads in the area and final optimized geotrail was prepared. Making use of the abovementioned information, infrastructures (gas stations, clinics, etc.), villages, and rivers were plotted on final map and output map was prepared to introduce the proposed geopark and geotourism of the study area.

2.5 Study Area

The study area is located at longitude 46° 56′ 26″E, 37° 2′ 23″ N to 47° 50′ 58″ E, 36° 25′ 54″ N.

Fig. 2.2 The Ancient relics of Takht-e Soleyman (World Heritage Site) located 45 km from the northeast of Takab in Western Azerbaijan, Iran. It covers an area of 12 hectares and is considered a significant historical site. Traces of living have been discovered in the area which go back to 1000 B.C to eleventh century. During the Sassanid period, Azar Goshnasb fire temple was built as the main temple of Iran in this place. It is said that the fire was burning ceaselessly for seven centuries in this temple which was called "the immortal fire" all through the country. This site has been inscribed on UNESCO's World Heritage List (*Source* Google Earth, Archive of GSI)

It covers an area of 5500 km^2, lying between three provinces in the northwest of Iran: East Azerbaijan (Hashtrood city: Qareh-Aghaj district), West Azerbaijan (Takab city: Ahmad-Abad, Chaman and Saroogh district) and Zanjan province (Mahneshan city, Anguran, and Markazi district). The region is one of the geological sheets 1:100,000, Takab–Shahin Dej quadrangle. These regions are very rich in terms ofthe diversity of geological domains (geologically and geomorphologically), while completely different units are observed in a close distance (see Figs. 2.1, 2.2, 2.3, 2.4, and 2.5). This morphological diversity is due to climatic factors, petrological characteristics, erosional processes, weathering, and tectonic movements. Additionally, the existence of Angouran Protected Area, the Takht-e Soleyman World Heritage Site, and the Baich-e Bagh abandoned mine (copper, lead and zinc, gold, tin, tungsten, and molybdenum), which date back to the Reza Shah Pahlavi reign (Khoie et al. 2000), have added to the importance of the region. According to Amrikazemi and Mehrpooya (2006, p. 87) "Takht-e Soleyman World Heritage Site which is a geosite itself (see Fig. 2.2), comprises a natural lake, a complex of ancient buildings, the Dragon Stone (Sang-e Ezhdeha) and Zendan-e Soleyman Mountain .... and around the main site there are other hot water and travertine springs, some of which are used for bathing and water therapy both by local people and tourists."

Fig. 2.3 Ahmad Abad-Olia hot spa and recreation complex is located 2 km to the northwest of Ahmad Abad village. There are some hot spas in the south of Zendan Mountain, which arise from the Earth. The groundwater temperature is 40-degree centigrade which is said to have therapeutic effects and is known as a tourism destination with favorable facilities for visitors throughout the year (*Source* Ali Haji-Moradi)

The study area is located in a mountainous region, covered with snow in winter with a mild weather during summer. The Peak of Belgheis Mountain is 3330 m and the lowest elevation point is 1464 m above the sea level (in the north of Qareh-Aghaj village).

Generally, the northern part of the study region is covered with metamorphic rocks, gneiss, marble rocks, amphibolites with rough topography, and steeped valleys while the southern part includes Takht-e Soleyman World Heritage Site consisting of Neogene Marl deposits and Miocene sandstone and calcareous depositS having a smooth topography than the north part. There are five mountains in the area with an altitude of 2800 m and important rivers flowing into the region such as Ai-DuGh-mush, Pari, Ghaleh khan, Anguran Chai, Donge, and Cham-Agh-Dareh. The average annual precipitation is about 300–400 mm. Takab—Ahmad Abad asphalt road connects Takht-e Soleyman World Heritage Site and Zendan-e Soleyman. The area is accessible through Zanjan—Tabriz highway to Mahneshan, Tabriz—Mianeh highway to Hashtrood and Tehran—Zanjan highway to Bijar. Urmia Lake is the closest water to the region and the study area has 44 disadvantaged villages nearby to be considered for the determination and feasibility of becoming a geopark (see Figs. 2.3, 2.4, 2.5 and 2.6).

Fig. 2.4 Arze-Khoran Castle, one of the Mahneshan castles which is a military one. Many archeologists are attracted to this place (*Source* Azar Mohammadi-Aragh)

2.6 Findings of the Research

In the study area, 48 important sites including geological sites, related cultural sites, and eco-sites were recognized (see Table 2.3) (Also see Fig. 2.6). Based on the aforementioned final formula (Eq. 2.3), in the Fassoulas et al. (2012) method, numerical index of protection necessity is calculated separately for each site which is indicated in Table 3.3. Regarding the educational potential value, the Numerical value for educational activities varied from 1 to 38. In the study region, some sites have educational value higher than 20, including Mine sites of the region, travertine fans, badlands, Chamali moving grass, hot and cold water springs, castles, hoodoo, different erosional forms, colored formations, lakes, and Nabikandi, Berenjeh, and Soleyman prisons (see Table 2.3).

Regarding the Touristic value, Numerical index for this value varied from 1 to 32. In the study region, some sites have touristic value higher than 20, including travertine fans, Chamali moving grass,[2] hot and cold water springs, badlands, hoodoo, different erosional forms, colored formations, lakes, other castles, and Nabikandi, Berenjeh, and Soleyman prisons. As for Protection necessity, to determine the numerical index of protection necessity (Eq. 2.3), the mean of scientific and integrity values

[2]Recently listed as national natural heritage by the government.

Fig. 2.5 A cultural geomorphology site, Behestan natural and manmade castle, is located 11 km from Mahneshan. According to the excavations in this area, the history of the castle goes back to the early and middle years of the Islamic period in Iran. However, some attribute it according to the terminology. The castle has 64 rooms and two staircases (*Source* AzarMohammadi-Aragh)

is considered. If this value is high, then the protection necessity is minimum. In this research, we used the formula Eq. 11 minus integrity value. Therefore, the resultant value will be from 1 to 10 which is compatible with the scoring system. In addition, there is another factor named the ecologic risk factor. The ecologic risk factor (F_{ecol}) is calculated based on the ecologic factor score (Parameter 1-2) divided by the protection necessity score (Parameter 2-2) (F_{ecol} = Ecologic effect score/protection condition score). If the ecologic effect and protection condition scores are high and low, respectively, the ecologic risk factor will be even higher.

Based on this formula, the highest protection necessity score is related to the region mines, travertine fans, badlands, Chamali moving grass, cold and hot springs, hoodoos, different erosional forms, colored formations, lakes, castles, and Nabikandi, Berenjeh, and Soleyman prisons.

1. Table containing criteria to evaluate potential educational use of geosites,
2. Table containing criteria to evaluate potential touristic use of geosites,
3. Criteria to evaluate potential educational and touristic uses of geosites and related scores for final evaluation,
4. Evaluation of degradation risk of a geosite (it is of great importance for management objectives and determining priorities of action plans).

Fig. 2.6 Up to down and left to right, respectively: 1—Baderlou erosional forms, 2—Shakourchi colored formations, 3—Nabikandi Prison, 4—Behestan Castle, 5—Erosional forms, 6—Zendan-e Soleyman (Soleymanprison), 7—Ghinerje sulfur hot spring, 8—Takht-e Soleyman World Heritage site, 9—Ahmadabad-e Olya hot spring 10—Erosional forms, 11—Laklakha castle, 12—Ahmadabad-e Olya hot spring, 13–Takht-e Soleyman World Heritage Site (*Source* Ali Haji-Moradi)

Criteria evaluation used in this method are comparable and in harmony with the Fassoulas et al. (2012) method. In this respect, numerical indices were obtained and based on these values; the following results have been concluded.

Figure 2.7 indicates the results of site evaluation in which the resultant scores are shown as circle diagrams in front of the sites. Circle diagrams and their valuation have been carried out based on educational, tourism, and protection necessity values

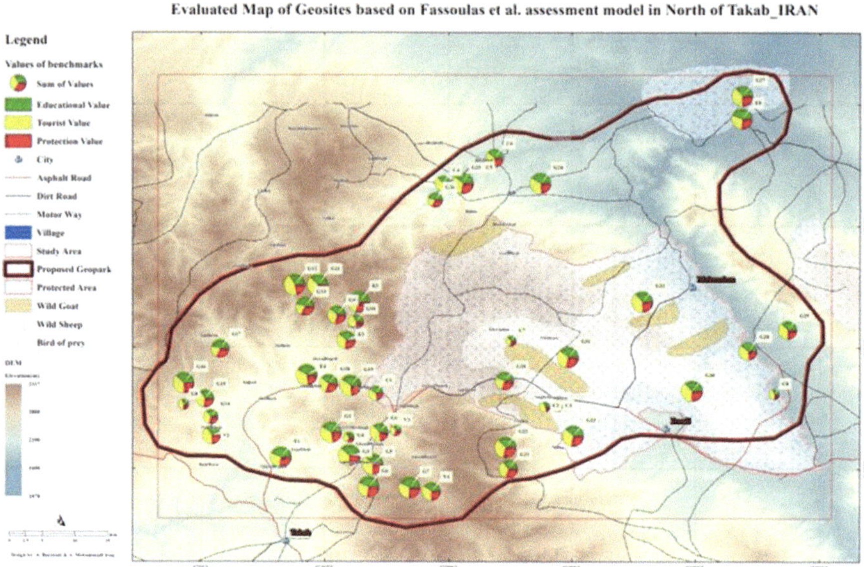

Fig. 2.7 Evaluation of criteria in the sites of the study area based on the Fassoulas et al. (2012) method. In the Lima et al. (2010) method, four separate identity certificates are prepared for each site

formula mentioned in the methodology. The size of each diagram is designed based on integrity criteria.

As mentioned in the research methodology, there is a protected area in the study region. In the wildlife region of Angouran located in the study, diverse variety of animals live including 33 species of mammals, 107 species of birds, 7 species of amphibians, and 78 species of insects. Typical animals of the region include brown bear, wolf, wild goat, wild sheep, rabbit, hog, etc.; and typical birds include eagle, stork, and so on. The area also is representative of a diverse flora that is not going to be discussed here. There is biodiversity besides geoheritage both in the protected area and out of it which needs further investigation. However, to find out and access the basic data and information for the present research, flora and fauna zoning, geology, and geomorphology mapping of the region by some state organizations or research institutions were used. Some of these organizations include Forests, Range and Watershed Management Organization (2004), and East Azerbaijan Environmental Protection Bureau (1993). Basic geological and geomorphological map of Takht-e Soleyman provided by Geography Research Institute (2006) and Babakhani and Ghalmash (Babakhani 1998) in GSI and Zarei-Nejad (Nejad and Mojgan 2009) were also used which were considered and presented in the final determination of the geopark territory determination (with biotic, abiotic, and cultural diversity) and by drawing a preliminary geotrail (see Fig. 2.9). According to the resultant numerical indices, 98% of the sites in the study area have high touristic, educational, and

Table 2.3 Socio-economic, cultural, ecological and scientific value of the various sites

Raw	Name	Integrity value	Ecologic effect value	Protection condition	Scientific value	Ecology value	Cultural value	Aesthetic value	Economic value	Potential use value	F_{ecol}	V_{edu}	V_{tour}	V_{prot}
1	Travertine terraces	7.5	1	2.5	13	3.5	4	10	11	13	0.4	8.7	9.6	5.633333
2	Qezel Gheshlagh travertine fans	10	1	10	33.5	11	17	12.5	12.5	40	0.1	21.5	18.9	11.53333
3	Berenjehpond	10	1	10	26	11	4	4.5	12.5	40	0.1	14.3	13.1	9.033333
4	Berenjeh spring	2.5	1	7.5	14.5	8.5	4	5	11	18.5	0.1333	9.3	8.7	7.711111
5	Baderlou road badlands	10	1	1	37.5	2	4	20	16	37.5	1	20.2	19.5	13.16667
6	Chamalimoving grass	10	10	10	47.5	520	4	11	20	40	1	26	17.2	16.5
7	Chahartagh hot and cold springs	10	5	10	32.5	5	4	8.5	15.5	30	0.5	18.5	13.3	11.33333
8	Chahartagh geo-village	7.5	1	2.5	35	3.5	21	8.5	15	35	0.4	20.6	17.6	12.96667
9	Ghezel-ghapan castle Bardiyeh castle	2.5	1	1	17.5	2	8	8.5	11	37.5	1	10.7	14.7	9
10	Nabikandi geo-village	7.5	1	2.5	18.5	3.5	4	8.5	11	30	0.4	10.6	12.4	7.466667
11	Alghadir Dam Googerdchi dam	10	5	10	37.5	15	4	11	12.5	31	0.5	21	13.9	13
12	Hampa geo-village Afshar-Khancastle	2.5	1	2.5	22	3.5	4	7.5	13.5	27.5	0.4	11.8	12	10.3

(continued)

Table 2.3 (continued)

Raw	Name	Integrity value	Ecologic effect value	Protection condition	Scientific value	Ecology value	Cultural value	Aesthetic value	Economic value	Potential use value	F_{ecol}	V_{edu}	V_{tour}	V_{prot}
13	Ghinerjeh Entrance waterfall	7.5	5	2.5	32.5	7.5	4	12.5	17.5	32.5	2	17.8	15.8	12.66667
14	Ghinerjeh waterfall and hot spring	10	5	2.5	45	7.5	4	10	20	32.5	2	22.3	15.3	16
15	Ahmad Abad storage dam	10	2.5	10	25	12.5	4	12.5	8.5	31	0.25	15.8	13.7	8.75
16	Ghinerjeh sulfuric hot spring	7.5	5	2.5	35	7.5	4	10	20	35	2	18.3	15.8	13.5
17	Ghinerjeh spring Baladeh	5	2.5	2.5	17.5	5	4	10	12.5	35	1	10.8	14.3	8.166667
18	Shirinsou spring	10	1	1	30	2	4	5	11	38.5	1	14.2	12.7	10.66667
19	Bekheir-Bolaghi spring	10	1	1	20	2	4	10	11	38.5	1	11.2	14.7	7.333333
20	Zarshouran gold mine	7.5	2.5	10	47.5	12.5	10.5	5	12.5	23.5	0.25	24.6	11.3	17.08333
21	Bikandi prison	5	2.5	1	37.50	3.5	4	12.5	20	42.5	2.5	19	18.3	15.33333
22	Ghaz-ghachankarstic craters & valley	7.5	2.5	1	25	3.5	4	7.5	13.5	40	2.5	13	14.5	10.33333
23	Balakhaneh castleGhez-ghapan	10	1	1	37.5	2	12	15	17.5	37.5	1	20.8	19.4	13.16667
24	Agh-Darreh gold mine	7.5	7.5	10	47.5	17.5	4	10	12.5	23.5	0.75	25.3	12	17.25

(continued)

Table 2.3 (continued)

Raw	Name	Integrity value	Ecologic effect value	Protection condition	Scientific value	Ecology value	Cultural value	Aesthetic value	Economic value	Potential use value	F_{ecol}	V_{edu}	V_{tour}	V_{prot}
25	Ahmad Abad hot spring complex	7.5	5	2.5	40	7.5	4	7.5	20	37.5	2	19.8	15.3	15.16667
26	SoleymanPrison	10	1	10	47.5	11	14.5	20	31	42.5	0.1	28.1	25.6	16.2
27	Salman terrace	5	2.5	10	42.5	12.5	31	20	21	42.5	0.25	29.7	26.9	16.25
28	Angouran lead-zinc mine	7.5	2.5	10	47.5	12.5	4	12.5	17.5	26	0.25	24.8	14.5	17.08333
29	Mianj cave-1	7.5	2.5	1	20	3.5	4	5	13.5	22.5	2.5	10.5	10	8.666667
30	Mianj spring & cave-2	10	2.5	1	30	3.5	4	5	13.5	37.5	2.5	14.5	13	11.16667
31	Shoura-gol lake & hot spring	10	2.5	1	35	3.5	4	10	13.5	36	2.5	17.5	14.7	12.83333
32	Ganj Abad castle	2.5	1	1	14.5	2	18.5	10	13.5	42.5	1	11.9	18.9	8
33	Ganj Abad old house	2.5	1	1	12	2	11	12.5	13.5	42.5	1	9.9	18.4	7.166667
34	Khandagh-lou or Pari lake	10	7.5	10	35	17.5	4	17.5	12.5	38.5	0.75	21.8	18	12.25
35	Hezar-darreh Arze-khoran	10	1	1	40	2	4	12.5	13.5	40	1	19.7	16.5	14
36	Arzeh-khoran castle	5	1	1	37.5	2	9.5	20	17.5	42.5	1	21.3	21.9	14.83333
37	BaichehBagh Mine	5	1	10	45	11	12	5	15	18.5	0.1	23.6	11.1	17.03333
38	Church	2.5	1	2.5	24.5	3.5	18.5	10	8.5	40	0.4	16.2	17.4	11.13333
39	Arzeh-khoranHusainiyah	7.5	1	2.5	29.5	3.5	21	10	12.5	35	0.4	18.7	17.7	11.13333

(continued)

Table 2.3 (continued)

Raw	Name	Integrity value	Ecologic effect value	Protection condition	Scientific value	Ecology value	Cultural value	Aesthetic value	Economic value	Potential use value	F_{ecol}	V_{edu}	V_{tour}	V_{prot}
40	Lak-lak-ha castle	10	10	10	40	20	4	17.5	15	25	1	24.3	15.8	14
41	Shakour-chi colored formations	10	1	1	32.5	2	4	12.5	17.5	38.5	1	16.7	17	11.5
42	Behestan castle	7.5	1	2.5	40	3.5	25	15	22.5	40	0.4	24.7	23.5	14.63333
43	Illi-bolaghhoodoo	7.5	1	2.5	37.5	3.5	4	20	17.5	40	0.4	20.5	20.3	13.8
44	Talzeh-kand erosional forms	10	1	1	27.5	2	4	10	13.5	36	1	14.2	14.7	9.833333
45	Honeycomb erosional forms	10	1	1	37.5	2	4	10	13.5	37.5	1	18.2	15	13.16667
46	Yasti castle	2.5	1	1	26	2	18.5	5	13.5	31	1	15.5	14.6	11.83333
47	BorounGheshlaghCastle	2.5	1	1	14.5	2	9.5	10	6	35	1	10.1	14.1	8
48	Mahneshan colored formations	10	1	1	35	2	4	17.5	20	38.5	1	18.7	19.5	12.33333

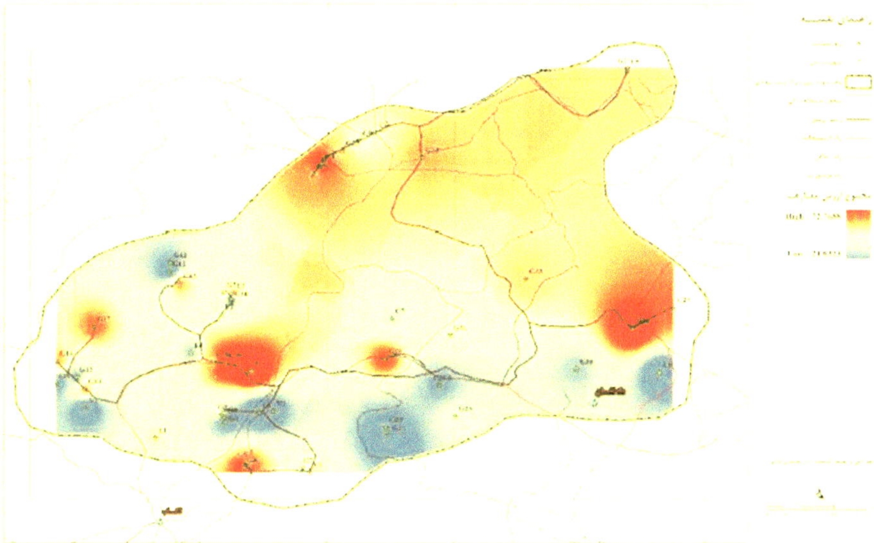

Fig. 2.8 Creating geotrail based on interpolation of the total values of the sites in the study area

protection necessity values. In order to determine the proposed area of the geopark, values of the criteria especially the protection necessity value indicates that all sites are located in the proposed area.

Figures 2.9 and 2.10 indicate the determined area of the geopark. Geotrails work as leverages of geo-conservation; special routes for visiting could be utilized in UNESCO Biosphere Reserve, national parks, and geoparks, or even individual scattered sites (Khoshraftar Khoshraftar and Reza 2011). As indicated in Fig. 2.9, to determine preliminary geotrail, the sum of three criteria is required to be calculated for each site; these criteria include education, tourism, and protection. Then, the point file of the sites is transformed to the raster file making use of the IDW method and finally, high-value pixels are used to determine geotrails of the study area (see Fig. 2.8). It must be noted that the final geotrail modified based on the existing road map of the region (Fig. 2.8). In general, in Fig. 3.9, the best preliminary geotrail for future development of geotourism was concluded and is indicated on the final geopark map of the study area.

2.7 Conclusion

The study area consists of considerable geoheritage sites and other precious heritage sites, and it seems necessary to be nationally protected.

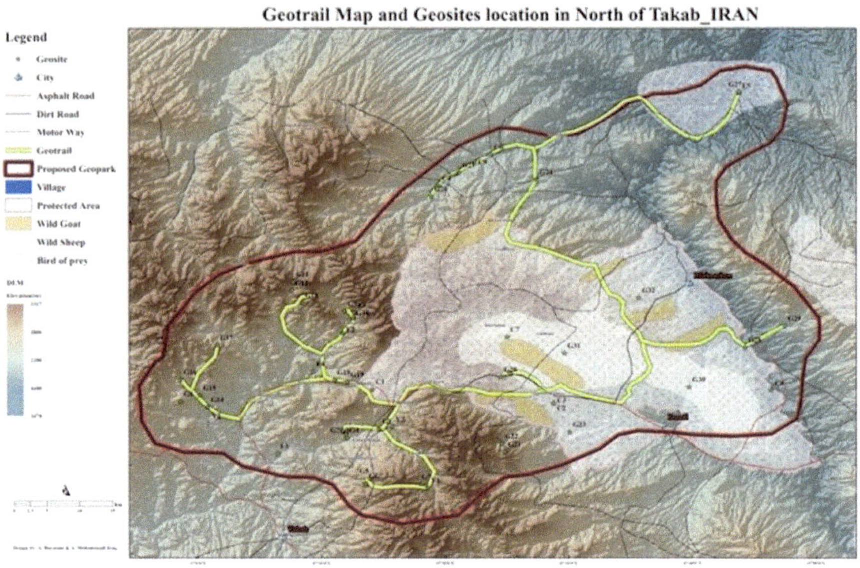

Fig. 2.9 Geotrail map and geosite location and wildlife zoning sites within the study area

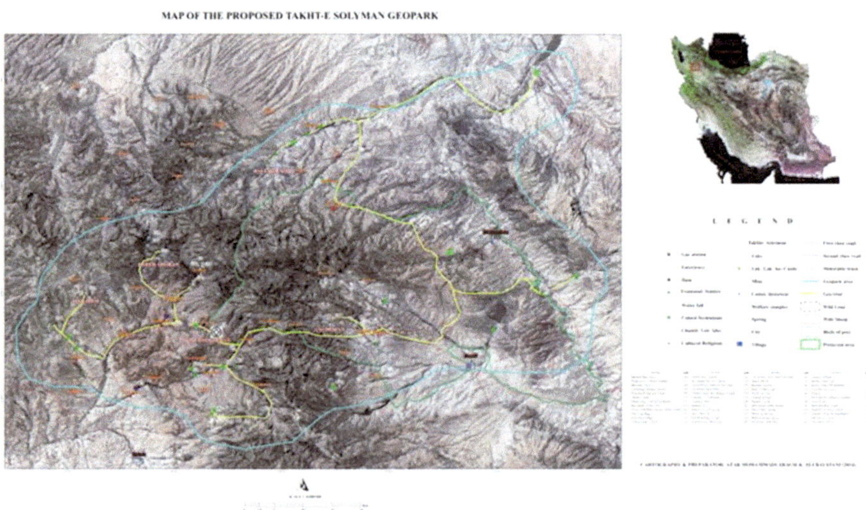

Fig. 2.10 Proposed Geopark map of Takht-e-Soleyman region in Takab

Sustainable utilization of the aforementioned heritage is potentially considered as geotourism development, and it is susceptible to be a national and/or international geopark.

Geotourism value is determined based on abiotic environmental features, biotic attributes as well as cultural–historical sites. It proves the high importance of the region from geological and geomorphological points of view. Geoheritage landscapes selected have made use of the evaluation as a geosite and a geomorphosite. They have the highest value for the protection and geotourism development that indicates their scientific, protection, and economic importance for the geotourism of the region (see Fig. 2.10).

In addition, easy access to the studied geosites is among the most important advantages for the development of geotourism in the region. However, there is still no geotourism infrastructure to make the region accessible for touristic and educational and recreational potential use.

Finally, regarding the fact that the main factor in the establishment of geoparks is geosites and local people, the region has a considerable potential for establishing geoparks, and the other issues depend on future studies by scientists and complementary geological studies. The most important outcome of the current study is introducing the Takht-e-Soleyman geopark area to conserve the geoheritage of the region, raise social awareness among the indigenous people, increase the popularity of geology, and lessen poverty through making use of geotourism development that may be applicable in the future. We hope that the writing of the present chapter could contribute to the proposed region to become a national and later a UNESCO Global Geopark in the future. We also hope that all developing countries including Iran could come up with special evaluation/assessment techniques related to the geographical and cultural features of their country and monitor every single corner of their land in search of any heritage potential, Especially geoheritage, toward developing this valuable strategy (establishing Geoparks) and mitigate poverty on the planet.

Acknowledgments We hereby would like to thank Dr. S. Saeed Hashemi, the Former Dean of Tourism faculty at the University of Science and Culture, for providing official requirements of the research project for us. The present study has been done as a master dissertation in eco-tourism by Azar Mohammadi-Aragh, under the supervision of her principal mentor, Dr. Sadry, at the University of Science and Culture. Since she was also a senior expert at Geological Survey & Mineral Explorations of Iran (GSI), simultaneously, the project has been used and named as a research project to facilitate support by her organization and consequently, the field visits and camping of the project have been supported financially and spiritually by adding human resource to this huge project by the GSI. And also, it was appointed as among the top research projects of her organization (GSI). We hereby thank the Geological Survey & Mineral Explorations of Iran (GSI) for financial support of the filed visits of this project. Also, finally, we are thankful to Mr. Amir Hekmat for providing us with some final graphic amends of the maps to be used in this chapter.

References

Amrikazemi A, Mehrpooya, A (2006) Geotourism resources of Iran. In: Dowling RK, Newsome D (eds) Geotourism, Oxford-Burlington (Elsevier Butterworth-Heinemann), pp 78–92

Amrikazemi A (2013) Email personal communication with Alireza Amrikazemi

Babakhani A, Ghalmash K (1998) Geological Map 1:100000, Takhte-Soleyman sheet, Geological & Mineral Explorations Survey of Iran (GSI)

Chen A, Lu Y, Ng CY (2015) The principles of geotourism. Springer Geography, Springer-Verlag, Science Press Ltd, Berlin, Heidelberg

Dowling RK, Newsome D (eds) (2018) Handbook of geotourism. Edward Elgar Publishers, Cheltenham

Dowling RK, Newsome D (eds) (2006) Geotourism. Oxford, Burlington (Elsevier Butterworth, Heinemann)

East Azerbaijan Environmental Protection Bureau (1993) Investigating past and present state of East Azerbaijan and the image its future environment. The report of investigating the state of East Azerbaijan province, Thorough environmental investigation of East Azerbaijan Province(in Persian)

Forests, Range and watershed Management Organization (2004) Mapping project of Iran's vegetation coverage via Landsat5-TM in blocks of 1:250000 in the present study area (Engineering office) (in Persian)

Fassoulas C, Mouriki D, Dimitriou-Nikolakis P, Iliopoulos G (2012) Quantitative assessment of geotopes as an effective tool for geoheritage management. Geoheritage J 4:177–193. https://doi.org/10.1007/s12371-011-0046-9

Ghasemi M (2010) Foreword: about the author. In: Amrikazemi A (ed) Atlas of geopark and geotourism resources of Iran: Geoheritage of Iran (P:11). GSI Publishing, Tehran

GGN Guidelines (2010) Guidelines and criteria for National Geoparks seeking UNESCO's assistance to join the Global Geoparks Network (GGN). http://www.unesco.org/new/fileadmin/MULTIMEDIA/HQ/SC/pdf/sc_geoparcs_2010guidelines.pdf. Accessed 1 Dec 2018

Hajalilou B, Sadry BN (2011) Geotourism. Payame Noor University Press (Geotourism textbook in Persian)

Hose TA (2006) Geotourism and interpretation. In: Dowling RK, Newsome D (eds) Geotourism. Oxford, Burlington (Elsevier Butterworth, Heinemann), pp 221–241

IRNA (2017) Submitting the Qeshm geo-park case to UNESCO to be listed in GGN again—Islamic republic of Iran News Agency online link: http://www.irna.ir (In Persian)

Khoshraftar R (2011) The role of geotrails in geoconservation. In: The proceedings of the 9th conference of Geology in Iran. Geological & Mineral Exploration Survey (GSI) of Iran (In Persian)

Khoie N, Qorbani M, Tajbakhsh P (2000) Copper deposits in Iran. Geological Survey & Mineral Explorations of Iran

Lima FF, Brilha J, Salamuni E (2010) Inventorying geological heritage in large territories: a methodological proposal applied to Brazil. Geoheritage J 2(3–4):91–99

Maghsoudi M, Barzkar M, Abbasi M, Moradi A (2014) Evaluation of geotouristic potentials of geomorphosites in Mahabad County. [Iranian] J Tour Stud. Spring No. 25. https://www.magiran.com/paper/1418249, Accessed 15 Nov 2019 (In Persian)

Mojgan Z-N (2009) Mapping geomorphology of Takht-e-Soleyman Sheet 1:100000, Master thesis of geomorphology, Geography Department. Islamic Azad University (Branch of Reyh),Tehran (In Persian)

Nabavi M-H (2000) Geotourism (geological tourism). In: The proceedings of the 18th Symposium on Earth Sciences. Geological Survey &Mineral Exploration (GSI) of Iran (In Persian), pp 1039–1049

Nabavi M-H (2011) 'Geologist is responsible for introducing geological attractions of Earth to the community'; Geoscience &Mining Monthly Magazine (Geoheritage Special Edition:

Geodiversity-Geoconservation-Geopark–Geotourism); Geological Survey & Mineral Exploration of Iran (GSI), pp 94–95

Panizza M, Piacente S (2009) Cultural geomorphology and geodiversity. In: Emmanuel R, Paola C, Géraldine R-B (eds) Geomorphosites. München, Verlag Dr. Friedrich Pfeil, pp 35–48

Patzak M, Eder W (1998) UNESCO GEOPARK, a new programme—a new UNESCO label. Geol Balc 28(3):433–35

Research Institute of Geography (2006) Geomorphologic researches in Takht-e Soleyman, Faculty of geography, University of Tehran (In Persian)

Reynard E, Brilha J (eds) (2018) Geoheritage: assessment, protection and management. Elsevier Inc, Amsterdam

Reynard E, Coratza P, Regolini-bissig G (eds) (2009) Geomorphosites. PfeilVerlag, München

Sadry BN (2006) Nature [Geotourism & Ecotourism]: a new approach for tourism industry. In: Proceedings of the first conference of the opportunities and threats of promoting tourism industry in East Azarbaijan Province, The Cultural Heritage, Handicrafts and Tourism administration of East Azarbaijan Province, 25 December, held in Tabriz(Hotel Pars), Iran (In Persian)

Sadry BN (2008) Zoning of Iran's geotourism potential Areas. In: The proceedings of the first geotourism conference in Iran, the Geological Survey & Mineral Exploration of Iran(GSI), October, 12 (In Persian)

Sadry BN (2009) Fundamentals of geotourism: with special emphasis on Iran, SAMT Organization publishers, Tehran, p 220. (English Summary http://physio-geo.revues.org/4873?file=1)

Sadry BN (2009a) Nomonation of suitable area for geopark and geotourism develoment in Jolfa and Varzeghan region in Iran. In: Brilha J, Pereira D, Pereira P (eds) Intensive course on geodiversity and geological heritage assessment workbook. University of Minho, Portugal, p 19

Sadry BN (2011) The role of geo-interpretation in the development of geotourism in Iran. In: The proceedings of the first national conference on tourism and sustainable development, Faculty of Geography and environmental planning, Islamic Azad University of Zahedan, Iran, May 24–26. (In Persian), pp 523–534

Sadry BN (2012a) Geodiversity & geoheritage education through geotourism textbooks in Iran. Geophysical Research Abstracts, EGU General Assembly, Vienna, 22–27 April, Austria, vol 14, EGU2012-14008. https://ui.adsabs.harvard.edu/abs/2012EGUGA..1414008S/abstract. Accessed 07 Nov 2019

Sadry BN (2012b) Geotourism industry at a glance in Iran: wrong beliefs and definitions. In: A presentation at the 2nd conference of geomorphology studies in developing and managing, Islamic Azad University Campus of Central Tehran, Iran (In Persian)

Sadry BN (2012c) Iran's geodiversity, a massage of wealth and welfare to Iranians, Iranian Mining and Engineering Organization. Magazine 13:46–59 (In Persian)

Sadry BN (ed) (2020a) The geotourism industry in the 21st century: the origin, principles and futuristic approach. Apple Academic Press Inc, USA

Sadry BN (2020b) An introduction to the geological heritage of Iran: Geosites & Geomorphosites; Tehran: Mahkameh Publishers & Iranian Association of Geomorphology Publishing (In Persian, forthcoming)

Safavi B (2011) Travel to our geoscience heritage, Jame Jam Newspaper, No. 3493, p. 12. http://www.magiran.com/npview.asp?ID=2570083 (In Persian)

Turner S (2018) Email personal communication with Dr. Susan Turner

Turner S (2008) Geotourism: where wise birds stay a while (Qeshm Island: Geoheritage Pearl versus Gasfields in the Persian Gulf?). GeoExpro Magazine. 60:56–58. www.GeoExpro.com

Turner S (2006) Forging a Geopark Network—Geopark Models, Planning and the Role of local Minorities and Women. In: Geotopge—Bausteine der Regionalentwicklung, Schriftenreihe der GesellschaftfürGeowissenschaften, Heft XX, S.-S

Turner S (2005) Assessment Report on Qeshm Geopark, Qeshm Island, Qeshm Free Area, Iran. Qeshm Geopark—the Geoheritage Pearl in the Persian Gulf. Report to UNESCO and Global Geoparks Network. 98 p., PrinteShoppe; Brisbane

UNESCO (2000) UNESCO Geoparks Programme Feasibility Study, UNESCO, Paris

Vafadari K, Cooper MJM (2020) Community engagement in Japanese geoparks. In: Sadry BN (ed) The geotourism industry in the 21st century: the origin, principles and futuristic approach (pp 357–373). Apple Academic Press, Inc., USA

Yamani M, Negahban S, Harabadi S, Alizade M (2012) Geomorphotourism and a comparison of geomorphosites' evaluating methodology in tourism development: a case study of Hormozgan province. Tour Plan Dev J 10–83 (In Persian)

Zouros N, Martini G (2003) Introduction to the European geoparks network. In: Zouros N, Martini G, Frey M-L (eds) Proceedings of the 2nd European geoparks network meeting, Lesvos, Natural History museum of the Lesvos Petrified forest pp 17–21

Chapter 3
Geoheritage: A Strategic Resource for the Society in the Anthropocene

Francesca Romana Lugeri, Piero Farabollini, and Vittorio Amadio

Abstract What geoheritage means for contemporary society? Socioeconomic developments pushed us in a new era, defined postmodern. Technological development requires a knowledge that goes hand in hand with its speed. The break between progress, society, environment, could be causing unpredictable effects on the planet: this is the Anthropocene paradigm. Knowledge, provided by science, seems to be the only weapon that leads us, troughs the shared awareness, toward changes. New and unconventional ways of communicating sciences, particularly earth sciences, are requested to provide a responsible interaction with the environment, which should massively involve societal tissue for a sustainable future.

Keywords Geoheritage · Geotourism · Knowledge · Soceity · Landscape · GIS

3.1 Introduction

The variety of our planet physiographic structures, are due from physical, biotic, and anthropic dynamics, interacting both in space and in time: an abundance of different landscapes are offered by a particular morphological configuration and geographical distribution of our planet's surface.

However, in many cases, the beauty of natural scenery, made of different special land settings, is a sign of extreme vulnerability.

The deterioration of landscapes characterizes most of the areas of elevated environmental value and high vulnerability, especially the touristic ones.

F. R. Lugeri (✉)
ISPRA Institute for Environmental Protection and Research, Roma, Italy
e-mail: francesca.lugeri@unicam.it

P. Farabollini
School of Sciences and Technology, University of Camerino, Camerino, Italy
e-mail: piero.farabollini@unicam.it

V. Amadio
OASI Department, University of Reggio Calabria Mediterranea, Reggio Calabria, Italy
e-mail: Vittorio.Amadio@unirc.it

R. B. Singh et al. (eds.), *Global Geographical Heritage, Geoparks and Geotourism*, Advances in Geographical and Environmental Sciences,
https://doi.org/10.1007/978-981-15-4956-4_3

Loss of landscape, loss of soil, and loss of biodiversity: an equation, whose solution is exponential, and whose effects are irremediable if we don't intervene promptly and appropriately. Furthermore, it is worth underlining that landscape loss has negative effects even on local and global cultures. Levy Strauss (1957), in his "Tristes Tropiques", shows us how dangerous is the loss of the landscape, referring to the symbolic value that landscape encompasses in every culture.

Landscape is a crucial element to set the quality of life and for the development of local cultures: this is what the European Landscape Convention determinate in 2000, adding that quality of landscapes has positive effects on social well-being.

The main causes of landscape's degradation are identified in the following crucial factors: climate change, anthropic pressure, lack of protection, prevention absence. Currently, human activities managed to influence the atmosphere and change its stability, and this is the first time in the planet history: the Anthropocene (Crutzen 2006). These factors are the result of the insufficient social awareness about what landscape is, about its importance for the life quality both of human beings and of Earth.

The landscapes that make up the background to our lives, are part of cultural heritage, when recognized, understood, and decoded, referring to their natural history and value. A virtuous circle can be activated by a mental process, that could empathize the link between humankind and environment, and encourage the participation of everybody in the management and development of territory, a successful strategy in these time of crisis. Communicating scientific concepts, as the explaining of what landscape is, why it reveals multiple shapes, why it is continuously evolving and mutation, which resources can be offered to us and which kind of danger could be hidden, is the leading step to reach this aim. All this in an easy, comprehensible, sustainable way.

An essential objective is to reach the widest possible audience, especially considering that time we are living in, the so-called Postmodernity, is characterized by an excess of tools and sources, that offer everyone the possibility to participate in social dynamics, however without proper guidance: the effects are often superficiality and knowledge fragmentation (Bauman 1997).

3.2 Sharing the Knowledge: A Geoethic Objective

Humankind must become aware of the environment: the knowledge of the environment in which we live and we are part of is fundamental for our survival. Territorial Science sets a new goal, represented by the diffusion of scientific heritage, through topics well known and appreciated. Future of our planet appears to be linked to humankind behavior, now more than ever: a new subject, very topical nowadays, is geoethic. A mix between philosophy and geoscience, geoethic arouses a new strong interest from scientists, aiming to transform into actions the result of those reflections (Peppoloni and Di Capua 2012). Earth sciences finally are worthy to be investigated through new ways of thinking, which involve ethical, social, and cultural implications of geoscience research, practice, and education. Analyzing the meaning, ethics is defined as the behavioral rules, recognized concerning a peculiar class of human beings, or group or culture. Thus, if nature is the existence and its foundation, with physical and biological settings, or different characteristics, in a region not yet modified by civilization, it is easy to understand how clear is and innate the integration between Earth Science and Ethics. There are new challenges for geoscientists: new skills and task that connect different disciplines have been developed, through different methodological procedures and technologies, facing different social, cultural, and scientific challenges, from micro to macro-scale studies (Martinez Frias 2011).

Referring to the prevention of risk (one of the main objectives in each concrete applied science's project) we can affirm that knowledge, consciousness, and action are fundamental in prevention. Referring to environmental risk, preventive attitudes and behaviors can be considered as elements of public health.

3.3 Knowledge and Communication

How to strategically organize information is a crucial point in scientific communication. It is important to know how to effectively communicate contents to the public. Landscape could become as a vector in the communication of earth science to the people: here, natural and cultural components interact, giving rise to the expression of the geological process that molds the Earth's surface. Landscape plays a fundamental role in the sharing of Earth science, being the object of human perceptions, and which the population interact whit (Council of Europe 2000). Aristotle, through the expression "shape is synthesis" teaches us to use an approach in the study of nature, based on reading the features of the land, as an expression of geological and environmental settings. New approaches in the study of the landscape, position it as an element of a system where geology leads, operating as an activator in the inspiration of emotions that otherwise would not be appreciated.

3.4 Geotourism: A Way Toward Sustainable Development

In tackling sustainable management of territory and landscape issues, it is a mandatory methodological approach, both scientifically and culturally deemed, that integrates all environmental and social elements and contains, as primary elements, awareness, participation, and public instruction. The management of an ideal equilibrium between using and respecting the territory, is fundamental; this approach is currently a way to reach a sustainable development of nature ecosystem (Wyss and Peppoloni 2014).

What is geotourism, and how can it help us with the territory management in a balanced, sustainable way? Tourism can be defined as "Geo" if it respects the canons of sustainability while protecting the geological characteristic of Earth in a way that encourages cultural and environmental understanding, and is locally beneficial (Dowling and Newsome 2006). It is a complementary form of tourism where the elements of the landscape and landforms, cause a more complete experience, whose result is more than the sum of the parts (Dowling and Newsome 2010; Allan 2012). The landscape is the integration between physical, biotic, and anthropic components, that operate in different space–time scales (Catton and Dunlap 2003; Forman and Godron 1986). As mentioned above, landscape shapes are the result of geological phenomena that molded the Earth's surface (Amadio and Lugeri 2016).

By considering the different land settings that give a region its own characteristics, it is consequential to observe how geological dynamics link all the territorial components. Risk and resources are two sides of the same coin, both regulated by the interaction between humankind and planet Heart: hazard or risk, exploitation, or development. Human behavior establishes the result of a sensitive balance. It is urgent the need to address individual/community/society towards a common aim, the identification and achievement of strategical ethical objectives: primarily, survival itself.

Why can a so easy concept not trespass the cognitive-behavioral borders of most of society? The message does not reach the whole society: it is prerogative of a part of it, the already sensitized one, the aware one. Despite the abundance of technology and media tools, the risk mentioned above of fragmentation of the information is high and continuous.

Dealing with the prevention of risk and protection from danger is really hard (Farabollini et al. 2014a, b). The traditional way of dealing with problems, so far, has not given the desired results, with no noteworthy result. The transformation of disasters into shows, diverts the audience from possible future implications. Talking about the same catastrophic event turns the scenario into a surreal one, which can lead to anger and confusion. Neither of these two modes of communication helps in the adoption of prevention and preservation measures. However, the public must be informed about the characteristic of territory through adequate communication strategies, which include understanding cultural and natural environment. In addition, dynamic operations on the territory must be clear to all, to better protect the population against any risk (Lugeri et al. 2012). It is necessary to tackle the theme by tactically enhancing the positive aspects in adopting a virtuous behavior regarding environmental responsibility, and successively introducing the concepts of vulnerability and risks as soon as the responsivity of the audience is "on".

3.5 New Semantic Tools

The scientific community has recently activated a scientific debate on the enhancement of geological heritage, providing sites of geological interest ("geosites") with instruction regarding methods and tools for census and knowledge (Prosser et al. 2018; Piacente 2005; Brilha 2016). The "geosite" is a locality, territory, or area identified for its peculiar geomorphological features, with the intent of conservation (Wimbledon 1996): all is defined by the IUGS (International Union of Geological Sciences) and applied in the GEOSITES projects. A "geomorphosite" is a geomorphological landform is a with scientific, cultural, and socioeconomic value (Panizza 2001), which in turn can be considered as a link between scientific research, cultural integration, and artistic suggestion (Panizza and Reynard 2005; Panizza and Piacente 2005). It often happens that "geotope" is used incorrectly to describe "geosites" causing confusion. A "geotope", following the guidelines of German scientists, is a basic element of the landscape. Further, recent definition agrees upon the term "geotope" (Troll 1950; Poli 1999) as the meaning of geological manifestations, the demonstration of physical elements of a territory, or particularity of a landscape with scientific value, in the expression of the process that transformed and shaped it, and at the same time, that is worthy of protection, for its aesthetic, naturalistic, historical, touristic, and educational attributes (Gray 2004; Ruban 2010). The "geotope" can be considered as the smallest part of the planet geographically circumscribable and having homogeneous characteristics (Brilha 2002). There is a deep link between geodiversity and biodiversity, and the connection between the components of the landscape and the environment are superimposable (Brilha 2016). The integration of biodiversity and geodiversity as integration concepts, lead us to consider geology as a part of a bigger natural system, where is no clear distinction between biological and geological process: this turns out to be crucial in the study of landforms, climate, and biodiversity (Amadio et al. 2002). In this perspective, it is appropriate to refer to environmental diversity, seen as the integration between biodiversity and geodiversity. Environmental heterogeneity is fundamental in evaluating and examining indicators to carry out a synthetic index that expresses the complex systemic functions of each portion of the landscape, considered in its identity (Brilha 2016; Gray 2008).

Moving on to the synthetic scale of the landscape analysis (1:250.000) it is useful to refer to the concept of "geomorpholandscape" defined as "a spatial object or component of a geological landscape, whose geological settings are easily understandable as results of its geomorphological evolution, thus allowing to link spatial patterns to geological processes" (Farabollini et al. 2014a, b).

3.6 Landscape Approach

The holistic approach seems to be the best one, regarding the analysis of the territory and the environment, taking into consideration all those aspects that constitute the complexity of nature (Farabollini et al. 2014a, b). That is the sense of the systemic approach, in contrast with the systematic one, this one based on a merely descriptive, taxonomic analysis of environmental components, often losing sight of the concrete objectives of the research.

Every environmental analysis must have as its basis the landscape ecology paradigm, which must have as its objective a significant representation of landscape and the assessment of its ecological value and vulnerability (Amadio 2003).

The interrelationship between structure and function is at the basis of the paradigm. Each landscape is studied individually at different scales, at which it shows peculiar elements. Structural elements, which are a function of physical form and specific spatial organization; functional element, which depend on the relationship between abiotic and biotic elements: dynamic elements, which are determined by the successive evolution of the structure (Forman and Godron 1986, Naveh and Lieberman 1994).

The universe, following a further "Landscape Ecology" principle, is not homogeneous but dynamic, multi-scale, and hierarchically organized. (Forman and Godron 1986). The related methodology, in studying landscapes, is functional to identify the criteria for analysis, evaluation, diagnosis, and control of the "systems of ecosystem" in which structure and function are integrated. A significant step forward is the possibility of developing indicators and models, scientific and technical tools suitable for a balanced land management.

The generic term of landscape indicates an intuitive notion deriving from the perception of a diversified physical environment. In this respect, it represents the visual manifestation of particular organizations of elements and structures (Forman and Godron 1986; Naveh and Lieberman 1994; Odum 1961). Researchers have been guided by the holistic approach to a concrete point of view, thanks to which the landscape can be interpreted. At the same time, the communicator has the possibility of providing the public with a complete vision of a complex system in a clear structure (Badiali and Piacente 2012).

3.7 GIS and Maps

The mapped representation of the territory, thanks to GIS and WebGIS, allows the description of different landscapes and related images, and allows user to take advantage of a landscape recognition process, according to its elements, which must be described, recognized, and explained.

Modern Cartography, as a GIS product, is no longer just a descriptive representation of phenotypes: it is now enriched with ecological–functional meanings, becoming an interpretative tool of socioecological systems. In this sense, it is a ground-breaking discipline and has many different fields of application, such as spatial planning at different scales, environmental analyses, environmental impact assessment, nature conservation, sustainable development. A natural evolution of GIS is given by the PPGIS (Public Participation Geographic Information System). The multipurpose function makes the instrumental use highly very powerful, for being made up of a different kind of information, referred to various disciplines and sectors, integrated with appropriate GIS structures (Sieber 2006; McKinster et al. 2013). The extreme usefulness of the 3D model is now widely demonstrated, while new analysis methodologies have been developed to expand themes of research. Displaying significant images represents the best aim of the 3D system (Martinez-Graña et al. 2013; Lugeri et al. 2015). In the model, the quantity of information and data intrinsically present, allows user to configure the system evolution. It is possible, in the field of Earth's sciences, through these innovations, to detect possible changes, form past to present, of those portion of the planet taken into analysis. The information given by the change over time of the observed area, are very important sets of integrated data that allow the modeling of the possible future changes. 3D models can play an important role in risk prevention practices. It is of fundamental importance to read the information not only in the spatial dimension but also through the diachronic one (Nettleton 2013; Lugeri and Muciaccia 2016). PPGIS can be considered as the natural and democratic evolution of GIS. The multidisciplinary, multi-sectoral and integrated functions make this tool very useful (Sieber, 2006; McKinster et al. 2013).

3.8 Mapping Landscapes

One of the most relevant studies realized in Italy, in the field of landscape analysis, is the Map of the Italian Physiographic Units at 1:250.000 scale", a GIS realized by ISPRA (Italian Institute for Environmental Protection and Research) following the abovementioned holistic approach (ISPRA 2015). Such map allows an interpretation of territory in a dynamic sense, by monitoring landscape settings evolutions in space and time. The Map identifies, within the Italian territory, 2155 areas characterized by homogeneous physiography, and by a characteristic geographic connotation as well. Each of these physiographic units classified is described using morphological, lithological, hydrographic, and soil coverage, to highlight the territorial specificities, referring to a list of 37 types of landscape, identified and reported at the national level (Amadio et al. 2002).

Those geographically characterized portions of territory, are defined units, mapped using the typical Italian classification of landscapes, created ad hoc to recognize and identify both the surface of the country and its evolution over time. These Units are described thus by typological properties, referred to the above-described landscape approach, by following an inductive process. In parallel, the unit has a precise and univocal denotation, about the geographical context in which it is located: such topological property is highlighted by following a deductive process. A series of criteria are used to define the typologies, based on the observation of the specific constituent characteristics that inform the structure of the landscape at the scale of analysis. In this case (1:250.000), the main component is referred to the lithology of the substrate. It is not of secondary importance to identify the relationship that a landscape has with the surrounding ones (Amadio et al. 2002). A database, part of GIS, was created by the digital conversion of related maps and tables. Two processes are involved in the description of the landscape unit: a brief description of the peculiarities of the polygon identified in the map, and a more detailed one (referred to a more detailed scale of analysis), both to be visualized in the personalized project of GIS. This paper refers to the scale 1:250.000, to report specific phenomena emerging from the landscape's representation, particularly the geomorphological settings (Fig. 3.1).

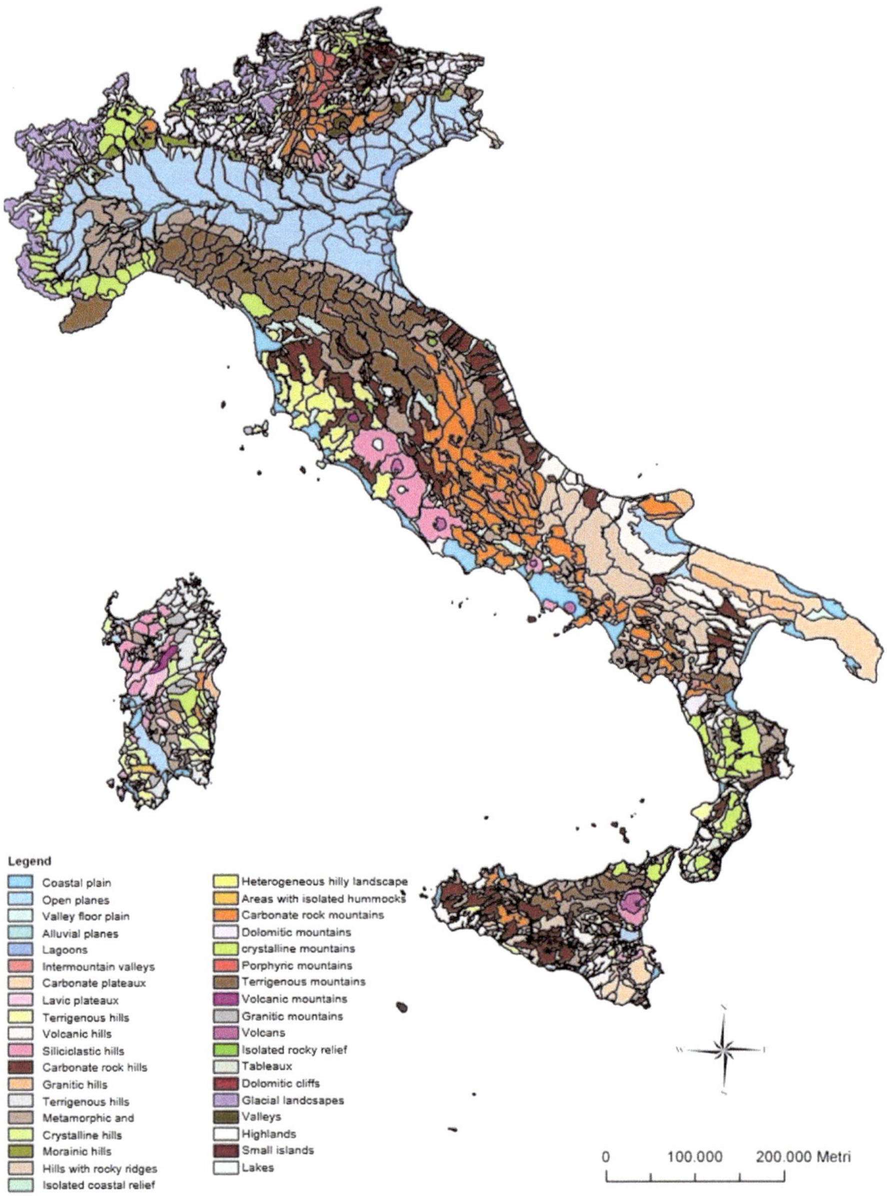

Fig. 3.1 Map of Italian physiographic units at 1.250.000 scale

The physiographic units have been evaluated regarding indexes: "Ecological value," "Ecological sensibility," "Human pressure," calculated by using specific indicators for each unit. The set of indicators refers to the unit composition, to its structure and the anthropic pressure stressing the unit.

Further indexes are calculated by combining the main three indexes: an overall index, "Environmental Profile," provides an approximate measure of the risk of environmental degradation for each physiographic unit.

Thanks to its characteristics, the Map of the Italian Physiographic Units, easily understandable and available online, is an effective tool in communicating to a wide public the most important information about landscape. In parallel, the information extractable from the data, organized in a GIS and graphically represented in the Map, supplies a broad range of instruments to territorial planners, or commonly to the final users (ISPRA 2015).

A new edition of the Map is in the project, analyzing the Italian Landscape by following an integrated geomorphological criterium, aiming at focusing those areas characterized by a high territorial vulnerability, and at explaining the landscape dynamics causing the fragility of an area. Referring to the previous considerations, the redefined Units are renamed as "Geomorpholandscapes".

3.9 From Theory to Practice

All the data and the results of the landscape analysis shouldn't hopefully be used only in the field of scientific research: as told before, there are new ways to give society information about territorial dynamics, aiming at starting a shared awareness of risks and resources, and promoting virtuous behaviors, in terms of environmental protection. In this perspective, geotourism represents a powerful strategy to invite people to perform pleasant, environmentally correct activities, by enjoying the landscape, geosettings, geoforms, while knowing it.

Geosciences are able to communicate and transmit a sense of environmental identity, derived from the consciousness of being inside an ecosystem; this thanks to the learning and knowledge of the environment. But geotourism is not tourism. How to make it appealing to most of the people? It is necessary to use some expedients. Justified by its nature, geological tourism is deeply correlated to outdoor sports, but this has not yet brought any benefits for development. Skiing, trekking, climbing, diving are just some of the many sports and disciplines that allow their players to discover different landscapes, from different points of view (Lugeri et al. 2011). Rocks, shapes, vegetation are some of the many components of the landscape that play a leading role in these sports (Fig. 3.2).

Our experience focuses on cycling, thanks to its rampant popularity and its deep relationship with landscape. Cycling is successfully practiced within the landscape, and is increasing as a meaning of integration and between impaired people. Its zero-impact transport value represents an important resource in plans that envisage possible smart cities (Lugeri et al. 2018). Cycling gathers a large number of people

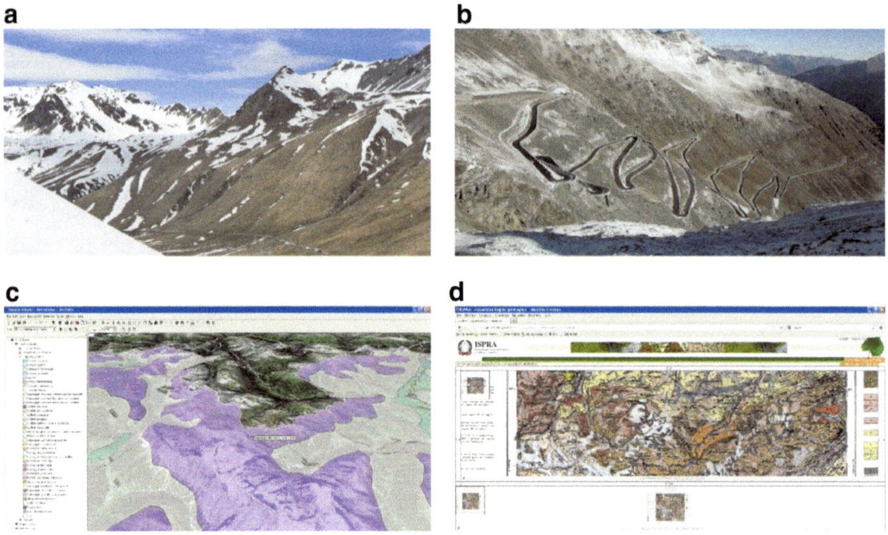

Fig. 3.2 Mountain glacial landscape and its representation. **a** East Alpine area. **b** Stelvio area. **c** Italian Physiographic Units' Map, Stelvio area. **d** Stelvio Geologic Map

at a competitive level, and has intrinsically the ability to inform the public about environmental scientific information (Lugeri and Farabollini 2018).

Moreover, it encourages contacts between cultures, peoples, allowing fans to know different realities, different landscapes, different lifestyles: it can help harmonize socioeconomic policies. Shape and position of the areas crossed by cycling routes, as well as by competitive races, entails a large number of different landscapes, both natural or artificial, particularly in the Italian country, where landscapes are developed, in a narrow, elongated area. The Giro d'Italia is one of the most important European cycling race, here identified as a versatile and powerful tool for scientific popularization.

The morphological characteristics of the area become of absolute importance in the race context: hill, mountains, slopes can offer the public a new vision of the landscape, a new point of view, combining scientific information with competitive competition values. The project came to life in 2012, and is continuously developing, being extremely effective in fair play and clean sport accessible to everybody (Fig. 3.3).

The knowledge of our country—of our Planet—performed thanks to a simple and popular code in communicating Earth sciences, represents a purpose of vital importance for the growth of an area, its culture and products.

Moreover, it is the first step in the implementation of risk prevention. According to the project, information regarding the geomorphological settings of an area are connected to cultural innovations, art and tradition (viscerally connected to geology as well as environmental conditions). In this way, it could help in enhancing the land as an anti-crisis resource. Through the presence of UNESCO and its world heritage

Fig. 3.3 The social participation in the Giro d'Italia

sites and protected area, the link between nature and hearth, as well as its own culture, can be analyzed more carefully. Many geologists participated in the Giro d'Italia, especially during the stages that winded through the Dolomites, offering a special occasion to observe the pink mountains, UNESCO WHS. The cognitive process of recognizing landscapes, activates the recognition of those who have the familiarity with the abovementioned places, and at the same time aroused curiosity in those who had never observed such beauties WHS (Fig. 3.4).

The basic cognitive tools were provided by the Map of Physiographic Units. GIS technology and WebGIS are means of communication with high educational potential and absolute versatility in use. Integrated by the 3D modeling technology,

Fig. 3.4 Dolomitic landscapes. **A** The "Tre Cime di Lavaredo". **B** The Girod' Italia Dolomitic stage. **C** The 3D geological map of Dolomites, featuring the Girosauro

they are flexible and useful tools either in educational plans, or in the promotion of territory, suitable for inclusion in sports programs, as confirmed by the Geologiro experience. The Geologist aims to inform about the geological features of the territory crossed by the tour, with the aim of spreading scientific knowledge: such project has been realized thanks to the collaboration between Camerino University and the ISPRA (Geological Survey of Italy). Rai sport (Italian state TV company), agreed to include in its schedule space, called "AnteprimaGiro", dedicated to providing the public with views of landscapes and sites, implementing scientific information on the competitive nature of the race. The presence of geologist at the event, triggered a collaboration between the various professional figures and allowed to participate in other sports (Gran Fondo, Cycling Marathons) aiming at encouraging the knowledge of the territory. The results have been very positive, and the connection between sport, nature, and culture provided the public with both sporting and scientific information.

Created for the youngest but also dedicated to a wider audience, is the "Girosauro" (Lugeri et al. 2018), a cartoon where a pink (as the color of the jersey given to the winner of the Giro) dinosaur (Girosauro) and a little girl accompany the public through the discovery of breathtaking landscapes, toward geological secrets (Fig. 3.4).

3.10 Further Initiatives: Science and TV Series

A further development of the project aimed at enhancing the social awareness of landscape, resources and its risks, is the Earth Sciences communication through films and fictions. Continuous communication has been allowed by television programs, transmitted subsequently to films and TV series. The function of this communication is to provide users with an operational chart of the world (Grasso 2002). It has been widely demonstrated that TV has not only a physical value (occupying a space) but a strong imaginary and cultural role.

Our proposal is to show and describe the complex aspects of the landscape of some areas particularly relevant to the series' plot, and to share their natural history to the public.

A format proposes the scientific reinterpretation of the environmental aspects of the movie's natural sceneries, immediately after its end. Guided by a speaker through the visualization of the natural environment, the audience is sensitized through a cognitive-recognitive process. This can be defined as "interpreter" in the process of mediating between nature and culture in a new landscape perspective approach. A first experience of the project was promoted in the week of geoscience 2014 (Lugeri et al. 2015) referring to the famous Italian series: "Commissario Montalbano", based on the novels of Andrea Camilleri. The episodes of the series are shot in Sicily, an island in the South of Italy, where most of the locations are WHS UNESCO. Sicily can be considered as a symbol of the beauty and fragility of the Italian country (Lugeri et al. 2015). Particularly, the popular actor Cesare Bocci, who plays an important role in the series, is a geologist: thanks to this special appeal, the proposed format makes the scientific themes more agreeable, producing a strong contribution to the communicative potential of natural and cultural heritage, toward the public.

The identification process in the common choice of themes to be used and language to be selected, are basic in applying the so-called storytelling method. (Lugeri and Farabollini 2016).

3.11 Conclusive Remarks

The starting point in achieving a shared knowledge of the planet is the promotion of a preventive culture aimed at avoiding environmental risks; it can be represented by the revelation of geosciences in an easy way, by using a proper language (Farabollini et al. 2014a, b). The consciousness of being within an ecosystem is accomplished through a full comprehension and experience of the environment (Amadio 2003; Lugeri et al. 2012). Cognitive process is crucial in prevention, leading society to an active and virtuous awareness in practice. A series of projects and activities, with the aim of making science accessible, must be carried out to ensure a useful dialog between people and science. New communication tools and codes are the basis of a new innovative process that connect communities and culture.

An essential objective is the enlargement of the audience of the scientific communication: it is necessary to involve the whole society into an educational process, reaching those ranges of society that had no tools to approach themes such as territorial balances, risks, and resources. Knowledge is intelligence, it is wisdom, it is a path toward "Civil self-protection". Many new initiatives have been started in recent times, in various parts of the planet, most of which have a schooling purpose. Unfortunately, many important figures such as policy-makers, stakeholders, and territorial planners are more interested in contingent and short-term problems, and must still be made aware of prevention, territory protection and enhancement, as vital concepts.

The scientific community has to meet real life; science has to walk the streets, to join the whole society, to communicate in new ways by using topics more appealing to the public.

Living the landscape is a process that involves multiple senses and generates emotions; it provides us with the essential topics in sharing key concepts of geological sciences, especially those that can lead society to adopt more appropriate behaviors, aiming at creating increasing welfare and a healthy planet.

References

Allan M (2012) Geotourism: toward a better understanding of motivations for a geotourism experience: a self-determination theory perspective. LAP Lambert Academic Publishing, Saarbrucken, Germany

Amadio V (2003) Analisi di Sistemi e Progetti di Paesaggio; Franco Angeli, Milano, Italy, p 23

Amadio V, Lugeri F (2016) The peri-urban natural areas: methodological proposals for the study area and the environmental assessment. In: Proceeding of international conference SIEP-IALE 2016 challenges of anthropocene and the role of landscape ecology. Asti, Italy
Amadio V, Amadei M, Bagnaia R, Di Bucci D, Laureti L, Lisi A, Lugeri FR, Lugeri N (2002) The role of geomorphology in landscape ecology: the landscape unit map of Italy, Scale 1: 250,000 ("CartadellaNatura" Project). In: Allison RJ (ed) Applied geomorphology: theory and practice. Wiley, London, UK, pp 265–282
Badiali F, Piacente S (2012) The study of the landscape: from a holistic approach to a social concept of knowledge. Ann Geophys 55:481–485
Bauman Z (1997) Postmodernity and its discontents. Polity, Cambridge
Brilha J (2016) Inventory and quantitative assessment of geosites and geodiversity sites: a review. Geoheritage 2016(8):119–134. https://doi.org/10.1007/s12371-014-0139-3
Brilha J (2002) Geoconservation and protected areas. Environ Conserv 29:273–276
Catton WR, Dunlap RE (2003) Environmental sociology. A new paradigm. Am Sociol 1978 13:41–49, 32
Council of Europe (2000) European landscape convention; European Treaty Series, No. 176; Council of Europe, Strasbourg, France, pp 2–4
Crutzen PJ (2006) The "Anthropocene." In: Ehlers E, Krafft T (eds) Earth system science in the anthropocene. Springer, Berlin/Heidelberg, Germany, pp 3–6
Dowling R, Newsome D (2006) Geotourism. Elsevier Butterworth Heinemann, Oxford
Dowling R, Newsome D (2010). Geotourism: the tourism of geology and landscape. In: Dowling RK (ed) Edith Cowan University, Australia David Newsome, Murdoch University, Australia
Farabollini P, Lugeri F, Amadio V (2014) Geomorpho-landscapes. EGU general assembly 2014, held 27 April–2 May 2014 in Vienna, Austria, id.4954
Farabollini P, Lugeri FR, Amadio V, Aldighieri B (2014) The role of earth sciences and landscape approach in the ethic geology: communication and divulgation for the prevention and reduction of geological hazards. In: Lollino G, Arattano M, Giardino M, Oliveira R, Peppoloni S (eds) Engineering geology for society and territory. Springer, Heidelberg, Germany, vol 7, pp 115–120
Forman RTT, Godron M (1986) Landscape ecology. Wiley, New York, NY, USA, p 620
Grasso A (2002) Enciclopedia della televisione. Milano, Italy, Garzanti, pp 375–382
Gray M (2004) Geodiversity: valuing and conserving abiotic nature. John Wiley & Sons, Chichester, UK
Gray M (2008) Geodiversity: developing the paradigm. Proc Geology Assoc 119:287–298
ISPRA (2015) Web GIS Carta della Natura. http://geoviewer.isprambiente.it/index_CdN.html?config=config_CdN.xml. Accessed 8 Jan 2015
Levy Strauss C (1957) TristesTropiques. Librerie Plon, France
Lugeri F, Farabollini P (2018) Discovering the landscape by cycling: a geo-touristic experience through Italian badlands. Geosciences 8:291. https://doi.org/10.3390/geosciences8080291
Lugeri FR, Farabollini P, Amadio V, Greco R (2018) Unconventional approach for prevention of environmental and related social risks: a geoethic mission. Geosciences 8:54. https://doi.org/10.3390/geosciences8020054
Lugeri FR (2018) https://www.youtube.com/watch?v=ge6K0kMTLMw
Lugeri FR, Muciaccia M (2016) Landscape approach and GIS: new ways in teaching territorial sciences. In: Conference proceedings. The future of education; Pixel LibreriaUniversitaria. It Edizioni: Padua, Italy, pp 216–221
Lugeri FR, Amadio V, Cardillo A, Bagnaia R, Lugeri N (2011) Landscapes and wine production areas: a geomorphological heritage. Geoheritage 3:221–232
Lugeri FR, Farabollini P, Graziano G, Amadio V (2012) Nature and culture in a landscape approach. Eur Geol 34:23–28
Lugeri FR, Farabollini P, Greco R, Amadio V (2015) The geological characterization of landscape in major TV series: a suggested approach to involve the public in the geological heritage promotion. Sustainability 7(4):4100–4119

Lugeri FR, Farabollini P (2016) Science and society: new proposals for an unconventional scientific communication. In: Innovation in Environmental Education: ICT and intergenerational learning International conference proceedings. Published by IBIMET-CNR, Firenze, Italy. ISBN: 9788895597287
Martínez-Frías J, González JL, Pérez FR (2011) Geoethics and deontology: from fundamentals to applications in planetary protection. Episodes 34:257–262
Martínez-Graña AM, Goy JL, Cimarra CA (2013) A virtual tour of geological heritage: Valourisinggeodiversity using Google Earth and QR code. Comput GeoSci 61:83–93
McKinster J, Trautmann N, Barnett M (2013) Teaching science and investigating environmental issues with geospatial technology: designing effective professional development for teachers. Springer Science & Business Media, Berlin, Germany, p 353
Naveh Z, Lieberman A (1994) Landscape ecology theory and application; Series on Environmental Management. Springer, New York, NY, USA, pp 99–103
Nettleton KF (2013) Cases on 3D technology application and integration in education. IGI Global, Hershey, PA, USA, p 359
Odum EP (1961) Fundamentals of ecology. Cengage Learning, Boston, USA
Panizza M (2001) Geomorphosites: concepts, methods and example of geomorphological survey. Chin Sci Bull 1:4–5
Panizza M, Piacente S (2005) Geomorphosites: a bridge between scientific research, cultural integration and artistic suggestion Geomorphological sites and geodiversity. Il Quaternario—Italian J Q Sci 18:3–10
Panizza M, Reynard E (2005) Géomorphosites: définition, évaluationetcartographie. Géomorphologie: relief, Processus, Environnement 3:77–180
Peppoloni S, Di Capua G (2012) Geoethics and geological culture: awareness, responsibility and challenges. Ann Geophys 55:335–341
Piacente S (2005) Geosites and geodiversity for a cultural approach to geology. Il Quaternario. Italian J Quat Sci 18(1) - Volume Speciale, 11–14
Poli G (1999) – Geositi, testimoni del tempo, Regione Emilia-Romagna
Prosser C, Díaz-Martínez E, Larwood J (2018) The conservation of geosites principles and practice. Book: Geoheritage, Elsevier. Chapter 11
Ruban DA (2010) Quantification of geodiversity and its loss. Proc Geolog Assoc 121:326–333
Sieber R (2006) Public participation geographic information systems: a literature review and framework. Ann Assoc Am Geogr 96:491–507
Troll C (1950) Die GeografischeLandschaft und IhreErforschung. Springer, Heidelberg, Germany, pp 163–181
Wimbledon WAP (1996) Geosites—A new conservation initiative. Episodes 19:87–88
Wyss M, Peppoloni S (2014) Science geoethics: ethical challenges and case studies in earth sciences. Elsevier, Amsterdam, The Netherlands, pp 3–13

Chapter 4
UAV's Multimedia Technology and Augmented Reality (Geointegration): New Concept and New Paradigm of Geodiversity Presentation

Ivaneide Santos, Renato Henriques, Gorki Mariano, and Edjane Santos

Abstract The search for interactivity and a greater degree of contextualization within the geographical space have fostered the integration of several technological mechanisms already applied in other areas of knowledge. The advent of multimedia technologies and augmented reality brings new challenges that arise in detriment of new forms of representation and dissemination of geodiversity and geological heritage. In this way, the objective of this work is to consider the main challenges when using image data acquired with Unmanned Aerial Vehicles (VANTs), the georeferenced information processed in Geographic Information Systems (GIS), the adaptation of the techniques of digital photogrammetry and multimedia technologies to existing static resources while maintaining user interactivity. The combination of these features can be called geointegration. This methodology also seeks to promote better results of the interval between the user, the computational visualization and what is illustrated and interpreted about the object of study, which in this case are the elements of geodiversity and geosites. These technologies provide results, on the one hand, excellent and versatile technological solutions, which allow not only the construction of a platform provided with several elements such as panoramic image (which allows a global view of the surroundings of the object of study, in this case the natural landscape), as well as capturing portions of terrain that can be converted, using photogrammetric techniques, into interactive 3D objects (which allow for detailed image data with various points of view of other aspects of the object of study). These results configure effective geointegration. This use also allows to

I. Santos (✉) · R. Henriques
Institute of Earth Sciences, Pole of University of Minho, Braga, Portugal
e-mail: Iva.Oliveira.Geo@outlook.com

R. Henriques
e-mail: rhenriques@dct.uminho.pt

G. Mariano · E. Santos
Universidade Federal de Pernambuco (UFPE), Recife, Brazil
e-mail: marianogorki@gmail.com

E. Santos
e-mail: ems_geo@yahoo.com

R. B. Singh et al. (eds.), *Global Geographical Heritage, Geoparks and Geotourism*,
Advances in Geographical and Environmental Sciences,
https://doi.org/10.1007/978-981-15-4956-4_4

obtain a new form of geovisualization, since all the information is aggregated in a static base that allows access to the represented multimedia contents of different forms. This base is built using the technology currently available, including solutions such as maps, the WebCL platform, finally, the interaction of other actors of the Society and the representation of specialized scientific content is produced. This whole workflow involves costs as well as repetition of the rendering processes in the search for the best result of the products (three-dimensional objects), besides some legal restrictions of the use of the VANT's in certain places and due to some meteorological conditions. The paradigm is that even when faced with several functionalities in some cases, this technology is not always adequate considering the available representation resources, some aspects of the structure to be imaged and the conditions of scale.

Keywords Augmented reality · Geointegration · Representation of geodiversity

4.1 Introduction

The search for interactivity and a greater degree of contextualization within the geographical space has the contribution of a large amount of technological resources already existing in other areas of knowledge. The advent of multimedia technologies and reality increases the new challenges that represent new forms of representation and dissemination of geodiversity and geological Heritage. Thus, the objective of this work is to conceptualize the integration of static and interactive elements for the representation and promotion of elements of geodiversity and geological heritage. The result is an interval between the computational visualization and the aspects that make up the imaged object of study, which are the elements of geodiversity and geosites.

These technologies promote, excellent and versatile, technological solutions, which allow that there is not a great variety of panoramic images (allowing a global point of view to be used as the object of study, if it is natural), as well as that the capture images can be converted, using photogrammetric techniques, into interactive 3D objects (giving access to detail image data, with various points of view from other aspects of the study object). This is activated from all geological aspects. Since information is aggregated to a site using currently available technology, including a web language, providing 3D objects online, favoring interaction with other actors of the Society and the representation of specialized content.

In areas where network access is limited, the information can be downloaded in small parts using "QR codes" available in the interest sites or in strategic locations nearby. The entire work process involves costs, as well as one of the rendering processes in search for the best product result, as well as some legal restrictions on the use of UAVs in certain locations and due to a series of meteorological conditions.

The paradigm is that, in some cases, attention is not always limited in the account for some aspects of the structure to be imaged. Augmented reality emerges as a new form of competition and also brings new challenges and particularities to an object of study.

The use of the term visualization in the cartographic literature began with Philbrick (1953), who pondered under the following aspects: 1—the real world is indivisible, 2—all visualization of phenomena is generalized, 3—all generalization is proportional to scale, 4—the form of the visualization is a suggestion, 5—the visualization depends on the contrast, 6—the contrasts are gradations of change, 7—the visualization of parts in relation to the whole depends on the balance, 8—all phenomena are not of equal importance, 9—all phenomena are repeated with variations, 10—the ideal of every expression is to say the maximum with the greatest saving of meanings.

MacEachren and Ganter (1990), in a parallel effort, developed a simple cognitive model to identify the fundamental parts of the interaction between the user and the product displayed on the computer screen, which occurs during map-based visual analysis. The International Cartographic Association (ICA 2003) defines "visualization" as the creation of a visual image of something, mentally, or physically, using graphics, photos, or other means. Display data may have been generalized or not. The term "Scientific Visualization" was adopted by computer scientists to refer to the exploration and analysis of data and information graphically.

According to ICA (2003), the "Cartographic Visualization", is considered as a subset of "Visualization", since the term incorporates the unique characteristics of a cartographic product (the map), that is, it is generalized, symbolized, and measurable in order to fulfill the purpose for which it was conceived.

In cartography, the term visualization became used more frequently four decades ago. Cartographic evolution stems from the evolution of mapping and technology techniques in visual information, including scientific visualization and computer graphics associated with geographic information systems. In 1987, when the U.S. National Science Foundation publishes a new meaning of visualization in cartographic scientific research, several cartographers come up with research related to visualization and communication. The authors used different nomenclatures: geographic visualization, cartographic visualization, geovisualization, or simply GVis (Santana 2014).

According to the U.S. National Science Foundation, "GVis (Geographic Visualization)" can be defined as a form of visualization based on maps that emphasizes the development and evaluation of visual methods to facilitate the exploration, analysis, synthesis and presentation of georeferenced information.

For Dykes et al. (2005), "Geo-Visualization" is an emerging field, which draws on approaches from a variety of disciplines that provide theory, methods, and tools, although the interactions between them are fluid. The art and science of Cartography have evolved to embrace and support Visualization. The increasing importance and use of spatial information and map metaphor make "Geo-visualization" an essential element and a genuine opportunity for Cartography in the twenty-first century.

Bleisch (2012) considers "3D Geo-visualization" to be a very generic term, used for a range of 3D visualizations representing the real world, real-world parts, or other data with a spatial reference. Especially with the advent of virtual globes or geobrowsers, they become increasingly popular and many people are aware of 3D geovisualizations even without referring to them with that technical term.

The twenty-first century brought changes of several paradigms of communication in detriment of the Internet and consequently in the manipulation and representation of data georeferenced through geovisualizers.

The exploration of visual communication, or what has been termed as visualization process in geotechnologies, and especially in Geographic Information Systems and its unfolding in geovisualizers, is defined as a discipline that allows "see the unseen" (see the invisible) (McCormick et al. 1987).

According to Santana (2014), the innovations brought by the cartographic visualization require that new research be carried out with new computational techniques applied within the context of the geoinformation. A new context of possibilities of manipulating data has arisen, which together with computational techniques and graphical interfaces, allow to make analyses and simulations of scenarios.

According to Sandercock (2000), there are three types of techniques that can be used by geovisualizers that allow the integration of multiple users of different profiles: animation, multimedia, and virtual reality. Added to this group is the augmented reality. With the animation, the variations of the characteristics of the information can be observed by the movements generated with the sufficiently rapid presentation of a series of pictures, in which the alterations of the phenomena are represented.

The multimedia techniques allow texts, graphics, animations, sound, and video to be used in a complementary way, to present the different aspects of the information. With virtual reality, the spatial characteristics of information can be represented three dimensionally (Santana 2014). The concept of Virtual Reality is the junction of three basic ideas: immersion, interaction, and involvement.

Immersion is the possibility of feeling inside the environment; the interaction is related to the possibility of the computer reacting to user actions and changing the virtual world in the function of this recognition; and the involvement may be passive (the user is the only viewer) or active (the system responds to commands).

When the virtual reality system is nonimmersive and only displays screen, keyboard, and mouse devices, it is called the "desktop." In the latter, which is the case here proposed, the effects are constructed by 3D and features of light and shadow (Moura 2003, p. 63). Zhou et al. (2008) define augmented reality (RA) as a variation of virtual reality (RV). Virtual reality technology completely immerses the user within a modeled synthetic environment.

While immersed, the user cannot see the real world around them. On the other hand, AR (Augmented Reality) allows the user to see the real world, with virtual objects overlapping or composing with the real world. For this work, AR completes the reality, instead of replacing completely, through the representation of geosites (Brilha 2016), as is the case of Serra das Russas, in northeast Brazil.

The concept of virtual geointegration is based on virtual navigation. According to Moura (2003), "virtual navigation is the first model that allows the exploration of the space environment in the third dimension through computer and multimedia. The observer can have the sensation of being in the place, moving around the landscape through focal points and axes of vision according to their interpretative interests.

Adapting this concept, "virtual geointegration consists of the insertion of scientific content in the specialty of geosciences, which can always be updated".

The base has a starter shape, that is, a base to promote a geointegration. This boot can be a static map material or even another online resource (Fig. 4.1).

The aggregation of elements already in use, such as static ethics, tables, interpretive panels, diagrams, maps, added to the use of geodiversity (Brilha 2016), sketches, contents, geosciences in conjunction with geovisualization composes "geointegration" (Fig. 4.2).

4.2 Methodology

In order to produce geointegration with reality for integrated representation and promotion of the geodiversity sites, three existing methodologies were integrated: Structure for motion, (for the production of three-dimensional objects); Virtual panoramas online (for the visualization in reality increased by one global landscape point of view and for the dissemination of interactive content multiplatform—mobile and desktop); WebCL technology (for online prospecting scientific content of GIS).

These three integrated methodologies are called integrated geointegration. The entire process starts with planning, which has three groups: Gather the characteristics of the object to be photographed and collect the images, information collection, validation, initialization of the flow of the imaging scale of the objects, whether detail or at various scales that define the choice of the imaging device (UAV), having specific characteristics).

Once the imaging device is specified, planning multimedia solutions are generated. Depending on the planned multimedia solutions, it is possible to optimize the planning of the collection of image data and flight plans of the collection device. The systematization of the methodological procedures consisted of the following steps:

I. **Selection of objects (structures) to be imaged**:
   This stage obeyed two criteria: (a) the objects were located in sites of geodiversity (Brilha 2016) and in sites of scientific interest (Brilha 2016); To be imaged, the objects do not necessarily need to have scenic beauty or have geotouristic relevance. Eventually, geotourism interest is considered, however, it is a secondary criterion.
   The 3D models configure a universal tool to enable access to field information for any public and in any location, preserving details and—often—allowing even broader viewing angles than in real field conditions. An example of this is an image work in the "Milonitos da Serra das Russas", state of Pernambuco, northeastern Brazil.
   It is a geosite of international value, which is configured in a section where it can be seen that the folds are synforms and antiforms, with amplitude varying from 1 to 10 m in milonites developed in an important transcurrent shear zone (Lineamento Pernambuco) that crosses the Borborema Province (Santos 2016).

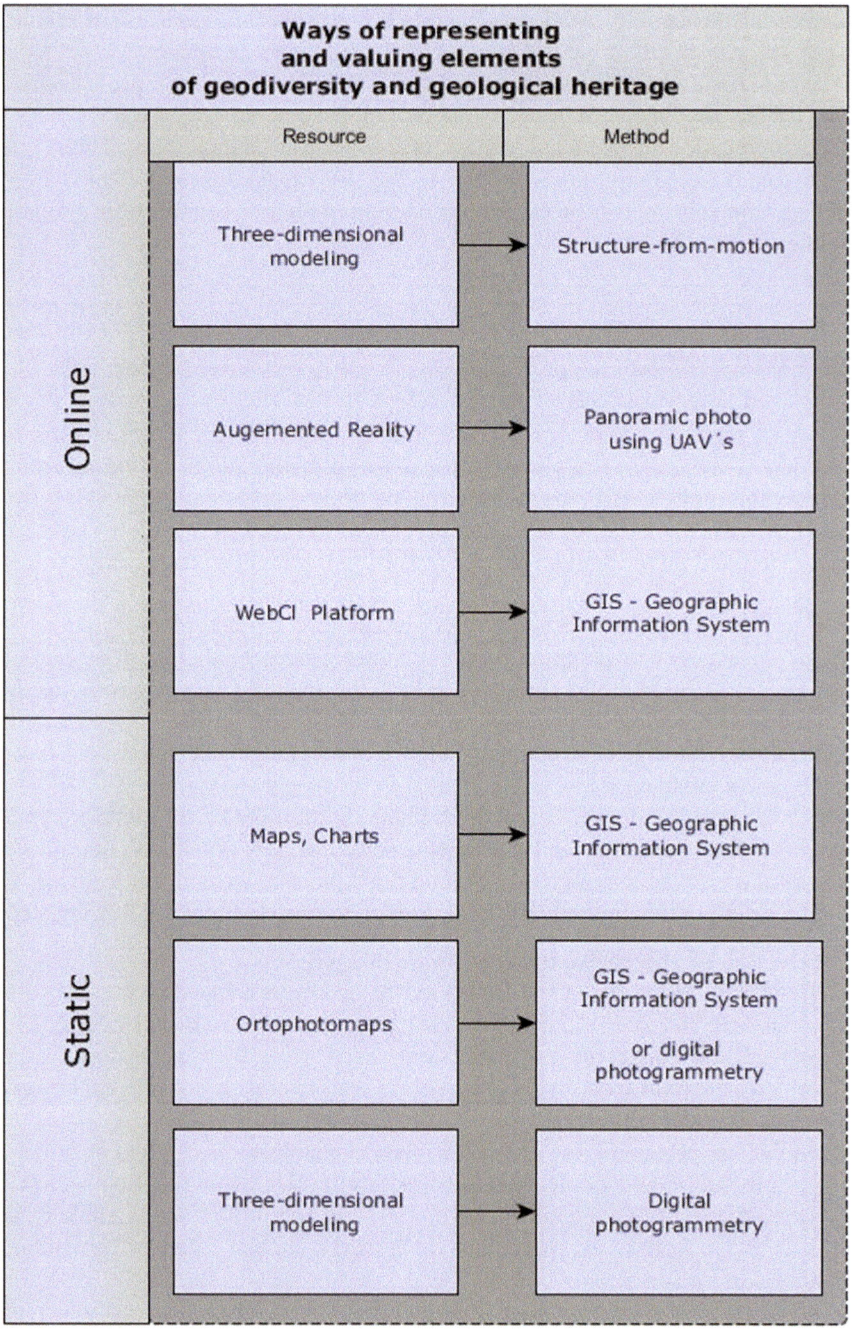

Fig. 4.1 Some forms of representation of the elements of geodiversity and geological heritage

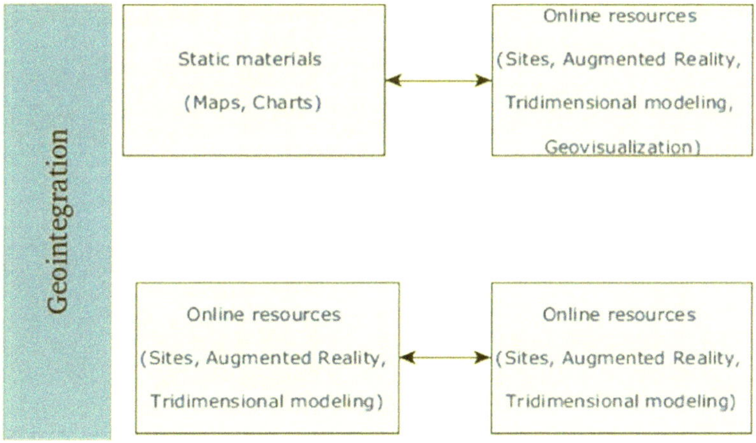

Fig. 4.2 Resources that integrate the geointegration

Through the creation of a 3D model of these forms, obtained by the digital treatment of images collected by VANTs, it was possible to solve some of the main impasses that make difficult the fieldwork in such places (Fig. 4.3).

The first problem is its accessibility, since the outcrop is located at the margins of a high-traffic highway (BR232) that only has a shoulder as a place for field

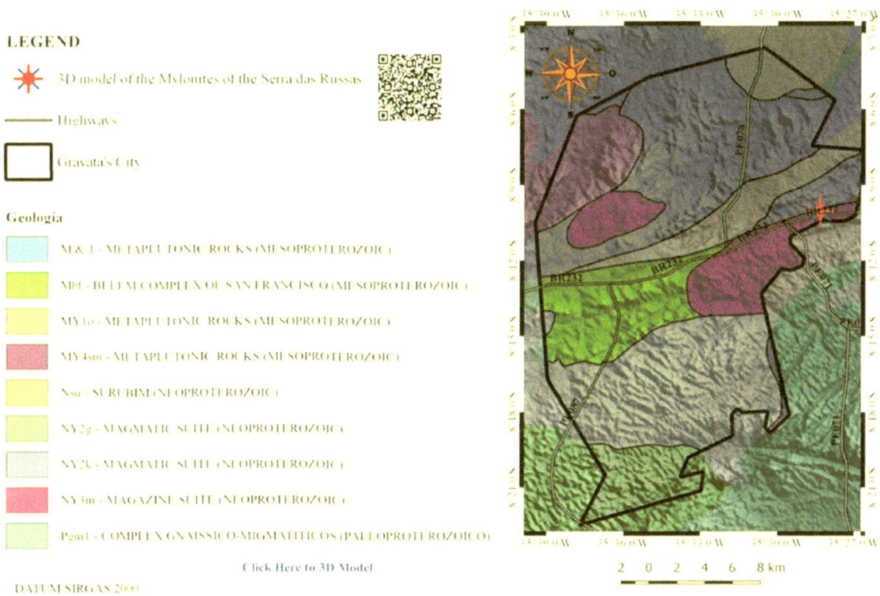

Fig. 4.3 3D model of the mylonites of the Serra das Russas

Fig. 4.4 Panorama Russas's Serra

lessons and data collection for research. The risk of accidents and they are related to the proximity of the highway, and the fall of blocks that can occur at any moment. In addition, natural elements that make difficult the work in this place during some periods of the year, such as rainfall and fog, as well as the caatinga vegetation covering the site. These problems can be eliminated by the creation of a model, provided that data collection takes place in the best conditions of the terrain. Another factor that brings advantages to the 3D model for the analysis of geological heritage for didactic, scientific, and tourist purposes is a possibility of new angles of vision that accompany a trajectory of the VANTs with panoramas that would be impossible to be carried out by the visitors in the locus.

II. **Planning of flight plans**:
Recognition of the surrounding area from the aerial point of view was first achieved. In this first procedure, a first grid flight was used to obtain a georeferenced orthophotomap and a digital surface model. This information allowed to make a previous analysis of the place and the structures with potential interest. From this, nine photogrammetric rows were carried out from an 11-min mission in the surrounding geoform, with the purpose of recognizing the other structures and better planning of subsequent missions (or mission).

III. **Obtaining image data**
These missions were executed in nodal flight mode, in which the UAV is centered, and the images are obtained in small angular intervals, guaranteeing an adequate overlap, covering 360° around the object.

IV. **Image data processing**

(a) Three-dimensional modeling:
In the specific case of the Serra das Russas, the flight was done in orbital mode, with manual control of the UAV. The use of the manual mode is justified, in this case, by the presence of some obstacles next to the object to be imaged, capable of intercepting an automatic trajectory. The manual control makes it possible to avoid these obstacles.

In the case of the other imaged structures, an automatic orbital flight was used, with two orbits positioned at different heights and with different radii. Photogrammetric techniques are supported by the adaptability of the application to the treatment of data from georeferenced images, obtained through unmanned or manual aircraft control, making them suitable for precise shapes and dimensions.

This procedure basically consisted of three distinct phases: In the first phase, the determination of the orientation parameters of the images (internal and external) and obtaining the products (coordinates, orthophotos, bidimensional and three-dimensional models).

In this phase, photogrammetric procedures are used for the three - dimensional restitution of structures, treatment of clouds of points and hyper-realistic textural image in mosaic, calculated from all the photographs obtained. Image editing applications are also used to correct errors in model textures. In this case, Adobe Photoshop was used.

For the georeferenced processing and treatment of the data obtained in the field, GIS software was used, mainly Quantum GIS (QGIS).

All the images obtained in the fieldwork contain georeferenced information in the metadata, this is possible because the coordinates of the center point and orientation data of each image are obtained from the positional drone sensors and included in the images by the firmware of the device.

Thus, included in each photograph are the X, Y coordinates of the center point, the altitude (obtained by combining the GPS sensor and the barometric sensor) and the azimuthal orientation (obtained from the magnetic compass).

These data are then used by the Agisoft PhotoScan application for initial georeferencing and camera positioning in the first photogrammetric alignment procedure. According to Westoby et al. (2012): One of the limitations of this technique is the determination of the simultaneous location of corresponding characteristics in several photographs, taken from different angles.

The initial processing step in solving this problem is the identification of characteristics in individual images that can be used for image matching. A solution used through the methods popularized by Snavely (2008) is the object recognition System of the Invariant Feature Transform SIFT scale. This solution is implemented in SFMToolkit3, through the incorporation of the algorithm SiftGPU (Lowe 1999, 2004).

This identifies the features in each image that are invariant to the scale and rotation of the image and partially invariant to changes in the lighting conditions and the view of the three-dimensional camera.

In this way, points of interest, or key points, are automatically identified scales and locations in each image, followed by the creation of a descriptor resource, calculated by transforming local image gradients into a representation that is largely insensitive to variations in illumination and guidance (Lowe 2004). These descriptors are sufficient to allow resources to be combined into large data sets. The number of key points in an image depends primarily on the texture and image resolution.

The density, sharpness, and resolution of the combined set of natural scene textures will determine the quality of the cloud point data. Likewise, the distance between the camera and the feature of interest, increasing the spatial resolution of the photograph, the spatial density, and the resolution of the cloud points. Variations in complexity, lighting, materials in all images influence the texture of the image. The minimum requirement is to obtain corresponding characteristics visible in a minimum of three photographs; However, using as many images as possible is highly recommended as the maximum number of key-point matches and system redundancy favors the final quality of the texture and 3D model.

The internal parameters were obtained from the complete calibration process of the chamber coupled to the UAV, which allowed the knowledge of its geometric characteristics, such as the main distance, geolocation of the central point and lens distortion values, were also obtained. These data are determined by the Agisoft Photoscan application during the process of correlating the various photographs obtained and also from the EXIF information of the image files generated by the camera itself.

In the second phase, the parameters of external orientation were determined, which allowed reconstituting the positioning of each camera (or each photo) at the moment of obtaining each of the images in relation to the coordinate system of the object. This operation was divided into two phases: in the first relative orientation, the relative position between the images was determined, from the identification of the homologous points in the same ones.

In the second operation, called absolute orientation, the orientation axis was established. In this way, the 192 captured photographs are better associated with obtaining a “pseudo-model". From this, it is possible to visualize the pre-positioning that will compose the three-dimensional model.

(b) Panoramic Image

In order to comply with this step, the following procedures were performed:

- Organization and cataloging of image data obtained in thematic folders;
- Processing of image data in Autopano Giga and Kolor Panotour Software;
- Processing images obtained using Adobe Photoshop;
- Search and selection of contents (sites, texts, photographs) for interpretation of the landscape and structures;
- Insertion of “HotSpots” to enable the insertion of contents through JavaScript language;
- Integration of panoramic and other multimedia elements through the HTML5 language.
- Assemblage of image data in Autopano Giga and Kolor Panotour Software.

In order to ensure better framing of the image data obtained during the data processing process, this phase is essential. In the database itself, it is possible to observe the organization of the image data obtained as a mosaic. This happens due to the positioning of the UAV during the process of obtaining images at a fixed point (Nodal), rotating 360° and shooting at angular intervals that allow at least 60% overlap between images. In addition, the images are collected already georeferenced by the UAV's firmware.

After being selected automatically by the Autopano Giga software, or by the user, these images are processed through common point correlation techniques to produce a panoramic image (Fig. 4.4). Later, these images are transformed into static panoramas.

https://www.dct.uminho.pt/geoviz/srussas/srussas.html

To achieve this result, 360° interactive panoramas, post-processing and conversion to interactive object is accomplished using the Kolor Panotour software. In both situations, the whole process of graphic editing and, in some cases, insertion of graphic elements of landscape interpretation, is processed through Adobe Photoshop software.

It is possible, for example, to promote accessibility for individuals with special needs through the representation of sites of scientific, didactic, and geotouristic value. Panoramic photographs can be interpreted according to the content you wish to represent and the audience you wish to attend (general public, scientists, students, photographers, for example, the panoramic images Serra das Russas.

(c) WebCL Platform

An innovative navigation interface allows access to three-dimensional terrain information based on the WebCL frameworks, compatible with the most modern browsers. For the production of this solution, cartographic bases are produced in a GIS environment and using the ThreeJS technology. This solution allowed to represent the inventory of places of scientific interest in the spatial canyon of San Francisco river, providing visual and interactive appeal, allowing access to inventory data not only through tables but with the use of graphical and cartographic resources innovative approach.

In this case, the vector layers of Geology, Hydrography, of the inventory of places of scientific interest according to Brilha (2016) and of the altimetric information that generated the digital model of elevation of the spatial cut of the Russian mountain range, Pernambuco, were superimposed.

On the vector information, it is possible to represent all the information of the attribute tables generated and compiled still in the cartographic production phase. The editing of this information is performed in a GIS (Quantum Gis) environment. WebCL (Web Computing Language) is an integrated online platform through JavaScript language binding to WebCL. First announced in March 2011, this framework features heterogeneous parallel computing in any browser without the use of plug-ins, This platform allows web applications to update with high speed using CPUs and multi-core GPUs.

This technology has emerged in the face of the growing demand for applications that require parallel processing, such as image editing, augmented reality applications, and sophisticated games. Faced with this reality, a nonprofit group, the Khronos Group, designed through portable kernel programming.

Therefore, this platform consists of two parts, one being the programming of the Kernel, which runs on the processors (devices) and the other is JavaScript, which connects to the OpenCL Web application (WebCL).

Open Computing Language (OpenCL) is a graphically structured language for writing program commands that run on heterogeneous platforms consisting of central processing units (CPUs), graphics processing units (GPUs), digital signal processors (DSPs), arrays field-programmable ports (FPGAs), and other processors or hardware accelerators.

It is a language that provides a standard interface for parallel computing using task and data-based parallelism. OpenCL is an open standard. For the layout adjustment (font, size, color, language) the JavaScript language was used.

This information in the attribute table is accessed on the WebCL platform with just a "click" on the "layer" of interest (Geology or Hydrography, in this specific case). A similar situation occurs by clicking on the icon representing the geosites.

https://www.dct.uminho.pt/webcl_per/russas.html

V. Insertion of contents of interest (geological, didactic, etc.);

Providing the interaction of specialized and didactic content based on a single starting static point, in this case a location map, requires the insert of multimedia resources. In the case of the insertion of geological-based didactic content, three-dimensional modeling is one of the multimedia resources used to represent elements of geodiversity. This feature allows the user to observe and interpret the image structure.

Once this step has been completed, it is possible to delineate elements (faults, rock paintings, erosion marks, etc.) of the structure (block, fossil, outcrop, etc.) through other image processing applications (Adobe Photoshop) and web applications (Adobe Dreamweaver CC).

4.3 Results

The representation of the geodiversity sites requires a simple and accessible language, as part of the toolkit for the construction of devices aimed at the dissemination and exploitation of scientific content.

This context involves different sectors of the society such as the scientific community and the non-specialized public. To meet these profiles, we used the JavaScript programming language and the representation language online that allowed the creation of customized multimedia products, through these languages it was possible to insert scientific content developed specifically for this platform.

Through these languages, it was possible to edit the access buttons to these contents, the updating of the contents and the insertion of images, graphics or sounds and other multimedia elements:

https://www.dct.uminho.pt/geoviz/Geonavigation.html

These elements are a platform of virtual reality that can be accessed using mobile computing (tablets, smartphones) that have position sensors (compass, gyroscope, GPS, and accelerometers). In these devices, the pointing azimuth and the inclination angle of the device.

In this work, the information was processed in georeferenced mode, so there is a great correlation between what is being shown on the screen and the very landscape to which the device is pointed. All layers with augmented reality can be enabled or disabled by the user.

Access to panoramic images, after corrections and processed in the immersive panoramic 360° panorama, follows a user-defined roadmap, where hotspots with links are placed over the image to allow the access to various information items and, consequently, the interpretation of geology. Each hotspot allows the access (to web pages, maps, text, photographs, tables, diagrams, etc.) contextualized to the geosite and can also set the navigation to other panoramic images of nearby places. (Fig. 4.5) They also allow access to animated evolution models that favor reconstruction of the geological history of that point of interest. The map is interactive and geographically contextualized, synchronized with any cursor movement.

The access to the base panoramic images can be made through links (HTML interfaces) with additional information, both accessible via strategic points manually marked on the topology or hotspots build from points shapefile format, implanted on an interactive map.

Fig. 4.5 Integration elements

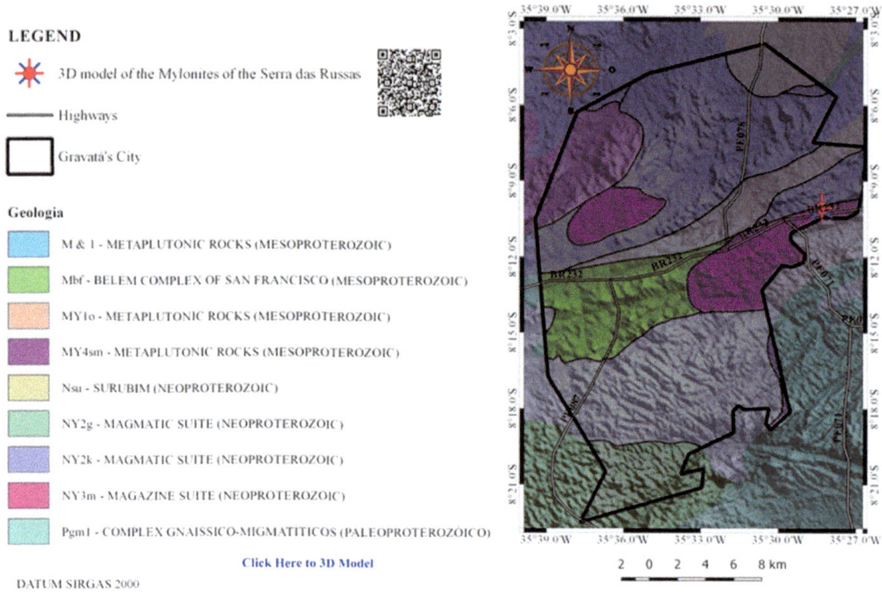

Fig. 4.6 Geointegration 's Serra das Russas, Brazil

These panoramic images are then used to navigate through the surrounding landscape and allow the access to deeper levels of information such as webpages or external websites, other panoramic images or detailed 3D models of landscape features (Fig. 4.6).

4.4 Conclusion

The advent of new technologies and the possibility of using interactive media stimulate the substitution of other existing static products, as is often the case, from the use of google maps in place of maps from traditional cartography.

These features are commonly used separately because of the integration of interactive features into the traditional cartographic representation. This work integrated the (static) location map of the Serra das Russas together with other forms of representation, such as the QR Codes for access to interactive panoramic photographs and three-dimensional modeling. This representation paradigm is innovative by using the location map of the Serra das Russas with interactivity elements to present geological aspects of the localized area. This new form of presenting geological aspects can be denominated as geointegration that represents the ample possibility of visualization of the landscape of diverse forms through the integration of several technologies in a single element.

The fact that the static cartographic resources are used in most cases through images, allow other forms of exploitation, as was the case with the insertion of QR codes. To fulfill this purpose, the use of VANTs is an important tool for acquiring images to compose panoramic images and 3D models.

This type of material has a high potential in search of geodiversity sites and new association of specialized scientific information. Through the integration of the generated products (3D models, panoramic view, data of images with privileged visions of vision), few provided a benefit in the dissemination of knowledge in geosciences.

The possibility of including elements of augmented reality in interactive images greatly increases the educational potential of the information provided. Interactive, geointegration although efficient in issues related to the geological/geomorphological structures of different interests (scientific, geotouristic, protective, among others) have, from a micro-scale to a macro, several options to represent geological, geomorphological, landscaping, etc., contributing significantly to the geodiversity and reusing static cartographic resources, maintaining quality interactivity of georeferenced information.

Acknowledgments This work was co-funded by the European Union through the European Regional Development Fund, based on COMPETE 2020 (Programa Operacional da Competitividade e Internacionalização), project ICT (UID/GEO/04683/2013) with reference POCI-01-0145-FEDER-007690 and national funds provided by Fundação para a Ciência e Tecnologia

CAPES - Coordenação de Aperfeiçoamento de Pessoal de Nível Superior.

References

Bleisch S (2012) 3D geovisualization –definition and structures for the assessment of usefulness. In: ISPRS annals of the photogrammetry, remote sensing and spatial information sciences, vol I-2, pp 129–134. https://doi.org/10.5194/isprsannals-I-2-129-2012

Brilha J (2016) Inventory and quantitative assessment of geosites and geodiversity sites: a review. Geoheritage. https://doi.org/10.1007/s12371-014-0139-3,pp1-16

Dykes J, MacEachren A, Kraak MJ (2005) Exploring geovisualization

Lowe DG (1999) Object recognition from local scale-invariant features. In: International conference on computer vision, Corfu, Greece, pp 1150–1157

Lowe D (2004) Distinctive image features from scale-invariant keypoints. Int J Comput Vision 60:91–110

Maceachren AM, Ganter JH (1990) A pattern identification approach to cartographic visualization. Cartographica Int J Geogr Inf Geovisualization 27(2):64–81. https://doi.org/10.3138/M226-1337-2387-3007

McCormick B, DeFant T, Brown M (1987) Visualization. Sci Comput Comput Graph 21:i-E-8

Moura ACM (2003) Geoprocessing in urban planning and management. In: Author (ed) Belo Horizonte, Brasil, p 294

Philbrick AK (1953) Toward a unity of cartographical forms and geographical content. The Professional Geographer 5(5):11–15. https://doi.org/10.1111/j.0033-0124.1953.55_11.x

Sandercock M (2000) Cartographic visualization relies on the use of maps to presentcartographic ideas, is this visualization metaphor relevant in today's digitalworld? [online]: https://www.gisca.adelaide.edu.au/~msanderc/

Santana SA (2014) Geoprocessing in parameterized modeling of territorial landscape: applications of geovisualization in the simulation of the urban landscape. Phd thesis, Federal University of Minas Gerais - School of Architecture - Postgraduate Center in Architecture and Urbanism

Santos Edjane Maria dos (2016) Geoconservation as a tool for sustainable development in semi-arid regions: a study applied to the harsh mesoregion of Pernambuco, northeastern Brazil. PhD Thesis - Geosciences. Federal University of Pernambuco, Brazil. [online]: https://repositorio.ufpe.br/handle/123456789/17896

Santos I, Henriques R, Mariano G et al (2018) Geoheritage 10:143. https://doi.org/10.1007/s12371-018-0305-0

Snavely N (2008) Scene reconstruction and visualization from Internet photo collections, unpublished PhD thesis, University of Washington, USA

Westoby MJ, Brasington J, Glasser NF, Hambrey MJ, Reynolds JM (2012) Structure-from-Motion' photogrammetry: a low-cost, effective tool for geoscience applications. Geomorphology 179:300–314

Chapter 5
Virtual Heritage: A Model of Participatory Knowledge Construction Toward Biogeocultural Heritage Conservation

Pablo Mansilla-Quiñones, Hermann Manríquez, and Andrés Moreira-Muñoz

Abstract Virtual heritage has recently received attention as a novel path to better conserve geoheritage values and sites by means of the use of advances in digital imaging technology to synthesize, reproduce, represent, and display information. Traditionally, there have been difficulties in the inventory, quantification, and consolidation of relevant geoheritage sites, especially as concerns the limited tools for an adequate understanding of its complex nature and multiple connections to other landscape values, including biodiversity, cultural values, and the more integrative concept of biogeocultural heritage. This point of contention has been particularly felt in Latin America: although relevant heritage sites are in peril of disappearing, the application of geoheritage and geoconservation concepts has been relatively slow and there continue to be salient difficulties in reaching audiences beyond academic circles. Extant Chilean biogeocultural geoheritage is remarkable not just due to its immense and impressive geoforms, running the gamut from arid to cold–humid climates with their respective geodiversities; it also harbors natural values regarding the same geological and geomorphological richness. The remarkable cultural and historic contributions from original communities are essential to understanding the dynamically evolved Andes geosystem over vast geological periods. Based on the above, this chapter seeks to advance, first, effective action to protect geoheritage in Chile; and second, to reframe that geoheritage with other landscape, biological, and cultural values toward the more holistic concept of biogeocultural heritage, and do

P. Mansilla-Quiñones
Laboratorio Territorios Alternativos, Instituto de Geografía, Pontificia Universidad Católica de Valparaíso, Valparaíso, Chile
e-mail: pablo.mansilla@pucv.cl

H. Manríquez (✉)
Laboratorio de Geografía Física, Instituto de Geografía, Pontificia Universidad Católica de Valparaíso, Valparaíso, Chile
e-mail: hermann.manriquez@pucv.cl

A. Moreira-Muñoz
Instituto de Geografía, Pontificia Universidad Católica de Valparaíso, Valparaíso, Chile
e-mail: andres.moreira@pucv.cl

R. B. Singh et al. (eds.), *Global Geographical Heritage, Geoparks and Geotourism*, Advances in Geographical and Environmental Sciences,
https://doi.org/10.1007/978-981-15-4956-4_5

so by integrating aspects of ancient and current communities with deep historical ties to this diverse landscape. This goal is forwarded through the implementation of the novel languages and practices made available by geovisualization. Integrally, this is the Virtual Heritage process and includes the use of mobile apps in which wider audiences may interact across critical aspects of environmental and social conflicts.

Keywords Virtual heritage · Geoheritage sites · Biogeocultural heritage · Geoconservation · Geovisualization

5.1 Introduction

Currently, the inventory, conservation, and formal declaration of relevant geoheritage sites (such as geoparks and geosites) are increasingly sustained by specific technological tools that enable adequate understanding of the complex nature of geoheritage and its multiple connections. This trend has also had beneficial effects for geotourism and in generating meaningful senses of place in geosites. Ongoing developments in 'Virtual Heritage' have facilitated the synthesis, reproduction, representation, digital reprocessing, display, and conservation of different aspects of cultural heritage through advances in virtual reality (VR) imaging technology (Roussou 2009). However, it still has some shortcomings that need to be addressed, like lack of meaning or sense of place, as well as those stemming from technological limitations (Tan and Rahaman 2009).

Addressing these challenges is especially critical in Latin America, where environmental conservation faces severe conflicts and territorial disputes regarding divided visions for development: there are discrepancies in social appropriations of nature between the traditional values assigned by local communities versus the exchange value promoted from the perspective of governments and capitalist companies and industries.

These challenges represent a constant threat to geoheritage and underscore the urgency in addressing geoconservation and sustainability. Furthermore, other aspects of landscape values remain to be combined with geographical and geological values, such as biologically and culturally relevant aspects of the territory. This interplay within different spheres has been called biogeocultural heritage as a term that encompasses the diverse aspects of these complex landscapes (Ray and Gregg 1991; Gordon et al. 2018; Manríquez et al. 2019a).

In this sense, virtual reality has the potential of integrating all these aspects in an attractive and meaningful virtual space. Local communities today need to be more active and engaged, i.e., actors in the conservation of all the different aspects of heritage.

Instead of passive observation, freely accessible geographic information tools and web 2.0 platforms facilitate the development of citizen science by non-experts and can promote the development of local environmental conservation strategies. This scenario demonstrates that it is important to promote—beyond scientific circles

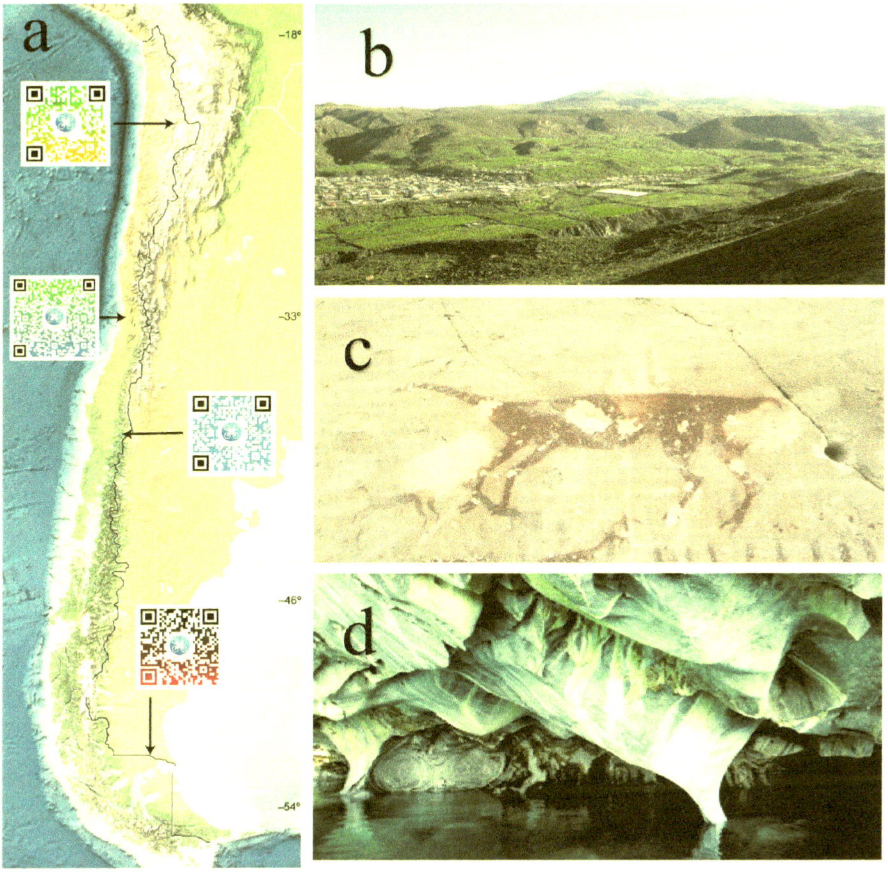

Fig. 5.1 Remarkable features of Chilean biogeocultural heritage: **a** examples of virtual journeys, by drone; **b** traditional terrace agriculture in the northern *Altiplano*; **c** Detail of pictography along with the "Inca Road System" (*Qhapaq Ñam*); **d** The "Marble Cathedrals" in the humid south. (Photographs by A. Moreira-Muñoz and Gaspar Gálvez)

or public and private institutions—environmental conservation through the development of citizen science. This would open transdisciplinary knowledge dialogues among these actors, and thus advance toward stronger social cohesion.

As a case in point, Chile—from the northern Atacama desert toward the southern temperate forests, channels, fjords, and ice fields (Fig. 5.1)—encompasses a remarkable diversity of landscapes and forms that constitute a rich biogeocultural heritage. Indeed, the Chilean Andes range is home to the highest concentration of active volcanoes; in the north, the highest plateau ("*Altiplano*"); remarkable longitudinal tectonic reliefs, such as the Coastal Cliff along the north, the Coastal *Cordillera*, and the intermediate depression; and in the south, dense outcrops of small islands, fjords, and glacial forms, and the largest ice field of the southern hemisphere.

Those shapes—related to Nazca and Antarctic subduction beneath the South America Plate—generate remarkable and distinctive landscapes along latitudinal and altitudinal gradients. Additionally, Chile harbors a long littoral of over 4,000 km, with many important morphological and paleontological features, and is recognized worldwide for its large cliffs, dune fields, wave-cut platforms at different heights above sea level, and macrofauna fossil sites.

Additionally, the above features are usually inseparable from traditional cultural elements of the landscape, such as "*bofedales,*" traditional agriculture terraces in the north (Fig. 5.1), pictographs in rock caves (Fig. 5.1), or ancient routes like the "Inca Road System" (*Qhapaq Ñam*) that unite Andean biogeocultural heritage. These elements encompass a diversity of forms, which further adds to the challenge for adequate communication and visualization at different scales (Manríquez et al. 2019b).

In practice, designing biogeocultural heritage conservation and geosites creates an opportunity to reunite relevant geological landmarks with both biotic values and those of culture; this acquires special relevance in times of global and local climatic and social changes. The valuation and conservation of different aspects of heritage have implications for future and current generations of students and the general public.

Today, significant pressures are associated with consequences from mining and energy extractivism and the deployment of transnational infrastructure works that consider the Andes Mountains a mere barrier to economic integration and capitalist development (Mansilla et al. 2019). The urgency of actions regarding such heritage continues to increase as its importance dwindles in society; for example, tourists damaged recently the most emblematic geoglyph from northern Chile, the "*Tarapacá* Giant." This is not merely a recent phenomenon: mineral activity and rallies like Dakar are known to generate profound damages to Atacama heritage. In spite of fees and national and international concerns, there continues to be neglect toward the importance of our heritage. In this sense, the use of new technologies and tools opens a world of possibilities for participation, social integration, and interaction with biogeocultural heritage.

5.2 Virtual Heritage and Dialogue Among Knowledges

Virtual heritage is the application of information and communication technologies to cultural heritage (Cayla and Martin 2018). It is capable of representing all elements of Geoheritage, scientific knowledge, local knowledge, and non-human knowledge, merged, integrated, and exposed to wider audiences through Digital media, Geonarratives, GIS, UAV, and audiovisual methods (Fig. 5.2). The connections among all these tools and cultural, biological, and geological heritage converge into the complex concept of Geoheritage. Adequate applications of Virtual Heritage for heritage conservation, however, require avoiding several pitfalls, for example,

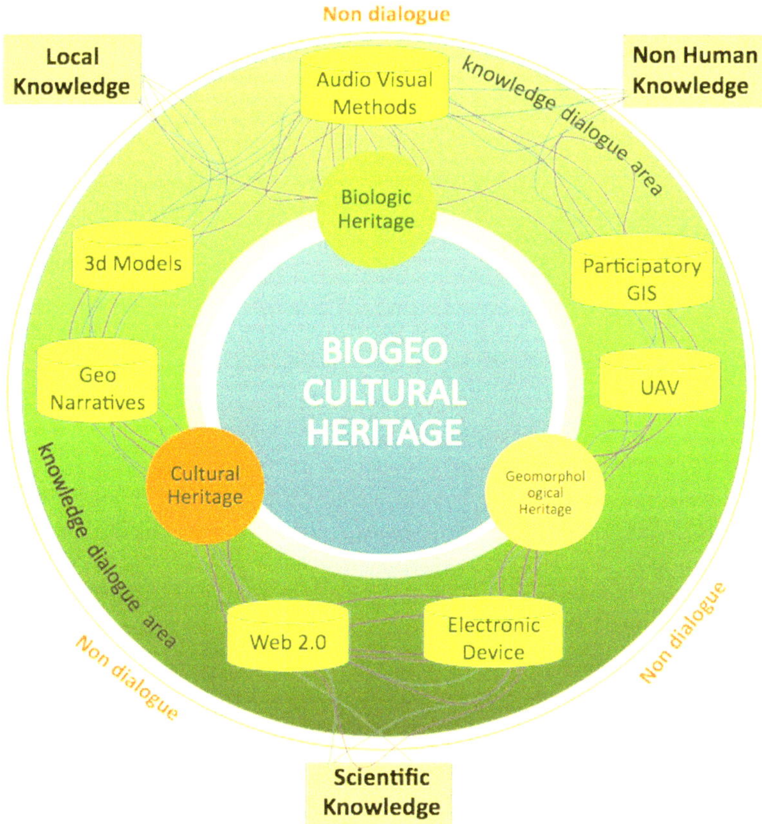

Fig. 5.2 A Virtual model for the inventory and valuation of biogeocultural heritage (based on concepts by Tan and Rahaman 2009; Escobar 2014; Cayla and Martin 2018)

focusing only on (touristic) products and not heritage valuation processes themselves (Tan and Rahaman 2009).

As shown in Fig. 5.2, the notion of Biogeocultural Virtual Heritage that we intend to promote focuses on transdisciplinary knowledge dialogue (Escobar 2014) among scientific, local and non-human knowledge—the latter of which has not been had an effectively representative dialogue.

Modern scientific knowledge has commonly maintained an arrogant perspective, discarding experience produced from local knowledge (De Souza Santos 2007). At the same time, its hierarchical position with respect to other social actors has impeded the production of constructivist knowledge. In contrast, the knowledge of local inhabitants is based on spatial practices located in the context of ecological relationships with their living spaces, in a process known as territorial ontologies (Blaser 2013). This experience allows them to have tools to generate creative strategies for environmental conservation. Likewise, recent perspectives in postcolonial geographies, animal geographies, and posthuman geographies point to the need to recognize non-human actors, such as rivers, mountains, flora, fauna, etc., beyond their status as objects, and rather as entities with the capacity to generate knowledge with an agency, and from which it is possible to learn and design other worldviews (Panelli 2009).

The dialogue among these multiple knowledges allows reconstruction of relational territorial ontologies that have been lost to the horizon of the modern development project that has laid the foundations of the current socio-environmental crises as manifested in the framework of climate change and globalization.

In this context, emerging technologies for geovisualization and the consolidation of Virtual Heritage are presented as devices to communicate knowledge and to cross borders of knowledge. These technical tools facilitate the communication of geographic information on biogeocultural heritage, as well as the multiple knowledges involved in its management.

The development of virtual environments offers immense possibilities; however, adequate development requires addressing several aspects or shortcomings, including interactive engagement with the public (game-based); constructivist interpretations; and the building of a meaningful 'virtual' sense of presence and place through hermeneutic virtual environments (Champion and Dave 2002, cited by Tan and Rahaman 2009). Some of the emerging tools in the field of Biogeocultural Virtual Heritage, also shown in Fig. 5.2, are listed below.

(a) Participatory GIS: Geographic Information Systems (GIS) had previously only been made available to experts; however, they are increasingly becoming available to widespread use. This trend is closing social gaps in communities or local institutions with low budgets. Platforms such as QGIS or gvSIG allow free access for the visualization, creation, and analysis of spatial data. Additionally, recent developments in web platforms such as Google Maps, Google Engine, or Mapbox have facilitated many geovisualization tasks.
(b) UAV, unmanned aerial vehicles: Growing access to aerial images through UAVs (drones) facilitates monitoring current states of biogeocultural heritage. At the same time, these devices are used for qualitative methodologies like guided tours of drone flights through biogeocultural heritage areas or regions (Ray and Gregg 1991). Here, inhabitants, through the lenses connected to the camera of the device, comment on their meanings and memories about these places.
(c) Web 2.0: Virtual platforms that facilitate communication among multiple actors who, in this case, interact with biogeocultural heritage. Web 2.0 refers to the capacity of a network to provide for the development of information through interaction and collaboration, in this case, geographic and geovisualization projects. An example of this comes from support provided by Wikimapia.
(d) Electronic Devices: Devices through which actors connect several of the tools indicated above. For example, to collect georeferenced information in real time, and incorporate it into collaborative mapping projects or for tracking paths in the pursuit of heritage recognition or monitoring, devices may include cameras,

augmented reality telephone applications, and QR codes to allow a greater degree of interaction with the geovisualization of data.

(e) 3D-Printed Terrain Models: Digital manufacturing tools currently provided by 3D printers make it possible to represent relief models easily and economically. Printing 3D terrain models also aids local communities in approximating geographical representations: whereas traditional map formats require knowledge of reading charts, 3D terrain models facilitate the recognition of geomorphological formations and biogeocultural heritage sites.

(f) Geonarratives: The life histories and biographies of inhabitants, investigated through spatiotemporal geovisualization (Kwan and Ding 2008). Although geonarrative prototypes have been supported by 3D models made in Geographic Information Systems, there is potential for expansion in open platforms like ESRI Story Maps or Google Earth.

5.3 Examples of Visualization for Environmental Assessment and Site Monitoring

Spatial information acquisition at different scales is currently quite feasible (Table 5.1). For the last five decades, aerial photographs were the most important media for acquiring spatial information and making maps. Although satellite imagery and digital processing have fully replaced aerial photographs, historical photogrammetric surveys still provide information on large swaths of territory. Similarly across Latin America, aerial photographs facilitate image overlap and stereographic visualization (3D vision). In Chile, some of the aerophotographic surveys still available are Trimetrogon (1944–45), Hycon (1955), OEA (1960), USAF (1970), SAF (1980), and Geotec (1980). Digitally treated historical aerial photographs enable stereographic and time modeled changes of any place and give a different value to biogeocultural heritage (Table5.1).

Table 5.1 Examples of geoheritage sites and features based on different scales of reference, as proposed by (Brocx and Semeniuk 2007)

Scales	Frame of reference	Geovisualization tool	Examples
Regional	100 km × 100 km or larger	Satellite images, Google Earth	The Atacama erg
Large	10 × 10 km	Fine Satellite images	Debris flows
Medium	1 km × 1 km	Aerial photographs, UAV	Aspects of Biogeoheritage at the National Park
Small	10–100 m × 10–100 m	UAV	Los Dedos paleontological site
Fine	1 m × 1 m	Photography	Taffonis, Granito orbicular
Very fine	1 mm × 1 mm or smaller	Microscope	Mineral crystals

Another important source of spatial information has been satellite images, which cover large surfaces over wide temporal frames. Depending on spatial and temporal resolution, a diversity of images taken by different satellites can be applied to different necessities, such as finer resolution urban planning (e.g., Sentinel images) or regional studies (NOAA images). Indeed, dramatic land use changes occurring in the most populated areas of Central Chile have been reported by means of remote sensing (Landsat images) (Schulz et al. 2010), as well as the impacts of immense wildfires from recent summers related to increases in global temperatures (Modis images) (Bowman et al. 2019).

But for detailed aspects of reality, even most accurate satellite images do not reach a resolution. It is here that the rapid increase in the use of UAV drones in environmental and geographic studies, especially in coastal environments, provide a valuable niche (Mancini et al. 2013; Gonçalves and Henriques 2015; Turner et al. 2016) (Table 5.1). The coastal interface of contact between the lithosphere, atmosphere, hydrosphere, and biosphere is a perfect example for detailed studies and is also where many environmental conflicts arise (see next section for case studies).

While traditional forms of research, representation, and exhibition of heritage are still in use, there have been significant advances in linking heritage values through Augmented Reality (Noh et al. 2009). Several examples of recent advances in geovisualization are presented below, as well as paths toward reinforcing Virtual Heritage in Chile.

5.4 Case Studies

5.4.1 Atacama Virtual Heritage

Among several features in this region, Atacama is the driest place on earth (Fig. 5.3). Interaction between the Nazca and the South America plates leaves units of relief, oriented north–south, which rises from sea level to over 6,000 m high, in a horizontal east–west average width of 250 km.

Of these, one of the greatest structural geological geoforms is the Atacama megafault, whose system has a 1,000 km N–S distance. From the western view, there is a large cliff named "*farellón costero*" that reaches 1,000 m above sea level (Paskoff 1989). From the eastern view, the tectonic fault scarp can be seen.

The coastal zone also has interesting geoheritage elements. The tafoni near *Caldera*, for example, show the natural weathering of a coastal saline environment on granitic rocks. Another site hosts orbicular granite, a Jurassic granitic rocky outcrop with spherolithic cores ("orbicules") originated by deep magmatic processes. These spherolithic cores are compound inclusions of tonalite mineral cores with many other minerals in a porphyritic matrix (Aguirre et al. 1976). This geoheritage, located mainly at the coast, is periodically (every 7 years) dressed by an ephemeral botanical cover associated with ENSO (El Niño Southern Oscillation), known as the "blooming

Fig. 5.3 Remarkable features of Atacama biogeocultural geoheritage: **a** the inland dune system; **b** tafonis at the Atacama coast; **c** blooming desert; **d** *Los Dedos* paleontological site at *Caldera* (Photos by A. Moreira-Muñoz; with permission of the IGM Military Geographic Institute of Chile)

desert" (Chávez et al. 2019), which is a marvelous example of the combination of geo- and biological heritages.

The Atacama coast is also home to one of the most important fossil deposits in Chile, *Los Dedos* paleontological site. Nearby paleontological sites include "*Cerro Ballena*" with more than 50 perfectly preserved whale fossils. North of *Copiapó* is an extensive sand dune system. The Atacama Erg, covering 240 km^2 (Araya 2001), has sands from the beaches and old wave-cut platforms since the Pleistocene carried inland to a distance of more than 60 km, following relief corridors

(Paskoff and Manríquez 2003). The complex dunes include transverse, seif, ankle-pattern, pyramidal, or barchans forms, visible from the air with detailed morphologies, physiognomy, and internal dynamics through high-resolution satellite images or UAV.

5.4.2 **Kütralkura** ***Geopark, 3D Model of*** **Llaima,** ***and Mapuche Culture***

The *Kütralkura* geopark is the first Chilean Geopark (Schilling 2014). It encompasses around 8,000 km^2 and some 50,000 people, mainly pertaining to the Mapuche-Pewenche ethnic groups. *Kütralkura* means "stone of fire" in Mapudungun language, referring to the *Llaima* volcano—within the geopark (*Conguillío* National Park) in the *Araucarias* Biosphere Reserve—that watches over most of this originally indigenous territory (Fig. 5.4).

Participatory mapping and new visualization tools have allowed the recent publication of the first "Mapuche Atlas" (Melin et al. 2018), a collaborative and intercultural work for the reconstruction and territorial defense from the Mapuche perspective, including that of the natural territory surrounding *Curacautín*. Through this project, the construction of at least two hydroelectric plants has been stopped by community action. The Atlas intends to deepen and expand the territorial and cultural knowledge of the area through cartographic models of reconstructed ancestral Mapuche territory. It does so from the orality that emanates from the Mapuche's own knowledge in conjunction with information available from historical sources and UAV technology support (QR in Fig. 5.1). This product integrates an approach from the perspective of Participatory Action Research, which not only addresses problems and alternative solutions from a local perspective but also allows dialogue with the cultural perspectives and knowledge of the Mapuche people. It is a clear advance in the construction of interculturality through practice and action.

The geopark designation looks to improve the quality of life of its inhabitants through the development of geotourism, education in geosciences, and geoconservation (Schilling 2014). *Kütralkura* geopark harbors a great geodiversity; a geological history that spans more than 200 million years, including recent notable active volcanic processes; and a remarkable biodiversity centered on the figure of the sacred "*pehuén*" tree (*Araucaria araucana*). The species, an iconic "living fossil," has only recently been confirmed as living as long as a thousand years (Aguilera-Betti et al. 2017). Additionally, the *pehuén* is an important element of biogeocultural heritage in a remarkable relationship with the Mapuche culture (Fig. 5.4).

Fig. 5.4 Remarkable features of the Biogeocultural Mapuche heritage: **a** Araucaria forest in *Conguillío* National Park; **b** traditional *nguillatün*, Mapuche ceremonial ritual; **c** terrain model of the geoheritage of the Araucanía region; **d** *Volcán Llaima* at the core of *Kütralkura* Geopark. (Photos by P. Mansilla)

5.4.3 Un-Habited Spaces of Patagonia

The cultural and geological landscape of mythical Patagonia and southernmost Chile (*Magallanes*) is shaped by a rich array of landscapes encompassing wonderful features of the physical geography (Coronato et al. 2017), rich biodiversity and impressive botanical landscapes, which have been documented in the stories of travelers and explorers since the times of Hernando de Magallanes and Antonio Pigaffeta.

The most striking feature of the western edge of the southern tip of South America that faces the Pacific Ocean is its great territorial fragmentation, a complex maze of multiple channels and islands of varying size.

In this regard, its unique views are due to two essential processes. Firstly, a tectonic collision process between the Antarctic and the South American plates, which manifests itself at 46° South latitude. This interaction, though poorly understood due to lack of seismic records, consists of a slow subduction process on the order of 20 cm/year (Charrier et al. 2007). Additionally, the Scotia plate in the extreme south of South America lies between the South America and Antarctic plates, contributing to further doubts in the context of interaction.

Secondly, Quaternary glaciations have structurally contributed to this territorial fragmentation, the high density of fractures and faults in the crust, and the deepening of ancient river valleys today transformed into fjords that take on the appearance of narrow, high-walled rocky corridors. Research suggests that quaternary glaciation covered the entire territory with inland ice (Laugenie 1982; Heusser 2003), of which only three ice cores remain as evidence: the Northern Ice Field, the Southern Ice Field, and the ice field located on the Darwin Mountain Range known as the Austral Ice Field (Fig. 5.5).

Next, the foremost morphological and structural feature of the Andes mountain range at these latitudes is undoubtedly the Patagonian Mountain Range. This great structural axis, formed by intrusive rocks that are part of the Patagonian batholith or the Andean *Aysén* batholith (Skármeta 1978), is found from 42° to 56°S and is more than 1500 km long.

Regarding culture, the population of Patagonia is intermittent, with marked stages of development and decline in relation to economic exploitation of natural resources associated with cycles of livestock, oil, and methanol production. These create a rural under population subjected to processes of depopulation and repopulation (Martinic 2006). This is in addition to the mobile practices of its inhabitants, characteristics of the territoriality and coping with the geographical and environmental characteristics of extensive Patagonia (Bascopé 2018).

All these features represent a paradoxically contested heritage, termed by Arqueros et al. (2015) a "heritage in extinction." In this context, one of the greatest challenges for the valuation and preservation of biogeopatrimony in Patagonia is related to its systematization and communication in highly fluid territorial contexts.

To address this challenge, we have developed work that rescues geographical geonarratives of the inhabitants of this territory, who, through their stories, name, and identify their places of cultural significance and the spatial practices in relation to these places. Based on these accounts, Geographic Information Systems are used to register places and routes.

Subsequently, fieldwork activities include audiovisual recordings of these places through film equipment and drone devices. These materials are edited and published to contribute to the continued narrative-building regarding the biogeopatrimony

Fig. 5.5 Remarkable aspects of southern Patagonia biogeocultural heritage: **a** 360° photography recorded by means of drone devices, inside an ancient volcano crater that served as a settlement for ancient indigenous peoples, in *Pali Aike* National Park; **b** Audiovisual record of biogeocultural heritage sites affected by oil extraction projects; **c** Records of soundscapes associated with biogeocultural heritage sites, in this case, the sounds of birds and fauna associated with bodies of water inside *Pali Aike* park; **d** Fluvio glacial sedimentary deposits near the glacial arc of terminal moraine at Magellan Strait, as viewed from satellite images (Photos by P. Mansilla)

present in these spaces. Such video support allows for more effective communication of territorial memories and local heritage and extends the diffusion of the results.

Indeed, further proposals regarding geosites and aspects of geoheritage have been undertaken in the extreme south of Chile and Argentina. Emblematic geosites already recognized include the Magallanes drumlins, the Mylodon cave, the *Torres del Paine* intrusive complex, and the *Pali Aike* volcanic field (Fig. 5.5), among others. Briefly, the Mylodon cave, a paleontological site, shows evidence of ancient human intermingling with prehistoric mylodon—an extinct Pleistocene herbivore—dating back to 12,000 years BP. *Pali Aike*, additionally, is considered a keystone of regional geoarqueological heritage (Barberena et al. 2006). Proposals to integrate geosites with tourism have been already undertaken by Mazzoni (2017) and Mazzoni et al. (2016).

5.4.4 *High-Resolution Imaging by UAV at a Sacrifice Zone (***Quintero-Puchuncaví***)*

The "*Loncura* dune field" (32°45′S), in Quintero bay, is a natural site in maximum degradation (Fig. 5.6). The 1955 Hycon aerial photo survey show a triangular dune field, bordered by a small river and the *Puchuncavi* wetland, with a surface of 182 ha, to the west.

Currently, the field is a narrow beach with a foredune running parallel to the coast. There are deflation corridors—longitudinal dunes orientated NE–SW—transversal dunes and scattered vegetation. The development of "*Las Ventanas*" complex industrial, currently the most polluted place in Chile, which began construction in the mid-twentieth century, has constantly ablated the riches of this place. Today, this dune field has been reduced to 12.6% of its original surface, which is now rife with pollutants and has altered aeolian morphologies. Some kilometers to the west of the *Las Ventanas* complex is the interesting paleontological site, "*Los Maitenes de Puchuncavi,*" at grave risk of disappearing due to pollutants affecting the fossils there?

Changes in historical land uses show the disappearance of sand dunes and the associated geological and biological values. However, even though there are still surviving paleontological sites of unquestionable geoheritage (Andrade et al. 2009), they have been classified a "sacrifice zone." Since the installation of industries in the 1960s, the area has undergone spoliation from the respective pollutants, and the air, soil, and water have been systematically poisoned (Muñoz et al. 2019). This has occurred to such an extent that nearby schools have been shuttered, and there are periodic environmental emergencies that sharply and negatively affect the quality of life of the people, with no end in sight. In one of the worst years to date, hundreds of cases were recorded in 2018 due to noxious chemicals, resulting in demonstrations against the "sacrifice zone" determination for industrial use, the low regulations, and the socio-environmental considerations (Pasten 2019).

Fig. 5.6 Integration of different visualization tools (aerial photographs and UAV imaging) at a coastal "sacrifice zone": **a** *Las Ventanas* industrial complex, UAV photo; **b** 1955 Hycon aerial photo showing the original *Loncura* dune field; **c** 2018 image showing the *Las Ventanas* complex industrial above the old dune field; **d** Natural arch, *Quintero* coast next to *Las Ventanas* complex (Photos by P. Mansilla and with permission of the IGM Military Geographic Institute of Chile)

5.4.5 *An Integrated Geobiocultural Approach to Virtual Heritage*

To address the above concerns, the authors are integrating all features mentioned throughout this paper into a mobile phone application, tentatively called "Geoscopio." This app is a platform for generating interactive channels in an information space, collected by a diversity of actors participating in Geoheritage and Environmental conservation (Fig. 5.7). Such applications—be they strictly virtual heritage, or virtual and augmented reality—can help the design of "Virtual Geographic Environments (VGEs)," considered the "new generation of geographic analysis tools"

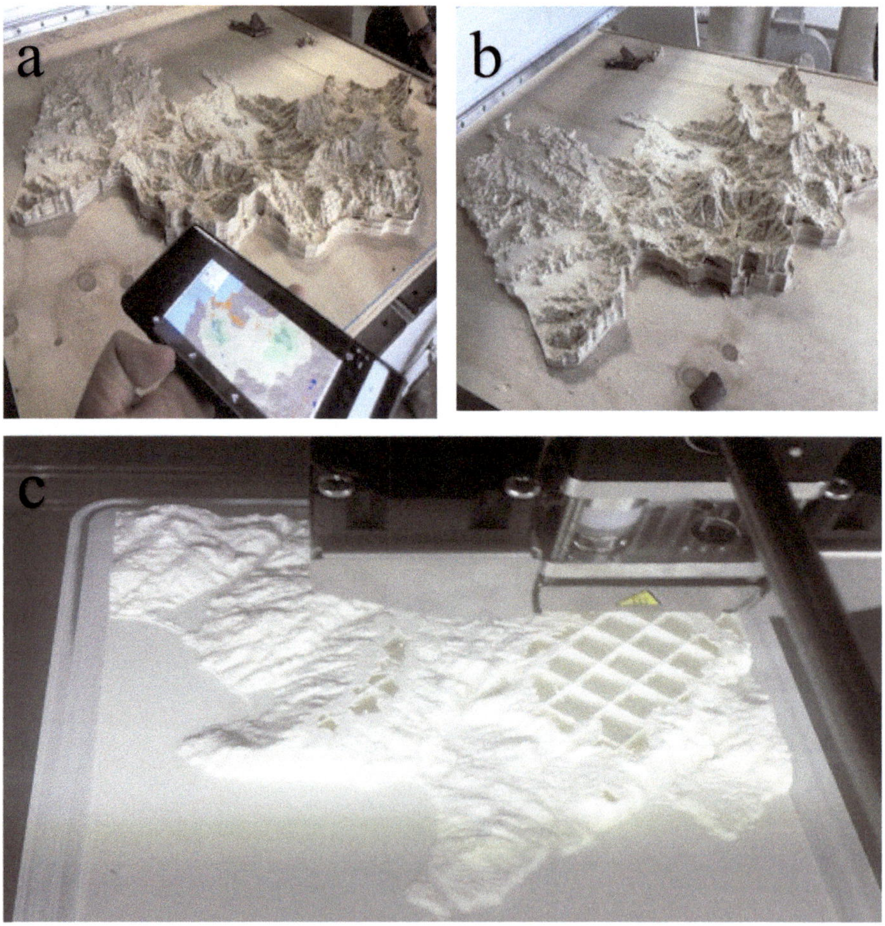

Fig. 5.7 3D model impressions: **a**, **b** in wood, for outreach and participatory actions toward the recognition of biogeocultural heritage; **c** through 3D Ultimaker 2+ printer, with PLA filament

(Lin et al. 2013) and serve as primers for future Geography and Geology teaching (Lv and Li 2016).

This will provide students with better comprehensions of the functioning of geosystems and the importance of conserving geoheritage and geodiversity (Brocx and Semeniuk, 2007; Gordon et al. 2018). It also opens the real possibility of building a future ethical framework for protecting geoheritage and its related fields in Latin America (Acevedo and Martínez Frías 2018).

5.5 Conclusions

This contribution proposes to advance and consolidate the concept of 'Biogeocultural Virtual Heritage' through the enhancement of geovisualization tools, toward a better understanding and conservation of biogeocultural heritage in Chile throughout its diverse landscapes and biogeocultural regions. This project implies the integration of traditional aspects of geoheritage with other important values of the landscape, such as the biological and cultural values of ancient and current communities with deep historical roots in these diverse landscapes. This goal is forwarded through the implementation of the novel languages and practices made available by geovisualization, which provides specific protocols and methods for effective participation and societal relationships to environmental problems and their possible solutions.

Although these devices do not themselves overcome the challenges of biogeocultural heritage conservation, they do reduce imbalances in the cultural capital of local actors living in territories where geoheritage is most evident. Advances toward the conservation of biogeocultural heritage will only be possible if we develop techniques and protocols for a better societal comprehension of these elements as parts of a rich and complex array of geosystems, landscapes, and heritages throughout Chile and the southern Andes. This requires encouraging interactive engagement and constructivist interpretation in citizens, scientists, and non-human actors to give sense to natural spaces and to environmentally devastated spaces, like *Las Ventanas* industrial complex.

Acknowledgments *Proyecto ANILLOS SOC* 180040 "Geohumanities and creative Biogeographies," funded by ANID—*Agencia Nacional de Investigación y Desarrollo* www.biogeoart.cl

Proyecto Fondecyt Iniciación N° 11181093: Dinámica Biogeomorfológica de las dunas costeras de Chile central: De la transgresión a la estabilización [Biogeomorphological dynamics of the coastal dunes of central Chile: From transgression to stabilization]

Fondecyt de Iniciación Científica N° 11181086: Deshabitar los extremos: Transformaciones de los modos de habitar lo rural en Magallanes.[Uninhabiting the Extremes: Transformations in Rural Living in Magallanes]

References

Acevedo RD, Martínez Frías J (2018) Geoethics in Latin America. Springer, Cham, Switzerland

Aguilera-Betti I, Muñoz AA, Stahle D, Figueroa G, Duarte F, González-Reyes A, Christie D, Lara A, González ME, Sheppard PR, Sauchyn D, Moreira-Muñoz A, Toledo-Guerrero I, Olea M, Apaz P, Fernandez A (2017) The first millennium-age Araucaria araucana in Patagonia. Tree-Ring Res 73(1):53–56

Aguirre L, Hervé F, Del Campo M (1976) An orbicular tonalite from Caldera, Chile. J Fac Sci, Hokkaido University, Japan 17(2):231–259

Andrade V, Schilling M, Mourgues FA (2009) Cuatro Sitios de Interés Paleontológico en la comuna de Puchuncaví: experiencia en educación escolar. In: Proceedings XII Congreso Geológico Chileno. Facultad de Ciencias Físicas y Matemáticas, Universidad de Chile, Santiago, pp 1–4

Araya JF (2001) Los Ergs del desierto marginal de Atacama, Chile. Investigaciones Geográficas 35:27–66

Arqueros G, Azúa A, Hidalgo J, Menard A, Morales H, Quiroz L, Radjl G, Uribe M, Urrutia F (2015) Patrimonio como extinción: Magallanes en el imaginario chileno. Sophia Austral 16:15–40

Barberena R, Blasi A, Castiñeira C (2006) Geoarqueología en Pali Aike: Cueva Orejas de Burro (Patagonia, Argentina). Magallania, (Chile) 34(1):119–138

Bascopé J (2018) En un área de tránsito polar. Desde el establecimiento de líneas regulares de vapores por el estrecho de Magallanes (1872) hasta la apertura del canal de Panamá (1914). CoLibris, Villa Tehuelches

Blaser M (2013) Notes towards a political ontology of 'environmental' conflicts. In: Green L (ed) Contested ecologies: nature and knowledge. HSRC Press, Cape Town, pp 13–27

Bowman DMJS, Moreira-Muñoz A, Kolden CA, Chávez RO, Muñoz AA, Salinas F, González-Reyes A, Rocco R, de la Barrera F, Williamson GJ, Borchers N, Cifuientes LA, Abatzoglou JT, Johnston FH (2019) Human-environmental drivers and impacts of the globally extreme, 2017 Chilean fires. Ambio 48:350–362

Brocx M, Semeniuk V (2007) Geoheritage and geoconservation—history, definition, scope and scale. J Roy Soc W Aust 90:53–87

Cayla N, Martin S (2018) Digital geovisualisation technologies applied to geoheritage management. In: Reynard E, Brilha J (eds) Geoheritage. Elsevier, Netherlands, pp 289–303

Charrier R, Pinto L, Rodríguez M (2007) Tectonostratigraphic evolution of the Andean orogen in Chile. In: Moreno T, Gibbons W (eds) The geology of Chile. The Geological Society, London, pp 21–114

Chávez RO, Moreira-Muñoz A, Galleguillos M, Olea M, Aguayo J, Latín A, Aguilera- BI, Muñoz A, Manríquez H (2019) GIMMS NDVI time series reveal the extent, duration, and intensity of "blooming desert" events in the hyper-arid Atacama Desert, Northern Chile. Int J Appl Earth Obs Geoinform 76:193–203

Coronato A, Mazzoni E, Vázquez M, Coronato F (2017) Patagonia: una síntesis de su geografía física, 1a edn. Universidad Nacional de la Patagonia Austral, Río Gallegos, Argentina

De Souza Santos B (2007) Para além do pensamento abissal: das linhas globais a uma ecologia de saberes. Revista Crítica De Ciências Sociais 78:3–46

Escobar A (2014) Sentipensar con la tierra. Nuevas lecturas sobre desarrollo, territorio y diferencia. Ediciones Universidad Autónoma Latinoamericana UNAULA, Medellín

Gonçalves JA, Henriques R (2015) UAV photogrammetry for topographic monitoring of coastal areas. ISPRS J Photogramm Remote Sens 104:101–111

Gordon JE, Crofts R, Díaz-Martínez E, Woo KS, Gordon JE (2018) Enhancing the role of geoconservation in protected area management and nature conservation. Geoheritage 10:191–203

Heusser CJ (2003) Ice age southern Andes. A Chronicle of paleoecological events. Developments in Quaternary Science 3. Series Editor. J. Rose. Elsevier Science Publishers B.V. 240 p

Kwan M-P, Ding G (2008) Geo-Narrative: Extending geographic information systems for narrative analysis in qualitative and mixed-method research. Prof Geogr 60(4):443–465

Laugenie C (1982) La región des Lacs, Chili Meridional. Recherches sur l'evolution géomorphologique d'un piémont glaciaire andine. Thése de doctorat dètat (unpublished). Université de Bordeaux III, 822 p

Lin H, Chen M, Lu G, Zhu Q, Gong J, You X, Wen Y, Xu B, Hu M (2013) Earth- Science Reviews Virtual Geographic Environments (VGEs): a new generation of geographic analysis tool. Earth-Sci Rev 126:74–84

Lv Z, Li X (2016) Virtual reality assistant technology for learning primary geography. In: Gong Z, Chiu D, Zou D (eds) Current developments in web based learning, pp 31–40

Mancini F, Dubbini M, Gattelli M, Stecchi F, Fabbri S, Gabbianelli G (2013) Using Unmanned Aerial Vehicles (UAV) for high-resolution reconstruction of topography: the structure from motion approach on coastal environments. Remote Sens 5(12):6880–6898

Manríquez H, Mansilla P, Figueroa-Sterquel R, Moreira-Muñoz A (2019a) Geodiversity meets Biodiversity: a landscape approach for biogeocultural conservation and governance in Mediterranean central Chile. Eco Mont—J Protect Mt Areas Res Manag 11:43–48

Manríquez H, Mansilla P, Moreira-Muñoz A (2019b) Valoración e inventario del Geopatrimonio del Norte Grande Chileno: Hacia una conservación integrada del paisaje biogeocultural de Atacama. Diálogo Andino 60:141–152

Mansilla P, Panez-Pinto A, Ponce-Hille M (2019) Discursos geopolíticos de desarrollo y reestructuración territorial IIRSA en el eje Mercosur-Chile. Diálogo Andino 59:37–53

Martinic M (2006) El poblamiento rural en magallanes durante el siglo xx.: realidad y utopia. Magallania (Punta Arenas) 34(1):5–20

Mazzoni E (2017) Landscapes of southern Patagonia, Argentina, their scenic characteristics and their importance for geotourism. In: Rabassa J (ed) Advances in geomorphology and quaternary studies in Argentina, Proceedings of the sixth argentine geomorphology and quaternary studies congress. Springer Series on Earth System Sciences, Springer Verlag, pp 140–179

Mazzoni E, Ferrari S, Navarro V (2016) Paisajes y modalidades turísticas en el Sur de la Patagonia continental argentina. In: Vieira A, Gonçalves A, Costa F, Cunha L, Lima Troleis A (eds) A Geografia Física e a Gestão de Territórios Resilientes e Sustentáveis. Portugal, Universidade do Minho, pp 743–754

Melin M, Mansilla P, Royo M (2018) Cartografía Cultural del Wallmapu: Elementos para descolonizar el mapa en territorio Mapuche. LOM Editores, Santiago

Muñoz AA, Klock-Barría K, Sheppard PR, Aguilera-Betti I, Toledo-Guerrero I, Christie DA, Gorena T, Gallardo L, González-Reyes Á, Lara A, Lambert F, Gayo E, Barraza F, Chávez RO (2019) Multidecadal environmental pollution in a mega-industrial area in central Chile registered by tree rings. Sci Total Environ 696:133915

Noh Z, Sunar MS, Pan Z (2009) A review on augmented reality for virtual heritage system. In: Chang M, Kuo R, Kinshuk G-DC, Hirose M (eds) Edutainment, 2009: learning by playing. Game-based education system design and development: 4th international conference on E-learning and games, pp 50–61

Panelli R (2009) More-than-human social geographies: posthuman and other possibilities. Prog Human Geogr 34(1):79-87.

Paskoff R (1989) Zonality and main geomorphic features of the Chilean coast. Essener Geogr. Arbeiten Bd. 18:237–267

Paskoff R, Manríquez H (2003) Origen de las arenas dunares de la region de Copiapó, desierto de Atacama. Chile. Revista Geológica De Chile 30(2):355–361

Pasten F (2019) El Paisaje del Conflicto: Estudio de las Representaciones Espaciales en la localidad de Ventanas. Tesis Geografía, Instituto de Geografía, Pontificia Universidad Católica de Valparaíso

Ray GC, Gregg WP Jr (1991) Establishing biosphere reserves for coastal barrier ecosystems. Bioscience 41(5):301–309

Roussou M (2009) A VR playground for learning abstract mathematics concepts. IEEE Comput Graph Appl 29(1):82–85

Schilling M (2014) Proyecto Geoparque Kütralkura. In: Moreira-Muñoz A, Borsdorf A (eds) Reservas de la Biosfera de Chile: Laboratorios para la Sustentabilidad. Academia de Ciencias de Austria, Pontificia Universidad Católica de Chile, Instituto de Geografía, Santiago, serie Geolibros 17, pp 178–181

Schulz JJ, Cayuela L, Echeverria C, Salas J, Rey Benayas JM (2010) Monitoring land cover change of the dryland forest landscape of central Chile (1975–2008). Appl Geogr 30:436–447

Skármeta J (1978) Carta Geológica de Chile N°29 Escala 1: 250.000. Región Continental de Aisén entre Lago General Carrera y la Cordillera Castillo. Instituto de Investigaciones Geológicas, Santiago de Chile

Tan B-K, Rahaman H (2009) Virtual heritage: reality and criticism. In: Tida T, Dorta T (eds) Joining languages, cultures and visions: CAAD futures, PUM

Turner IL, Harley MD, Drummond CD (2016) UAVs for coastal surveying. Coast Eng 114:19–24

Chapter 6
Communitarian Ethics, Environmental Conservation and Development

Sibonakaliso Shadrack Nhlabathi and Brij Maharaj

Abstract This paper, being conceptual in nature, arises out of the imperatives of sustainable usage of both natural and man-made heritage resources, particularly, monuments, parks and other natural resources which are generally consumed for the purposes of tourism. The paper observed the obligation of centralising the planet Earth in the consciousness of global development. In this regard, the paper argued for an adoption of the communitarian ethics as a frame of reference in any development project. Communitarian ethics has been found on the principle of common good. Thus, following this framework any development would meet the sustainable development obligation only if was predicated, not on individual or unilateral interests but on the interests of the global community at large. This is what the concept of communitarianism is about.

Keywords Communitarianism · Conservation · Sustainable development · Kantian ethics

6.1 Introduction

Common to all the emerging concepts of conservation and development is the prefix '*geo-*' which is derived from the Greek prefix '*geō-*' which combines 'earth, land, and country'. Further, the origin of '*geō*' could be traced from '*gaîa*' meaning 'earth'. Therefore, in view of this, the processes of utilising natural heritage called for the recognition of the centrality of planet Earth. Thus, this paper maintains that it is only by being ethical that planet Earth's resources would be preserved and conserved as they are exploited. Essentially, there are three categories of ethics, that is, Kantian (deontological) ethics, Teleological/Consequentialist ethics and Virtue/Aristotelian ethics. A brief word about each of the three types of ethics. Central to the Kantian/deontological ethics is the concept of moral duty of an individual,

S. S. Nhlabathi (✉) · B. Maharaj
Department of Anthropology and Development Studies, University of Zululand, Richards Bay, South Africa
e-mail: bona.nhlabathi@outlook.com; bona.nhlabathi@gmail.com

R. B. Singh et al. (eds.), *Global Geographical Heritage, Geoparks and Geotourism*, Advances in Geographical and Environmental Sciences,
https://doi.org/10.1007/978-981-15-4956-4_6

which suggests that an individual has a moral duty in every situation that arises. Kant called this moral duty categorical imperatives. Thus, people as rational beings have an obligation to act according to the directive of the morals. People have a duty to act morally. The Kantian ethics is governed by the maxim that one needed to treat humanity always as an end never as a means to an end (Bynum 2012; Lee 2017). The consequentialist/teleological moral systems are characterised primarily by a focus on the consequences which might arise from any action. Thus, in order to make correct moral choices, there have to be an understanding of what would result from choices made. In short, consequentialists define the ethical as whatever maximises the good (White 2009). The Virtue/Aristotelian ethics maintains that an object becomes good by achieving it purpose (*telos*). The *telos* function of a human being is to reason as it is through reason that human beings achieve good life (Dimmock and Fisher 2017).

However, this paper is predicated on communitarian ethics. Communitarian ethics bears some similarities with the Kantian/deontological ethics; however, communitarians reject the isolated self with duties, rights, interests, values, and ends which are independent of the social context. Instead, communitarian ethics posits that people have duties and responsibilities as citizens within a social context. Communitarians work to make sure that all ships rise by acting in the best interest of others. So, this paper argues that the communitarian ethics framework provided a sound basis to guide activities to exploit the rich terrestrial geo-heritage and communitarian ethics provided a lens through which the world and the people's place in it can be understood.

Before the paper explores the communitarianism philosophy in depth, the next section turns to what currently typifies the relationship between mankind and the natural environment.

6.2 Mankind and Their Relationship with the Natural Environment

The mankind's drive over centuries to be dominant over nature has created serious ecological and environmental problems (Tsekos and Matthopoulos 2009; Liu et al. 2018). These problems have become conspicuous in the twenty-first century, a period which could be described as the age of the global environmental crisis where myriad environmental and social issues challenge societies' and the whole ecosystem's ability to survive (Schild 2016). At the root of the global environmental crisis are ontological and epistemological worldviews which have determined how human kind related to their surrounding environment. The positivist science, which belongs to the realm of mechanistic concepts has exerted a considerable influence as mankind relates to the environment (Acott and McGibbon 2007; Cirillo 2014; Cox 2015).

Notwithstanding the denialist stance, concerning climate change and global warming, by some who hold positions of power in the global arena, for example, the former president of the United States George Bush and the current president Donald J. Trump, who express doubt about some central tenets of climate science (Powledge

2009; Björnberg et al. 2017), and that more than 50% of US citizens resolutely refuse to accept that emissions are changing the climate (Kool 2013), there is an urgent need to forge new viable modes of existence in order to bring harmony in the relationship with the environment. In order words, the business-as-usual way of dealings with the Earth has to be replaced, as soon as possible, by deliberate novel approaches and strategies to help in the better understanding of complex systems and of the scale on which the interactions between natural and human systems play out (Stafford 2010; The Global Environmental Change Programmes 2001). As the global environmental problems are highly complex, interrelated and interdependent, they demand novel approaches in understanding, managing and conceiving humans' relationship to the natural world (Schild 2016). There is no single way to face the environmental problems; however, the holism, interdisciplinary framework which entails the intermeshing of disciplines from the natural sciences, social sciences, engineering and management is being increasingly emphasised in the drive to address today's environmental challenges (Acott and McGibbon 2007; Cirillo 2014; Sauvé et al. 2016; Gibert et al. 2018). Interdisciplinarity and postmodern science is proposed in order to come to grasp with environmental problems. *The Condition of Postmodernity by* Lyotard celebrated the demise of grand narratives of modernisation and opened up 'difference' in science (Kwa 2005). In this regard, geographers could play a critical role in forging new modes of existence.

There are numerous models and concepts which have been put forward in order to interrogate and address environmental problems, for example, circular economy (Sauvé et al. 2016); the modern-day prophecy concept based on the 1972 book *Limits to Growth* (Kool 2013); the environmental stewardship concept (Cirillo 2014; Mathevet et al. 2018) and environmental citizen science concept (Owen and Parker 2018; Turrini et al. 2018). The governance for sustainability concept (O'Riordan 2004) which entails greater cooperation between nation states is also emerging to provide a framework for sustainable use of natural resources. The discussion below turns to the communitarian ethics framework, a framework which has not received the attention it deserves, with regard to the understanding of the relationship between humankind and the environment. The paper maintains that this framework further adds more understanding to the evolving body of knowledge on sustainable usage of natural and human kind made resources.

6.3 The Communitarian Ethics Framework

John Goodwyn Barmby first used the term communitarianism in 1841, out of a concern with the development of intentional and experimental communities, a concept which at the time was not being accorded the necessary status (Etzioni 2014). But the communitarian concept was not to become widely used up until the 1980s when Michael Sandel revived interest in it. Michael Sandel used this concept to counteract the apparent accentuation of individualism and autonomy by those who subscribed to liberalism. In the 1990s, Amitai Etzoini, a sociologist, emerged as a guru in the communitarianism intellectual movement (Golby 1997). During this time, communitarianism was to develop as a response to the Rawlsian liberalism, following John Rawls, who was a proponent of individualistic liberalism (Golby 1997; Etzoini 2014). Liberalism understood that the good resided in the individual rather than in the community; however, Michael Sandel and, subsequently, Amitai Etzioni were critical of this point of view and asserted that it is the communities rather than individuals who ought to be arbitrators of the good (Golby 1997; Arthur 1998; Etzioni 2014). Liberalism was further understood to be grounded on a male anthropology and thus sexist since it presumed that the natural self was solitary, independent, alone and self-centred, qualities which typify the male gender (Shields and Serna 2011).

Further criticism of liberalism, which supported the course of communitarians, is the liberal belief in the ethics of unfettered market and the laissez-faire economics (Golby 1997; Arthur 1998; Shields and Serna 2011; Etzioni 2014). The observation is that in a capitalist state liberalism had resulted in substantial gains to small number of communities at the expense of the larger masses; in this way, liberalism ignored the great importance of interpersonal relationships in sustaining everyone's life (Brown-Scott 1994; Shields and Serna 2011).

The above does not suggest that communitarianism was a perfect philosophical point of view. Critics of the communitarian philosophy argued that some non-liberal societies such as those grounded in fascism or caste system accorded a higher standing to their conceptions of a common good (Etzioni 2014). Thus, authoritarianism and repression could be disguised under the concept of common good, a concept which communitarians extol. As a consequence, there could also be a contingent danger that communitarianism comes to be used as a vehicle for dominant interests to repress those of minorities (Golby 1997). The communitarian philosophy has also been criticised that its position is anti-intellectual, as it holds that moral behaviour is principally a matter of conforming to established norms (Golby 1997).

Be that as it may, according to this paper, communitarianism serves as a systematic framework which provides tools to deal with ecological and environmental problems and enhance geo-conservation in a highly globalised world. As a postmodern philosophy, communitarianism rejects grand narratives such as those associated with liberalism in favour of interdisciplinary approaches (Brown-Scott 1994). As it was stated above, this entails the intermeshing of disciplines from the natural sciences, social sciences, engineering and management in so far as today's natural and social environmental challenges are concerned.

Communitarian ethicists emphasise the connection between an individual and his/her community. This serves to reconcile individualism and communal concerns. Communitarianism is based on the principle that there has to be common formulations of the good rather than leaving it to be determined by individual by him or herself, for themselves. Communities ought to secure a balance between individual autonomy and rights, on the one hand, and social responsibilities as they relate to the larger good of a community, on the other hand (Painter 2017). While individualism implies an atomistic conception of society, in which the individual is the ultimate source of value and meaning, communitarianism, on the other hand, views society organically, and the community as more than the sum of its individuals (Leeds 1998). According to Garfinkle (1997), communitarians believe that with every right there is a responsibility and that rights and responsibilities can enhance one another if they are in proper balance. Communitarians argue that the design of a good society requires drawing on: multiple normative principles, principles that conflict with each other at least in part, a careful balancing of these principles and principles whose point of equilibrium changes as the historical conditions change (Etzioni 2014). The communitarian ethic asserted that human beings were naturally depended, connected to one another in a web of moral obligation (Shields and Serna 2011).

Although bearing some similar features to Kantian deontological hypothesis, nevertheless, communitarianism represented a departure from the strict Kant's deontological theory. While the Kantian deontological hypothesis suggested that 'being right or correct' is given priority over the 'good', communitarianism brings the common good to the centre and rejects the isolated rational self with duties, rights, interests, values and ends which are independent of social context (White 2009). What personal choices are made are to be made not principally through moral reasoning (a principle which the deontological theory upholds) but through the recognition of duties, emulation of moral exemplars and followership. Communitarianism has an explicit and strong emphasis on morality and diverse expressions of morality (van Staveren 2009). Further, principles of communitarianism are at variance with the Kantian theory that an individual possesses dignity (an absolute inner worth) by which that individual command respect for himself from all other rational beings in the world (Rolf 2012). Being guided by their basic observation that the modern society has lost its sense of social solidarity, communitarians highlight the embeddedness of persons to their communities (Arthur 1998).

Communitarianism is based on the principle that forges an equilibrium between rights and responsibilities (Arthur 1998). According to Garfinkle (1997), communitarians seek to mediate the tension between these two forces of extreme autonomy and extreme centralised authority based on their understanding that societies remain healthy only so long as they effectively provide a balance between the centrifugal forces of autonomy and the centripetal force of centralised authority. This philosophy is therefore about taking on board civic responsibility in exchange for an individual opportunity (Arthur 1998). Communitarianism is a movement for moral renewal. What morality is not, as far as communitarians are concerned, is a personal conduct of individual choice (Golby 1997), but communitarians understand morality as emerging in communities through the interaction between agents in practices.

Communitarians deny the market as a location of morality (van Staveren 2009). According to communitarians, economic policies depend critically on the common and not on individual purposes to be achieved. This principle, however, need not be confused with the teleological or consequential ethics which informs utilitarianism which has its pillar that the moral value of actions depends entirely on their consequences/ends/goals for human welfare (Rolf 2012). These common purposes must be found on the core values of the citizens of the community. Put differently, while the criterion of common purpose defines communitarianism, teleological ethics insists that the ultimate standard for moral duty or moral value is the non-moral value it brings about, that is, the good in a non-moral sense, such as joy, happiness and utility (Lee 2017).

In terms of their economic philosophy, communitarian's philosophy holds a centrist position of the social order that mediates between libertarian and totalitarianism (Garfinkle 1997). Thus, communitarian economics seeks to mediate the tension between the thrust toward a laissez-faire economy characterised by decentralised autonomous individual behaviour and the thrust toward a command economy managed by a highly structured centralised authority. Communitarian economics adopts the positive elements from both conservative economics, with its emphasis on individual responsibility and wealth creation, and liberal economics, which is underpinned by an emphasis on community objectives and government responsibility for their achievement (Garfinkle 1997). According to Leeds (1998), communitarians attempt to find a means of combining the most desirable attributes of both individualism and collectivism.

Though the communitarian framework is not popularly known, it has had a considerable influence on the policies of both the Democrats in the USA and the New Labour Party in Britain (Arthur 1998). Thus, the above has been a rundown on this concept of communitarianism; the next section discusses how communitarianism would reflect on conservation and development practice.

6.4 A Communitarian Grounded Conservation and Development Project

Sustainable development and conservation have become topical global concerns. However, globally there appears to have been differences in terms of how to deal with the vexing issues of global warming and climatic change which can be traced from the unsustainable use of resources. Within major economic powers themselves, for example, in the USA there have been differences in terms of policy positions adopted by the Republicans and the Democrats. The Republicans, both during the Bush administration and now during the Trump administration, appear to have adopted a denialist position (Powledge 2009; Björnberg et al. 2017). It is the Democrats who during the Barack Obama administration appeared to have been amenable to climatic change and global warming problems (Powledge 2009). Recently, in 2018, President Trump pulled the USA out of the Paris Agreement on climate change and global warming. These global differences in the understanding of global warming and climatic change also seem to be rooted in nationalism which appears to be on the rise. National interests are taking precedence over multilateralism with dire consequences for the environment and conservation.

In view of the above, what would be the relevance of the communitarian concept to the preservation and conservation of the terrestrial heritage? In order to be able to respond to this question, it would be proper at this stage to explore the two key principles that define communitarianism. These are defined as an adopting of a centrist position which would mediate between liberal ideas and centrally planned government environmental programmes and reconcile individual and communal concerns (Garfinkle 1997; Painter 2017). And, secondly, the recognition that communitarianism is a postmodern concept (Brown-Scott 1994).

With regard to the centrist position, a communitarian grounded conservation and development project would adopt positive elements from both liberal system and a centralised system. In terms of liberal values, a communitarian grounded system would be conscientised about the community concept and communal ideals. It would be cognisant of the effects of its environmental policies and decisions on the communities and would reconcile the needs of the communities and the demands of the market.

Communitarianism being a postmodern grounded concept (Brown-Scott 1994) denotes that a communitarian grounded conservation and development project would be consistent with postmodern attributes which include intertextuality, lack of universal laws, irrationality and a rejection of grand narratives (Batkin 1997; Goneos-Malka, et al. 2013). Intertext from intertextuality suggest that 'text' lacks stability but is dynamic in nature, in that it is versatile as it is capable of tearing itself away from its historical and personal place (Batkin 1997). This suggests that environmental conservation needed to use text (knowledge) from other field of study in order to be able to make a meaningful contribution to environmental conservation. Related to the concept of intertextuality is that of diversity. This suggests that environmental sciences would benefit from the multiplicity of inter-meshing ideas.

6.5 Conclusion

There is no denying that global warming and climate change are a reality that deserved urgent attention. However, there has not been an agreement as to how to handle this phenomenon. This paper argued that sustainable usage of resources is an ethical issue. On the whole, there are three categories of ethics, namely, the Kantian/Deontological ethics, the Consequentialist/Teleological ethics and the Virtue/Aristotelian ethics. This paper argued for the less written about but according to this paper a pivotal concept in sustainable development, that is, communitarian ethics. Communitarianism regards the global nations as a community. Thus, any project of development ought to secure a balance between individual rights and country autonomy, on the one hand, and social responsibilities as they relate to the larger good of the global community, on the other hand.

References

Acott TG, McGibbon M (2007) Environmental science and sustainability: some critical thoughts. Environ Sci 4(4):199–207

Arthur J (1998) Communitarianism: what are the implications for education? Educ Stud 24(3):353–368
Batkin L (1997) On postmodernism and "postmodernism." Russ Stud Lit 33(3):62–85
Björnberg KE, Karlsson M, Gilek M (2017) Climate and environmental science denial: a review of the scientific literature published in 1990–2015. J Clean Prod 167:229–241
Brown-Scoot W (1994) The communitarian state: lawlessness or law reform for African-Americans. Harvard Law Rev 107(1209):1209–1230
Bynum GL (2012) Immanuel Kant's account of cognitive experience and human rights education. Educ Theor 62(2):185–201
Cirillo MC (2014) Science and environmental stewardship. Global Bioeth 25(2):114–124
Cox M (2015) A basic guide for empirical environmental social science. Ecol Soc 20(1):63–67
Dimmock M, Fisher A (2017) Aristotelian virtue ethics. Ethics for A-level. Open Book Publishers, London, pp 50–63
Etzioni A (2014) Communitarianism revisited. J Polit Ideol 19(3):421–260
Garfinkle N (1997) Communitarian economics. J Socio-Econ 26(1):1–24
Gibert K, Izquierdo J, Sànchez-Marrè M, Hamilton SH, Rodríguez-Roda I, Holmes G (2018) Which method to use? An assessment of data mining method in environmental data science. Environ Model Softw 110:3–27
Golby M (1997) Communitarianism and education. Curric Stud 5(2):125–138
Goneos-Malka A, Grobbler A, Strasheim A (2013) Suggesting new communication tactics using digital media to optimise postmodern traits in marketing. Communication 39(1):122–143
Kool R (2013) *Limits to Growth*, environmental science and the nature of modern prophecy. Ecol Econ 85:1–5
Kwa C (2005) Interdisciplinarity and postmodernity in the environmental science. Hist Technol 21(4):331–344
Lee M-H (2017) Confucianism, Kant, and virtue ethics. In: Jones D (ed) Its roots and global significance. University of Hawaii Press, Honolulu, pp 92–101
Leeds CA (1998) A study of communitarianism as a feature of contemporary capitalist societies and management. Int Bus Rev 7:51–67
Liu G, Yin X, Pengue W, Benetto E, Huisingh H, Schnitzer H, Wang Y, Casazza M (2018) Environmental accounting: in between raw data and information use for management practices. J Clean Prod 197:1056–1068
Mathevet R, Bousquet F, Raymond CM (2018) The concept of stewardship in sustainability science and conservation biology. Biol Cons 217:363–370
O'Riordan T (2004) Environmental science, sustainability and politics. Trans Inst Br Geogr 29:234–247
Owen RP, Parker AJ (2018) Citizen science in environmental protection agencies. In: Hacker S, Haklay M, Bowser A, Makuch Z, Vogel J, Bonn A (eds) Citizen science: innovation in open science, society and policy. UCL Press, London. https://doi.org/10.14324/111.9781787352339
Painter C (2017) All the game. J Stud 18(1):11–27
Powledge F (2009) Environmental science after bush. Bioscience 59(3):200–204
Rolf A (2012) Humanity as an object of respect: Immanuel Kant's anthropological approach and the foundation for morality. Hyethrop J 53:594–605
Sauvé S, Bernard S, Sloan P (2016) Environmental sciences, sustainable development and circular economy: alternative concepts for trans-disciplinary research. Environ Dev 17:48–56
Schild R (2016) Environmental citizenship: what can political theory contribute to environmental education practice. J Environ Educ 47(1):19–34
Shields JA, Serna S (2011) The demise of feminist communitarianism. Perspect Polit Sci 40(1):9–15
Stafford SG (2010) Environmental science at the tipping point. Bioscience 60(2):94–95
The Global Environmental Change Programme (2001) Earth system science: an integrated approach. Environ Sci Policy Sustain Dev 43(8):21–27
Tsekos CA, Matthopoulos DP (2009) Ethics, science and environment: the need for a new environmental worldview. Int J Environ Stud 66(6):679–687

Turrini T, Dörler D, Richter A, Heigl F, Boon A (2018) The threefold potential of environmental citizen science—Generating knowledge, creating learning opportunities and enabling civic participation. Biol Cons 225:176–186

Van Staveren I (2009) Communitarianism and the market: a paradox. Rev Soc Econ 67(1):25–47

White MD (2009) In defence of deontology and Kant: a reply to van Staveren. Rev Polit Econ 21(2):299–307

Chapter 7
Ecological and Socio-Economic Vulnerability to Climate Change in Some Selected *Mouzas* of Gosaba Block, the Sundarbans

Nabanita Mukherjee and Giyasuddin Siddique

Abstract This paper is an attempt to recognize ecological and socio-economic vulnerability to climate change in some selected *mouzas* of Gosaba Block, Indian Sundarbans. Ecological vulnerability has been identified by intensive studies on plant morphology (plant height, leaf area, leaf count, branches/plant) and stomata index to probe into the variation between the plants growing in salt-affected areas and those growing in the areas free from such phenomena. GPS survey has been conducted to perceive stretches of embankment breaching to perceive physical vulnerability. Socio-economic vulnerability has been worked out from the information and facts extracted from the primary household survey. Data collected on socio-demographic profile, livelihood strategies, health, food, water, social networks, natural disaster and climatic variability, indicators for Livelihood Vulnerability Index (LVI), and Livelihood Vulnerability Index–Intergovernmental Panel on Climate Change (LVI–IPCC) have been used to measure and evaluate the vulnerability of individual *mouzas* suffering from recurrent flooding, coastal erosion, and embankment breaching. The findings may assist the government and non-government groups to improve sectors for better coping and adaptation strategies for the poor households located in the marginal areas.

Keywords Climate change · Ecological vulnerability · Socio-economic vulnerability · Adaptation strategies

---

N. Mukherjee (✉) · G. Siddique
Department of Geography, The University of Burdwan, Bardhaman, West Bengal, India
e-mail: nabanita.mkj@gmail.com

G. Siddique
e-mail: gsbu2008@gmail.com

R. B. Singh et al. (eds.), *Global Geographical Heritage, Geoparks and Geotourism*, Advances in Geographical and Environmental Sciences,
https://doi.org/10.1007/978-981-15-4956-4_7

7.1 Introduction

Climate change has been inextricably linked with the rising sea and air temperature, the rise of sea level, increasing frequency and intensity of storm surges, surges with tide, fierce cyclones, untimely incessant rainfall events, flooding, etc. (Alston 2013; Resurreccion 2013; IPCC 2014; Hondula et al. 2015). It directly threats biophysical and socio-economic systems of a community (Yunlong 1997; Milly et al. 2005). Climate change leads to the intensification of atmospheric hazards, higher levels of flooding, and sea-level rise (McInnes et al. 2003; Nicholls and Tol 2006; Kleinen and Petschel-Held 2007); disruption of agricultural production (Tubiello 2005); destruction of biodiversity and natural resources (Thuiller et al. 2005); deterioration of human health and quality of life (Rogers and Randolph 2000; Reiter et al. 2004); and, finally, fluctuation of economy and polity (Kumar and Gautam 2014). The tropical developing countries of the world are among the most vulnerable to climate change because their limited human, institutional, and financial capability force them to accept a greater share of the global cost of climate change (IPCC 2007; Welp et al. 2009).

Such hydrometeorological hazards are very much evident in the world's largest trans-boundary repository of mangrove—the Sundarbans (Das and Das 2015; Samanta et al. 2017). The coastal tracts of the Sundarbans, washed in and out by sea waves, are exposed to the atmospheric hazards like storms, cyclones, and tidal surges that cause constant erosion and accretion of the river banks. Embankments made to protect the cultural areas of the Islands in the Sundarbans have been a result of the premature reclamation and severe coastal erosion leads to the frequent breaching of them. It allows saltwater intrusion into land, surface and sub-surface water sources which in turn affect the growth of plant and animal species endemic to the area along with damage to agriculture, fishing, and other subsistent occupations followed by the coastal people. The livelihood of the communities rooted in such coastal areas thus becomes vulnerable and insecure. Thus, it is necessary to assess the livelihood vulnerability of the people of the area for the sake of developmental planning, adaptation, and mitigation of the negative impact of changing climate.

It has always been a challenge for the policymakers, ecologists and other stakeholders to identify measures to meet the impact and risk from climate change (Velepucha et al. 2016). There are various methods of vulnerability assessment but the problem is that vulnerability in one sector is repeatedly linked with the vulnerability of other sector which is the foremost reason behind lack of execution of strategies. Vulnerability emerges out of multifaceted causal networks with diverse background conditions and dynamics such as differences in environmental conditions, pre-existing stresses to ecosystems, current resource use patterns, etc. (IPCC 2014). It instigates the adaptation measures to be far more complex (Magrin et al. 2007; FAO 2013; Merino et al. 2018).

Following IPCC, geographical and scientific studies use systematic methods of risk assessment and management to identify socio-economic and biophysical impacts (National Research Council 1997; Demeritt 2001; IPCC 2007; Yuehong et al. 2008;

Hahn et al. 2009; Pandey and Jha 2012; Etwire et al. 2013; Shah et al. 2013; Toufique and Islam 2014; Panthi et al. 2016; Alam 2017) related to changing climate. Here, Livelihood Vulnerability Index (LVI) and Livelihood Vulnerability Index–Intergovernmental Panel on Climate Change (LVI–IPCC) have been structured to estimate differential impacts of climate change on communities living in seven *mouzas* of Gosaba Block, the Indian Sundarbans. The LVI and LVI–IPCC indices use diverse indicators like socio-demographic profile, livelihood strategies, health, food, water, social networks, natural disaster and climatic variability to assess exposure, sensitivity, and adaptive capacity, the three prerequisite indicators of vulnerability (Hahn et al. 2009; Mukherjee and Siddique 2019a, b; Mukherjee et al. 2019). Natural disaster and climatic variability indicators are under exposure; health, food, water indicators are under sensitivity and socio-demographic profile, livelihood strategies, social network indicators are under adaptive capacity.

The mangrove species of the Sundarbans have greater ecological importance and uniqueness but climate change poses a threat on them. Studies reveal salinity-induced mangrove migration will have a strong regressive impact on the mangrove species, especially on *Heretierafomes* (*Sundari*) (Dasgupta et al. 2016, 2017). Thus, the identification of ecological vulnerability becomes very much necessary. Here, in this study, plant morphology (plant height, leaf area, leaf count, branches/plant) and stomata index have been worked out to differentiate the condition of growth and development of plants thriving in the core (comparatively low salinity areas) and marginal (high salinity zones) areas.

The findings of the study might be useful to the developmental organizations and policymakers for framing necessary adaptation strategies on the basis of varying degrees of vulnerability of community to climate change. It will assist the government and other organizations too to recognize the areas need particular intervention for future development.

7.1.1 The Study Area

The area under review includes seven *mouzas* of Gosaba Community Development Block: Bali, Sonagram, Hamilton Abad, Dayapur, Satjelia, Hetalbari, and Kumirmari. The Gosaba CD Block is situated in the Canning Sub-Division of South 24 Parganas District of West Bengal State. It is located at 22°09′55″N 88°48′28″E and is encircled by the Matla River in the South West, the Bidya River in the West, the Gosaba River in the South, and there are also numerous other small rivers (*khal/gang*). The Haribhanga River and the Raimangal River lies in the South East and the Eastern part, respectively, while the Bay of Bengal lies in the South. It is located in the south-eastern part of South 24 Parganas District. North 24 Parganas District is situated in the northeastern part of the Gosaba Block. The block has an area of about 296.43 km^2 and a population of 246,598 according to the 2011 Census (Fig. 7.1).

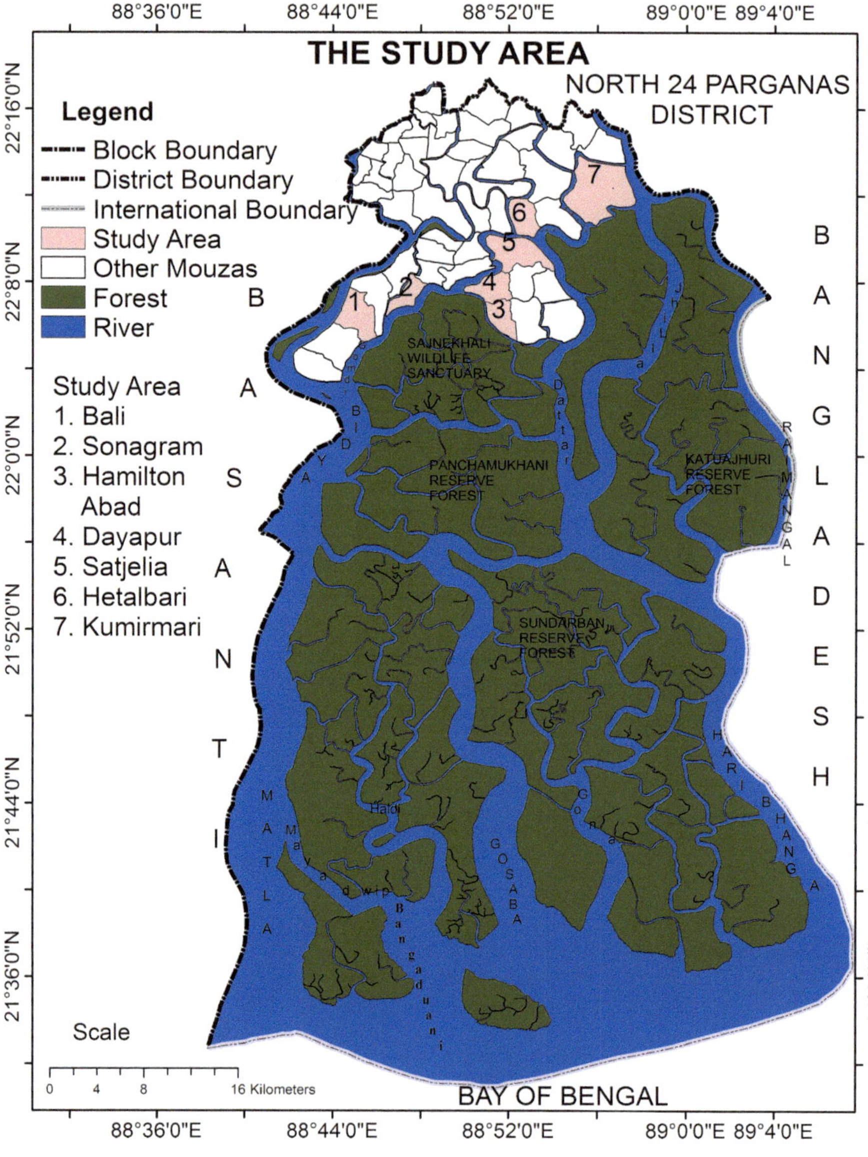

Fig. 7.1 The study area

7.1.2 Database and Methodology

Seven out of 51 *mouzas* of Gosaba have been selected as the area under review on the basis of the severity of erosion that was identified through a review of literature, field investigation, and discussion with experts and respondents. A primary survey has been conducted to recognize the livelihood vulnerability of marginal rural people to climate change and related problems. Random sampling technique has been applied to collect household data for its inherent quality of being unbiased owing to the homogeneity of the population. A statistically significant sample size of 572 households (92 households from Bali/Bally, 75 from Sonagram/Sonaga/Sonagao, 72 from Hamilton Abad, 76 from Dayapur, 78 from Satjelia, 87 from Hetalbari, 92 from Kumirmari) has been selected following the National Statistical Service Organization led by the Australian Bureau of Statistics (2018). Structured questionnaires have been prepared according to the need of the study. The survey was conducted from December 2017 to October 2018. These seven *mouzas* have been chosen for vulnerability assessment because they have recently reported coastal erosion, embankment breaching, and flood events on an annual basis. The interviewers followed computer-generated random number table to survey the required number of household and when a household refused to reply, surveyors moved to the next one. The heads of the households were asked to participate in group discussion and, according to their willingness, ten household heads were selected for the discussion. The information obtained from group discussion helped us to understand the phenomena much better. GPS survey has been conducted to find out the current embankment breaching spots (December 2017–October 2018). 6 edible plant species (Alternanthera phyloxeroides, Alternanthera sp, Azadirachta Indica, Colocasia Esculants, Amaranthus sp, Dolichos Lablab) have been chosen (respondents were asked to name few plant species and green leafy vegetables whose availability has been decreasing) in the study area to analyze the difference in stomata index and plant morphology (Plant height, leaf area width, length, leaf count/branch) in core and marginal zones. Five plants under each species, both from core and marginal areas, are collected to authenticate the information. Salisbury (1927) introduced the concept of stomata index (SI) which is now widely used in several researches related to plant growth under salinity condition (Villiers et al. 1996; Royer 2001). The formula is

$$\begin{aligned} \text{SI} = {} & \text{No. of Stomata per unit area} * 100/\text{No. of Stomata per unit area} \\ & + \text{No. of epidermal cell per unit area} \end{aligned}$$

7.1.3 Calculating LVI: Composite Index Approach

The present study has employed the LVI approach of Hahn et al. (2009), Alam (2017) following vulnerability framework of IPCC; however, a slight modification of the approach has been made to comply with the local reality of the specific area under study (Mukherjee and Siddique 2019a, b; Mukherjee et al. 2019). The LVI is used to measure the risk of each village to climatic anomaly and comparison of vulnerability among the villages (inhabited *mouzas*) to find out the causes which make those villages more vulnerable. The major components selected for estimating

LVI are socio-demographic profile, livelihood strategies, health, food, water, social networks, natural disaster, and climatic variability. Each of the major components is sub-divided into sub-components that contribute equally to the overall index. The sub-components are selected in such a way that they have a specific contribution to each major component. The scores range from 0 to 1, where 0 is least vulnerable and 1 is most vulnerable. A balanced weighted approach has been employed for the calculation (Table 7.1). The calculation of LVI follows a number of steps. At the first step, values are standardized using the formula:

$$\text{Index } s_d = (s_d - s_{\text{Min}}/s_{\text{Max}} - s_{\text{Min}}), \tag{7.1}$$

where Index s_d is the original sub-component for an area d and $s_{\min}$ and $s_{\max}$ represent the minimum and maximum values for each sub-component, respectively. The sub-components were averaged after being standardized using the following formula to calculate the value of each major component:

$$M_d = \sum_{i=1}^{n} \text{Index } s_d\, i/n, \tag{7.2}$$

where M_d is one of the eight major components for an area d, Index $s_d i$ denotes the sub-components, indexed by i, which constructs each major component, and n indicates each major components' number of indicators. Once the calculation of the values of each of the eight major components has been done, they were then averaged to obtain the LVI using the formula:

$$\text{LVI}_d = \sum_{i=1}^{8} Wm_i Md_i / \sum_{i=1}^{8} Wm_i, \tag{7.3}$$

where LVI_d is the LVI score of area d, Wm_i is the weight which equals to the number of sub-components that make up each major component.

Calculating LVI–IPCC, IPCC framework approach: LVI–IPCC approach has also been adopted in this study which takes into account IPCC vulnerability definition. The LVI–IPCC differentiates from the LVI as the major components are combined following the three elements of IPCC vulnerability mentioned earlier. Exposure of the target population is measured by the perception of people on natural disaster and climatic variability. Sensitivity is measured by assessing the present status of food, water, and health security. Adaptive capacity is quantified by assessing socio-demographic profile, livelihood strategy, and social safety. LVI–IPCC score varies from −1 to +1, where −1 is least vulnerable and +1 is most vulnerable. LVI–IPCC has been calculated using the following formula:

Table 7.1 Major components and sub-components of livelihood vulnerability index

Major component	Sub-component	Explanation of Sub-component	Survey question
Socio-demographic profile	Dependency	Percent of HH members below 15 and above 60 years	Please mention the age and sex of people who live in this HH
	Percent of female-headed HH	Percent of female members to total HH members	whether you are the head of the HH
	Percent of HHs where the head of family did not attend school	Percent of HHs where head did zero years of schooling	Whether the head of HH attended to the school
	Percent of HHs where head is the only earning member	Percent of HHs where head is the only earning member	Whether the head of HH is the only earning member
	Average number of family members in a HH	Average number of family members in a HH	How many people live in the family?
	Percent of HHs having non-climate-resilient home	Percent of HHs having *kaccha* (mud) house type	Mention your house type? How many times you have reconstructed your house in your lifetime when it got damaged by hazards?
Livelihood strategy	Percent of HHs where family member migrate for work	Percent of HHs where family members migrate for work	How many people from your family migrate as they don't get enough job opportunity here? Where do they go? How much they earn? How long they stay away from home?
	Percentage of HHs dependent upon natural resources	Percentage of HHs dependent upon natural resources	Do you or anybody else from your HH collect anything from forest, rivers or any other natural resources?
	Percent of HHs where agriculture is the main source of income	Percent of HHs where agriculture is the main source of income	Is your family completely dependent upon agriculture for livelihood?

(continued)

Table 7.1 (continued)

Major component	Sub-component	Explanation of Sub-component	Survey question
	Percent of earning member in a HH	Percent of members who earn in a HH	How many people of your HH are engaged in any sort of work for earning livelihood?
Health	Percentage of HHs who find it difficult to reach health facilities	Percentage of HHs who find it difficult to reach health facilities because of the greater distance of HH from health center	How long it takes to reach health facility?
	Percentage of HHs whose family members died/injured without treatment	Percentage of HHs whose family members died/injured without treatment	How many people of your family got injured or died during the period of climatic perturbation?
	Percent of HHs without proper sanitation facilities	Percent of HH without sanitary latrine or sanitation facilities	Do you have a proper sanitary latrine in HH? Do you use proper sanitation?
	Percent of HHs where members suffer from illness	Percent of HH-reported diseases	How many members of the family remain chronically ill throughout the year?
	Percent of HHs not visiting doctors during illness	Percent of HH not visiting doctors during illness due to lack of awareness	How often do you visit a doctor? Does the hospital have adequate doctors or you depend upon local unscientific quack doctors?
Food	Percentage of HHs that get partial or insufficient food from the family farm	Percentage of HHs that do not get sufficient food from the family farm	From where does your HH get daily food mostly?
	Percent of HHs reported decreasing agricultural production	Percent of HH reported decreasing agricultural production	Have you observed any change in the agricultural production post-*Aila*? Does salinity play any role to it?

(continued)

Table 7.1 (continued)

Major component	Sub-component	Explanation of Sub-component	Survey question
	Percent of HH losing agricultural land	Percent of HHs reported a decrease of the agricultural field due to coastal erosion	Have you lost your agricultural land owing to erosion? How much area have you lost?
	Percent of HHs reported increasing food insecurity during a natural disaster	Percent of HH reported increasing food insecurity during a natural disaster	How many months a year the HH is in trouble to get enough food after a natural disaster?
	Percent of HHs reporting decreasing availability of freshwater fishes for consumption	Percent of HH reporting decreasing size and amount of fresh water fishes due to salinity effect	Do you find that that saline water affects availability of good catches in the study area?
	Percent of HHs reporting decreasing regeneration of green leafy vegetables	Percent of HH reporting decreasing regeneration of green leafy vegetables	Do you find that that salinity or any other condition has been affecting regeneration of green leafy vegetables in your locality?
Water	Percent of HHs whose members walk more than 500 m–2 km to reach the water source	Total distance to reach the safe drinking source	How long it takes to reach to the source of water?
	Percent of HHs use unsafe water for drinking, cooking, bathing, and washing	Percent of HH use unsafe water for drinking, cooking, bathing, and washing (river, pond, hole, contaminated water)	How many people use unsafe water (saline pond water)?
	Percent of HH reported increasing effect of saline water	Percent of HH believes that there is greater effect of saline water in the area	Do you think that there is any effect of salinity in the water that you use on a daily basis? State the effect
	Percent of HHs reported water conflict	Percent of HH reported conflict due to using same water sources	Is there any conflict over water? How many tube well/other source is there in the village?

(continued)

Table 7.1 (continued)

Major component	Sub-component	Explanation of Sub-component	Survey question
Social safety	Percent of HHs do not receive assistance from social network	Percent of HH do not receive assistance from social network	During the period of disaster or during flood did your relatives help you financially or physically to cope up with the situation?
	Percent of HHs do not receive assistance from Government after hazard/disaster	Percent of HH do not receive assistance from governtment	Whether you receive any help from Government during yearly flood hazard or during disaster period? Are you an *Aila* beneficiary?
	Percent of HHs do not receive assistance from other sources after hazard/disaster	Percent of HH do not receive assistance from NGOs (It may be a shawl, tarpaulin, cloth, or any other item that the community needs during hazard or disaster)	Whether you receive any help from non-government organizations during yearly flood hazard or during disaster period?
	Percent of HH do not get awareness	Percent of HH do not get awareness to reduce the risk of disaster through training provided in schools or local clubs	Is there any training programme arranged in society to teach people how to deal with hazards?
	Percent of HHs do not use mobile phone for communication	Percent of HH do not use mobile phone for communication	Do you have a mobile phone? Do you get information regarding onset of any hazard by it?
Natural hazard, disaster	Percentage of HHs reported increased frequency and intensity of cyclone and storm surges	Percentage of HH reported increased frequency and intensity of cyclones and storm surges	How many times this area is affected by cyclone in the past 10 years?
	Percentage of HHs with injury or death as a result of natural disasters in the last 10 years	Percentage of HHs with injury or death as a result of natural disasters in the last 10 years	How many members of the family died or got injured in the past 10 years?

(continued)

Table 7.1 (continued)

Major component	Sub-component	Explanation of Sub-component	Survey question
	Percentage of HHs with injury or death to their livestock as a result of natural disasters that occurred in the last 10 years	Percentage of HHs with injury or death to their livestock as a result of natural disasters that occurred in the last 10 years	How many livestock got died or washed away in the past 10 years?
	Percentage of HHs with losses to physical assets	Loss of assets like (homestead/agricultural equipment/machinery)	Have you lost your assets during period of hazard or disaster?
	Percentage of HHs that do not receive warning before a natural disaster	Percentage of HH that do not receive warning before a natural disaster	Did you receive warning by the local authority before any hazard took place?
Climatic variability	Percentage of HHs reported change in summer temperature	Percentage of households reported increase in summer temperature	Do you find any change of summer season in the past 10 years?
	Percentage of HHs reported change in winter temperature	Percentage of households reported increase of temperature during winter	Do you find any change of winter season in the past 10 years?
	Percentage of HHs reported decreasing span of winter season	Percentage of households reported decrease of span of winter	Do you find any change of span of winter season in the past 10 years?
	Percentage of HHs reported vagaries of monsoon	Percentage of households reported variability in monsoon rainfall	Do you find any change of monsoon rainfall in the past 10 years?
	Percentage of HHs reported change in Winter and post-monsoon rainfall	Percentage of households reported increase in winter and post-monsoon rainfall	Is there any sort of change in winter and post-monsoon rainfall in your locality?
	Percentage of HHs reported change in the frequency of floods	Percentage of households reported increase in the frequency of floods	How many times a year this area gets flooded?

*HH = Households

$$\mathrm{CF}_d = \sum_{i=1}^{n} Wm_i Md_i / \sum_{i=1}^{n} Wm_i, \tag{7.4}$$

where CF_d is an IPCC-defined contributing factor (exposure, sensitivity, or adaptive capacity) for *mouza d*, Md_i are the major components for area d indexed by i, Wm_i is the weight of each major component, and n is the number of major components in each contributing factor. Once exposure, sensitivity, and adaptive capacity were calculated, the three contributing factors were combined using the following equation:

$$\mathrm{LVI{-}IPCC}_d = (e_d - a_d) * s_d, \tag{7.5}$$

where $\mathrm{LVI{-}IPCC}_d$ is the LVI for *mouza d* expressed using the IPCC vulnerability framework, e is the calculated exposure score for area d (weighted average of Natural Disaster and Climate Variability major component), a is the calculated adaptive capacity score for area d (weighted average of the Socio-Demographic, Livelihood Strategies, and Social Networks major components), and s is the calculated sensitivity score for area d (weighted average of the Health, Food, and Water as major components). Hahn et al. (2009) and Alam (2017) has been followed for both the calculation, but following experts' opinion, here adaptive capacity calculation has been treated same as exposure and sensitivity as all the sub-components express vulnerability.

7.2 Results and Discussions

7.2.1 Biological Vulnerability

Table 7.2 shows the average number of epidermal, stomata cell and stomata index of six plants collected from the core is higher than that of the marginal zones. Similarly, the average height of plant; width and length of leaf area; number of leaf per branches are found to be higher in core areas compared to marginal areas. Previous researches state that soil salinity is a major constraint affecting crop yield and plant growth (Qiu et al. 2007; Munns and Tester 2008; Gupta and Huang 2014; Negrao et al. 2017; Sandhu et al. 2017). Moisture stress generally affects stomatal openings of a plant, which leads to closure of stomata in order to reduce the amount of carbon dioxide assimilation. Closing of the stomata allows plants to slow down the rate of transpiration (limit water loss) and helps to prevent the wilting effects of moisture stress (Solmaz et al. 2013; Osakabe et al. 2014; Acosta-Motos et al. 2017). Thus, it becomes clear that plants grow under more stressful condition (Kerstiens et al. 2002) in marginal areas (where the intrusion of saline water takes place) compared to lower salinity areas in the core.

Table 7.2 Stomata index and plant morphology analysis

Name of plant	No. of epidermal cell	No. of stomata cell	Stomata Index	Height (m)	Width of Leaf Area (m)	Length of Leaf Area (m)	Leaf/Branches
Alternanthera phyloxeroides							
CORE (22.205 N 88.936 E)	19	12	38.709	0.100	0.004	0.006	20
MARGIN (22.192 N, 88.957 E)	19	9	32.142	0.100	0.004	0.005	17
Alternanthera sp							
CORE (22.205 N 88.936 E)	12	9	42.857	0.150	0.003	0.005	14
MARGIN (22.192 N, 88.957 E)	10	8	44.444	0.140	0.003	0.004	13
Amaranthus sp(Amaranthus gangeticus)							
CORE (22.115 N, 88.87 E)	39	10	20.408	0.166	0.016	0.036	1 (8 leaves per plant)
MARGIN (22.117 N, 88.850 E)	36	9	20.000	0.138	0.013	0.018	1 (6 leaves per plant)
Azadirachta indica							
CORE (22.115 N, 88.87 E)	410	51	11.062	1.830	0.014	0.522	10
MARGIN (22.1162 N, 88.849 E)	408	44	9.7345	0.860	0.010	0.500	8
Colocasia esculanta							
CORE (22.189 N 88.926 E)	76	15	16.483	0.457	0.063	0.112	1
MARGIN (22.1162 N, 88.849 E)	76	12	13.636	0.451	0.053	0.102	1
Dolichos lablab							
CORE (22.189 N 88.926 E)	42	16	27.586	6.096	0.085	0.100	3

(continued)

Table 7.2 (continued)

Name of plant	No. of epidermal cell	No. of stomata cell	Stomata Index	Height (m)	Width of Leaf Area (m)	Length of Leaf Area (m)	Leaf/Branches
MARGIN (22.1162 N, 88.849 E)	40	12	23.076	2.896	0.050	0.075	3

7.2.2 LVI Analysis

Table 7.3 shows the values of component, sub-component chosen for LVI analysis and also the composite value of LVI for each of the *mouzas*. The highest vulnerability is observed in Kumirmari (0.631) followed by Hamilton Abad (0.596) and Satjelia (0.566), while the lowest has been observed in Bali (0.496) followed by Sonagram (0.516). Dayapur (0.574) and Hetalbari (0.543) recorded moderate LVI values (Fig. 7.2). The following table signifies that higher the index value score, higher the vulnerability, and vice versa.

Table 7.3 LVI components, sub-components, and analysis

	Bali	Sonagram	Hamilton Abad	Dayapur	Satjelia	Hetalbari	Kumirmari
1. *Socio-Demographic Profile*							
Dependency	0.335	0.267	0.400	0.393	0.734	0.224	0.218
Percent of female-headed HH	0.050	0.069	0.130	0.012	0.075	0.044	0.050
Percent of HH where head of family did not attend school	0.230	0.243	0.273	0.212	0.116	0.234	0.278
Percent of HH where head is the only earning member	0.900	0.627	0.815	0.837	0.775	0.733	0.925
Average number of family members in a HH	0.390	0.458	0.484	0.464	0.472	0.402	0.405
Percent of HH having non-climate resilient home	0.675	0.714	0.842	0.864	0.775	0.600	0.900
Index Value for component	0.430	0.396	0.490	0.464	0.491	0.372	0.462
2. *Livelihood Strategy*							
Percent of HH where family members migrate for work	0.175	0.182	0.552	0.324	0.375	0.422	0.375

(continued)

Table 7.3 (continued)

	Bali	Sonagram	Hamilton Abad	Dayapur	Satjelia	Hetalbari	Kumirmari
Percentage of HH dependent upon natural resources	1.000	1.000	1.000	1.000	1.000	1.000	1.000
Percent of HH where agriculture is the main source of income	0.300	0.461	0.321	0.297	0.500	0.511	0.575
Percent of earning member in a family	0.303	0.461	0.268	0.292	0.280	0.411	0.311
Index Value for component	0.444	0.526	0.535	0.478	0.538	0.586	0.565
3. *Health*							
Percentage of HH who find it difficult to reach health facilities	0.915	0.945	0.926	0.897	0.875	0.222	0.975
Percentage of HH whose family members died without treatment	0.050	0.023	0.157	0.105	0.100	0.088	0.075
Percent of HH without proper sanitation facilities	0.250	0.465	0.447	0.540	0.625	0.422	0.775
Percent of HH where members suffer from illness	0.425	0.488	0.552	0.567	0.500	0.644	0.600
Percent of HH not visiting doctors during illness	0.500	0.465	0.011	0.567	0.450	0.400	0.500
Index Value for component	0.428	0.477	0.418	0.535	0.510	0.355	0.585
4. *Food*							
Percentage of HH that get partial or insufficient food from the family farm	0.975	0.860	0.631	0.702	0.550	0.866	0.750
Percent of HH reported decreasing agricultural production	0.100	0.674	0.447	0.486	0.250	0.511	0.725
Percentage of HH losing agricultural land	0.350	0.790	0.789	0.573	0.525	0.400	0.725

(continued)

Table 7.3 (continued)

	Bali	Sonagram	Hamilton Abad	Dayapur	Satjelia	Hetalbari	Kumirmari
Percent of HH reported increasing food insecurity during natural disaster	0.700	0.101	1.000	1.000	0.825	0.822	0.975
Percent of HH reported decreasing fish production	0.250	0.674	0.368	0.594	0.500	0.511	1.000
Percent of HH reporting decreasing regeneration of green leafy vegetables	0.550	0.860	0.605	0.783	0.700	0.955	0.900
Index Value for component	0.487	0.660	0.640	0.690	0.558	0.677	0.845
5. *Water*							
Percent of HH who walk more than 500 m–2 km to reach the water source	0.075	0.023	0.447	0.054	0.025	0.022	0.400
Percent of HH use unsafe water for drinking, bathing, cooking, and washing	0.875	0.829	0.910	0.890	0.880	0.891	0.945
Percent of HH reported increasing effect of saline water	0.650	0.558	0.921	0.648	0.650	0.800	0.475
Percent of HH reported water conflict	0.025	0.255	0.917	0.378	0.025	0.088	0.450
Index Value for component	0.406	0.416	0.799	0.492	0.395	0.450	0.567
6. *Social Safety*							
Percent of HH do not receive assistance from social network	0.175	0.325	0.105	0.405	0.275	0.311	0.675
Percent of HH do not receive assistance from Government after natural hazard/disaster	0.300	0.232	0.552	0.434	0.275	0.488	0.200
Percent of HH do not receive assistance from NGO after natural hazard/disaster	0.300	0.302	0.578	0.459	0.600	0.622	0.700
Percent of HH do not use mobile phone for communication	0.100	0.116	0.078	0.123	0.050	0.045	0.025

(continued)

Table 7.3 (continued)

	Bali	Sonagram	Hamilton Abad	Dayapur	Satjelia	Hetalbari	Kumirmari
Percent of HH do not get awareness	0.725	0.627	0.921	0.729	0.625	0.911	0.350
Index Value for component	0.320	0.320	0.447	0.430	0.365	0.477	0.390
7. *Natural hazard, disaster*							
Percentage of HH reported increased frequency and intensity of cyclones and storm surges	0.800	0.790	1.000	0.918	0.950	0.800	0.950
Percentage of HHs with injury or death as a result of natural disasters in the last 10 years	0.002	0.053	0.157	0.012	0.100	0.044	0.006
Percentage of HHs with injury or death to their livestock as a result of natural disasters that occurred in the last 10 years	0.750	0.017	0.815	0.810	0.875	0.715	0.750
Percentage of HHs with losses to physical assets	0.750	1.000	1.000	0.810	0.875	0.933	1.000
Percentage of HH that do not receive warning before a natural disaster	1.000	0.837	0.894	0.702	0.750	0.733	0.950
Index Value for component	0.660	0.539	0.773	0.651	0.710	0.645	0.731
8. *Climatic Fluctuations*							
Percentage of HHs reported increasing summer temperature	0.975	0.860	0.947	0.942	0.950	0.977	1.000
Percentage of HHs reported increasing winter temperature	0.650	0.511	0.684	0.810	0.475	0.555	0.900
Percentage of HHs reported decreasing span of winter	0.850	0.883	0.947	0.810	0.950	1.000	0.950
Percentage of HHs reported variability in monsoon rainfall	0.975	0.930	0.763	0.756	0.950	0.844	0.800

(continued)

Table 7.3 (continued)

	Bali	Sonagram	Hamilton Abad	Dayapur	Satjelia	Hetalbari	Kumirmari
Percentage of HHs reported increase of rainfall during winter and post-monsoon	0.300	0.418	0.421	0.540	0.925	0.288	0.825
Percentage of HHs reported change in the frequency of floods	0.625	0.744	0.368	0.785	1.000	0.800	0.500
Index Value for component	0.729	0.724	0.688	0.774	0.875	0.744	0.829
LVI SCORE	0.496	0.516	0.596	0.574	0.566	0.543	0.631

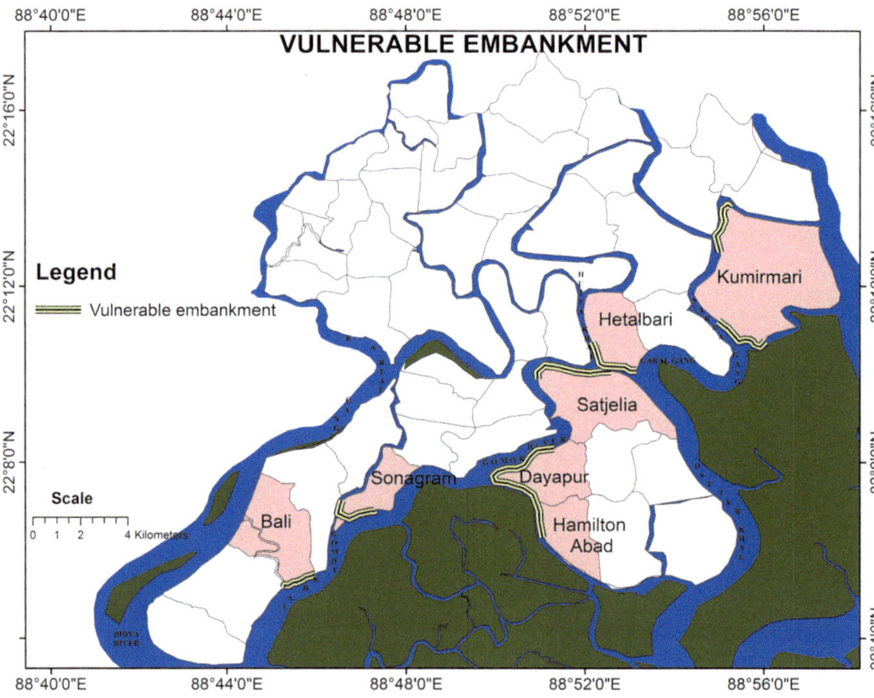

Fig. 7.2 Vulnerable Embankment Sites

7.2.3 Socio-Economic Vulnerability Analysis in the Light of Component Specific Discussion

A higher number of reliant family members, greater family size, poverty, illiteracy, and a greater percentage of people having non-climate-resilient *kaccha*/mud house are the foremost reasons behind the poor socio-demographic profile of households. Such cases have been observed in Kumirmari (46.27% HH), Hamilton Abad (49.08% HH), Dayapur (46.41% HH), and Satjelia *mouza* (49.13% HH).

Under the livelihood strategy component, all of the *mouzas* are more or less vulnerable. They have a greater dependency on natural resources (100% HH), dependency on agricultural income (highest in Kumirmari *mouza*—57.5% HH), a lower percentage of earning member in a family (1 member work in a 4 member family/2 member work in 8 member family in most cases), and a greater degree of migration to industrial centers (highest in Hamilton Abad *mouza*—55.26% HH) which in addition lead to their vulnerability. Since natural resources and agricultural activity are very much climate sensitive in nature, any kind of extreme weather condition directly distresses it.

Under the health component, the highest vulnerability score has been found in Kumirmari (58.5% HH) *mouza*. The main health center of the block is present in Gosaba Island, it is 20 km away from Kumirmari but one has to cross Sarsa *Gang*, Garal *Gang*, and Hutta *Khal* to reach Gosaba. Improper development of infrastructure in terms of transport and communication retards people from visiting health facilities. Greater distance to main health center compels people to visit quack doctors for treatment which is insufficient and poor in nature. People of the study area also have lack of awareness to visit health facility during illness (40% HH on an average) and the greater number of people has no proper sanitation facilities (highest in Kumirmari—77.5% HH).

The fourth issue is food and here highest vulnerability score has been found in Kumirmari (0.845). Severe coastal erosion during natural hazards causes embankments to breach and consequent intrusion of saline water takes place into the study area. It deteriorates the fertile character of soil and directly affects agricultural production. Households have also reported a loss of cropland (59% HH on an average) and decreased agricultural production (45% HH on an average) due to saline water intrusion. Before *Aila* of 2009 (a severe cyclonic storm which devastated the Sundarbans), the yield was much higher compared to the present situation. Certain species' phenotypic plasticity cannot adjust with high temperature and salinity conditions which often lead to their near extinction (Sengar and Sengar 2015; Mukherjee and Siddique 2018) which has been the case in the study area. The local people have observed decreased regeneration of certain green leafy vegetables (*Marsilea minuta, Amaranthus spinosus, Alternanthera phyloxeroides, Alternanthera sp.*) owing to increased temperature and high salinity condition over the years. Rural people who used to consume these leafy vegetables claimed that nowadays few of these vegetables are very less available and few are not at all available because of the high salinity of the soil. The number of households dependent upon the yield of their private farms seems to be less. Their farmland supports them partially. The higher percentage of people buys all sorts of food items from the market. All such facts turn the local people to be food insecure (77% HH on an average) during hazardous events. Even people of the study area have informed that ever since there is saline water intrusion into their pond, there is a massive decrease of fishes which thrive in freshwater system. After saline water intrusion, few local people drain out saline water from the pond and added fresh water into the pond for fishing activity whereas others started prawn culture. But people are of the opinion that there is decreasing availability of good catches post-*Aila* period. Both the quality and the number of catches have decreased. Thus changing climate has a greater impact on agriculture and fishing activity. Rural people who are becoming landless or jobless are thus compelled to migrate to different other places such as Andaman Nicobar Island, Kerala, Tamilnadu, Chennai, Gujarat, Hyderabad, Punjab, Maharastra, Puruliya, Burdwan, Asansol, Kolkata, Delhi, and Bangalore in search of jobs for their livelihood. They earn Rs. 250–400/per day but their jobs are irregular in nature. So after just 3–9 months or even 1 year, they again return back to their place and live with the vulnerable situation. Migration is based on agent network hence destination of migration are same. In most of the instances, migration to Kerala and Gujarat has been found.

Saline water intrusion also lowers down the quality of potable water sources. So female members of the household have to walk for kilometers to reach water sources where water is available for a fixed time period and the queue is quite long. Availing water for drinking and cooking for the household from the same source creates an extra burden to the female members. For washing clothes and bathing, saline pond water is used in most of the cases which leads to skin diseases of household members. Especially women are more susceptible to such diseases because they often use saline water for varied household activities and they are also mostly engaged in prawn seed collection activity which demands greater contact of their body to saline water sources for a long time period. Several researches on shrimp fry (*meen*) farmers of Sundarban Mangrove Forest states that direct and prolonged contact with salt water during farming activity exposes women more to skin infections and other waterborne diseases (Chowdhury et al. 2008; Das et al. 2016; Mukherjee and Siddique 2019a, b).

Under Social safety component, people of all of the *mouzas* expressed disappointment on the role played by government, and other non-Government organizations in terms of insufficient post-hazard mitigation strategies. Figure 7.2 shows that in the study area there is 16.46 km stretch of the vulnerable embankment which is prone to be breached every 2–3 years because of severe coastal erosion (1.1 km in Bali, 2.1 km Sonagram, 1 km Hamilton Abad, 3.8 km Dayapur, 2.41 km Satjelia, 1.85 km Hetalbari, and 4.2 km in Kumirmari). Villagers opined that the kind of materials that the officials use for construction and management of embankment is poor and uncohesive in nature which leads to frequent breaching. Embankment in the study area is considered to be the lifeline, because whenever it gets breached, saline water enters and there is a total menace already discussed above (Fig. 7.2).

The seventh and eighth components attempt to understand the perception of people on natural hazards and climate change. It is noticed that majority of the people of all

Table 7.4 LVI–IPCC contributing factors for the study area

	Bali	Sonagram	Hamilton Abad	Dayapur	Satjelia	Hetalbari	Kumirmari
Exposure	0.697	0.640	0.727	0.718	0.8	0.699	0.785
Sensitivity	0.446	0.534	0.608	0.586	0.498	0.509	0.685
Adaptive Capacity	0.397	0.405	0.488	0.456	0.461	0.463	0.465
LVI–IPCC	0.134	0.125	0.146	0.153	0.168607	0.119	0.218

mouzas perceive that there is increasing effect of storm surges and natural hazards (>79% HH); increasing temperature during summer (>95% HH), winter (65.5% HH); decreasing span of winter (91.31% HH); greater anomaly of monsoon rainfall (>85% HH) (which affects the agricultural community) and greater extent of rainfall during the post-monsoon and winter season (>53.1% HH) (rainfall in October–November 2017, 2018 affected *Aman* rice production). Monsoon rainfall, higher storm surges, and tidal ingression cause breaching of embankment and inundation of the area from June to September. An area prone to such hazards must organize awareness generation programme to educate people on how to be safe during such incidents but more than 69.8% HH opined that there is no such programme organized.

7.2.4 LVI–IPCC Analysis

Table 7.4 shows LVI–IPCC scores which states that Kumirmari *mouza* (0.218) is highly vulnerable followed by Satjelia (0.168); Dayapur (0.153) and Hamilton Abad (0.146). Lowest vulnerability is observed in Hetalbari (0.119), Sonagram (0.125), and Bali (0.134). The LVI score matches with LVI–IPCC score, it deviates only in case of Hetalbari. Under LVI analysis, Hetalbari *mouza* is moderately vulnerable, but under LVI–IPCC analysis, it is least vulnerable. Table 7.4 clearly shows that the adaptive capacity of Hetalbari *mouza* is quite high compared to Bali, Sonagram, Satjelia, and Dayapur, which lowers the effect of exposure, sensitivity, making the LVI–IPCC score to be low. All the highly vulnerable *mouzas* experience a higher rate of coastal erosion, embankment breaching, and flooding phenomena that differentiate the levels of vulnerability among them.

7.3 Conclusion

Emphasis has been given in this study to explore ecological and socio-economic vulnerability of the communities living under the threats of changing climate. Analysis related to stomata index and plant morphology of six species of plants helped us to understand that plants grow under the stressful condition in marginal areas where salinity intrusion is more. Comparison of LVI among seven villages helped us identify the aspects that require special consideration to develop resilience for coping up with the perils related to climate change in each of the *mouzas*. The study also supports the notion that socio-economic and biophysical attributes of the communities are related to climate change vulnerability. Climatic fluctuation, natural hazards, etc., make the communities exposed to climatic stimuli while improper access to food, health, water, sanitation, etc., make them extremely sensitive to climate change phenomena and finally lowers their adaptive capacity in terms of poor livelihood status. In addition, social security does not help the communities to get adapted to the situation. Thus, the ultimate result sorts out the vulnerability of local marginal human groups.

Acknowledgments The first author is thankful to UGC, New Delhi, India, for providing fellowship to carry out research work. We are immensely thankful to Dr. Avipsita Chatterjee, Research Fellow, University of Calcutta for helping us with LVI and LVI–IPCC calculation and Bhabatosh Roy, Assistant Teacher in Geography, Bhawani High School, Jalpaiguri, for helping in the collection of data. Equally, we are thankful to Dr. Moumita Chatterjee, Project Fellow, Institute of Wood Science and Technology, Bangalore, Karnataka, and Mr. Adwaita Das, CSIR-JRF, Department of Botany, University of Burdwan, for providing necessary help in biological vulnerability analysis. We also thank the anonymous reviewers and the editor for their valuable comments and assistance in the improvement of the paper.

References

Acosta-Motos JR, Ortuno MF, Bernal-Vicente A et al. (2017) Plant responses to salt stress: adaptive mechanisms. Agronomy 7(18):1–38. https://doi.org/10.3390/agronomy7010018
Alam GMM (2017) Livelihood cycle and vulnerability of rural households to climate change and hazards in Bangladesh. Environ Manag 59:777–791. https://doi.org/10.1007/s00267-017-0826-3
Alston M (2013) Gender mainstreaming and climate change. Women's Stud Int Forum 47:287–294. https://doi.org/10.1016/j.wsif.2013.01.016
Census (2011) District Census Handbook: South 24 Parganas District
Chowdhury AN, Mondal R, Brahma A, Biswas MK (2008) Eco-psychiatry and environmental conservation: study from Sundarban Delta, India. Environ Health Insights 2:61–76
Das K, Das K (2015) Embankment breaching and its management in Gosaba and Basanti blocks of Sundarbans, West Bengal. Indian J Spatial Sci 6(1):1–73
Das P, Das A, Roy S (2016) Shrimp fry (meen) farmers of Sundarban Mangrove Forest (India): a tale of ecological damage and economic hardship. Int J Agri Food Res 5(2):28–41
Dasgupta S, Sobhan I, Wheeler D (2016) Impact of climate change and aquatic salinization on mangrove species and poor communities in the Bangladesh Sundarbans. Policy Research Working Paper 7736, World Bank Group
Dasgupta S, Sobhan I, Wheeler D (2017) The impact of climate change and aquatic salinization on mangrove species in the Bangladesh Sundarbans. Ambio: 1–15
Demeritt D (2001) The construction of global warming and the politics of science. Ann Assoc Am Geogr 91(2):307–337
Etwire PM, Al-Hassan RM, Kuwornu JKM et al (2013) Application of livelihood vulnerability index in assessing vulnerability to climate change and variability in Northern Ghana. J Environ Earth Sci 3:157–170

FAO (2013) Resilience of natural resources & climate change. https://www.fao.org/climatechange/35936060c2ab16ecebffe3bd024dee829faa2e.pdf. Accessed 22 March 2017

Gupta B, Huang B (2014) Mechanism of salinity tolerance in plants: physiological, biochemical, and molecular characterization. Int J Genom. https://doi.org/10.1155/2014/701596

Hahn MB, Riederer AM, Foster SO (2009) The livelihood vulnerability index: a pragmatic approach to assessing risks from climate variability and change-A case study in Mozambique. Global Environ Change 19(1):74–88. https://doi.org/10.1016/j.gloenvcha.2008.11.002

Hondula DM, Balling RCJ, Vanos JK et al (2015) Rising temperatures, human health, and the role of adaptation. Curr Clim Change Rep 1:144–154. https://doi.org/10.1007/s40641-015-0016-4

IPCC (2007) Climate change 2007: impacts, adaptation and vulnerability. https://www.ipcc.ch/pdf/assessment-report/ar4/wg2/ar4_wg2_full_report.pdf. Accessed 29 August 2017

IPCC (2014) Climate change 2014: Impacts, adaptation and vulnerability. Part A: global and sectoral aspects. https://www.ipcc.ch/report/ar5/wg2/. Accessed 29 Aug 2017

Kerstiens G, Tych W, Robinson MF et al (2002) Sodium-related partial stomatal closure and salt tolerance of Aster tripolium. New Phytolog 153(3):509–515

Kleinen T, Petschel-Held G (2007) Integrated assessment of changes in flooding probabilities due to climate change. Clim Change 81:283–312

Kumar R, Gautam J (2014) Climate change and its impact on agricultural productivity in India. J Climatol Weather Forecast 2(1):1–3. https://doi.org/10.4172/2332-2594.1000109

Magrin G, Gay García C, Choque DC et al (2007) Latin America. Climate change 2007: Impacts, adaptation and vulnerability. In: Parry ML (ed) Contribution of working group II to the fourth assessment report of the intergovernmental panel on climate change. Cambridge University Press, Cambridge, UK, pp 581–615

McInnes KL, Walsh KJE, Hubbert GD et al (2003) Impact of sea-level rise and storm surges on a coastal community. Nat Hazards 30:187–207

Merino MV, Jost DSF, Berger U (2018) Archetypes of climate vulnerability: a mixed-method approach applied in the Peruvian Andes. Climate Dev. https://doi.org/10.1080/17565529.2018.1442804

Milly PCD, Dunne KA, Vecchia AV (2005) Global pattern of trends in stream flow and water availability in a changing climate. Nature 438(7066):347–350

Mukherjee N, Siddique G (2018) Climate change and vulnerability assessment in Mousuni Island: South 24 Parganas District. Spatial Inf Res 26(2):163–174. https://doi.org/10.1007/s41324-018-0168-0

Mukherjee N, Siddique G (2019a) Assessment of climatic variability risks with application of livelihood vulnerability indices. Environ Dev Sustain. https://doi.org/10.1007/s10668-019-00415-3

Mukherjee N, Siddique G, Basak A, Roy A, Mandal MH (2019) Climate change and livelihood vulnerability of the local population on Sagar Island India. Springer: Chinese Geogr Sci 29(3):417–436. https://doi.org/10.1007/s11769-019-1042-2

Mukherjee N, Siddique G (2019b) Gendered vulnerability of climate change: experiences of women in the sundarbans. In: Chattopadhyay et al (eds) Gender and development: aspects of social and economic change. New Delhi Publishers, New Delhi, pp 15–29

Munns R, Tester M (2008) Mechanisms of salinity tolerance. Annu Rev Plant Biol 59:651–681. https://doi.org/10.1146/annurev.arplant.59.032607.092911

National Research Council (1997) Rediscovering geography: new relevance for science and society. National Academy Press, Washington DC

National Statistical Service (2018) Sample size calculator. Australian Bureau of Statistics. https://www.nss.gov.au/nss/home.nsf/NSS/0A4A6.42C71.2719D.CCA25.71AB0.0243D.C6?opend.ocume.nt. Accessed 12 Jan 2017

Negrão S, Schmöckel M, Tester M (2017) Evaluating physiological responses of plants to salinity stress. Ann Bot 119(1):1–11. https://doi.org/10.1093/aob/mcw191

Nicholls RJ, Tol RSJ (2006) Impacts and responses to sea-level rise: a global analysis of the SRES scenarios over the twenty-first century. Philos Trans Royal Soc London 364:1073–1095

OsakabeY, Osakabe K, Shinozaki K et al (2014) Response of plants to water stress. Front Plant Sci 5(86). https://doi.org/10.3389/fpls.2014.00086
Pandey R, Jha S (2012) Climate vulnerability index-measure of climate change vulnerability to communities: a case of rural lower Himalaya, India. Mitig Adapt Strat Glob Change 17:487–506. https://doi.org/10.1007/s11027-011-9338-2
Panthi J, Aryal S, Dahal P, Bhandari P, Krakauer NY, Pandey VP (2016) Livelihood vulnerability approach to assessing climate change impacts on mixed agro-livestock smallholders around the Gandaki River Basin in Nepal. Reg Environ Change 16(4):1121–1132. https://doi.org/10.1007/s10113-015-0833-y
Qiu DL, Lin P, Guo SZ (2007) Effects of salinity on leaf characteristics and CO_2/H_2O exchange of Kandeliacandel (L.) Druce seedlings. J Forest Sci 53(1):13–19
Reiter P, Thomas CJ, Atkinson P et al (2004) Global warming and malaria: a call for accuracy. Lancet Infect Dis 4:323–324
Resurrección PB (2013) Persistent women and environment linkages in climate change and sustainable development agendas. Women's Stud Int Forum 40:33–43. https://doi.org/10.1016/j.wsif.2013.03.011
Rogers DJ, Randolph SE (2000) The global spread of malaria in a future, warmer world. Science 289:1763–1765
Royer DL (2001) Stomatal density and stomatal index as indicators of paleoatmospheric CO_2 concentration. Rev Palaeobot Palynol 114(2001):1–28
Samanta B, Das S, Hazra S (2017) Micro level vulnerability assessment of a community living in Mousuni Island in the Indian Sundarban: an integrated study employing geoinformatics. In: Hazra S et al. (eds) Environment and earth observation, Springer Remote Sensing/Photogrammetry. Springer International Publishing, Switzerland, pp 195–213. https://doi.org/10.1007/978-3-319-46010-9_13
Sandhu D, Cornacchione MV, Ferreira JFS et al (2017) Variable salinity responses of 12 alfalfa genotypes and comparative expression analyses of salt-response genes. Sci Rep. https://doi.org/10.1038/srep42958
Salisbury EJ (1927) On the causes and ecological significance of stomatal frequency, with special reference to the woodland ora. . Philos Trans Royal Soc London Series B, Biol Sci 216:1–65
Sengar RS, Sengar K (2015) Climate change-effect on productivity. Taylor and Francis Group, New York
Shah KU, Dulal HB, Johnson CA, Baptiste A (2013) Understanding livelihood vulnerability to climate change: Applying the livelihood vulnerability index in Trinidad and Tobago. Geoforum 47:125–137. https://doi.org/10.1016/j.geoforum.2013.04.004
Solmaz I, Sari NZM, Dasgan HY et al (2013) The effect of salinity on stomata and leaf characteristics of dihaploid melon lines and their hybrids. J Food Agric Environ 9(3):1–5
Thuiller WS, Lavorel MB, Araujo MT et al (2005) Climate change threats to plant diversity in Europe. In: Proceedings of national academy of sciences, USA
Toufique KA, Islam M (2014) Assessing risk from climate variability and change for disaster-prone zones in Bangladesh. Int J Disaster Risk Reduct 10:236–249. https://doi.org/10.1016/j.ijdrr.2014.08.008
Tubiello FN (2005) Climate variability and agriculture: perspectives on current and future challenges. In: Knight B (ed) Impact of climate change, variability and weather fluctuations on crops and their produce markets. Cambridge, UK
Velepucha P, Eguiguren A, Chamba JAM et al (2016) Tropical ecosystems vulnerability to climate change in southern Ecuador, pp 1–16. https://doi.org/10.1177/1940082916668007
de Villiers AJ, Teichman IV, van Rooyen MW et al (1996) Salinity induced changes in anatomy, stomatal counts and photosynthetic rate of AtripZexsemibaccata R. Br South Afr J Bot 62(5):270–276
Welp M, Battaglini A, Faeger CC (2009) Defining dangerous climate change: the Bejing experience. In: Patt AG, Schroter D, Klein RJT and Leinert VAC (eds) Assessing vulnerability to global environmental change. Earthscan, London

Yunlong C (1997) Vulnerability and adaptation of chinese agriculture to global climate change. Chin Geogr Sci 7(4):289–301

Yuehong Z, Shaohong WU, Erfu DAI et al (2008) Identification and categorization of climate change risks. Chin Geogr Sci 18(3):268–275. https://doi.org/10.1007/s11769-008-0268-1

Part II
Geosites and Geoparks

Chapter 8
An Example of Geosite Evaluation of Fossils: Zonguldak Coal Basin (Turkey)

Selin Isik Cakti, Okan Pulat, and Hulya Keskin Citiroglu

Abstract Today thanks to having knowledge about the geology, it is necessary to protect the areas of geological heritage and geosite and inform the people about them. In this study, the Zonguldak hard coal basin and fossils and pollens of coal in the basin were aimed to be evaluated as a geosite. Fossils and pollens of coal as a geosite have been studied from geological point of view in detail and aimed to draw attention of the researchers and local authorities Zonguldak hard coal basin's historical, cultural sides and also create awareness to the geosite and geosite areas and bring these basin as an alternative for geotourism. Hard coal which is the most important underground wealth in the Zonguldak basin, and coal fossils, Carboniferous, aged between 290 and 350 million years, are found in the basin. Zonguldak hard coal basin on the Black Sea coast is 160 km in length through the E-W line between Eregli and Inebolu and has 1,100,357,359 tons of geological coal reserves in total. Alacaagzi, Kozlu, and Karadon formations in the region contain Carboniferous aged coal fossils. The samples of coal fossils are gathered from the sample archive of Turkish Hard Coal Enterprises, Department of Plan-Project and Facility as well as the personal fossil collections. It is determined that fossils of the coal formations in Zonguldak basin are the elements of geological heritage and the region containing those coal formations with fossils is the geosite area. Therefore, the detected geosite area should be taken under protection not only to understand the coal formation in the region but also to be able to transfer this information to the next generations. To protect and to introduce the geosite to the next generations, coal fossils, which were formed millions of years ago and which are of great significance in terms of geological heritage, should be protected under appropriate conditions; a museum which contains a collection of coal fossils should be established in the region. The museum will be also important

---

S. Isik Cakti
Kemalpasa District, 112 Street, Nisanci Buildings No: A/4 Serdivan, Sakarya, Turkey

O. Pulat
MTA General Directorate of Mineral Research and Exploration, West Black Sea Regional Office, Zonguldak, Turkey

H. Keskin Citiroglu (✉)
Directorate of Investment Monitoring and Coordination, YIKOB, Aydin, Turkey
e-mail: keskinhc@yahoo.com

R. B. Singh et al. (eds.), *Global Geographical Heritage, Geoparks and Geotourism*, Advances in Geographical and Environmental Sciences,
https://doi.org/10.1007/978-981-15-4956-4_8

institution that will help to the presentation, discovery, and protection of these geosite fields. Creating a museum containing fossil collection, introduction, protection, and transmission of the geosite to the next generation is important for the development of geotourism and the sustainable regional development.

Keywords Geosite · Hard coal · Fossil · Zonguldak

8.1 Introduction

For a sustainable regional development and geotourism, the detection, utilization, and protection of the geosite areas are of great significance. Geosite is a rock, mineral, fossil collection or a part of a land describing a geological process, event, and property (Jemirko 2018). The set of fossils, rocks, and minerals which tell about the features of any geological event, processes, or a part of land is called a geosite. There is no dimension limit in geosite description; very large areas or very small areas can also be considered as geosite. If the area is small, a single geological formation is considered and that is the geosite itself. In a narrow space, two or more geosites are not defined. If there is more than one geological formation in a wide area, this time not only a specific geological feature but the area is considered as geosite area itself (Inan 2008). Geoparks are composed of a combination of same or different types of geosites (Jemirko 2018). It is a known fact that the world continuously changes under the influence of the geological events. However, if we see or not, the events occurring on the earth are recording that day to some places. For example, a fossil helps us to get knowledge about millions of years ago. The earth can give us information about the past by recording the tectonic stages from the formation to the present to the rocks. A mineral, a rock, a cave, or a fossil may shed light on today's events. However, many geological formations on earth are faced with the threat of extinction due to unplanned urbanization and ignorance. The biggest reason of this is caused by the people who see these materials only as stones. Today where people have information about geology, it is required to protect geological heritage and geosite areas and inform people. The geosites which are under the danger of becoming extinct for natural or man-made reasons are called geological heritage. The first aim of geo-conservation is the protection of geosites as member of the geological heritage (Henriques et al. 2011).

Natural formations that have given evidence to earth's geological history are diverged from resembling formations because of their unordinary visual character and they may not be built again and are under danger of extinction, considered as geological heritage. Geological heritage elements are being cultural, touristic, and natural sources of wealth and also belong to the whole world (Inan 2008). However, some geological formations on earth are facing the risk of extinction along with urbanization and unawareness.

Geoparks are the areas which provide the sustainable development and protection of geological heritage (Wang et al. 2015). It is necessary to educate people about environment and geology, the realization of sustainable development, and safeguarding of geological inheritance for future generations when a geopark is being formed (Yilmaz 2002). Geosite is a place elucidating an area which helps to clarify a certain geological feature (Inan 2008). The aggregated show of geosite or geological heritage element without classifying in itself is called geo-inventory (Jemirko 2018). Geotourism is trips organized to investigate the geological heritage elements, geosites, or nature (Kazanci 2010). Martín-Duque et al. (2012) defined as process of organizing geoheritage information for areas of geotourism. Hose explained modern geotourism as geo-interpretation, geo-history, and geo-conservation. Hose and Vasiljević (2012) described the modern geotourism approaches and its further practices within Europe. The plurality of the geotourism activities is realized in geoparks (Yildirim 2012). Pamukkale, the red fairy chimneys of Narman, Cappadocia, Salt, and Meke Lakes, the volcanoes of Kula, the caldera of Nemrut, and natural caves are among the most widely known leading values of geotourism in Turkey (Gungor 2009). Anand et al. (2014) studied geotourism assessment of Varkala geopark in India. Alonso-Almeida (2013) emphasized that the tourism industry can only operate with adequate energy, water, and waste management facilities. Roads are important assets for society for mobility and enable economic growth (Butt et al. 2015). The routes and paths which to be followed in order to visit the geological heritage elements and geosites are called under the heading georoute (Boyraz and Yedek 2012).

Geodiversity approaches are required for highlighting sustainable future (Gray et al. 2013). Aras (2012) has investigated on the benefits of the geological diversity for sustainable development in the central Anatolia, Kazanci et al. (2012) have studied the meteorite impact findings on the Neogeneclastic sediments of the region and also Gullu and Kazanci (2013) applied a study on the identification of the Pelitcik fossil forest located in Kizilcahamam-Camlidere geopark in Ankara in Turkey. Atici et al. (2013) investigated on Hatila Lava Lake in the Hatila Valley and Artvin as a potential geopark area in the Eastern Black Sea Region, Gungor et al. (2012) studied on geopark inventory of Levent Valley in Akcadag-Malatya Region in the Eastern Anatolian Region, Vural et al. (2013) researched on geotourism-geosite potential of the abandoned mine areas in the province of Gumushane in Turkey. Gogin and Vdovets (2014) studied geosites located in Nature Park in Russia which have stratigraphic, paleontologic, paleoecologic, and geomorphologic properties. Singh and Anand (2013) investigated geodiversity, geographical heritage, and geoparks in India. Shin et al. (2008) studied disciplines such as geology, ecology, chemistry, sociology, economics, engineering, and politics for sustainability problems.

In this study, Zonguldak coal basin and coal fossils and pollens in the basin are aimed to be evaluated as geosite and also coal formations and coal fossils located within the study area were investigated. Keskin Citiroglu et al. (2017) emphasized that the caves, mineral waters, landslides, cliffs, faults, quartz sand, column andesite, and coal and coal fossils have been defined as elements of geological heritage and Zonguldak hard coal basin should be utilized as a large geopark with its geological diversities each of which is a geosite in their study. In this study, recommendations were presented in order to protect, introduce, and transmit the geosite which has great importance as a geological heritage for sustainable development and sustainable geotourism regarding the region. Two of the important wealth that worth to be protected in terms of geological heritage are coal and coal fossils. Fossils which are the stoned remains of the creatures left in the sediments where they were after the fossilization of the creatures lived in various geologic times after they die are located only in sedimentary rocks. Plants can be fossilized by being buried in water environments like sea, lake, or swamps. Plants are becoming partially or completely coal as a result of carbonization event so called coalification (Inan 2009). Coal, mostly formed with plant parts, is defined as a sedimentary rock containing carbon in the form of free or composition with a rate changing between 55 and 95% (Nakoman 1971; Karayigit and Koksoy 1988).

8.2 Zonguldak and Coal

Zonguldak Province which has an important place on the production of coal is located in the Western Black Sea Region of Turkey at 41°–27′ N latitude and 31°–49 E longitude (Fig. 8.1).

As in all Black Sea regions, mountains and hills are rising parallel to the sea in Zonguldak, but unlike the other places, topography rises immediately from the sea front in Zonguldak. Because of this corrupted topography, no settlement was observed in the geography where the city center of Zonguldak is located until the recent times. Zonguldak city started to be formed in the beginning of 1800s when coal is found in the region, and started to be produced in 1848 officially. Rapid industrialization started with the hard coal processing caused the population density to increase. As a consequence, Zonguldak has been registered as the first city of the Republic of Turkey. Meanwhile, with a law known with a name as Havza-iFahmiye enacted in 1910, immovable properties in coal basin was registered on behalf of treasure (Tamzok 2008; Quataert 2009). Zonguldak coal basin shows spread approximately 160 km in EW direction between Eregli-Inebolu in West Black Sea coast. In the basin, total of five manufacturing establishments, Kozlu, Uzulmez, Karadon, Amasra, and Armutcuk, are located. In Zonguldak coal basin where total of 1,100,357,359 tons of geological reserves of coal are found (Table 8.1), total of 451 of drillings have been carried out in 323,775,610 m depth (TTK 2012) so far.

8.3 Formations Containing Coal

In the region, Alacaagzi, Kozlu, and Karadon are the formations containing coal. Alacaagzi formation is formed with sandstone, claystone, siltstone, shale, and coal alterations. Its age is Namur according to the palynological analysis performed in the coal level in the stack. Kozlu formation is found as conglomerate, sandstone, siltstone, claystone, and coal sequenced. The thickness of the coal vessels varies

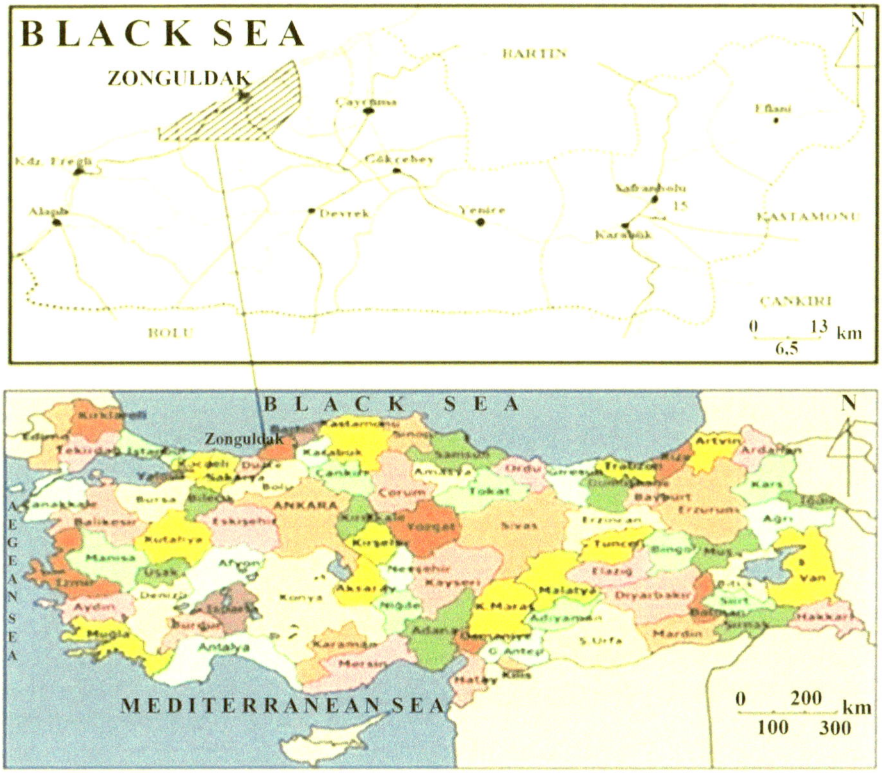

Fig. 8.1 The geographical location of the province

Table 8.1 Zonguldak hard coal basin reserves (TTK 2012)

Establishments	Apparent reserve	Probable reserve	Possible reserve	Total (Ton)
Armutcuk	8,045,551	15,859,636	7,883,164	31,788,351
Kozlu	66,744,799	40,539,000	47,975,000	119,258,799
Uzulmez	135,794,982	94,342,000	74,020,000	304,156,982
Karadon	81,852,172	93,179,000	63,134,000	238,165,172
Amasra	170,401,055	115,052,000	121,535,000	406,988,055
Total	462,828,559	424,783,636	368,621,164	1,100,357,359

between 0.5 and 6 m. The age of the unit is Westphalian A according to palynological estimation made on coals. Karadon formation is composed of conglomerate, sandstone, siltstone, claystone, and coal (Yergok et al. 1987). Carboniferous age coal fossils are found in the region in the ages between 290 million year and 354 million years. Ralli (1933) divided Zonguldak carboniferous fossils into four main groups as Sphenopterides, Pecopterides, Odondopterides, and Nevropterides in his book "Le

BassınHouillerD'Heraclee Et La Flore Du Culm Et Du Houiller Money" published in 1933 as a result of the researches he made in Zonguldak coal basin.

8.4 Palynological Analysis Method

The science branch allowing us to estimate the age estimation of coal units via spores and pollens is called as palynology. When palynological age estimation is performed, samples taken from the coal units are powdered and left for oxidation. This process separates plant remains into spores and pollens. Samples in liquid form are heated up with the help of gelatin and after sticking on the lam, it is closed with lamellae (Fig. 8.2). After the lamella is ready for examination, it is checked with a polarizing microscope with top and bottom backlit. On the samples archived in TTK General Directorate, the microscope from the Department TTK Research Plan Projects Facility and a LEICADM2500P brand microscope illuminated from top and bottom. By looking at the spores inside the coal samples in 10 × 25 and 10 × 40 objectives with single polar and bottom illuminated (Fig. 8.3).

Microscopic views of the spores obtained from this investigation are given below. Lycospora in Westphalian A aged sample taken from Amasra, Zonguldak (Fig. 8.4a), Lophotriletes (Fig. 8.4b), Granulatisporites (Fig. 8.4c), Tripartites, and Schulzospora (Fig. 8.4d) which are characteristic fossil of Namurwith Namur age taken from Armutcuk-532 elevation and Tripartites (Fig. 8.4e) are seen. Furthermore, commonly Convolutispora (Fig. 8.4f), Punctatisporites (Fig. 8.4g), Densosporites (Fig. 8.4h),

Fig. 8.2 Prepared microscope samples

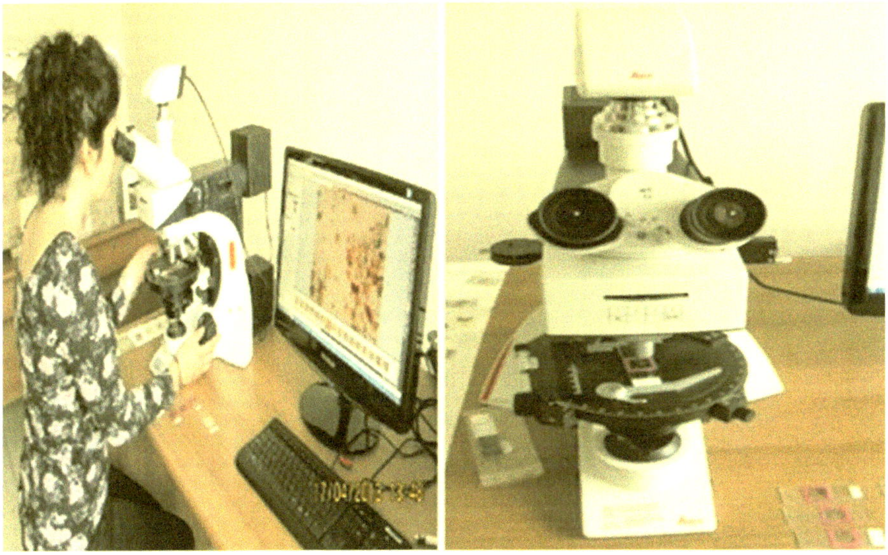

Fig. 8.3 Image of the microscope for review

and Cyclogranisporites (Fig. 8.4i) are seen. Schulzospora is important because it is a rare type of pollen seen in Namur and Westphalian A-B.

Spores observed in the coal samples in Karadon, Uzulmez, Amasra, and center of Zonguldak region are (reproductive cells of flowerless plants): Sporinites, Chaetosphaerites, Laevigatosporites, Punctatisporites, Leiotriletes, Punctatisporites, Calamospora, Cyclogranisporites, Granulatisporites, Verrucosisporites, Convolutispora, Planisporites, Apiculatisporites, Anapiculatisporites, Pustulatissporites, Lophotriletes, Ibrahimisporites, Acanthotriletes, Cristatisporites, Raistrickia, Tuberculatisporites, Camptotriltes, Microreticulatisporites, Egemenisporites, Dictyotriletes, Reticulatisporites, Knoxisporites, Crassisporakosankei, Stenozonotriletes, Lophozonotriletes, Anguisporites, Simizonotriletes, Bellisporites, Sinussporites, Callisporites, Rotasporites, Procoronasporites, Lycospora, Densosporites, Cirratriradites, Reinschospora, Tholisporites, Triquitrites, Yahsimanisporites, Tripartites, and Ahrensissporites. Pollens are (reproductive cells of flowering plants): Perisaccus, Florinites, Endosporites, Microsporites, Velosporites, and Schulzospora (Akyol 1983; Agrali and Konyali 1973). Plants remaining and macrofossils found in Alacaagzi formation; some plants remaining were determined from the samples gathered and evaluated by various researchers around Kandilli. These are Sphenopteris Bermudensiformis Schlot, Sphenopteris (Diplothmema) dissectumBrgt, Sphenopteris Divaricate Goepp, SphenopterisElegans-Brgt, SphenopterisLinkiliGoepp, SphenopterisSchutzeiStur, SphenopterisStangeri, SphenopterisAdiantoidesSchlotheimi, CelamitesCistiformisStur, CelamitesRamiferStur, SphenophyllumTenerrimumEtinghauss, SphenophyllumSewardiZeiller,

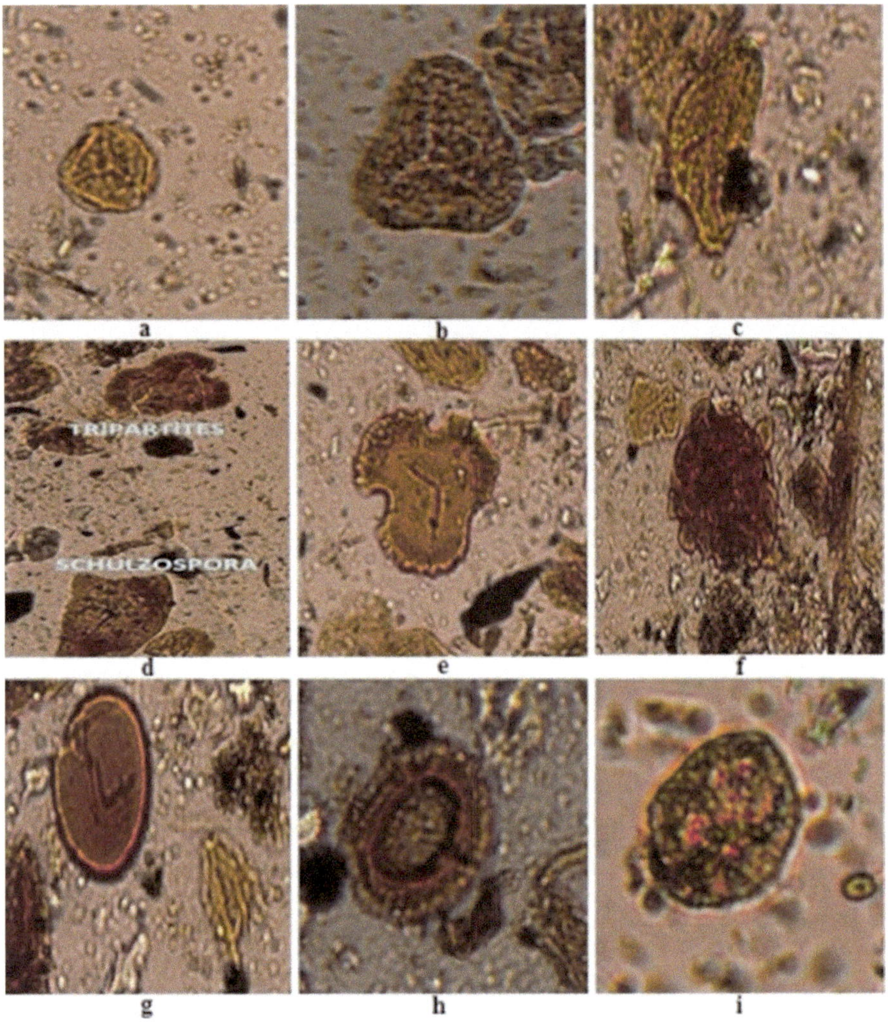

Fig. 8.4 Microscopic views of the **a** Lycospora, **b** Lophotriletes, **c** Granulatisporites, **d** Tripartites and Schulzospora, **e** Tripartites, **f** Convolutispora, **g** Punctatisporites, **h** Densosporites, and **i** Cyclogranisporites

DiplothmemaMladikinin, DiplothmemaDissectumBrgt, CardiopterisPolymorpha-Goepp, StigmariaFicoidesBrgt, ArcheaocalamitesRadiatusBrgt, AsteroclamitesS-crobioutatusSchloth, Lepidodendron sp. like plant remaining, Gnathodus sp., Peleksygnathus sp., Hindeodella sp., Ligonodinasp.,Roundya sp., Neoprioniodus sp. like conodonts, Semiplanus sp. and Delepinas sp. were found as Brachiopod from the samples taken from marine parts of Tarlaagzi and their ages were found to be Namur (Yergok et al. 1987).

8.5 Samples From Zonguldak Coal Fossils

Fossil samples were reached from fossil collection located in TTK etude, Plan-Project and Facility Department Management, and from the personal fossil collection of Ekrem Murat ZAMAN who is a mining engineer serving in Zonguldak TTK Kozlu facility (Zaman 2013). Obtained fossils are observed generally to be in Kozlu formation. Sphenopteris fossil obtained from Kozlu formation in Westphalian age can be seen in Fig. 8.5a and Lepidophloios fossil can be seen in Fig. 8.5b. Sigillaria Elegans Sternb (Fig. 8.6) were identified by Alpan (1968) and published in his work named as macrofossils (Alpan 1968). Sigilari leaf traces taken out from −320 m elevation again from Kozlu formation are seen in Fig. 8.7.

The image of calamites undulates fossil obtained from −560 m depth in Lepidostrobus body fossil obtained from Uzulmez region, the image of Cordoitine fossil

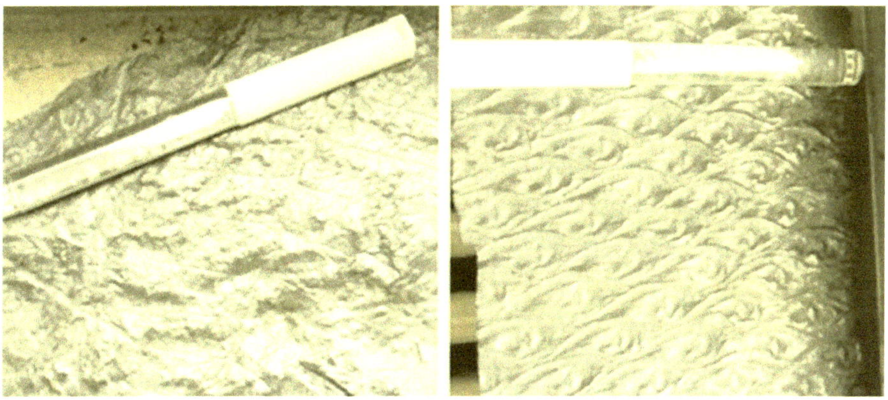

Fig. 8.5 Sphenopteris and Lepidophloios

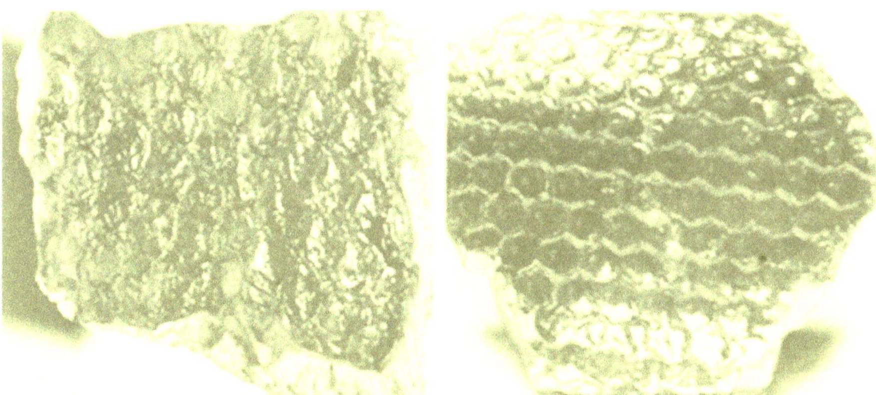

Fig. 8.6 Sigillaria Elegans Sternb (Alpan 1968)

Fig. 8.7 Sigilari leaf traces

taken from Uzulmez region of Kozlu formation, Pecopteris taken from -320 elevation of Uzulmez obtained from Kozlu formation, leaf fossil, Artisia from coal fossils located in Kozlu formation are seen in Fig. 8.8a, b, c, d, e, f, respectively.

8.6 Conclusions

Scientific studies were realized about the geology of the zone and the coal fossil which is one of the elements of geological inheritance for the sustainable development of the region within the framework of this study. If the study area is thought as a geopark containing various geological features, the rocks, mineral water springs, caves, landslides, faults, quartz sands, column andesites, sea cliffs, and coal and coal fossils form a geosite area. Zonguldak hard coal basin has various geosite elements that attract attention of both scientific investigators and nature-lovers.

Zonguldak coal basin is situated on the E-W axis along the Black Sea coast for 160 km. In the study area, the formations of Alacaagzi, Kozlu, and Karadon are the units which contain coal reserves. The most important underground wealth of Zonguldak basin is coal and Carboniferous coal fossils in the age of 290 million years and 350 million years are located in the basin. It is decided that fossils of coal formation in Zonguldak basin are heritage items and can be considered as geosite. For the purposes an old coal mine, which has stopped operating, should be turned into a museum and opened to visitors in order to educating the students, foreign and domestic tourists, the people of the region whose main livelihood is from coal mining. Protecting the coal fossils which have great importance in terms of geological heritage in proper storage conditions, creating a museum for the fossils, and opening for visiting are concluded to be necessary. An old coal mine which is no longer in operation must be transformed into a museum and must be opened to visits. Also, it is thought that the coal which is left on the pillar part would be beneficial

Fig. 8.8 **a** Calamites undulatus, **b** Lepidostrobus body, **c** Cordoitine, **d** Pecopteris, **e** leaf fossil and **f** Artisia (Zaman 2013)

for the researchers. Including a microscope inside the museum suggested for the coal fossils and displaying pollen and spores via microscopes in thin sections of the current coal samples taken from the same formation in addition to the fossil sample and also preparing introductory brochures are recommended. These fossils which are of utmost importance for the science of geology must be preserved and maintained in suitable conservation conditions, and a museum must be established and opened to visits. Creation of a museum containing fossil collection, and introduction, protection, and transmission of the geosite to the next generation are important for the development of geotourism and the sustainable regional development.

Acknowledgments The authors thank Turkish Hard Coal Enterprise (TTK), Biologist Zeliha CETIN working in TTK Etude, Plan-Project Facilities Department for helping in microscope examination and Mining Engineer Ekrem Murat ZAMAN who works in TTK Kozlu Institutions for allowing to use his coal fossil collection. This study was presented as an oral presentation at the 1st South East Europan Conference On Sustainable Development Of Energy Water And Environment Systems–SEE SDEWES in Ohrid, Republic of Macedonia between 29 June–3 July, 2014.

References

Agrali B, Konyali Y (1973) Investigation of Microsporidia of Amasra Carboniferous Basin. MTA Bull. Ankara 53:50–143

Akyol E (1983) Palynology investigations of Westphalian A and Namur vessels cut by southern and eastern elevations in -50 Roadway, Zonguldak, Uzulmez Area, Asma, MTA Bull. Ankara 59:47–108

Alonso-Almeida MM (2013) Environmental management in tourism: students' perceptions and managerial practice in restaurants from a gender perspective. J Clean Prod 60:201–207

Alpan S (1968) Macrofossils. MTA Institute and Museum of Natural History, Ankara, pp 41

Anand S, Saluja V, Singh RB (2014) Varkala as a national geopark of India: ggeographical realities, geotourism and community participation. Int J Geoherit Parks 2(1):65–81

Aras A (2012) Use of geodiversity for the sustainable development of the regions. In: A Kocbay et al (eds) 65th geological congress of Turkey Abstract Book. TMMOB JMO, Ankara, pp 328–329

Atici G, Kandemir O, Pehlivan S, Cobankaya M, Kanar F, Tok T (2013) Hatila Lava Lake, a potential geopark field in Eastern Black Sea Region, Hatila Valley, Artvin. In: B. Ozmen (ed) 66th Geological congress of Turkey Abstract Book. TMMOB JMO, Ankara, pp 342–343

Boyraz S, Yedek O (2012) Kizilcahamam-Camlidere Geopark. Mag Newslett 2:21–24

Butt AA, Toller S, Birgisson B (2015) Life cycle assessment for the green procurement of roads: a way forward. J Clean Prod 90:163–170

Gogin IY, Vdovets MS (2014) Geosites of international significance in the UNESCO WHS Lena Pillars Nature Park (Sakha Republic, Russia). Geoheritage 6:173–184. https://doi.org/10.1007/s12371-013-0089-1

Gray M, Gordon JE, Brown EJ (2013) Geodiversity and the ecosystem approach: the contribution of geoscience in delivering integrated environmental management. Proc Geol Assoc 124:659–673

Gullu E, Kazanci N (2013) Characterization of fossil forest PelitcikKizilcahamam-Camlidere Geopark. In: Ozmen B (ed) 66th geological congress of Turkey Abstract Book, TMMOB JMO, Ankara, pp 344–345

Gungor Y (2009) To tell the story of nature: geotourism. Blue Planet Mag 14:4–8

Gungor Y, Iskenderoglu L, Azaz D, Gungor B (2012) Geopark inventory study of Levent Valley, Akcadag-Malatya. In: Kocbay A et al (ed) 65th geological congress of Turkey Abstract Book, TMMOB JMO, Ankara, pp 322–323

Henriques MH, Pena dos Reis R, Brilha J, Mota T (2011) Geoconservation as an emerging geoscience. Geoheritage 3:117–128. https://doi.org/10.1007/s12371-011-0039-8

Hose TA (2012) 3G's for modern geotourism. Geoheritage 4:7–24. https://doi.org/10.1007/s12371-011-0052-y

Hose TA, Vasiljević DA (2012) Defining the nature and purpose of modern geotourism with particular reference to the United Kingdom and South-East Europe, Geoheritage 4:25–43. https://doi.org/10.1007/s12371-011-0050-0

Inan N (2008) Geological heritageand natural historymuseums. TUBITAK Sci Tech Bull 493:80–84
Inan N (2009) Paleontology, 2nd edn. Seckin Publishing, Ankara, p 205
JEMİRKO (2018) Terms of geological heritage. Geological Heritage Conservation Society. https://www.jemirko.org.tr/jeolojik-miras-terimleri/Accessed 28 Oct 2018
Karayigit AI, Koksoy M (1988) Coal formation and classification. In: Kural O (ed) Coal. Kurtis Press, Istanbul, pp 51–70
Kazanci N (2010) Geological conservation concepts and terms. Chamber of Geological Engineers. pp. 60
Kazanci N, Kibar M, Kadioglu YK, Erkoc H (2012) Evidences of asteroid impact within Neogeneclastic sediments of central Anatolia, Turkey. In: Kocbay A et al (ed) 65th Geological Congress of Turkey Abstract Book. TMMOB, JMO, Ankara, pp 324–325
KeskinCitiroglu H, Isik S, Pulat O (2017) Utilizing the geological diversity for sustainable regional development, a case study-Zonguldak (NW Turkey). Geoheritage 9:211–223. https://doi.org/10.1007/s12371-016-0196-x
Martín-Duque JF, Caballero García J, CarcavillaUrquí L (2012) Geoheritage information for geoconservation and geotourism through the categorization of landforms in a karstic landscape. A case study from Covalaguaand Las Tuerces (Palencia, Spain). Geoheritage 4:93–108. https://doi.org/10.1007/s12371-012-0056-2
Nakoman E (1971) Coal. MTA Institute of Broadcast Training Series No:8, Ankara, 358 p
Quataert D (2009) Miners and the state in the Ottoman Empire, Zonguldak Coal Basin. In: Gundogan AZ, Gundogan NO (Translators) Bogazici University publication, pp 1822–1920
Ralli G (1933) Le BassinHouillerD'HeracleeEt La Flore Du Culm Et Du Houiller Money. Istanbul, ImprimerieZellitchFreres-Pera, p 166
Shin D, Curtis M, Huisingh D, Zwetsloot GI (2008) Development of a sustainability policy model for promoting cleaner production: a knowledge integration approach. J Clean Prod 16:1823–1837
Singh RB, Anand S (2013) Geodiversity, geographical heritage and geoparks In India. Int J Geoherit Parks 1(1):10–26
Tamzok N (2008) The last period of the Ottoman Empire in Multi-Party Period, 1861-1948, Ankara University Faculty of Political Sciences, Issue: 63–64
TTK (2012) Reserves report of Northwestern Anatolia Hardcoal Basin. Turkey Hard Coal Enterprises Survey, Plan-Project and Facilities Department, Zonguldak, pp 5–29
Vural A, Sahin E, Guner S (2013) Geotourism-geosite potantial of abandoned mine sites, Gumushane, Turkey. In: Ozmen B (ed) 66th Geological Congress of Turkey Abstract Book, TMMOB JMO, Ankara, pp 348–349
Wang L, Tian M, Wang L (2015) Geodiversity, geoconservation and geotourism in Hong Kong Global Geopark of China. Proc Geol Assoc 126(3):426–437
Yergok AF, Akman U, Keskin I, Ipekci E, Mengi H, Karabalik NN, Umut M, Armagan F, Erdogan K, Kaymakci H, Cetinkaya A (1987) Geologic report of Western Black Sea Region I. Report No: 8273. MTA, General Directorate of Mineral Research and Exploration, Ankara, pp. 237
Yildirim G (2012) Rising trend in tourism: geotourism. Mag Newslett 2:18–20
Yilmaz A (2002) Geoparks. TUBITAK Sci Tech Bull 417:64
Zaman EM (2013) Personal interview. TTK Kozlu Facility, Zonguldak

Chapter 9
Gullies and Badlands as Geoheritage Sites

Wojciech Zgłobicki, Jean Poesen, Veena Joshi, Albert Sóle-Benet, and Sofie De Geeter

Abstract Permanent gullies and badlands are found on all continents and their spectacular nature and beauty has attracted scientists, artists and tourists. These landforms are elements of geomorphic heritage and development of geoeducation is the best way to transfer geomorphological knowledge to society. Gullies and badlands have been studied for many years, and their value for geoheritage is well recognized in many places. However, tourism promotion and interest in such sites vary a lot among tourists. Badlands in Africa, India, South Africa are rarely visited by local people or national citizens, in contrast to such sites in the European Mediterranean and North America, where badlands attract many national and international geotourists. This chapter discusses the worldwide importance of badlands and gullies as geoheritage sites and also the need for their conservation through the development of geotourism. The advantages and disadvantages of gullies and badlands as geosites at a regional and global scale are discussed.

Keywords Geo-education · Geotourism · Geomorphosites

9.1 Introduction

Geoheritage, being an integral part of global natural heritage, encompasses special places and features that are crucial for our understanding of earth's history (Wimbledon and Smith-Meyer 2012). The need for geoheritage conservation can

W. Zgłobicki (✉) · J. Poesen
Faculty of Earth Sciences and Spatial Management, UMCS, Lublin, Poland
e-mail: wojciech.zglobicki@umcs.pl

J. Poesen · S. De Geeter
Division of Geography and Tourism, KU Leuven, Leuven, Belgium

V. Joshi
SP Pune University, Pune, India

A. Sóle-Benet
Arid Zone Experimental Station, CSIC (National Research Council), Madrid, Spain

R. B. Singh et al. (eds.), *Global Geographical Heritage, Geoparks and Geotourism*, Advances in Geographical and Environmental Sciences,
https://doi.org/10.1007/978-981-15-4956-4_9

be promoted by taking advantage of a broad range of spectacular geo(morpho)sites, commonly occurring at the surface of our planet, in geotouristic developments and thus reaching larger audiences, including, for example, students. Such geomorphosites are found on all continents and include permanent gullies and badlands whose spectacular nature and beauty has attracted scientists, artists (e.g. painters, film producers) and tourists. These geotourists visit such sites for their spectacular and "surrealistic" character, diverse morphology, biodiversity and aesthetical appeal. Two types of badland and gully landscapes may be selected:

(a) Natural badlands—dissected topography was produced by intense surface and subsurface erosion due to the occurrence of erodible sediments, steep slopes, sparse vegetation and erosive climate. Such badlands are typically found in Mediterranean, semi-arid and arid regions of the globe.
(b) Human-induced badlands and gully regions—development of gullies and badlands was a result of past or present land use, which destroyed the natural land cover and intensified water erosion. Such regions can be observed in areas with a temperate climate.

Gullies and badlands, resulting from various geomorphological processes that are also considered to be soil degradation processes, may in particular environmental conditions result in large permanent gully channels and badlands. Gully channels are formed by intensely concentrated flow erosion, typically occurring in landscape positions where during rainfall runoff water accumulates and removes the soil from this narrow area to considerable depths (typically ranging from 0.5 m to as much as 25–30 m). If the gully channels are not filled in by tillage or land levelling operations, they become permanent (Poesen et al. 2003). Depending on their location in the landscape, gully channels are further classified as hillslope gullies, valley-bottom gullies, bank gullies, edge-to-field gullies, road gullies, or urban gullies (Poesen 2018). Gullies are often depicted to illustrate that soil erosion may have severe on-site consequences (e.g. large volumes of soil lost, terrain deformation with consequences for its trafficability).

Badlands are densely and deeply dissected erosional landscapes, formed in soft rock terrain, commonly but not exclusively observed in dry regions with sparse vegetation that have a very high drainage density of rill and gully channels (e.g. Figure 9.2) (Grove and Rackham 2003) described badlands as "the despair of cartographers and delights of artists". Badlands can be seen as miniature versions of larger river channel systems, but with the advantage that their spatial and temporal scale is strongly reduced compared to larger drainage network systems, which makes them such interesting research objects for better understanding landscape evolution. Soil erosion processes in badlands are most often intense because of their relative steep topography, soft lithologies, erosive rainfall and a small vegetation cover. As a consequence, badlands have received significant attention in the scientific literature as well as by (geo)tourists visiting such sites. This is illustrated in Fig. 9.1 showing badland sites around the world that are reported in research papers or that are shown on photos frequently posted on Google Images (retrieved by typing the word "badland" plus the

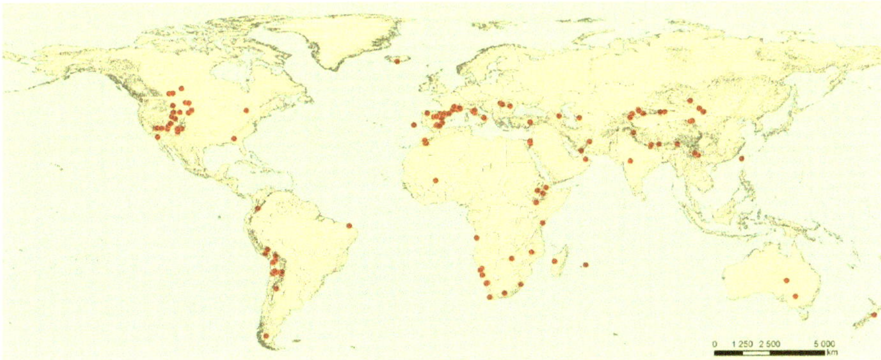

Fig. 9.1 Global overview of badland sites that have been frequently described in research papers or that were photographed many times by geotourists and shown on Google Images (after De Geeter 2018, modified)

name of continent or country, De Geeter 2018). However, badlands are not restricted to those countries and their presence is almost worldwide.

There are many badlands to be found around the world today: New Zealand has the Putangirua Pinnacles on its North Island, Italy has the Calanchi in Basilicata, Spain has the Bardenas Reales in Navarre and the Tabernas Desert in Almeria, among many others, Argentina has the Valle de la Luna in its midwestern regions, and Taiwan has the Gutingkeng Formation in its south, just to name a few. Badlands vary in their respective topographies and rock formations, but the most famous ones are located in the United States and Canada. These include the Big Muddy Badlands in Saskatchewan, Canada and the spectacular Dinosaur National Park, also in Canada, which was declared as a UNESCO World Heritage Site in 1979. In the United States, the Grand Staircase-Escalante National Monument has the Chinle Badlands in Utah, the Makoshika State Park in Montana, the Toadstool Geologic Park in Nebraska and the El Malpais National Monument in New Mexico.

Gullies and badlands may be regarded as geomorphosites—landforms or landscapes of high scientific value, also being attractive for tourists, i.e. sites having cultural, ecological or scenic values (Panizza 2001; Reynard 2008). They are elements of geomorphic heritage that is appreciated by society. Studies made by De Geeter (2018) indicated that more than 100 badland sites all around the world are recognized as tourist attractions (frequently posted pictures at Google Images). Applying the concept of geotourism to particular landforms is the best way to transfer geomorphological knowledge to society (e.g. Thomas 2012; Hose 2013; Zgłobicki et al. 2015). Besides being important for scientific researchers, geomorphosites also attract tourists. To capture the attention of geotourists, these sites must be of direct geotouristic interest or constitute convenient areas to other cognitive/tourist/recreational activities—hiking areas, viewing points (Warowna et al. 2016; Zgłobicki et al. 2017, 2018). The scientific potential of gullies and badlands helps: (1) to better understand past environmental changes (i.e. land use or climate change) and their impact on the landscape (through gully development), (2) to model the development of landscape dissection, erosion phenomenon and drainage network in landscapes and (3) to study regional geology through rock outcrops in gullies (walls and channel bottoms). The educational potential is especially important in the case of gullies, landforms formed by concentrated flow erosion and mass movements as a result of human activity (e.g. deforestation, overgrazing, poor land management) that constitutes an important geohazard in some regions. Furthermore, gullies and badlands have a significant touristic value: gullies can contribute to spectacular hiking experiences while badlands are much appreciated for their aesthetic value (see many photos of badlands posted on the Internet). Scenic values are among the most important reasons for selecting tourist destinations (Zgłobicki and Baran-Zgłobicka 2013). Especially badlands attract visitors thanks to the impressive landscapes, spectacular morphology, colours and particular features (e.g. earth pillars, hoodoos, flutes) (Zgłobicki et al. 2018). Because of these attractive aspects, the educational function (and conservation through education) can be developed through the establishment of geosites and georoutes.

The objective of this chapter is to explain the worldwide importance of badlands and gullies as geoheritage sites (geomorphosites), and therefore also the need for geoheritage conservation through the development of geotourism.

9.2 Geoheritage of Badlands and Gully Regions

Without a doubt, lithology is one of the main factors controlling gullies and badland formation. Gullies and badlands typically reflect the low resistance of particular lithologies to water erosion (sheet, rill, gully and piping erosion) and mass movements. As such, the presence of these geomorphic features are strong indicators of the outcrop of a particular geology (e.g. rock type or tectonic history).

Lithologies most susceptible to gullying and badland formation often consist of soft, unconsolidated sediments or barely cemented clay or silt materials: common examples are loess, marls, mudstones, shales, volcanic ash deposits, sands and sandy gravelly sediments. Due to a rather uniform particle-size distribution, the packing of particles is not optimal and the soil is less resistant to water erosion processes (Gallart et al. 2002). Favourable characteristics for weathering are furthermore the presence of soluble minerals such as halite, gypsum and carbonates or clay minerals that swell upon wetting and shrink upon drying, forming deep cracks (e.g. vertisols). Also, networks of cracks and fissures (joints, faults), as a consequence of unloading or tectonic activity, favour the breakdown of rocks. Of secondary importance are the structural properties of the material (e.g. aggregate stability) and the physical (e.g. porosity) and chemical properties (e.g. presence of cementing agents, clay dispersion, wetting–drying cycles, etc.) (Gerits et al. 1987; Solé et al. 1992; Regüés et al. 2000; Cantón et al. 2001b; Gallart et al. 2002).

Most badlands are found in marls because this lithology produces more runoff compared to sandy material. Sand-dominated badlands are more affected by splash, rill and interrill erosion while in clay-dominated badlands piping frequently occurs. It is unclear whether badlands in sandy material are scarcer worldwide or few publications have been reporting on such badlands. Lithologies of different gully and badland regions are presented in review papers by Zgłobicki et al. (2017, 2018).

Other important factors explaining the formation of gullies and badlands are topography and land use. For a gully channel to develop, a critical topographic threshold must be exceeded. This threshold is determined by (1) local slope gradient (S) of the soil surface, (2) area (A) that drains towards the gully head and (3) land use. Land use controls the runoff discharge that will be produced during rainfall in a particular environment. This, in combination with A and S, then controls the magnitude of the flow shear stresses (runoff erosivity) exerted on the soil surface. Once flow erosivity exceeds a critical flow shear stress, a deep gully channel can be eroded during a single rainfall event (Poesen et al. 2003). On bare soil surfaces, large runoff volumes are produced and hence small A-values are needed to initiate gully channels. This mechanism explains the high rill and gully density on bare soils (Fig. 9.2; Torri and Poesen 2014).

All regions around the globe that have the above characteristics in terms of topography and land use are prone to gully erosion and badland formation. Figure 9.1 illustrates some of these regions as far as they have been described in scientific publications or reported by geotourists that have posted photos of badlands on Google Images. Several example regions located in India, Poland and Spain characterized by different values of geoheritage are presented below.

India. There are four major areas of severe gully erosion and badlands in India and the largest one is the Chambal Ravine Zone (Fig. 9.3). The Chambal Badlands occupy a belt of ca.500 km length and 10 km width (ca. 0.5 million ha of land) along the banks of Chambal River (Haigh 1984). These gullies are deeply incised (up to 45 m) and widespread over the plain region. The basement for the alluvium consists of granite,

Fig. 9.2 **a** Badlands formed in highly weathered volcanic ash deposits (containing petrified wood fragments) at Bulbul (30 km east from Jimma, Southwest Ethiopia). Overgrazing is most probably the main cause for triggering the development of these badlands (photo: J. Poesen). **b** Badlands (locally named "New York") formed in sandy sediments (derived from highly weathered granite and gneiss, Williams 2016) 15 km west from Konso, South Ethiopia (photo: L. Vranken)

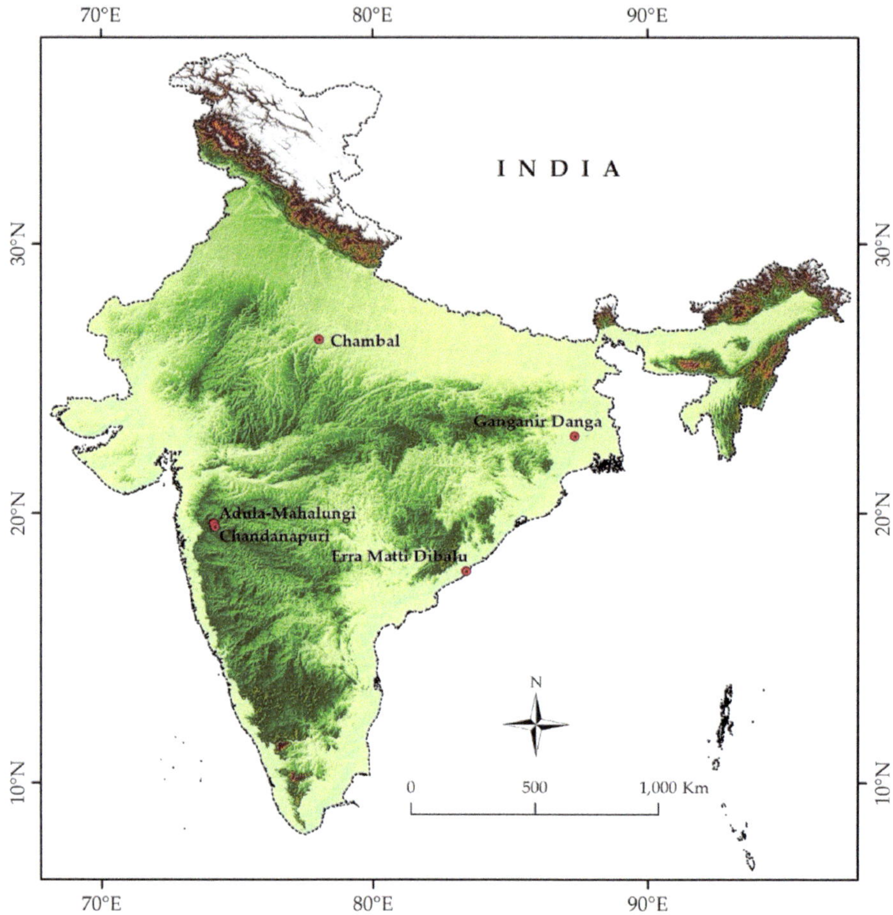

Fig. 9.3 Location of the 4 badland areas in India that are discussed in this chapter

with Precambrian rocks occurring at shallow depths (50–60 m) at some places. The alluvial sequence consists of a highly oxidized polycyclic sequence of fine sands, silt and clay in varying proportions. The sediments from these two contrasting lithologies eventually resulted in a large diversity of gully morphologies (Fig. 9.4a, b).

A second area characterized by severe gully erosion is "Erra Matti Dibbalu", a local name meaning "Red Sand Hills". It is one of the most spectacular badlands in the country, located on the eastern coast of Vishakhapatnam in the state of Andhra Pradesh. There are only three such badlands in South East Asia making this a very unique geomorphic feature/landscape/landform. These badlands formed on a 24 km stretch of coastal red sand with ca. 16 major gully systems cutting across the red sand mound. The possibility of neotectonics as a factor in the formation of these badlands was reported by Madabhushi (1995). The sedimentary formations reflect a polygenetic and multicyclic evolution, involving fluvial, fluvial-marine and aeolian

Fig. 9.4 **a** Chandanapuri site where the pediment slope deposits reach a thickness far beyond 20 m (which is rare in the Deccan Trap region). These deposits are characterized by intricate network of gullies: **b** Chambal badlands in the great alluvial plains of India—exceptionally unique in terms of their extent, unique flora and fauna (photo: V. Joshi)

processes modified by late Quaternary sea-level changes, weathering and neotectonic movements(Rao et al. 2006).

Thirdly, the Deccan Volcanic Province in India consists of an essentially rocky landscape. Godavari Basin in Maharashtra shows an exception with Quaternary alluvium underlain by basalts of Cretaceous–Eocene age (Kale and Rajaguru 1987). These deposits are densely dissected by gullies at many locations, e.g. the pediment hill deposits at the Chandanapuri site (CHP). The sediments are unconsolidated and poorly cemented silty sands. Associated with the colluvial deposits, there are strata with calcium carbonate nodules and tubules of pedogenic origin. Gully density is high near the foothills with V-shaped gully channels and when moving downstream, the gully floors become wider and the gully channels transform into a U-shaped morphology. The length of these gullies ranges from a few 100s m to more than 2 km. The sediments in which the V-shaped gullies developed have a high percentage of fine earth fractions with textures such as silty sand and sandy loam, while U-shaped gullies are observed in sediments with a relatively high coarse fraction. The deposits are rich in smectite clays that have a very high swelling capacity, which contributed to the development of a dense network of gullies in the area (Joshi and Kale 1997). Gully systems are believed to have initiated during the early Holocene humid phase but considerable modifications have occurred under present hydrologic conditions. In the same region, Adula Mahalungi (ADM) is another interesting site with true badland topography located along the banks of two tributaries of the River Pravara in Godavari Basin. The site provides one of the classic sites for studying colluvial deposits and badlands in the Deccan Volcanic Province. The alluvium is highly dissected by deep V-shaped gullies and it is hard to distinguish between a main gully channel and the tributaries. The formation of these badland gullies is reported to be the result of lineament induced minor episodic block uplifts, which are typical of the Deccan Trap Terrain (Joshi and Nagare 2013).

Finally, extensive badlands are also found in several parts of Chota Nagpur Plateau in India. These badlands are locally known as the Rarh Plain. It is a sub-hilly lateritic topography with flat to concave slopes. Ganganir Danga in the Garbheta block is one of the most extensive badland areas in this region. The area is covered by nearly 4 km^2 of Pleistocene lateritic upland and is known for its spectacular ravine development (Bandyopadhyay 1988). The gully channels are deeply incised and have both V- and U-shaped morphologies. Sidewalls are steep exceeding 85° at many cross sections (Shit and Maiti 2012). The gully network displays both trellis and dendritic patterns. The role of piping and sidewall retreat plays a major role in the mechanism of gully growth in this region.

Poland. The largest permanent gullies in E Poland dissect the loess cover and their floors reveal the underlying rocks: sand, clay, limestone and opoka (or gaize, i.e. a hard porous silica-based sedimentary rock from the Upper Cretaceous). Most channels are incised into the bottoms of older dry valleys (Fig. 9.5). Gullies usually form multi-branched systems, where the total length of all gullies can reach 30–40 km with the highest density in Poland—up to 20 km/km^2 (Zgłobicki et al. 2014). The main gullies are usually several hundred m long, while the length of the side gullies ranges from a few to a few dozen m. The average depth and width of gullies is between 10 and 20 m. Their cross section is close to V-shaped, gully sides are steep, or even vertical in some places. The floor is uneven in the upper reaches of the gullies and in the side gullies: numerous steps and collapsed pipes occur here. Gullies incised into valley sides occur more rarely; the origins of such gullies are

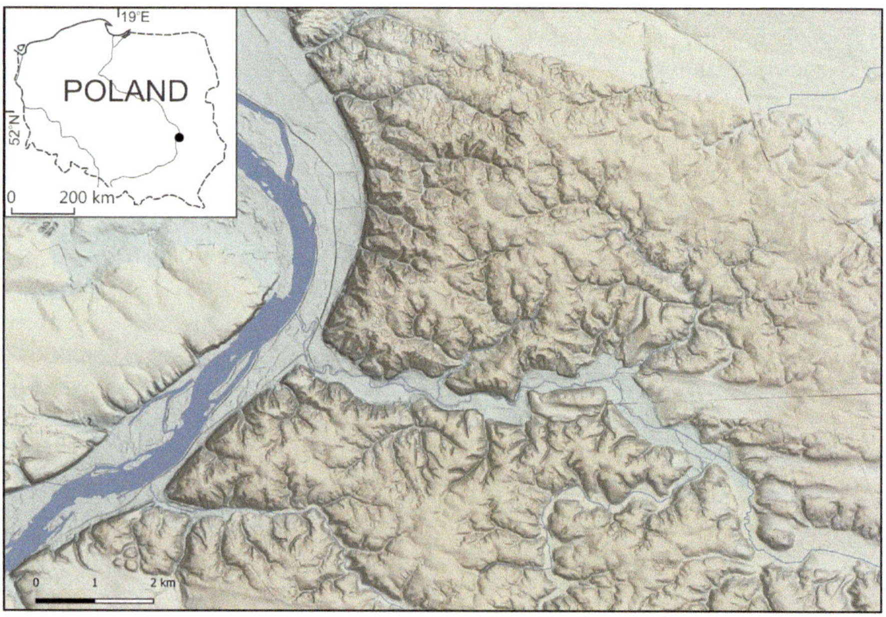

Fig. 9.5 Permanent gullies in the western part of Nałęczów Plateau (East of Vistula river, E Poland)

Fig. 9.6 Geoheritage features in gullies near Kazimierz Dolny (E. Poland). **a** Exposure of fluvioglacial sediments. **b** Infill of fossil gullies shown in the bottom of the present-day gully. (photo: W. Zgłobicki)

usually linked to roads (sunken lanes). The largest forms are straight, 6–8 m deep, and have a rectangular cross section (Fig. 9.6).

A profile of sediments infilling the fossil gully that began to develop as early as the Bronze Age has been preserved in the *Doły Podmularskie* gully (Fig. 9.5b). These sediments document several phases of gully erosion linked with periods of increased anthropogenic pressure (Dotterweich et al. 2012). In some places, fluvioglacial sediments are visible at the bottom of the gully (Fig. 9.5a). Located very close nearby, *Korzeniowy Dół* is the best-known sunken lane (road gully) among tourists and is regarded as a landmark of the region. The vertical loess walls of the gully are several m high and the trees growing on them are several decades old. The gully was named after the curious shapes of the tree roots. In this road gully, tourists can observe the characteristics of loess as a rock. The *Chałajowy Dół* gully is characterized by intense erosion processes (piping and headcut erosion). Their erosion intensity is closely related to the field and land use pattern in the catchment. In the *Kamienny Dół* gully, rocks of the region's older bedrock are exposed; i.e. a rock profile where the Mesozoic/Cenozoic boundary can be observed in the old quarry. The erosion channels represent different gully types and stages. The *Ośmiornica* gully is a multi-branched system drained by picturesque streams formed due to the dissection of aquifers by gully erosion. Also here, one can observe elements of an older topography, contemporary geomorphological processes as well as problems linked with the degradation of gullies caused by recreational activities such as quad riding) (Zgłobicki et al. 2015).

Spain. Gullies and badlands are very well represented in Spain because the country has a profusion of soft lithologies from different ages and climates dominated by the Mediterranean-type, which produces intense and erosive rainfalls. Gullies and badlands are present across the whole country but their number is higher in the Mediterranean arid regions (sub-humid, semi-arid and arid) (Fig. 9.7). Most gullies formed on Upper Tertiary sediments like marls, filling the main intermountain basins such as those of Sorbas-Tabernas in Almería and Mula in Murcia, but also in Quaternary fillings like those of Guadix in Granada. But other badlands have formed on

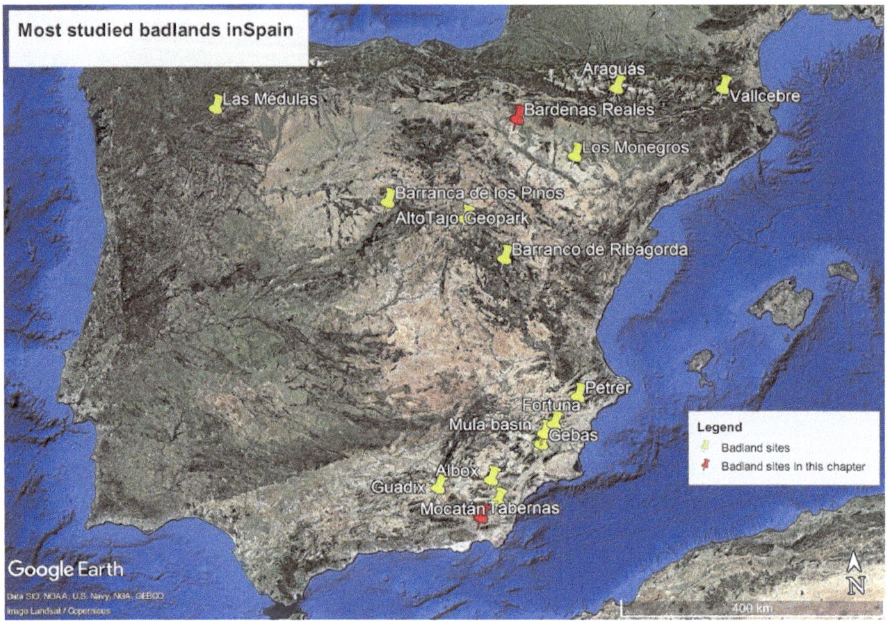

Fig. 9.7 Location of most studied Badlands in Spain

upper Cretaceous and Paleogene sediments (marls, clays and sands) like those of Alicante (Petrer, Montnegre) and also on Quaternary sediments like those of Guadix (Granada) and Bardenas Reales (Navarra). Only gullies and badlands showing extreme values of length, width, depth and/or special features have been attributed to a geoheritage and/or geotouristic value. Other well-studied sites are the Matagorda gully on poorly compacted sands and sandstones of the Utrillas Formation, East-Central Spain in the Iberian range (Martín Moreno et al. 2014) also mentioned in Zgłobicki et al. (2017), the badlands on Eocene marls in Central Pyrenees (Nadal-Romero et al. 2007) and those of Vallcebre, Eastern Pyrenees (Moreno-de-las-Heras and Gallart 2016).

In this chapter, only the two most conspicuous ones will be described in more detail: (a) those developed on grey Tortonian (Upper Miocene) gypsiferous-calciferous mudstones (marls) from SE Spain; and (b) those developed on reddish, both Miocene and Holocene, sediments (clays and silts) in Navarra, at the northern part of the Ebro depression (i.e. the Bardenas Reales). Here is a brief description of these two regions:

(a) Gullies and badlands from SE Spain are found on several interlinked Miocene basins (Tabernas-Sorbas, Mula-Gebas, Abanilla, Fortuna, Lorca) filled with relatively soft mudstones-marls and somewhat hard calcaric sandstones which, upon recurrent tectonic movements due to the African plate pushing towards the Iberian plate, and further erosion cycles, have formed the present erosional landscapes (Fig. 9.8a). These badlands are frequently associated to outliers and

Fig. 9.8 **a** Gullies and badlands from Tabernas Desert (S Spain). **b** Cabezo de Castildetierra (pinnacle), symbol of Bardenas Reales natural park (N Spain) (photo: A. Sóle-Benet)

cuestas. The badlands from these Miocene basins are very scenic and morphologically somewhat similar as they formed on relatively similar lithologies. From SW to NE, we find the remarkable badlands of the Tabernas desert (Calvo et al. 2014), Sorbas (barranco de Mocatán, Faulkner et al. 2000, 2004), Albox (Gurrea et al. 1995), the Mula basin (Romero-Díaz et al. 2007), Rambla de Algeciras (barrancos de Gebas Natural Reserve), Rambla Salada (Fernández-Gambón et al, 1996), Fortuna and Abanilla (Alonso-Sarriá et al. 2011) to mention the most studied ones. In these highly dissected areas, the main channels can be either oriented SW-NE or perpendicular, have a length of 40–200 m and a width from 15 to 55 m, with a quite variable depth. Secondary channels, usually perpendicular to the latter have a length between 40 and 50 m, a width of 5–15 m and depths not over 5–10 m. The amount of gypsum and/or other soluble salts in the parent material control processes which are related to clay dispersion and piping, with a maximum intensity at the centre of most depressions, e.g. Sorbas. Despite the scientific and scenic interest of all these areas, only two, "Tabernas desert" and "barrancos de Gebas", are "paraje natural" or natural reserves for their significant geological and geomorphological interest as erosional landscapes (badlands).

(b) The Bardenas Reales is a vast erosional depression of 425 km^2 located in the central-western sector of the Ebro Depression in the south-east of Navarra province (Desir and Martín 2007). Outstanding badlands and gullies can be observed in this protected area (natural park since 1999, world biosphere reserve since 2000, though two areas within it were declared a natural site in 1986). Four different lithologies from the Bardenas Reales (Desir et al. 2005): (1) Lerin Gypsums, Aquitanian in age (Early Miocene), are found 5 km away of the erosional depression, changing laterally to the Tudela Formation (2) composed of a 320-m-thick clay deposit mostly illite having different colours (red, grey, beige), interbedded with lacustrine limestones, sandstones and gypsum (Leránoz 1993 and Murelaga 2000, both cited by Desir et al. 2005, 2007), Buridigalian-Vindobonian in age (Lower-Middle Miocene) ; (3) Slightly Marly

Limestones, Vindobonian-Portian in age (Middle-Upper Miocene), capping the former formations and responsible of mesas, cuestas and stepped landscapes; (4) Quaternary clays and silts of Holocene age which mostly formed by the erosion of clays of the Tudela formation. Badlands are mostly found on the slopes of the Miocene clays where they develop spectacular morphologies of knife-edged interrill areas separated by a dense rill network and some monoliths of high altitude (Desir et al. 2005) (Fig. 9.8b). Besides badlands, gullies up to 10 km long and 8 m deep have developed during the Holocene and can be seen within the natural park.

9.3 Research Within Badlands and Gully Regions

Gully erosion, gullies and badlands have been the subject of intensive research during the last decades, and many studies on these geomorphic features have been published (see for example review papers on gully erosion by Poesen et al. 2003; Valentin et al. 2005; Castillo and Gómez (2016), and books on badlands by Bryan and Yair 1982; Nadal-Romero et al. 2018). Therefore, the knowledge base on the geoheritage of these geo-features is extensive. Research mainly focused on the following problems: (1) contemporary erosional processes and controlling factors (e.g. Solé-Benet et al. (1997); Nachtergaele and Poesen 1999; Cantón et al. 2001b; Nadal-Romero et al. 2007; Desir and Marin 2007; Rodzik et al. 2009; Boardman et al. 2015; Ranga et al. 2015; Su et al. 2015; Shellberg et al. 2016; Del Monte 2017; Guerra et al. 2017), (2) human impact on gully development in the past (e.g. Prosser et al. 1994; Vanwalleghem et al. 2006; Dottereweich et al. 2012; Nogueras et al. 2000), (3) recent land reclamation of badlands and gully regions (Paterson et al. 2013; Peng et al. 2013; Marzolff and Pani 2017; Shellberg and Brooks 2013).

The geoheritage of badlands and gully regions and their potential geotouristic values, however, have received much less research attention. Most studies on the latter topic are of local or regional character (Bruno and Perrotta 2012; Kushwah et al. 2015; Zgłobicki et al. 2015; Jorge et al. 2016, Palacio-Prieto et al. 2016; Warowna et al. 2016). The first synthetic studies devoted to the geotouristic values of gullies and badlands at the European and global scale have been recently published (Zgłobicki et al. 2017, 2018). From a broader perspective, the authors have emphasized the significance of research on geomorphological heritage and the ways of assessing its touristic values in the context of geotourism development (Panizza 2001; Pralong 2005; Pralong and Reynard 2005; Reynard et al. 2011; Coratza and DeWaele 2012; Zgłobicki and Baran-Zgłobicka 2013).

Badlands and gully regions, affected by human activities for centuries and millennia, have a high educational potential thanks to the diverse characteristics of their natural environment. Based on the investigation of 42 gully sites located in 11 European countries, Zgłobicki et al. (2017) indicate the following aspects that can be observed within such sites: (i) geological structure and properties of the bedrock; (ii) active geomorphological processes; (iii) past gully erosion phases; (iv) interactions between human activities and contemporary erosion (intensity of processes, mitigation, reclamation, soil and water conservation techniques). It appears that similar aspects can be the basis for educational activities also at gully sites and badlands in other continents.

Badlands in semiarid regions, located in bare landscapes, represent natural laboratories for the study of weathering of the bedrock (Cantón et al. 2001a) and different processes of erosion (soil particle detachment and transport by runoff, piping, fluting and sediment deposition (Nadal-Romero et al. 2007; Nadal-Romero and Regüés 2010). In temperate climate zones, permanent gullies are usually in a dormant state but, at some locations the geomorphological processes shaping them become more intense as a result of human activity (Zgłobicki et al. 2017).

Badlands offer good conditions for observing the bedrock (marls, mudstones, granites, limestones). In temperate zones, permanent gullies serve as a geological window on fossil and recently formed soil horizons and subsurface sediment layers (glacial and periglacial sediments). In some areas, these windows are among just a few locations where geology education is possible.

In many locations in Central Europe, present gullies are covered by forests, indicating that intensive agricultural use of the land in the past could lead to the initiation and development of gully erosion processes. The erosional phases of gully erosion processes were directly linked to increased human pressure on the environment. Such observations help to better understand the importance of soil and water conservation as well as sustainable management of the land.

The rapid development of gullies and badlands can significantly impact the livelihood of local communities through the decreasing land area suitable for agriculture, the lower suitability of eroded soils, the disruptions in the road network and, lastly, the sediment deposition on crops. However, in other badlands like those of the Tabernas Desert, receiving less than 250 mm annual rainfall, farmers managed to cultivate small plots in hanging valleys fed by hillslope runoff, fields nowadays abandoned.

At present, the reclamation of gully sites is conducted in locations where there is a need for increasing the agricultural land area or rehabilitating badlands which were the result of centuries of land abuse. Such efforts, including land levelling, have been made in Norway (Borselli et al. 2006), in India (Ranga et al. 2016), in Spain (e.g. Los Monegros in the Ebro valley, Mula and Fortuna basins in Murcia), which resulted in the destruction of valuable and distinctive badland landscapes (Zgłobicki et al. 2018). An example of ancient badlands restoration can be found in Saldaña, Duero basin, Spain (Navarro et al. 2014).

9.4 Geotouristic Value of Badlands and Permanent Gullies

As already mentioned above, badlands and gullies occur on all continents, and the latter landforms are particularly common. During their stay in the mountains, uplands or lake districts, tourists can also often visit these landforms. The tourism promotion and interest among tourists in such sites varies a lot. These landforms are regarded as great tourist attractions in countries such as Spain, Italy, China or the USA, and they define the identity of a particular territory. Some of them are even on the UNESCO World Heritage List (e.g. Australia, Italy, South Africa), which are part of the Global Geoparks Network (e.g. Italy) or are protected within nature reserves or national parks (Zgłobicki et al. 2018). Sometimes they do not have an official status as Geopark or natural park but still, they are part of the region's tourism brand development (e.g. *Land of Loess Gullies*, Poland). In other cases, gullies are mainly regarded as a threat to agricultural development (e.g. Romania, Slovakia, South Africa) (Fig. 9.9).

Badlands in Africa like in India, are rarely visited by local people or national citizens, in contrast to badlands in Europe and North America, where badlands attract many national and international geotourists. Based on this observation we can expect that, with improving living conditions in the (near) future, badlands in India, Africa will receive more attention from national and international tourists.

Except for a few cases, the desire to learn about the geoheritage of badlands and gullies is not the main reason why tourists visit the areas where these landforms occur. Instead, tourists are attracted by these geomorphic features mainly because of high scenic values and hiking possibilities. Some of these sites can also have other values: e.g. palaeontological (Dinosaur Provincial Park, Canada), cultural (Italy, Australia), biotic (Brazil, India). The touristic potential because of their high geoheritage values

Fig. 9.9 Gully erosion in the Iasi area (E Romania) (photo: W. Zgłobicki)

may not always be utilized. Thirty percent of the European badlands studied by Zgłobicki et al. (2018) received a low scoring for their geotouristic values, mainly due to their low accessibility (road network, tourist infrastructure and vicinity of bigger cities).

The successful development of geotourism requires, alongside the natural environment assets (primary touristic product), information on the external and internal accessibility, the existence of tourism infrastructure (roads, accommodation, car parks, restaurants) and specific coherent secondary products dedicated to tourists (guides, leaflets, georoutes, geosites). Moreover, efforts must constantly be made to raise the reputation of such geomorphosites. This reputation can be evaluated by, for instance, the number of photos from such geosites that are posted by local stakeholders and geotourists on the internet (these photos can be compiled via Google Images). Studies conducted by Zgłobicki et al. (2018) revealed that for only a small number of the European badlands and gully regions geotouristic products describing the most important geoheritage values (i.e. geology and geomorphology) were prepared. The accessibility of sites, security and potential demand should also be taken into account. Below, a description is given of three regions differing from each other in terms of their tourism development level: i.e. low level: Chambal Badlands and Erra Matti Dibbalu (India); medium to high level: western part of the Lublin Upland (Poland) and high level: the Tabernas Desert and Bardenas Reales (Spain).

India. Gullies and badlands are generally not touristic destinations in India. It has never been a social tradition or preferred destination for tourists as opposed to the hill stations, coastal beaches, historical sites and religious centres. However, recent initiatives have been taken for two badland areas to promote tourism in the country, though the attraction is not the landscape per se. Over centuries, the Chambal Badlands (Fig. 9.4b) have been famous for all the wrong reasons. The name "Chambal" has been associated with fear in the minds of Indians. Hence, though the striking beauty of the landscape has never escaped the amazement of all, Chambal Badlands has been anything but a tourist place. The situation is changing slowly with the government's current initiatives to bring the dacoits (bandits in Indian English) to the mainstream. India's biggest 'Birdfest' was held near the beautiful riverine National Chambal Sanctuary in 2016 that attracted many birding fans from within India and abroad. Realizing the potential Birdfest has for tourism, the state government has made this a regular activity every year and large numbers of visitors have started coming during this event period. Another attraction of Chambal is the Ganges River Dolphin, one of the world's most threatened mammals and fish-eating river crocodiles. Visitors come to enjoy the unique experience of these animals by means of sighting boat trips. Madhya Pradesh State Tourism Development Corporation has made arrangements for providing infrastructures for tourists and also to provide information regarding tourist facilities and attractions in Chambal Circuit, such as water sports, rafting, camping, boating, bird watching and an interpretation centre. Besides these, there are a number of significant archaeological sites located in the vicinity of the Chambal ravines. Tourist facilities and information centres have been constructed at all these places including arrangements for parking, signages and lodging.

Fig. 9.10 Spectacular badlands in the 'Red Sand Hills' ('Erra Matti Dibbalu') at the east coast of Vishakhapatnam (Andhra Pradesh, India). (photo: N. Rao)

'Erra Matti Dibbalu' in Andhra Pradesh is the only badland area in India that serves as a famous tourist destination in India. The awesome scenery of the landscape coupled with its location near beaches makes this a popular tourist destination for over decades (Fig. 9.10). The site is visited by the local population and visitors from outside, for family picnics, recreation, and sightseeing, especially in the cooler season. Currently, there is no legal protection and this fragile system is being jeopardized by quarrying and by trampling by tourists. The state tourist brochure refers to this location as a "picnic spot". Except for this casual announcement of the site in the tourist brochure, there is no published material for tourists that focuses on the understanding of this spectacular landscape. Much scientific literature is available on these coastal red sands but not presented in a format aiming at geotourists. The site is suffering from an unregulated tourist flow. Actions have been taken to declare this area a 'National Park' or an 'Eco Park' and to request UNESCO's recognition for this unique site.

Poland. Tourists visit gullies in eastern Poland mainly during hiking and bike trips. For visitors, these areas are an alternative for natural areas in agriculturally used uplands. In the case of some gully sites, an important role is played by aesthetic values that attract tourists (Figs. 9.5b, 9.10 and 9.11). Although tourist visits to gullies are not motivated by geoheritage, questionnaire surveys indicate that gullies constitute the most recognizable geomorphological feature in this region (Zgłobicki and Baran-Zgłobicka 2013). An important role is also played by the well-developed tourism infrastructure and the existence of cultural assets (e.g. historical buildings, local food) encouraging people to visit areas where the gullies occur. The region where Europe's densest system of permanent gullies (badlands) occurs, is a tourist

Fig. 9.11 Sunken lane (road gully) that attracts many tourists in Poland (Korzeniowy Dół, SE Poland) (photo W. Zgłobicki)

product of supra-regional significance. Tourist trails designated along some of the gullies are quite frequently described in guidebooks and are a destination for school field excursions. The most interesting sites are listed in the national register of geosites (https://geostanowiska.pgi.gov.pl/gsapp_v2/). Loess gullies offer an opportunity to present geological and geomorphological topics in conjunction with active hiking tourism. Geotourist trails can become a tool helping tourists to learn more about the very interesting geological history of this region and to understand its impact on the present-day landscape of the region. The rocks that can be observed here contain a record of a long and eventful geological history of the area, constituting part of the geological history of the European continent. For the eastern part of the Lublin Upland, several geotourist trail proposals have been developed but none of them have been implemented yet (Zgłobicki et al. 2015). The area is located within Kazimierz Landscape Park.

At present, the Natural Science Museum—branch of the Nadwiślańskie Museum in Kazimierz Dolny plays a leading role in the propagation of knowledge about loess relief and gullies. The Museum is visited by 20,000 to 30,000 tourists annually, and organized school trips are the dominant group. Since 2002, the Museum has featured an exhibition "*In the Land of Gullies*" that presents topics related to gully erosion and gully regions in an attractive way. The exhibition offers a possibility of interactive tours that involve nearly all senses, e.g. tourist can not only touch but also hold a lump of loess in their hand. Alongside a 3D map of the Grodarz catchment and models presenting phases of gully development, a diorama of a sunken lane attracts a lot of attention. Some gullies of East Lublin Upland are also mentioned in the Geotouristic Map of Poland 1:750 000.

Gully systems in eastern Poland are currently quite commonly used by people engaging in active and adventure tourism. In some cases, an uncontrolled development of these activities may pose a serious hazard to these gullies (badlands). A particular hazard is linked to the quad and all-terrain vehicle traffic which causes damage to the soil in the gullies (e.g. soil compaction and soil erosion) on a considerable scale, through the degradation of the vegetation cover.

Spain. Like in India, gullies and badlands are generally not touristic destinations in Spain as there are many more sites of greater interest: i.e. beaches, historical and religious monuments, museums, caves of all types (karstic areas, Palaeolithic and Neolithic dwellings, etc.). However, nature lovers find a few attractive badland sites which are part of Natural Parks other types of protected landscapes, e.g., Nature Sites.

Among sites in this last category, we find the Tabernas desert, in the Almeria province in SE Spain, in the municipality of Tabernas with an average annual rainfall below 250 mm and the *maximum* insolation in Europe. Such climatic characteristics explain why the area is known as the Tabernas desert, with well-developed badlands and scarce vegetation. As the area is crossed by two main roads connecting important touristic cities, people can enjoy this fantastic landscape and, depending on sunlight conditions, its beautiful colours. The resemblance with some parts of the American Far West has ensured that the area became the scenery of many Western-type movies. Badlands themselves and the movie industry gave rise to several *Far West villages*, as movie sets. Three of them are now touristic attractions where daily shows simulate typical western scenes (https://www.oasyspark.com/, https://fortbravo.com/, https://www.western-leone.es/). Within the town of Tabernas, an eleventh-century Moorish fortress, and an Interpretation Center that opens upon demand are becoming touristic destinations, still poorly known despite the ca. 1000 annual visits and the availability of leaflets, brochures and hiking maps. An enterprise is in charge of the guided tours through the badlands and they had ca. 1000 visitors in 2017. The desert hiking trail PR-AL-269 (Tabernas-Almeria) is also quite known among regional hikers. Likewise, near Alhama de Murcia, in the Murcia region, the badlands named *Barrancos de Gebas*, being also a *nature site*, is a touristic attraction where one can hike amid scenic gullies and badlands (in the tourism office of nearby Alhama de Murcia we can get information and brochures of hiking trail PR-MU-92).

The Natural Park Bardenas Reales in the province of Navarra has an Information Center open daily, visited daily by up to 700 people and a total record of ca. 70,000 visitors in 2017. Besides hiking (there are several hiking trails across the park) and mountain bike visits, a private enterprise (https://bardenasrealesnavarra.com/) is in charge of guided 4WD, buggies, quads and horse riding tours. Lodging in nearby troglodyte dwellings is also well organized.

9.5 Advantages and Disadvantages of Badlands and Gullies as Geosites

At a world scale, there are different levels of progress in the inventory of geosites and attention to geoheritage: for example, national inventories exist in some countries while in other countries they do not (Zgłobicki et al. 2017). Detailed information on the geotouristic values of gullies from Brasil, Spain and Poland has been published in international or national journals (e.g. Zgłobicki et al. 2015; Jorge et al. 2016; Palacio-Prieto et al. 2016). The physical characteristics and hence the touristic values of badlands and gully regions show a large variation. For example in China, US, Italy, Spain and Poland, these regions have a medium to high touristic value, a developed infrastructure and a high intensity of tourist traffic even if their geomorphology is not the main reason why people are attracted by these geomorphic features. On the other hand, badlands in South Africa, Romania and India lack tourist infrastructure and are rarely visited by tourists (Zgłobicki et al. 2017, 2018). Here, gullies are recognized as a threat to agricultural development and not as sites important from the geotouristic point of view. However, it should be underlined that even such anthropogenically induced gullies may play an important educational role in illustrating human–environment interactions. Such gullies clearly show the negative consequences of improper land use by humans for the environment and may, therefore, be used for formal education during the field classes.

The previously discussed diversification of selected countries and regions with gullies and badlands is confirmed by the results of a simple assessment of their values as geosites (Table 9.1). Sites in Spain are characterized by higher values of geoheritage (scientific values) while in the case of India and Poland they are generally lower. This may be explained by the fact that the Spanish geosites have been more studied by scientists, and are therefore better known by (international) tourists compared to the Indian and Polish geosites. On the other hand, the situation for educational and touristic values is more complicated. Sites in Spain have both high and moderate values, whereas in Poland moderate values prevail. Badlands in India received both high, moderate and low values but in most categories they are low. The latter is in line with observations made around the globe that gullies and badlands in the Global South have generally received no or very limited attention by the research community and by national tourists.

From the previous sections, it becomes clear that gullies and badlands are important geomorphosites (geosites): i.e. landforms of particular significance for a better understanding of the history of the Earth and landscape evolution. Such geosites should have a protective, scientific, educational and touristic function. In fact, many permanent gullies and badlands around the world attract tourists as these sites often also have ecological and aesthetic values. Gullies and badlands definitely have strengths as geosites and offer opportunities for geotouristic development. Unfortunately, they also have some weaknesses and are threatened in the long run. Table 9.2 summarizes the strengths, weaknesses, opportunities and threats of permanent gullies and badlands as geo(morpho)sites.

Table 9.1 Value ratings of gullies and badlands as geosites in India (I), Poland (P) and Spain (S); criteria according to Brilha (2016), evaluation made by the authors

	High value	Moderate value	Low value
Scientific value			
Representativeness	S	P	I
Key locality	S	I	P
Scientific knowledge	S		P, I
Integrity	S	P	I
Geological diversity	I, S		P
Rarity		I, S	P
Use limitations	P	S	I
Educational and touristic value			
Vulnerability	P	S	I
Accessibility	I	I, S	P
Safety	P	S	I
Logistics	S	P	I
Density of population	I	S	P
Association with other values	S	P	I
Scenery	I, S	P	
Uniqueness	I, S	I	P
Observation conditions		P, S	I
Didactic potential	S		P, I

Table 9.2 Overview of strengths, weaknesses, opportunities and threats of permanent gullies and badlands as geo(morpho)sites

Strengths	Weaknesses
Erosional landforms found around the globe (primary geotouristic product) In some regions, particularly in the Global North, these geosites are very popular among (geo)tourists Possibility of practicing various forms of tourism, e.g. hiking, biking)	Not always very spectacular landforms Lack or limited tourist infrastructure, roads and trails In most countries, particularly in the Global south, lack of evaluation of geoheritage potential of these geomorphosites
Opportunities	**Threats**
Growing demand for new types of tourism (i.e. geotourism) Development of adequate secondary geotourist products (e.g. educational trails, booklets, leaflets) Development of network of national and global geoparks	Destruction (e.g. by land levelling for agriculture or road and building construction, by bombturbation on military training areas, by dumping waste, by overuse by off-road vehicles) Lack of strategies for the development of geotourism

Gullies and badlands typically represent the "worst-case" scenario of areas degraded by water erosion. As such they are the best example to illustrate (1) the relative role of controlling factors (i.e. lithology, tectonics, topography, climate, land cover, land use and land management) and (2) the impact of rainfall and runoff processes on intense soil erosion rates that lead to the (rapid) formation of a (dense) network of erosional channels, earth pillars, sinkholes formed by piping collapses as well as colluvial and alluvial fans. Moreover, these geosites are ideal places to learn about landscape dissection through the development of a (dense) drainage network. On top of this, exploring gully channel walls and bottoms often offer a very cheap (and non-destructive) way to explore the characteristics of various soil horizons, weathering profiles, lithologies and sedimentological structures below the soil surface. Therefore, gullies and badlands have a large scientific and educational potential. On top of this, gullies and badlands often offer beautiful and spectacular scenery, which explains the large numbers of photos of these geomorphic features that are posted by visitors on the Internet. Hence, these geomorphosites have a large geotouristic potential, provided that external and internal accessibility is made easy, that secondary touristic products (explanation boards, leaflets, booklets) are available and that local and national authorities invest in the promotion of their reputation.

Some gullies and badlands may pose a threat to visitors because of unstable cliffs and gully walls that may collapse (rock and soil fall) and hence such geomorphosites may be hazardous and may require special measures or interventions to be taken by local authorities to avoid problems. On the other hand, large numbers of (geo)tourists visiting gullies and badlands may pose a threat to the conservation of these geomorphosites as they may damage the sites by, e.g. trampling or by the use of off-road vehicles. In addition, in some regions, gullies and badlands are threatened by damage or destruction due to particular land use types and changes: e.g. land levelling for expansion of agriculture or road and building construction, bombturbation on military training areas located within badlands, dumping waste). Establishing strategies for the development of geotourism in regions with gullies and badlands might support the willingness to conserve these geomorphosites.

9.6 Conclusions

Permanent gullies and badlands show large variations (in dimensions, morphology, colour, vegetation cover, accessibility) on our globe. There is a large contrast in attitude (by the local and national population) towards conservation and geotouristic exploitation of gullies and badlands between the Global North and the Global South (Fig. 9.1). In the Global North permanent gullies and particularly badlands are visited by many national and international tourists because of various reasons (scientific, educational, aesthetic) which is not or much less the case in the Global South. As tourism is developing worldwide, it may be expected that these geomorphosites in the Global South will receive more geotouristic attention.

In order to further develop these geomorphosites for geoeducation, geotourism, several aspects need to be addressed: apart from the primary touristic product (i.e. the geomorphosite itself), attention should be given to the development of secondary touristic products (e.g. hiking trails, explanation boards, leaflets, booklets), their external and internal accessibility. Moreover, efforts must constantly be made to raise the reputation of such geomorphosites. This reputation can be evaluated by, for instance, the number of photos from such geosites that are posted by local stakeholders and geotourists on the internet (these photos can then be compiled via Google Images).

In many cases, the first step in the tourist use of gullies and badlands is the inventory and assessment of their geotouristic and geo-educational potential. In most countries, gullies are not perceived as a tourist attraction. Only after investigating their potential, it will be possible to include features with the highest geoheritage values to the tourist offer of the area and to prepare secondary tourist products based on these values. The role of earth scientists is very important here, and they must be actively involved in the promotion, education and protection of gullies and badlands. The most effective way to use the values of gullies and badlands would be to include them in a network of geoparks. It could be the best strategy for economic development (tourism) of gully and badland regions together with education and protection of most valuable sites.

References

Alonso-Sarria F, Romero-Díaz A, Ruíz-Sinoga JD, Belmonte-Serrato F (2011) Gullies and badland landscapes in Neogene basins, region of Murcia, Spain. Landf Anal 17:161–165

Bandyopadhyay S (1988) Drainage evolution in Badland Terrain at Gangani in Medinipur District West Bengal. Geograph Rev India 50(3):10–20

Boardman J, Favis-Mortlock D, Foster IDL (2015) A 13-year record of erosion on badland sites in the Karoo, South Africa. Earth Surf Proc Land 40(14):1964–1981

Borselli L, Torri D, Øygarden L, De Alba S, Martınez-Casasnovas J, Bazzoffi P, Jakab G (2006) Land levelling. In: Boardman J, Poesen J (eds) Soil erosion in Europe. JohnWiley & Sons, Chichester, pp 643–658

Brilha J (2016) Inventory and quantitative assessment of geosites and geodiversity sites: a review. Geoheritage 8:119–134

Bruno DE, Perrotta P (2012) A geotouristic proposal for Amendolara territory (northern ionic sector of Calabria, Italy). Geoheritage 4:139–151

Bryan RB, Yair A (1982) Badland geomorphology and piping. Geo Books, Norwich, p 408

Calvo-Cases A, Harvey AM, Alexander RW, Cantón Y, Lázaro R, Solé-Benet A, Puigdefábregas J (2014) Badlands in the Tabernas Basin, Betic Chain. In: Gutiérrez F, Gutiérrez M (eds) Landscapes and landforms of Spain, Springer Science+Business Media, Dordrecht, pp 197–212

Cantón Y, Domingo F, Solé-Benet A, Puigdefábregas J (2001) Hydrological and erosion response of a badlands system in semiarid SE Spain. J Hydrol 252:65–84

Cantón Y, Solé-Benet A, Queralt I, Pini R (2001) Weathering of a gypsum-calcareous mudstone under semi-arid environment at Tabernas, SE Spain: laboratory and field-based experimental approaches. Catena 44:111–132

Castillo C, Gómez JA (2016) A century of gully erosion research: urgency, complexity and study approaches. Earth Sci Rev 160:300–319

Coratza P, DeWaele J (2012) Geomorphosites and natural hazards: teaching the importance of geomorphology in society. Geoheritage 4(3):195–203

De Geeter S (2018) Topographic thresholds for gully channel heads in Mediterranean Badlands and sunken lanes in loess areas. Unpublished MSc. Thesis, KU Leuven, Division of Geography and Tourism, 132 pp

Desir G, Marín C (2007) Factors controlling the erosion rates in a semi-arid zone (Bardenas Reales, NE Spain). CATENA 71(1):31–40

Desir G, Marín C, Guerrero J (2005) Badlands and talus flat irons in the Bardenas Reales region. Field Trip Guide B3. In: 6th International Conference Geomorphology, Zaragoza (Spain) 2005, 37 p

Dotterweich M, Rodzik J, Zgłobicki W, Schmitt A, Schmidtchen G, Bork H-R (2012) High resolution gully erosion and sedimentation processes, and land use changes since the Bronze Age and future trajectories in the Kazimierz Dolny area (Nałęczów Plateau, SE-Poland). CATENA 95:50–62

Faulkner H, Alexander R, Teeuw R, Zukowskyj P (2004) Variations in soil dispersibility across a gully head displaying shallow sub-surfacr pipes, and roles of shallow pipes in rill initiation. Earth Surf Proc Land 29:1143–1160

Faulkner H, Spivey D, Alexander R (2000) The role of some site geochemical processes in the development and stabilisation of three badland sites in Almeria, southern Spain. Geomorphology 35:87–99

Fernández-Gambón I, López-Bermúdez F, Alonso-Sarriá F., LeGoué P (1996) Papeles de Geografía 23-24:127–145

Gallart F, Solé A, Puigdefàbregas J, Lázaro R (2002) Badland systems in the Mediterranean. In: Bull M, Kirkby MJ (eds) Dryland rivers: hydrology and geomorphology of semi-arid channels. John Wiley & Sons, Ltd., pp 299–326

Gerits J, Imeson A, Verstraten JJ, Bryan RB (1987) Rill development and badlands regolith properties. Catena Supp 8:141–160

Grove AT, Rackham O (2003) The nature of Mediterranean Europe. An ecological history. Yale University Press, London, p 384

Guerra AJT, Fullen MA, Jorge MCO, Bezerra JFR, Shokr MS (2017) Slope processes, mass movement and soil erosion: a review. Pedosphere 27:27–41

Gurrea MM, Jiménez MA, Berrad F, García-Rossell L (1995) La vegetación en una zona de badlands (Albox-Almería). Studia Oecologica 12:173–182

Haigh MJ (1984) Ravine erosion and reclamation in India. Geoforum 15(4):543–561

Hose TA (2013) 3G's for modern geotourism. Geoheritage 4(1):7–24

Jorge MCO, Guerra AJT, Fullen MA (2016) Geotourism, geodiversity and geoconservation in Ubatuba Municipality, São Paulo State, Brazil. Geograph Rev 29:23–29

Joshi VU, Kale VS (1997) Colluvial deposits in northwest Deccan, India: their significance in the interpretation of late Quaternary history. J Quat Sci 12:391–403

Joshi VU, Nagare V (2013) Badland formation along the Pravara River, Western Deccan, India, Can tectonism be the cause? Zeitschrift fur Geomorphol 57(3):349–370

Kale VS, Rajaguru SN (1987) Late quaternary alluvial history of the northwestern Deccan upland region. Nature 325:612–614

Kushwah RS, Singh P, Dubey S (2015) A case study on gharial and tourism: role of national chambal sanctuary in conservation of gharials and promoting Tourism. In: Bansal SP, Walia S, Rizwan SA (eds) Tourism: present and future perspective. Kanishka Publishers, New Delhi, pp 75–83

Leránoz B (1993) Geomorfología y Geología Ambiental de la Ribera de Navarra. Ph.D. Thesis. Faculty of Science. University of Zaragoza

Madabhushi S (1995) Possible indicators of neotectonic activity near Bhimunipatnam Coast, Andhra Pradesh. Curr Sci 68:552–555

Martín-Moreno C, Fidalgo-Hijano C, Martín Duque JF, González Martín JA, Zapico AI, Laronne JB (2014) The Ribagorda sand gully (east-central Spain): sediment yield and human-induced origin. Geomorphology 224:122–138

Marzolff I, Pani P (2017) Dynamics and patterns of land levelling for agricultural reclamation of erosional badlands in Chambal Valley (Madhya Pradesh, India). Earth Surf Process Landf. https://doi.org/10.1002/esp.4266

Del Monte M (2017) The typical badland landscapes between the Tyrrhenian sea and the Tiber river. In: Soldati M, Marchetti M (eds) Landscapes and landforms of Italy. Springer, Cham, pp 281–291

Moreno-de las Heras M, Gallart F (2016) Lithology controls the regional distribution and morphological diversity of montane Mediterranean badlands in the upper Llobregat basin (eastern Pyrenees). Geomorphology 273:107–115

Murelaga X (2000) Estudio de las faunas de vertebrados del Mioceno Inferior de las Bardenas Reales y áreas colindantes (Cuenca del Ebro, Navarra). Ph.D. Thesis. Faculty of Science. University of País Vasco

Nachtergaele J, Poesen J (1999) Assessment of soil losses by ephemeral gully erosion using high-altitude (stereo) aerial photographs. Earth Surf Process Land 24:693–706

Nadal-Romero E, Martinez-Murillo JF, Kuhn NJ (2018) Badland dynamics in the context of global change. Elsevier, Amsterdam, p 320

Nadal-Romero E, Regüés D (2010) Geomorphological dynamics of subhumid mountain badland areas—weathering, hydrological and suspended sediment transport processes: a case study in the Araguás catchment (Central Pyrenees) and implications for altered hydroclimatic regimes. Prog Phys Geogr 34:123–150

Nadal-Romero E, Regüés D, Martí-Bono C, Serrano-Muela P (2007) Badland dynamics in the central Pyrenees: temporal and spatial patterns of weathering processes. Earth Surf Process Landf 32(6):888–907

Navarro J, de Araújo JC, Mongil J (2014) Assessment of 80 years of ancient-badlands restoration in Saldaña, Spain. Earth Surf Process Land 39:1563–1575

Nogueras P, Burjarchs F, Gallart F, Puigdefábregas J (2000) Recent gully erosion in the El Cautivo badlands (Tabernas, SE Spain). Catena 40(2):203–215

Palacio-Prieto JL, Rosado-González E, Ramírez-Miguel X, Oropeza-Orozco O, Cram-Heydrich S, Ortiz-Pérez MA, Figueroa-Mah-Eng JM, Fernández de Castro-Martínez G (2016) Erosion, culture and Geoheritage; the case of Santo Domingo Yanhuitlán, Oaxaca, México. Geoheritage 8:359–369

Panizza M (2001) Geomorphosites: concepts, methods and example of geomorphological survey. Chin Sci Bull 46(1):4–6

Paterson DG, Smith HJ, Van Greunen A (2013) Evaluation of soil conservation measures on highly erodible soils in the south-eastern free state province, South Africa. South Afr J Plant Soil 30:213–217

Peng SL, Chen AQ, Fang HD, Wu JL, Liu GC (2013) Effects of vegetation restoration types on soil quality in Yuanmou dry-hot valley, China. Soil Sci Plant Nutr 59(3):347–360

Poesen J (2018) Soil erosion in the Anthropocene: Research needs. Earth Surf Process Land 43:64–84

Poesen J, Nachtergaele J, Verstraeten G, Valentin C (2003) Gully erosion and environmental change: importance and research needs. CATENA 50(2–4):91–133

Pralong JP, Reynard E (2005) A proposal for a classification of geomorphological sites depending on their tourist value. Il Quaternario 18(1):315–321

Pralong JP (2005) A method for assessing tourist potential and use of geomorphological sites. Géomorphologie: relief, processes, environnement 3: 189–196

Prosser IP, Chappel JMA, Gillespie R (1994) Holocene valley aggradation and gully erosion in headwater catchments, shoutheastern highlands of Australia. Earth Surf Process Landf 19:465–480

Ranga V, Mohapatra SN, Pani P (2015) Geomorphological evolution of badlands based on the dynamics of palaeo-channels and their implications. J Earth Syst Sci 124(5):909–920

Ranga V, Poesen J, Van Rompaey A, Mohapatra SN, Pani P (2016) Detection and analysis of badlands dynamics in the Chambal River Valley (India), during the last 40 (1971–2010) years. Environ Earth Sci 75(183):1–12

Rao NK, Bhaskara Rao Udaya Ch, Vijaya PP, Thimma RK (2006) Morphostratigraphy and evolution of the quaternary red sands near Bhimunipatnam, east coast of India. J Geol Soc India 68:857–873

Regüés D, Guardia R, Gallart F (2000) Geomorphic agents versus vegetation spreading as causes of badland occurrence in a Mediterranean subhumid mountainous area. Catena, Special Issue: Badlands in Changing Environments 40:173–187

Reynard E (2008) Scientific research and tourist promotion of geomorphological heritage. Geogr Fis Dinam Quat 31:225–230

Reynard E, Coratza P, Giusti C (2011) Geomorphosites and geotourism. Geoheritage 3(3):129–130

Rodzik J, Furtak T, Zgłobicki W (2009) The impact of snowmelt and heavy rainfall runoff on erosion rates in a gully system, Lublin, Upland, Poland. Earth Surf Process Land 34:1938–1950

Romero-Díaz MA, Marín P, Sánchez A, Belmonte F, Faulkner H (2007) The causes of piping in a set of abandoned agricultural terraces in southeast Spain. CATENA 69:282–293

Shellberg JG, Spencer J, Brooks AP, Pietsch T (2016) Degradation of the Mitchell River fluvial Megafan by Alluvial Gully Erosion Increased by Post-European land use change, Queensland Australia. Geomorphology 266(1):105–120

Shellberg JG, Brooks AP (2013) Alluvial gully prevention and rehabilitation options for reducing sediment loads in the Norman by catchment and Northern Australia. Griffith University, Australian Rivers Institute, Final Report for the Australian Government's Caring for our Country - Reef Rescue Initiative, pp 312

Shit PK, Maiti RK (2012) Rill hydraulics—An experimental study on gully basin in Lateritic upland of Paschim Medinipur West Bengal. India J Geograph Geol 4(4):1–11

Solé A, Josa R, Pardini G, Aringhieri R, Plana F, Gallart F (1992) How mudrock and soil physical properties influence badland formation at Vallcebre (Pre-Pyrenees, NE Spain). CATENA 19:287–300

Solé-Benet A, Calvo A, Cerdà A, Lázaro R, Pini R, Barbero J (1997) Influence of micro-relief patterns and plant cover on runoff related processes in badlands from Tabernas (SE Spain). CATENA 31:23–38

Solé-Sedo J (1977) Mapa Geológico de España (Geological Map of Spain (Escala 1:50.000. Hoja 282 (Tudela). Madrid. IGME. 2ª Serie, 1ª Edición, 1 mapa. 18 p

Su ZA, Xiong DH, Dong YF, Yang D, Zhang S, Zhang BJ, Zheng XY, Zhang JH, Shi LT (2015) Influence of bare soil and cultivated land use types upstream of a bank gully on soil erosion rates and energy consumption for different gully erosion zones in the dry-hot valley region, Southwest China. Nat Hazards 791:S183–S202

Thomas M (2012) A geomorphological approach to geodiversity—Its applications to geoconservation and geotourism. Quaest Geograph 31(1):81–89

Torri D, Poesen J (2014) A review of topographic threshold conditions for gully head development in different environments. Earth-Sci Rev 130:73–85

Valentin C, Poesen J, Li Y (2005) Gully erosion: impacts, factors and control. CATENA 63(2–3):132–153

Vanwalleghem T, Bork HR, Poesen J, Dotterweich M, Schmidtchen G, Deckers J, Scheers S, Martens M (2006) Prehistoric and Roman gullying in the European loess belt: a case study from central Belgium. The Holocene 16(3):393–401

Warowna J, Zgłobicki W, Kołodyńska-Gawrysiak R, Gajek G, Gawrysiak L, Telecka M (2016) Geotourist values of loess geoheritage within the planned Geopark Małopolska Vistula River Gap, E Poland. Quat Int 399:46–57

Williams FM (2016) Understanding Ethiopia. Geology and Scenery. Spinger Geoguide. ISBN 978-3-319-02180-5

Wimbledon WAP, Smith-Meyer S (2012) Geoheritage in Europe and its conservation. ProGEO, Oslo, p 405

Zgłobicki W, Baran-Zgłobicka B (2013) Geomorphological heritage as a tourist attraction. a case study in Lubelskie province SE Poland. Geoheritage 5:137–149
Zgłobicki W, Gawrysiak L, Kołodyńska-Gawrysiak R (2015) Gully erosion as a natural hazard: the educational role of geotourism. Nat Hazards 79(Supplement 1):159–181
Zgłobicki W, Rodzik J, Superson J, Dotterweich M, Schmitt A (2014) Phases of gully erosion in the Lublin upland and Roztocze region. Annales UMCS, Sec B 69(1):149–162
Zgłobicki W, Poesen J, Cohen M, Del Monte M, García-Ruiz JM, Ionita I, Niacsu L, Machová Z, Martín-Duque JF, Nadal-Romero E, Pica A, Rey R, Solé-Benet A, Stankoviansky M, Stolz Ch, Torri D, Soms J, Vergari F (2017) The potential of permanent gullies in Europe as geomorphosites. Geoheritage. https://doi.org/10.1007/s12371-017-0252-1
Zgłobicki W, Poesen J, Daniels M, Del Monte M, Guerra AJT, Joshi V, Paterson G, Shellberg J, Solé-Benet A, Su Z (2018) Geotouristic value of Badlands. In: Nadal-Romero E, Martínez-Murillo JF, Kuhn NJ (red.) Badland dynamics in the context of global change. Elsevier, Amsterdam-Oxford-New York, pp 277–313

Chapter 10
Geoeducation and Geoethics Among the Children for Sustainable Tourism and Development of Aliage Geological Park in Spain

Julia Escorihuela

Abstract Most part of the population finishes their secondary studies without basic notions about Earth Sciences. For this purpose, the geoscientific community is promoting a reintroduction of this discipline and Geoethics, by means of field activities in geosites, geoparks, and geological parks. The survey conducted by the manager of the Geological Park of Aliaga (Teruel, Spain) aims to compare the geological, environmental, and ethical knowledge acquired by the children of the community attending summer courses on Geology and nature with regard to other children and adults. Results show that children who have been attending the summer courses had a higher knowledge than adults, not only about the environment, but also about the implications of the human activities in the territory. Children no students and young people have shown lower geological and environmental knowledge than adults, and therefore, they are falling behind in the basic education. This minor knowledge is inadmissible, due to the fact that they will become a predatory society incapable of judging the repercussions of the actions in the territory. On the other hand, the higher knowledge of the ex-students of the intensive courses shows the positive effect that these courses can have on the education of future professionals.

Keywords Geoethics · Geological knowledge · Geological park · Intensive courses

10.1 Introduction

Modern society development has been threatening and damaging relevant geosites for several decades; by means of open mining, building, highways, and intensive farming among other activities. Furthermore, environmental issues are being emerged in delicate as yet unknown areas in the countryside, due to the fact that nowadays

J. Escorihuela (✉)
Geologic Park of Aliaga, Calle San Antonio S/N 44150, Aliaga, Teruel, Spain
e-mail: jumidosiv@gmail.com

R. B. Singh et al. (eds.), *Global Geographical Heritage, Geoparks and Geotourism*, Advances in Geographical and Environmental Sciences,
https://doi.org/10.1007/978-981-15-4956-4_10

city dwellers are coming as tourists covering almost every holiday period of the year (Escorihuela 2017a).

On the other hand, several destructive examples caused by academic disciplines have made the scientific development impossible by eliminating relevant outcrops, now called "geovandalism" (Mansur et al. 2017).

For all the above reasons, not only is lack of geological knowledge of the population being highlighting by a lot of geologists (Escorihuela 2016), but also lack of geological knowledge of students who get to the universities of Geology. These facts show that the most part of population finishes their secondary studies without basic notions about Earth Sciences (Pedrinaci 2012). This ignorance also affects politicians, technicians, journalists, and scientific communicators that act and give their opinion in natural resources and environmental issues without a global comprehension, helping to increase the confusion in the public (Lacreu 2017).

It is not possible to suitably teach Natural Sciences if the Geology is absent, and for this purpose the geoscientific community is promoting a reintroduction of this discipline, using a new approach by means of field activities in geosites, geoparks, and geological parks (Escorihuela 2017b, c). But, at the same time, it is not possible to forget that the geologists have a relevant extra role in sustainability. Geoethics appears in this context; this new discipline investigates the values that support the behaviors and practices in any of the interactions of professional activity in the environment.

This way, it is believed by many teachers that moral issues related to Earth Sciences are better understood by means of Geoethics, and so Geoethics develops sustainability strategies (Vasconcelos et al. 2016). Nevertheless, there are not any practical studies focusing on this approach (Allan 2015). This is the reason why the manager of the Geological Park of Aliaga—Spanish pioneer in promoting Geotourism and geodidactics in relation to Geoethics at early ages—has been organizing intensive summer courses for 8 years with the children of the locality (Escorihuela 2017b, c).

The present article tries to outline the geological, environmental, and ethical knowledge acquired by the children of the community, by means of the study of the environmental–geological knowledge of the adults and children—population and visitors—and the ethical considerations underlying this knowledge. Results show that children who have been attending the summer courses have a major knowledge than adults, not only about the environment, but also about the implications of the human activities in the territory. This fact is allowing them to be able to develop a more critical analysis, and major ethical skills when they value the repercussions of projects affecting the municipality.

10.2 Methodology

10.2.1 Study Site: Geological Park of Aliaga and Its Geomorphological Context

The Geologic Park of Aliaga is situated in the Teruel Region of Spain (Fig. 10.1). Teaching about and promoting geology by guiding visitors along with several points of special interest —providing a sound overview of the past 200 million years—were the main objectives when it was founded in 1993. Since then, scientists have been evaluating the park as a unique example of geology due to its geologic structures and formations, offering impressive landforms (Soria de Miguel et al. 1996). In addition, stratigraphic record from Upper Triassic to the Quaternary is completely shown by means of its exceptional outcrops, while Cretaceous and Tertiary formations have an important relevance. Moreover, the diversity of tectonic structures allows us to appreciate the two fold systems superposition (North–South and East–West). This way, the research of the Earth history can be possible through the superposition of rock layers created by the history of the Planet Earth.

Fig. 10.1 Location of Geological Park of Aliaga within Spain, in the same region as Cultural Park of Maestrazgo. Being the first one the coordinator of the second one in the Red Ibérica de Espacios Geomineros

Fig. 10.2 Panoramic view of "La Olla", spectacular meandering fold with an international relevance because of its singularity and dimension

An old erosion surface limits the summits of hills and plateaus surrounding Aliaga, raising no more than 1500 m of altitude. Selective erosion due to different rock types has modeled the several degrees of the slopes, and so crest and cliffs are formed in the more resistant limestone, dolomitic and conglomeratic beds, whereas gentle slopes coincide with clay, marl, and sandy beds. This erosion process has enabled us to observe these complex folding structures, which have been completely leveled by the erosion surface (Fig. 10.2).

This geological framework gives the manager of the Geological Park of Aliaga the opportunity to lead 3 main axes of action: Earth science teaching and territorial awareness/protection by means of Geoethics, promotional activities, and developing the understanding of the Aliaga territory in a scientific way. The background of knowledge acquired by the manager of the Park through these activities (addressed to different public groups—of various age and career—and intensive summer courses), gives an important experience in adapted contents for various educational levels (Escorihuela and Dowling 2015) all these activities are created and coordinated by the local company charged of the management of the Geological Park from 2002 until the present, called "Jumdosiv".

10.2.2 Summer Courses Methodology

In summer courses, concepts as biodiversity and geodiversity of the municipality were introduced, and so as the idea of a possible sustainable development if we

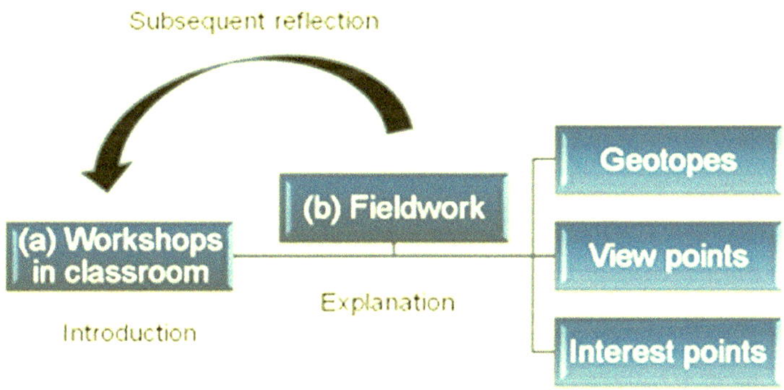

Fig. 10.3 Structure of Geologic Park of Aliaga method in intensive courses. *Source* Escorihuela (2016)

use the Geoethics principles in the management of the area. Classes (Fig. 10.3) were taught indoors (a) and outdoors (b) by means of several interest points: (b.1) outcrops, known in the scientific community as geotopes, (b.2) viewpoints in which different elements and processes can be observed, and (b.3) panoramic views that allow students think in a territorial scale.

Likewise, the holistic approach to enable the preservation of the land and its inhabitants is pointed in EGN (European Geopark Network) foundations. In the foundational statement of EGN, it is said that geological and cultural heritage must be joined in order to promote a recognized place (McKeerver et al. 2010). Furthermore, the inhabitants of an area with a relevant landscape own the real knowledge to correctly manage the territory. This important principle is shown by the manager of the Geological Park of Aliaga, because of the fact that research carried out in the Park has highlighted the importance of this local "savoir faire" in the Geosciences apprenticeship.

In the educational activity carried out in Geological Park, the analysis of the way of teaching the different concepts to different kinds of students is crucial for successful learning. The authors as Hamm (Ham 1992) have pointed out the idea of the necessity of translating the technical concepts of environmental sciences into a more comprehensible language to create the interest of children (Fig. 10.4).

10.2.3 Selected Groups Characteristics

Three groups were studied in this research. One of the three groups consisted of ex-students of intensive summer courses, aged 12–17, it served to assess their general environmental-geological knowledge acquired after these summer courses. The second group was composed of children age 12–17 years. The environmental–geological acquired knowledge was examined in this group to study children who

Fig. 10.4 Children in a field activity of the summer course in Aliaga

did not attend these intensive courses, and so they had just the environmental knowledge acquired from the school and family. On the other hand, the third group consisted of adults: parents, relatives of the children, people from Aliaga and other regions of Spain that visited the village during the summer. With this last group, we wanted to compare with respect of children, not only the differences between generations, but also the effect that the intensive summer courses had on children's environmental–geological knowledge.

First two groups—ex-students and no students—consisted of 60 children each. The study was approved by the families of the children and privacy was guaranteed.

The 60 adults of the third group were divided as follows: 20 of them aged 25–35, 20 were aged 35–45, and 20 were older (>45 year old). Figure 10.5 shows a group of young people doing an interview next to the visitors 'center of Geological Park of Aliaga. These three groups of adults were divided on the basis of a general behavior analyzed by the interviewer, depending on the number of successes and failures.

10.2.4 Studied Topics

The six topics evaluated in this research were

1. Environmental contamination: It happens when chemicals alter our environment, producing negative effects on living beings' lives. Pollution has reached its peak with modernization and development in our lives; contributing to climate change and worsening human health.

Fig. 10.5 Young people doing the interview

2. Earth resources: Our society based its energy and products on resources becoming from the Earth. Likewise, energy is highly dependent on these resources.
3. Hydrological cycle. This cycle makes possible the redistribution of water, and so the interactions among the different spheres of the Planet Earth. Groundwater, led by the terrestrial system is crucial to life and ecosystems.
4. Water erosion. Ecosystems are based on this process caused by the interaction of rocks, water, and plants. The material movement caused by water is the origin of landforms and system soil.
5. Animal's and plant's habitat. Habitat is the environment that has all the properties—nutrients and physical elements—for animals and plants to birth, grow, and reproduce. The good quality of a habitat is produced by a balanced mixture of the natural elements. Knowing these properties is crucial to understand the animals' and plants' requirements and how to act or manage them.

In order to value this general knowledge of population, it must be highlighted the fact that all these concepts were understood by everybody. So, the geology, plants, and animals of the interviews were present in the area, and due to this fact, we have called as general this type of environmental–geological knowledge, because it must be known by the population in a general way, like the general cultural knowledge.

On the other hand—and in a complementary way to the analyzed topics—all the interviewed adults were asked if they would attend a course on geology and nature.

Table 1 Index and items evaluated in interviews

Variables	
$\text{Generalenvironmental} - \text{geologicalknowledge} = \frac{e*16.6}{10} + \frac{er*16.6}{10} + \frac{ec*16.6}{10} + \frac{php*16.6}{10} + \frac{ahp*16.6}{10} + \frac{hc*16.6}{10}$	erosion (e)
	Earth resources (er)
	environmental contamination (ec)
	plant's habitat preservation (php)
	animal's habitat preservation (ahp)
	hydrological cycle (hc)

10.2.5 Procedure and Description of Interviews

The study follows the methodology proposed by Valles (2002) by planning a themed analysis in the surveys, clustering the achieved contents in meaningful topics, to finally highlight the results in ecological-geological understanding in an analytical and quantitative way. Furthermore, thanks to the qualitative interview, a discussion about Geoethics of all the interviewed could be carried out. Table 10.1 shows how the values for environmental–geological general knowledge were integrated and the related variables evaluated.

The methodology of the interviews was improved using additional procedures and interactions with the participants by means of observation and improvised conversations. The responsibility of this study lead all the interviews in order to prevent any interference.

10.2.6 Data Analysis

In order to study the differences in the general environmental–geological knowledge in ex-students and no students, the Mann–Whitney U test was utilized. This test is suitable to compare two independent groups with ordinal or continuous dependent variables without a normal distribution.

Furthermore, to identify differences between three groups (ex-students, no students and adults), we used the Kruskal–Wallis ANOVA. This test compares the variation between data sets among different groups, and lets us know when the differences of the study variables are significant.

10.3 Results and Discussion

To study the existence of differences between the general environmental and geological knowledge in the three groups of the study (25–35, 35–45, and 45–55), Fig. 10.6 shows a box and whisker plot. A significant difference was found between the three

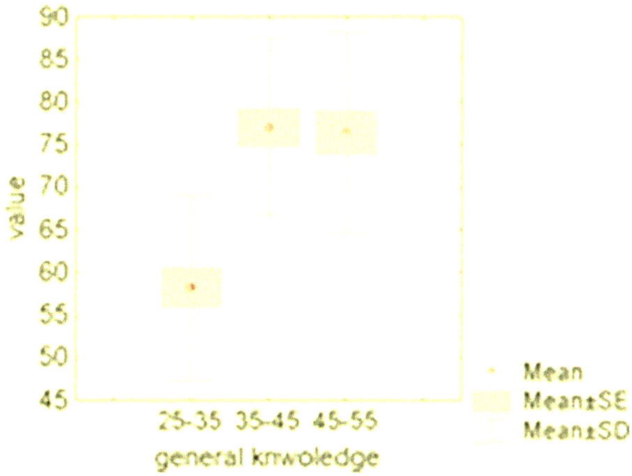

Fig. 10.6 Box and whisker plot for adults' geological and environmental knowledge. Adults of different ages showed the following data: 25–35 (58.27 ± 10.63, mean ± SD); 35–45(77.02 ± 10.16, mean ± SD) and 45–55 (76.44 ± 11.53, mean ± SD)

groups of adults ($H_{2,60} = 22.44$; $p < 0.01$), in which a lower knowledge was found in adults aged 25–35 (58.27 ± 10.63, mean ± SD). It must be underlined that we found significant differences between these two other of adults aged 35–45 and 45–55 (77.02 ± 10.16, mean ± SD and 76.44 ± 11.53, mean ± SD, respectively). So, there were common answers in adults aged >35 years, independently of their social condition and careers. These results endorse the idea of reduction of contents on Geology in Spanish schools, a reduction which is having implications in the present professionals, due to the fact that territorial management is dealing with people who have a minor geological and environmental knowledge than the previous generations.

Likewise, a minor knowledge involves a minor conscience of the implications of the actions. This way of acting was demonstrated when they were asked about the consequences of different territorial projects. This fact leaves the population defenseless and less critical facing decision making at territorial level (Macedo 1997).

Furthermore, it must be highlighted the fact that all the adults—without exception—would like to participate in courses on geology and nature to expand their knowledge, regardless of the age stratum and their professional career (85% have university degrees).

In order to compare the differences in the environmental knowledge between children and adults, Fig. 10.7 shows a box and whisker plot of three groups: children––ex-students and no students, and adults (>25).

This study reveals that there are significant differences between the three groups ($H_{2.180} = 115.07$ $p < 0.001$), and every group was different from the others. Children who had been attended to the summer course in Aliaga had a higher general knowledge on geology and nature than the rest of the population (85.57 ± 11.07, mean ± SD), being even higher than adults' knowledge (70.58 ± 13.86, mean ± SD). On the other hand, students who did not attend the summer course had a much lower general knowledge (46.59 ± 10.09, mean ± SD). This low level of knowledge—even in young people aged 17 years—coincides with the assertions of the authors as Pedrinaci (said in the introduction) who affirms that most of the population that ends up with secondary

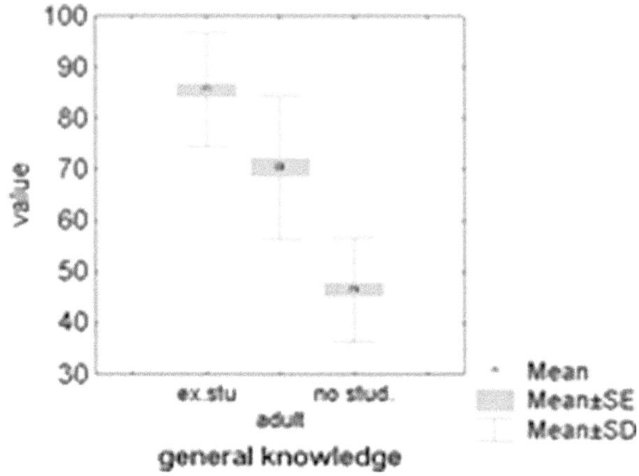

Fig. 10.7 Box and whisker plot for children's geological and environmental knowledge. Children (ex-students-no-students) and adults showed the following data: ex-students (85.57 ± 11.07, mean ± SD); adults (70.58 ± 13.86, mean ± SD) and no students (46.59 ± 10.09, mean ± SD)

education lack basic knowledge in Earth Sciences. This lower value, lower than people aged 25–35, could be showing a reductionist trend for decades, which means that we must put the spotlight on a generational problem of ignorance, that will worsen over time if no measures are taken to correct it.

However, not only was a higher degree of knowledge on ex-student proved but also this fact demonstrates the important educational work in geosciences that can be done in the geosites with a teaching infrastructure (such as geological parks and geoparks and geological–paleontological museums). Such work is being claimed from the *Red Ibérica de Espacios Geomineros*, recently created by Spain, Portugal, and Andorra (https://patrimonigeominer.eu/). This network has been founded by members such as the Geological Park of Aliaga, which highlight the recognition of the responsible role that geologists have in sustainable development, forcing them to carefully evaluate the existing educational structures to ensure more effective contributions to the population.

In this context, these children who have shown a higher knowledge, also have become more involved in territory, due to the fact that they have been taught notions on Geoethics in the summer courses. It is not possible to have a good notion of nature without basic geological notions, and geologists also have an important function in the world of ethics (Martínez-Frías et al. 2011) not only to protect the geosites from the geo-depredation, but also to achieve the sustainable development of the area.

Established educational courses on Earth Sciences can build good foundations, however, extracurricular initiatives can achieve the same goal (Gill 2016), like those which can be offered in geological sites with an adequate educational framework.

Tomorrow's land managers are the children of today, and they could start to understand the environment if an earlier acquisition of Geosciences knowledge is provided. This fact will contribute to giving children qualified skills to interpret the geosphere and biosphere and their diversity since earlier ages, contributing to create a more critical and combative society.

The satisfactory results in this earlier teaching in children make possible this initiation of early ages in geosciences concepts and issues.

10.3.1 The Geoethics' Role on the Education by Geotourism in Geological Sites

Natural hazards and anthropogenic activities have led to the rapid destruction of several geosites; what is putting emphasis on the need for geoconservation and Geoethics. The scientific community dedicated to Geoethics is strongly linked to the Geotourism activity, since it is one of the key tools to promote values of Geoethics among stakeholders and participants (Gill 2017). Moreover, the teacher community thinks that learning Geoethics contribute to successfully consider the ethical quandaries of Earth Sciences, and so students could develop sustainability approaches to face environmental issues (Vasconcelos and Almeida 2014).

Understanding Geoethics as "... research and reflection on the values which underpin appropriate behaviors and practices, wherever human activities interact with the geosphere" (Peppoloni and Di Capua 2015), teaching Geoethics provides an added-created value for a societal resilience for environmental justness defying the technological society (Bellaubi 2018). This way, this critical pedagogy becomes a useful tool to teach how to think in a conscious way.

Characteristics of the environment and school spaces have a decisive influence on the activities and behaviors of children and young people. The contact with nature minimizes the stress factors that today's childhood supports, this contact also increases their well-being, stimulates cognitive processes, facilitates emotional regimentation, and promote resilience. In addition, natural spaces offer opportunities for motivation and learning in all areas of knowledge, creating ecological awareness. Direct experience of natural processes is provided by this approach (Barberá and Valdés 1996), which develops a practical way of thinking, and general interpretation patterns of the planet Earth in the pupils' brains (Kirschner 1992). This outdoor experience is unique due to the fact that the geological processes complicatedness is re-established.

10.3.2 Implications of a Sustainable and Quality Tourism for the Sustainable Development of the Area

Due to the low geological knowledge showed on population, the scientific community is promoting the reintroduction of geosciences based on field activities, as well as ensuring the conservation of geological and geomorphological sites, which would be through Geotourism. The development that is being done from Europe, shows an effort to expand the physical interpretation of landscapes to tourists, promoting their conservation (Hose 1997).

Geotourism teaches tourists to better understand and appreciate cultural aspects, encouraging an effective geological understanding through education, which in turn provides tourist satisfaction (Dowling and Newsome 2006). Thus a geotourist can have a holistic experience that brings together knowledge, appreciation and ethical considerations. In addition, Geotourism is considered as an activity that adds significant economic value to the geological heritage, because of the fact that an infrastructure must be created and adapted to the exhibition and access to points of geological interest, as well as the establishment of jobs and personnel training for the maintenance of facilities and explanatory disclosure (Nieto et al. 2006).

In contrast to the new model of mass tourism that is being reported recently among the population of the cities—which goes to the rural environment to enjoy outdoor activities, and which is threatening the sustainability of host areas—the Geotourism is shown as a sustainable way to develop these areas of special geological and natural richness, using tourism as a vertebrate economic activity. Within this geotouristic activity, we highlight two main axes of action that are essential for the sustainability of the resource.

Firstly, in order to know the quantity of visitors who does not damage the environment of a place; touristic planning requires the implementation of the concept called "host capacity". Therefore, in sensible natural areas, a necessity to distinguish between touristic activities and good environmental quality appears as a key factor to protection, particularly if natural assets are the main pole of attraction (Edwards and Priestley 1996). Secondly, touristic guides related to the environment are essential to evaluate the natural resources of a place. In addition, geotouristic guides or geological guide-interpreters are the main factors of valorization and protection for special sites with a higher number of geological diversity elements. Moreover, Geoethics concept should be applied by these professionals, by teaching the commitment required to the correct protection of the environment. By means of these under-control tours, the touristic recreation based on a responsible satisfaction will be guaranteed, and so as the future viability of the activity itself.

Furthermore, stakeholders and touristic decision takers must equate geological heritage with cultural, archaeological and ethnographic heritage in the social and economic strategies of development in the special areas with high geological values (Hose 2013).

All the above mentioned about the need for Geoethics and geological knowledge, has a greater relevance if the focus of interest is the children's audience; because this experience would be part of a non-regulated education that will strengthen their background integrated by a balanced, respectful and realistic knowledge.

10.4 Conclusions

Children no students and young people have shown lower geological and environmental knowledge than adults, and therefore, they are falling behind in basic education. The children of today are the future managers of tomorrow, this minor knowledge—without basic notions of geology—is inadmissible, due to the fact that they will become a predatory society incapable of judging the repercussions of our actions in the territory. And, at the same time, defenseless against projects which can damage the environment in an irreversible way.

On the other hand, the entire population interviewed agreed that they would like to attend a course on geology and nature if they had the opportunity. The fact that the entire population demands this type of learning, highlights the need for public authorities to act in response to a unanimous social demand that is not being provided.

The higher knowledge of the ex-students of the intensive courses provided by the manager of Geological Park of Aliaga, shows not only a positive effect that these courses can have in the education of our future professionals, but also shows this type of courses as suitable for school and extracurricular activities. These activities are necessary to reintroduce the basic notions of Earth Sciences that have been lost in Spain for decades.

The scientific and geological community is reacting, calling for a more active and responsible action in the sustainability of the environment of the professionals of Earth Sciences through Geoethics. Promoting educational plans and didactic activities for the general population is being requested by means of Geotourism.

All the above mentioned knowledge and measures will not be effective, if society does not equate the geological and natural heritage with the cultural one. Public institutions should become aware of the need for education of the population, who are responsible for the preservation and the effective sustainable development of their own environment.

References

Allan M (2015) Geotourism: an opportunity to enhance geoethics and boost geoheritage appreciation. In: Peppoloni S, Di Capua G (eds) Geoethics: the role and responsibility of geoscientists. Geological Society, London, pp 25–29

Barberá O, Valdés P (1996) El trabajo práctico en la enseñanza de las Ciencias: una revisión. Enseñ Cienc Tierra 14(3):365–379

Bellaubi F (2018) Teaching geoethics as a form of eco-political resistance. https://iapgeoethics.blogspot.com/2018/11.html

Dowling RK, Newsome D (2006) Geotourism. Elsevier, Oxford

Edwards JA, Priestley GK (1996) European perspectives on sustainable tourism. In: Edwards JA, Coccossis H (eds) Sustainable tourism? European experiences. CAB International, Wallingford, pp 189–198

Escorihuela J (2016) Ineffectiveness in natural resource management in modern society: Geoparks proposed as possible tools for increasing awareness in the short to medium term. Episodes 40:57–62

Escorihuela J (2017) A new mass market outdoor tourism model threats the sustainability of rural environment in Spain: the necessity for regulation. Arch Anthropol Open Acc 1(3):1–3. https://doi.org/10.31031/AAOA.2017.01.000515
Escorihuela J (2017) The role of the geotouristic guide in earth science education: towards a more critical society of land management. Geoheritage 10(2):30–311
Escorihuela J (2017) The contribution of the geotourism to the sustainable development: the role of geoethics. Arch Anthropol Open Acc 1(3). https://doi.org/10.31031/AAOA.2017.01.000511
Escorihuela J, Dowling R (2015) Analysis of the geotouristic activity in the geologic park of Aliaga, Spain: progress, threats and challenges for the future. Geoheritage 7:299–306
Gill J (2017) Geology and the sustainable development goals. Episodes 1(40):70–76
Gill JC (2016). Building good foundations: skills for effective engagement in international development. In: Wessel GR, Greenberg JK (eds) Geoscience for the public good and global development: toward a sustainable future. Geol Soc Am Spec Pap 520. https://doi.org/10.1130/2016.2520(01)
Ham SH (1992) Environmental interpretation: a practical guide for people with big ideas and small budgets. North American Press, Colorado
Hose TA (1997) Geotourism—selling the earth to Europe. In: Marinos PG, Koukis GC, Tsiambaos GC, Stournaras GC (eds) Engineering geology and the environment. AA Balkema, Rotterdam, pp 2955–2960
Hose TA (2013) Geotourism in Almeria Province, southeast Spain. Tourism 55:259–276
Kirschner PA (1992) Epistemology, practical work and academic skills in science education. Sci Educ 1:273–299
Lacreu HL (2017) The Social Sense of Geological Literacy. Annals of Geophysics 60(7). https://doi.org/10.4401/AG-7558
Macedo (1997) La educación científica, un aprendizaje accesible a todos. In: UNESCO (ed). Proy Princ educ Am Lat Caribb 44: 5–7
Mansur KL, Ponciano LCMO, De Castro ARSF (2017) Contributions to a Brazilian code of conduct for fieldwork in Geology: an approach based on geoconservation a Geoethics. An Acad Bras Ciênc 89:431–444
Martínez-Frías J, González JL, Pérez FR (2011) Geoethics and deontology: from fundamentals to applications in planetary protection. Episodes 34(4):257–262
McKeerver P, Zouros, N, Patzak M (2010) The UNESCO global network of national geoparks. In: Bailey H, Hill W (eds) Geoparks: think outside the Park. George Wright Forum 27(1), pp 14–18
Nieto F, Guillén F, Díaz-Martínez E (2006) Estado actual de la legislación para la geoconservación en España. Trab Geol 26:187–201
Pedrinaci E, Ruiz de Almodóvar G, Alfaro P et al (2012) Alfabetización en ciencias de la tierra: propuesta curricular. Simposio sobre Enseñanza de la Geología. Paper presented at the 8th symposium on the Enseñanza de la Geología Huelva, pp 295–309
Peppoloni S, Di Capua G (2015) The meaning of geoethics. In: Wyss M, Peppoloni S (eds) Geoethics: ethical challenges and case studies in earth science. Elsevier, Waltham, pp 3–14
Soria de Miguel AR, Liesa-Carrera CL, Simón-Gómez JL (1996) El Parque Geológico de Aliaga. Geogaceta 20(6):1457–1459
Valles M (2002) Entrevistas cualitativas, Colección Cuadernos Metodológicos, vol 32. Centro de Investigaciones Sociológicas, Madrid
Vasconcelos C, Almeida A (2014) Sustentabilidade e Questões de (Geo)ética. Rev Ciênc Elementar 3(2):41–44
Vasconcelos C, Torres J, Vasconcelos J et al (2016) Sustainable development and its connection to teaching geoethics. Episodes 39:509–517

Chapter 11
"Location, Location, Location": Challenges of Effective Geoscience Education Within Geotourism Opportunities at Coastal US Fossil Park Sites

Renee M. Clary

Abstract In the United States, a unique geotourism venue exists with fossil parks. These sites provide informal geoscience education for the visitor while also permitting the sustainable collection and retention of a small number of personal fossils. US fossil parks bridge the protected and well-interpreted US National Park Service sites (where no fossil or rock collecting is allowed, but visitor education is developed through signage and visitors' centers) and public fossiliferous outcrops that local collectors may visit and find fossils to keep, but which do not offer any educational instruction for the uninformed or casual visitor. Because fossils illustrate geologic time, evolution and extinction, and Earth's changes in biodiversity, sea level, and climate, they can be used to improve visitors' geoliteracy by addressing several big ideas of the Earth Science Literacy Initiative (ESLI in Earth science literacy principles: The big ideas and supporting concepts of Earth science. National Science Foundation, Washington, D.C., 2010) that all people should understand in order to live harmoniously on our planet. Many variables contribute to visitors' enjoyment and improved geoliteracy at fossil parks, including the authenticity of collecting experiences, geological age of the strata, training opportunities for visitors with regards to fossil retrieval, availability of paleontology mentors, ease of fossil identification by visitors, the organization of the site, and site accessibility, including visitor safety and handicapped visitor access (Clary and Wandersee in Qualitative Inquiry in geoscience education research, GSA special papers, vol 474. Geological Society of America, Denver, CO, USA, pp. 113–134, 2011; Lessons from US fossil parks for effective informal science education. Geoheritage 6(4):241–256, 2014). Underdeveloped fossil park sites can be optimized through the fossil park model and targeting the Earth Science Literacy Initiative's 9 big ideas (Clary in Handbook of geotourism. Edward Elgar Publishing, Cheltenham, UK, pp. 244–253, 2018) However, this process is more challenging when the park contains marine fossils and

---

R. M. Clary (✉)
EarthScholars Research Group, 15 Degree Laboratory, Department of Geosciences, Mississippi State University, Mississippi State, MS, USA
e-mail: RClary@geosci.msstate.edu

R. B. Singh et al. (eds.), *Global Geographical Heritage, Geoparks and Geotourism*, Advances in Geographical and Environmental Sciences,
https://doi.org/10.1007/978-981-15-4956-4_11

is located near a coastal area. Three US Miocene-age coastal sites were researched in 2016: Calvert Cliffs State Park, Maryland, Purse State Park, Maryland, and Westmoreland State Park, Virginia. At each of the parks, extensive observations were recorded, fossils were collected, onsite signage and website information were coded and analyzed, and the park was scored with the fossil park model. All three sites faced challenges for promoting visitor geoliteracy. Visitors struggled to understand (1) fossils were *millions of years* old, (2) many fossils represented extinct organisms; and (3) sea level had been higher in the past leading to the deposition of the fossiliferous layers. Even with additional interpretive challenges, the parks can extend visitor geoliteracy for optimized geotourism opportunities.

Keywords Fossil parks · Geoliteracy · Informal science education · Geotourism · United States geotourism

11.1 Introduction

The United States of America (US) contains diverse geological sites for visitors to explore. The sites not only range in their geological content—from glacial, volcanic, coastal, and mountainous landscapes—but they also differ in their levels of protection and educational interpretation (Clary 2018; Clary and Wandersee 2011, 2014). The most familiar US geoheritage sites are those within the US National Park System, including the first protected US geosite, Yellowstone National Park. Established in 1872, Yellowstone may also have been the first national park in the world; it is now overseen by an organization that celebrated its centennial in 2016 (Clary 2016). Along with other national agencies (e.g., US Army Corps of Engineers, US Bureau of Land Management, US Fish and Wildlife Service), the US National Park Service oversees, protects, and manages national sites "for the benefit and enjoyment of the people." In addition to being protected—with no collecting or removal of geological, biological, or cultural specimens allowed by visitors—the US National Park sites are typically effectively interpreted for informal visitor education.

On the opposite side of the US spectrum are unmarked areas that are known to local fossil and rock enthusiasts, such as road cuts on public roads or borrow pits where the landowner allows collecting.[1] Local geoscience instructors also make use of these collecting sites, rallying around the slogan that "geology is best taught in the field." And while collecting is permitted on these properties, the sites have no educational interpretation for the visitor—and are often unknown to the public.

Fossil parks bridge the protected US parks and undeveloped field sites. Similar to the protected US parks, fossil parks offer educational instruction for the visitor through signage, brochures, exhibits, and/or paleontological mentors or interpreters. However, unlike the protected US park sites, fossil parks allow visitors to personally

[1]Rules differ between sites as far as what can be collected, and whether or not the landowner or public entity prohibits certain types or large numbers of items to be collected. All fossil and rock enthusiasts are responsible for verifying collecting rules at a site.

collect and retain a small number of fossils through sustainable site management. Through the 16 years of our fossil park research, we realized that the fossil park concept appears to be unique to the United States (Clary and Wandersee 2011, 2014; Clary 2018).

While all US fossil parks are similar in that they offer some type of visitor education and allow the collection of personal fossils, there is considerable variation between the sites (Clary and Wandersee 2011, 2014; Clary 2018). Some have the potential to be effective informal geoscience education sites, but are currently under-developed with regards to educational instruction for the visitor (Clary 2018). This research investigates underdeveloped US coastal fossil park sites that face unique challenges because of their locations, and analyzes how coastal parks with marine fossils can effectively interpret critical geological concepts to promote visitors' geoliteracy.

11.2 Geotourism in the United States Through Fossil Parks

According to a 2017 Gallup poll, 62% of Americans vacationed in 2017 (McCarthy 2018), and more than half of US tourists reported that learning during their travels augmented their experience (Stueve et al. 2002). In the early 2000s, the US states of Arizona and Oregon explored options to promote geotourism (Long 2005; Oregon Solutions 2005). Not only were states focusing upon tourism to geologic sites, but some areas began to investigate whether the personal collection of fossils might draw visitors—and in some situations, provide a safe, sustainable solution to collectors and geotourists who might otherwise attempt to collect at active quarry sites or illegally attempt to remove specimens from protected US National Park sites (Clary and Wandersee 2011).

Our research into fossil parks was inspired by an online article on The Fossil Park at Sylvania, Ohio, published by Cable News Network (CNN 2003). This led to the investigation of Sylvania's fossil park, Penn-Dixie Paleontological Park (New York), and Rockford Fossil and Prairie Park (Iowa) and the subsequent development of an optimized fossil park model for fossil park development and evaluation (Clary and Wandersee 2011, 2014). The second phase of our research investigated Trammel Fossil Park (Ohio) and Wheeler High School Fossil Beds (Oregon), and analyzed fossil park effectiveness as outdoor learning laboratories using the US National Research Council's *America's Lab Report* (Singer et al. 2005). Further investigation of Aurora Fossil Museum's (North Carolina) and Stonerose Interpretative Center's (Washington) fossil collecting sites resulted in an evaluation of all 7 fossil park sites with the fossil park model (Clary and Wandersee 2011, 2014).

In 2014, research led us to underdeveloped fossil parks in Ohio, namely, Hueston Woods State Park, Caesar Creek Lake, William Harsha Lake, and Oakes Quarry (Clary 2018). The parks were researched and analyzed for opportunities to advance the geoliteracy of visitors, and foster an appreciation for the geoheritage of the sites (Clary 2018).

11.2.1 The Fossil Park Model

The fossil park model evaluates US fossil park sites through fundamental variables (Clary and Wandersee 2011, 2014). These include the authenticity of the collecting site, geologic age of strata, fossil retrieving opportunities, onsite paleontological mentors, fossil identification processes for visitors, park organization, and accessibility/visitor safety. We documented how in situ collecting from strata facilitated visitors' understanding of the fossil record, whereas collecting from spoil piles resulted in visitor misconceptions, including that notion the jumbled fossils in the spoil pile resulted from a biblical "great flood." Paleozoic-aged fossils held a greater attraction for visitors than more recent Paleogene ones. Ease of collecting with a reasonable success rate for finding fossils resulted in visitor satisfaction. With regards to visitor education, the onsite mentors were valuable for enhancing visitors' experiences through individualized attention, but the fossil identification processes could also be successfully accomplished through displayed fossil examples, brochures, and perhaps most importantly, signage at the collecting site. Visitors should be able to easily find permitted collecting sites, and nearby washing stations helped park visitors clean their fossil finds for quick identification. Finally, we documented how safety at fossil park sites varied greatly, with some sites being more hazardous for families with small children. Public fossil park sites should also have some paved walkways and wheelchair height collecting opportunities so all visitors can participate in the experience (Clary and Wandersee 2011, 2014).

11.2.2 Developing Geoliteracy Through Earth Science Literacy Initiative

When visitors arrive at fossil parks, they have opportunities to collect and retain fossils, and learn a multitude of scientific concepts, including fossilization, geologic time, evolution and extinction of life forms, and changes in continents and sea level over time. Informal environments, such as fossil parks, are the venues in which we engage in the majority of our science learning—even for elementary, secondary, and college students (Falk and Dierking 2002). Since visitors are not obligated to read every sign or listen to every comment from an onsite mentor, informal education is often referenced as free-choice science learning (Falk 2001). Because fossils are not always intuitive to the general population, effective interpretation is needed at these informal sites for effective geoscience education (Larwood and Durham 2005).

Since visitors can choose which signs and brochures to read and attend to, and because the natural environment is competing for visitors' attention, the geologic concepts that should be explained and interpreted at fossil park sites should be chosen with care. In 2010, the Earth Science Literacy Initiative (ELSI), funded by the US National Science Foundation, identified nine core concepts that everyone should understand about the planet in order to be earth science literate and make wise decisions for the Earth's future (ELSI 2010). These core principles, or big ideas, include that scientists use repeatable observations and testable ideas for understanding Earth (Big Idea 1). The planet is 4.6 billion years old (Big Idea 2) and a complex system with interacting air, water, rock, and life (Big Idea 3) that is dynamic and constantly changing (Big Idea 4). The ELSI principles also acknowledge that we live on the water planet (Big Idea 5). Life has evolved—and continues to evolve—and shapes the planet (Big Idea 6), which provides us with natural resources (Big Idea 7), but also poses natural hazards to the human occupants (Big Idea 8). Conversely, humans impact and alter the Earth (Big Idea 9) (ELSI 2010). The Earth Science Literacy Principles distill important constructs that not only help interpreters design optimal informal geoscience education at fossil parks, but also provide a guideline to aid us in analyzing the effectiveness of the educational opportunities to learn at these sites (Clary and Wandersee 2014; Clary 2018).

11.3 Methods

Three eastern US coastal fossil parks were visited in Spring 2016: Calvert Cliffs State Park, Maryland; Purse State Park, Maryland; and Westmoreland State Park, Virginia (Fig. 11.1). All of the parks have Miocene-aged strata exposed at the surface, and all

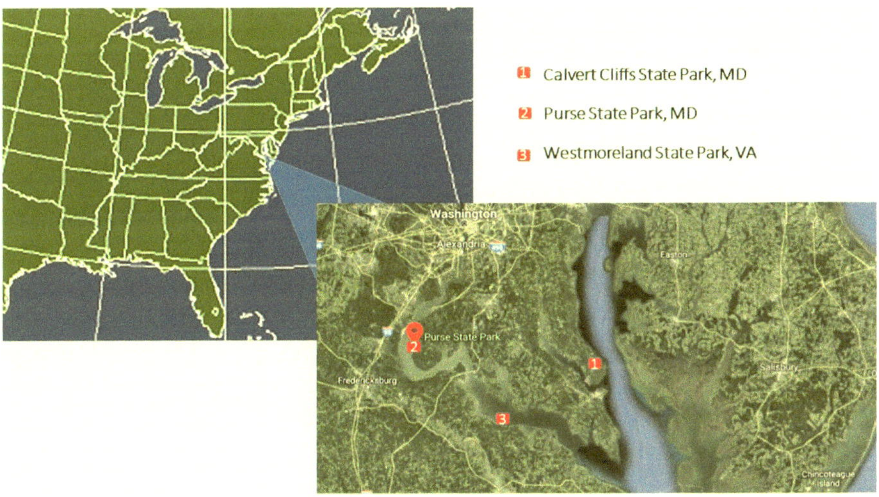

Fig. 11.1 The three eastern US coastal fossil parks investigated are Calvert Cliffs State Park (1), Purse State Park (2), both in the US state of Maryland, and Westmoreland State Park (3), in Virginia. Calvert Cliffs is located on the Chesapeake Bay, while Purse and Westmoreland are located on the Potomac River, which terminates in the Chesapeake Bay

parks have water frontage to Chesapeake Bay (Calvert Cliffs) or the Potomac River (Purse, Westmoreland) which empties into the Chesapeake Bay.

Similar to the earlier fossil park research, this investigation used a mixed-methods design (Creswell 1994; Tashakkori and Teddlie 1998) and case study analysis (Yin 2003) with a constant comparative method (Glaser 1978; Strauss 1987). Electronic sources were examined before site visits, and onsite investigations provided additional data that were used to assess the opportunities to successfully collect fossils and opportunities to learn about critical concepts related to them.

11.3.1 Identification of US Coastal Fossil Parks

Our research group became aware that the Calvert Cliffs site allowed fossil collecting when a colleague mentioned hunting for shark's teeth with her family. Upon further investigation, we found Internet posts from fossil enthusiasts who actively searched for the *Carcharocles megalodon* teeth along the park's beach. *C. megalodon*, a Cenozoic shark that lived during the Miocene through Pliocene epochs, possessed teeth that sometimes extended more than 17 cm in length.

We then uncovered information posted by the Maryland Geological Survey about fossil collecting sites at Calvert Cliffs, which inspired us to investigate other parks in the Maryland Park Service system. This resulted in the identification of the Purse State Park site as a location that allowed collecting, although the information about fossil collecting on the website was sparse. Another electronic search within nearby states revealed a "Fossil Beach" at Westmoreland State Park in Virginia.

11.3.2 Site Evaluation

Using a naturalistic lived learning experience, a minimum of 4 h was spent at each site reading signage, visiting with park personnel (if available), and collecting fossils in the allowed areas. Copies of all available brochures, maps, and other handouts were collected for use at the fossil site and for later analysis. All parks were photographically documented, including all posted information in visitor centers, information boards, and waypoint signage. Detailed notes recorded site perceptions and visitor activities, as well as personal success—or lack thereof—in fossil collecting.

11.4 US Fossil Parks in Coastal Areas

11.4.1 Calvert Cliffs State Park, Maryland, US

The Calvert Cliffs State Park was easy to locate. Using a map printed from the Maryland Geological Survey website (as well as the posted Fossil Collecting Sites article), I parked in a designated area near a junction of three trail paths. Not far from the parking area, a posted sign stated, "Due to numerous landslides, access to the cliffs has been closed. You may walk along the open portion of the beach to find fossils." The Red Trail sign also noted that "climbing upon and walking beneath the cliffs is prohibited" immediately below an image of two shark's teeth. The easy trail was almost 3 km (2.9 km; 1.8 miles) to the beach, and led through some wooded and wetland areas. One additional station on the path displayed a cross-section of a 112-year-old red oak trunk, with numbers pointing to various historic years (1901–2006) and various milestones (e.g., 1966—Calvert Cliffs state park opens). At the end of the trail, another board posted information that was specifically dedicated to the fossils (Fig. 11.2).

Although the trail path was quiet, several couples, family groups, and friends were enjoying the beach (Fig. 11.3). There were also several dogs with their owners. Some

Fig. 11.2 At the end of the Red Trail at Calvert Cliffs State Park, an information board displayed a brief summary of fossils and the Miocene Epoch (left panel), diagrams of fossil shark teeth and other fossil teeth of the Maryland Miocene (middle panel), and photographs/diagrams of shark teeth, other teeth and fish/reptile bones, and a few gastropods and bivalve fossils (right panel). *Photograph by EarthScholars Research Group*

Fig. 11.3 Several visitors enjoyed the Calvert Cliffs beach and searched for fossils along the water's edge. The cliff area was closed because of landslides, and visitors could only search for fossils along the shore. *Photograph by EarthScholars Research Group*

groups were looking for fossils as they walked along the water's edge, while other groups seemed to enjoy casual walks with friends and pets. One group was playing Frisbee.

As warned in the earlier signs, the cliff area of the park was closed since the 1980s because of landslides (Maryland Geological Survey n.d.). Through casual conversations, visitors who were searching for fossils stated they were looking for the "big shark teeth." I only managed to collect a few small shark's teeth fossils, but these were intermingled with modern shells that washed up on the shore. Conversations revealed that the other visitors did not seem to understand the difference between older, fossilized shark teeth and the modern remains they collected.

No park personnel was encountered on this visit, nor was there a museum/visitor center nearby with displayed fossils.

11.4.2 Purse State Park, Maryland, US

Unlike Calvert Cliffs State Park, the second Maryland fossil park site was more difficult to locate. An address listed on a website led to a maintenance office within another park. Fortunately, the GPS device picked up the park location and directed me to a dirt parking lot. Similar to the Calvert Cliffs trailhead parking lot, there was no visitor's center, but there was an information board with posted material. Fossils were not discussed or interpreted, but within the "non-hunting uses" the sign stated, "Fossil hunting and collecting is permitted." There was also a warning sign that "Trails leading to the shoreline may have drop-offs due to the erosion of the natural shoreline."

The 1.6 km (1 mile) trail to the beach began across the road. The trail is fairly straightforward through a wooded area, with a house slab with brick chimney jutting out from the trees immediately before the beach. There were the smoldering remnants of a campfire near the house slab. The descent to the beach from the trail was fairly steep.

There were no additional vehicles in the parking lot, and I didn't observe anyone else until I arrived at the beach and encountered an extremely angry woman accompanied by her two dogs. Between her anger exclamations (e.g., "I can't do this anymore!;" "I am so angry!"), the woman mentioned that the Purse State Park area was designated to become an NOAA (National Oceanic and Atmospheric Administration) marine preserve park, and she thought that the state of Maryland had not prepared the site in advance. She spoke of an uncapped well where a child could fall in, and having personally encountered several smoldering fires. She also pointed out a knife on the beach (e.g., a hunting/fishing knife) and bemoaned how this area would be inviting families when it wasn't safe for them to visit. She also spoke of *E. coli* and heavy metal contamination in the water—and stated that dogs had become poisoned and died from drinking the water. Our research group did not independently verify these stories.

The woman asked if I was searching for shark teeth and gave me a half dozen small ones she had picked up. She then pointed out a jutting point where she said the big *Turritella* gastropods will weather out of the banks, and stated that is the fossil for which the park is best known (Fig. 11.4). We had not seen that information on any fossil website, or Maryland Park Service website.

The woman then offered a quick lesson in searching for shark's teeth. She stated that since they are lighter (less dense) than the rocks, they tend to get carried a little higher than the pebbles and shells by the water action. Therefore, she advised me to look at the top of the shell/gravel line, and toward the bottom of that shell/gravel swath.

While I was walking to the point with the *Turitella* fossils, the tide was rising. Along the beach, there was a lot of exposed cliff, but not all of it was stable. And at Purse State Park, retaining fences were not in place. Accumulated driftwood and fallen trees also made the beach difficult to navigate. There was a fair amount of trash. Along the beach toward the point with the best fossils, fallen trees blocked the path and forced visitors to crawl through the tree limbs with the water rising. At that point, I opted to search for fossils weathering from the cliffs before reaching the target area. Fossiliferous blocks had fallen and rested against a cliff, and fossils were easily seen within the strata (Fig. 11.5). I found a *Turritella* weathering from the strata, and walked back to the entrance pathway. The woman had been joined by another woman and a uniformed man, who seemed to work for the park system or department of natural resources. I thanked her for her help.

Fig. 11.4 Purse State Park was deserted except for one other woman, and underdeveloped. Fallen trees and branches made traversing the beach difficult. *Photograph by EarthScholars Research Group*

Fig. 11.5 Purse State Park had many exposed cliff banks, with fossils weathering out of the limestone. *Photograph by EarthScholars Research Group*

11.4.3 Westmoreland State Park, Virginia, US

The last US coastal fossil park visited was Westmoreland State Park in Virginia. There were no issues locating the park with the online address and directions. This was the only park that charged a small fee; entrance cost was $6 USD for non-Virginia residents.

The park includes a Discovery Center, which was closed and locked the day I visited. However, when peering through the windows, I spotted a plastic replica of a *C. megalodont* tooth on a shelf and framed fossil displays on the back wall, indicating that the park must address fossils some of the time in their outreach programs. The Visitor's Center did not have a fossil display, but they offered several brochures, including the month's program guide, which listed "Fossil Hike—Search for the Past" as an event. Other brochures included a trail guide/map, a bird field checklist, and a Fossil Guide.

The Fossil Guide unfolded to reveal 19 grayscale photographs, primarily of fossil shark teeth. A few fossil collecting tips are listed at the top of the page. Other panels discuss a brief definition of fossils, and where the fossils at Westmoreland State Park originated. Types of fossils are listed (i.e., mold, permineralization [*sic*], groundwater [*sic*], petrification). Seven additional fossil collecting tips are itemized, as well as fossil stewardship. Notably, fossil collecting is permitted in the park, but any other resource collecting—including animals, plants, historical artifacts, or mineral—is prohibited.

Near the Visitor's Center, signage interpreting historical conflicts during the US War of 1812 were displayed. An additional sign in front of the Visitor's Center explained the center's designation as a LEED (Leadership in Energy and Environmental Design) Gold Facility, indicating that sustainability and energy conservation were carefully considered in its construction.

The trailhead to the fossil collecting area started across from the circular drive in front of the Visitor's Center, with "Fossil Beach" clearly marked (Fig. 11.6). A Big Meadows Trail information board, across from the "Fossil Beach" sign, had a map and information. The 0.6 km (0.4 mile) Beach trail is marked as "more difficult." The last item of the "Trail Guidelines" on the left of the large map is the only rule listed in all capital letters: "11. DIGGING TOOLS OF ANY KIND ARE NOT ALLOWED ON FOSSIL BEACH."

The path to the beach was a short walk, with a rather steep incline which made the return trip slightly challenging. Signs marking "Big Meadow Trail" assured hikers they were on the correct path. Multiple signage along the path provided an opportunity to learn about trees in the area. The signage, emplaced by the Virginia Department of Forestry, provided simple diagrams of leaves (and sometimes associated berries or flowers) for tree identification along with information about the species. The density of signage was greater at the beginning of the trail, though the signage continued more sporadically down the trail. The waypoint signs effectively communicated information through its use of simple diagrams, reading level, a limited number of words in the text, and focus upon one theme (Wandersee and Clary 2007).

Fig. 11.6 Westmoreland State Park's trail to Fossil Beach starts near the park's Visitor's Center. The larger "Big Meadows Trail" information board has information for hikers. Additionally, the pathway features waypoint signs with identifications of trees. One waypoint sign can be seen behind the Fossil Beach sign in the distance. *Photograph by EarthScholars Research Group*

At the trail's end, the beach area to the left side was bordered by a fence with a mounted sign informing visitors "Danger" and "No Trespassing," and information that this "beach closed due to hazard of cliff slides." However, one woman with two dogs ignored the sign and was navigating through the off-limit area, which was strewn with fallen trees and sticks.

Within the open public beach area, two family groups were present, one with 2 dogs (Fig. 11.7). A man and his son walked approximately 2–3 m into the water with a bucket and a shovel, and began to sieve for shark's teeth. The visitors mentioned "shark's teeth" as being the valuable find at this beach.

There was a cliff at the opposite end of the beach, and I took off my shoes and socks and rolled up my pants to cross a water inlet to get there. The outcrop wasn't fossiliferous like the one at Purse State Park, but had the appearance of the weathered chalk/limestone with some iron staining. While I found some small fossils in the beach area, I had less success at finding fossils in this park than the other two sites in Maryland.

When walking back along the trail to the parking area, I encountered three young men who asked if I had found any shark's teeth. They said that their mission was to find big shark's teeth. Back at the visitor's center, an employee told me that a woman had uncovered a 7.5 cm (~3 in) *C. megalodon* the previous weekend.

Fig. 11.7 On Westmoreland State Park's beach area families enjoyed the sand and water, and one father and son team sieved sediments looking for fossils. The larger "Big Meadows Trail" information board has information for hikers. Additionally, the pathway features waypoint signs with identifications of trees. One waypoint sign can be seen behind the Fossil Beach sign in the distance. *Photograph by EarthScholars Research Group*

11.5 Analysis of Coastal Fossil Parks: Opportunities and Challenges

For each of the coastal parks (Calvert Cliffs State Park, Maryland; Purse State Park, Maryland; Westmoreland State Park, Virginia), all researcher notes, park signage, web-housed information, and brochures were transcribed and coded. Fossils collected were also used as data. Through content analysis (Neuendorf 2002), stable themes emerged for coastal fossil parks. Each of the three parks was also analyzed on the optimal Fossil Park Model (Clary and Wandersee 2011, 2014).

11.5.1 Data Analysis

Calvert Cliffs State Park was relatively easy to access, and although some information was available at the site (with the most informative material posted on the information board at the trail's end at the beach), visitors did not recognize that there were fossil organisms present other than large shark's teeth. I had little success with fossil collection, and it was extremely easy for the uninformed visitor to pick up shell fragments along the beach and not realize that these were the remains of modern organisms. Visitors could easily come to the misconception that the 13-million-year-old fossil teeth from *C. megalodon* resulted from the extinct sharks living in the Chesapeake Bay area, with their teeth washing into the beach area. There was little opportunity for understanding that this area was once underwater, and ancient shark teeth were deposited within sedimentary layers that are now eroding to reveal the entombed contents. There is also no understanding of the biodiversity of the Miocene, or how the *C. megalodon* fit into a food web of this epoch.

Conversely, fossil collecting was moderate at Purse State Park's outcrops, with multiple gastropod shells eroding from the cliff wall on the beach. However, the information available to visitors was the sparsest of any of the coastal parks in this research. The best information came from the Maryland State Park website, and even this was nominal. The area was not well known, and other than one concerned resident—who called others to report a safety issue—no visitors were observed. While I have not independently verified issues with water quality and uncapped wells, I did observe a knife, an abandoned smoldering fire, and numerous debris on the beach.

Westmoreland State Park offered the best interpretation *at the park site*, but also the lowest fossil success among the three state parks researched in this study. Stratigraphic context was absent at this park, and visitors were observed sieving sands at distance from the beach to hunt for shark's teeth. While the brochures offer some information for visitors to use in identifying fossils, some visitor confusion will likely result in the inclusion of "groundwater" and the listing/description of "per mineralization" [*sic*] and "petrification" as fossil types. Through our past research, we observed how brochures are often discarded by visitors while signage offers more permanence with regards to informal education (Wandersee and Clary 2006). Similar to Calvert Cliffs State Park, visitors at Westmoreland sought the large *C. megalodon* teeth, but were unaware of any other potential fossils that might be found.

Through content analysis and comparison with the Earth Science Literacy Initiative big ideas (ELSI 2010), three stable themes emerged with regards to these coastal fossil parks' educational challenges in promoting visitors' geoliteracy. Visitors had difficulties understanding the geological context of the fossils, particularly that the large shark's teeth and other Miocene fossils were *millions of years* old. Visitors could also leave these fossil park sites failing to comprehend that many of these fossils represented extinct organisms. Finally, with the absence of stratigraphic context and explanation of deposition and lithification, visitors have few clues that sea level has fluctuated in Earth's history, and the eastern US coast was once covered in an ocean—and it was within ancient seas that fossiliferous layers were deposited, and are now exposed and eroding.

11.5.2 Fossil Park Model

Using the Fossil Park Model (Clary and Wandersee 2011, 2014), the three coastal fossil park sites were ranked according to key variables (Table 11.1). These coastal

Table 11.1 Coastal fossil park sites analyzed with the Fossil Park Model key variables

Fossil park sites	Authenticity	Geological age and ease of retrieval	Training and facilities	Onsite mentor	*Onsite* signage and brochures	Organization	Accessibility and safety
Calvert cliffs state park					*x*		X
Purse state park	X						
Westmoreland state park			X		*x*	*x*	X

fossil parks scored lower than the previously researched underdeveloped parks in Ohio (Clary 2018). Only Purse State Park site had obvious stratigraphic context, with fossils eroding from the cliffs for an authentic collecting experience. All these coastal fossil parks had fossils that were more recent in age (Miocene) compared to other fossil parks (Silurian, Ordovician). Our previous research documented that visitors are more attracted to the older fossils—perhaps because they represent organisms very different than those today, and are undeniably extinct (Clary and Wandersee 2011, 2014). Neither Calvert Cliffs State Park nor Purse State Park had a visitor's center nearby the fossil collecting site. However, it appears that Westmoreland's Discovery Center offers programs to engage visitors in understanding the fossils and shark's teeth found in the area. Westmoreland State Park's brochure of events also listed a fossil event. Since the fossils are found at beach areas in all three coastal park locations, no washing stations are needed to clean the fossils for identification.

None of the locations had an onsite mentor. The Westmoreland State Park Visitor's Center was staffed, but a paleontological mentor was not available at the Fossil Beach site to help visitors identify their find—as we experienced in previous fossil parks in Sylvania, Ohio and Fossil, Oregon (Clary and Wandersee 2011, 2014). Only Westmoreland State Park had a brochure to help with identification, and Calvert Cliffs posted fossil identification charts on the identification board where the trial ended at the beach. Westmoreland State Park offered the best organization with posted trail markers and a nearby Visitor's Center, but the trail was steeper than the others. However, both Westmoreland State Park and Calvert Cliffs State Park showed more concern about accessibility and visitor safety compared to Purse State Park. The closed-off beach at Westmoreland State Park was not a deterrent to determined visitors, though.

11.6 Optimizing Geoliteracy at Coastal US Fossil Parks

The data and analyses from three US coastal fossil parks in Maryland and Virginia indicate that these sites face greater barriers than the underdeveloped Paleozoic sites in Ohio for optimizing visitor geoliteracy (Clary 2018). There are some basic safety issue considerations with collapsing cliffs and shore debris. At both Calvert Cliffs State Park and Westmoreland State Park, personnel attempted to implement barriers and rules for visitor safety, though it appears that some visitors at Westmoreland paid no attention to the signage warning them to keep out of fenced-off areas. Purse State Park exhibited some unsafe conditions for the unwary visitor when I visited in 2016.

Even with safety optimized, all three parks face the obvious obstacle that visitors collect *marine* fossils near a modern body of water. Location of a fossil park with marine fossils near a modern marine environment is problematic for visitor geoliteracy, if the science behind the fossils and their current location are not explicitly interpreted for educational effectiveness. It is not readily apparent to geological novices that the fossils do not represent the same age, nor the same environment, as the present day. Explicit instruction noting the different environments—including comparative images of modern and Miocene fauna—could go a long way in increasing fossil collectors' understanding of the differences.

Additional information for stratigraphic context and the fossilization process could help illuminate how marine fossils end up on the shoreline. Visitors need explicit instruction through signage and brochures that the fossils found on the shore did not simply wash up from the ocean. Diagrams could illustrate the processes by which shark's teeth and other fossils settle on the ocean floor, are then covered with sediment, and the resulting layers lithified to be later exposed with falling sea level—resulting in rock layers being weathered and eroded to free the fossils from their matrix.

Visitors may also struggle to understand the differences between *fossils* and *modern* shells that wash on the shore. Demonstrating the difference with signage or brochures would help dissipate misconceptions that all the shelly remains on the beaches are fossils, or that anything other than shark's teeth on the beach are not fossils.

Optimizing informal education to focus upon the geologic age of the fossils, the difference between the modern shells and fossil materials, and the formation of the fossiliferous layers through sedimentation, lithification, and sea-level changes could help improve the geoliteracy of visitors with regards to several of the big ideas of the Earth Science Literacy Initiative (ELSI 2010). Simple signage or brochures could educate visitors about the millions of years that have passed since the Miocene Epoch (Big Idea 2) and the different types of animals that lived 13 million years ago (Big Idea 6). The complexity and changing nature of Earth (Big Ideas 3, 4) is easily demonstrated at these coastal park sites, especially how the area has experienced changes in sea level and erosion (Big Idea 5). These coastal fossil park sites also offer the potential to address modern climate change issues, and how human impact can result in future sea-level rise (Big Idea 9).

Although coastal face parks face greater challenges for effective informal education, the solutions are not difficult, nor expensive to implement. With safety issues addressed, and effective signage and brochures developed, these coastal fossil parks can be improved to extend visitor geoliteracy for optimized geotourism opportunities.

References

Cable News Network (CNN) (2003) Finding trilobites by the bucket, for free. https://www.cnn.com/2003/TECH/science/06/26/fossil.park.ap/index.html. Accessed 26 June 2003; now archived in an e-mail at https://dml.cmnh.org/2003Jun/msg00472.html

Clary RM (2016) Science and art in the National Parks: Celebrating the centennial of the US National Park Service. Sci Teach 83(7):33–38

Clary RM (2018) Geoheritage and public geoliteracy: opportunities for effective geoscience education within US parks (Chap. 18). In: Dowling R, Newsome D (eds) Handbook of geotourism. Edward Elgar Publishing, Cheltenham, UK, pp 244–253
Clary R, Wandersee J (2011) Geobiological opportunities to learn at US fossil parks. In: Feig A, Stokes A (eds) Qualitative Inquiry in geoscience education research, GSA special papers, vol 474. Geological Society of America, Denver, CO, USA, pp 113–134
Clary RM, Wandersee JH (2014) Lessons from US fossil parks for effective informal science education. Geoheritage 6(4):241–256. https://doi.org/10.10007/212371-014-01016-x
Creswell J (1994) Research design: qualitative and quantitative approaches. Sage Publications, Thousand Oaks
Earth Science Literacy Initiative (ELSI) (2010) Earth science literacy principles: The big ideas and supporting concepts of Earth science. National Science Foundation, Washington, D.C. https://www.earthscienceliteracy.org/
Falk J (2001) Free choice science education: how we learn science outside of school. Teachers College Press, New York
Falk J, Dierking L (2002) Lessons without limit: how free-choice learning is transforming education. Alta Mira Press, Walnut Creek, CA
Glaser B (1978) Theoretical sensitivity: advances in the methodology of grounded theory. Sociology Press, Mill Valley
Larwood J, Durham E (2005) Earth heritage: word heritage. involving people in geodiversity. Status Design, Spalding
Long L (2005) Arizona promotes 'geotourism.' The Arizona Daily Star. https://tucson.com/news/opinion/arizona-sonora-join-up-to-promote-geotourism/article_8f748ccf-1b8a-5367-965c-198c614d48d1.html. Accessed 9 Dec 2005
Maryland Geological Survey (n.d.) Fossil collecting sites at Calvert Cliffs. https://www.mgs.md.gov/geology/fossils/fossil_collecting.html
McCarthy J (2018) Six in 10 Americans took a vacation in 2017. Gallup. https://news.gallup.com/poll/224843/six-americans-took-vacation-2017.aspx
Neuendorf K (2002) The content analysis guidebook. Sage Publications, Thousand Oaks
Oregon Solutions (2005) The Oregon Paleo Project. https://orsolutions.org/osproject/the-oregon-paleo-project
Singer SR, Hilton ML, Schweingruber HA (eds) (2005) America's lab report: investigations in high school science. National Academies Press, Washington, D.C.
Strauss AL (1987) Qualitative analysis for social scientists. Cambridge Press, Cambridge
Stueve A, Cook S, Drew D (2002) The geotourism study: phase I executive summary. Travel industry Association of America, Washington, D.C., 24p
Tashakkori T, Teddlie C (1998) Mixed Methodology: combining qualitative and quantitative approaches. Sage Publications, Thousand Oaks
Wandersee JH, Clary RM (2007) Learning on the trail: a content analysis of a university arboretum's exemplary interpretative science signage system'. Am Biol Teach 69(1):16–23
Yin R (2003) Case study research: design and methods, 3rd edn. Sage Publications, Thousand Oaks

Chapter 12
Geoparks and Geo-Sites: Geological "Learning Objects"

Susanna Occhipinti

Abstract The paper presents a research starting from an acquired analysis of a general lack of sensitivity towards the Earth Sciences in Italian schools, also extended to the social and cultural background. This research has mainly analysed the Italian context, and it highlights that the problem resulting from a lack of sensitivity directly leads to a lack of attention to the territory which whereas requires to be protected and "geo-preserved". If the aim is to promote the culture of geoscience, but alike to enhance the sensitivity towards geo-heritage or geo-conservation, geo-sites can become a kind of learning objects, useful in the teaching–learning process. The use of geo-sites as learning objects has been experimented effectively in various other contexts both for their scientific interest or beauty and for their dramatic aspect, in the case of a catastrophic event; whatever is the reason, they represent an opportunity for cultural enrichment and awareness. To turn the "geological object" into an opportunity to deepen in the scientific contents and to develop skills, but also to promote sensitivity and passion, the use of effective methodological approaches is however necessary.

Keywords Learning objects · Unexpected connections · Investigative approach

12.1 Introduction

The paper presents a research that, starting from a proven lack of sensitivity in Italian schools towards geosciences, a subject rarely taught in schools with its own epistemology, highlights the need to search for new tools and to find new paths to increase attention towards a territory increasingly fragile and sensitive to geological problems.

This lack of attention to the geosciences is also widespread in society and in the common thought; the consequence is that the sensitivity to the issues of this branch of science only grows, unfortunately when a catastrophic event occurs, the scientific

S. Occhipinti (✉)
Secondary School Principal, IESO and IGEO Coordinator for Italy, Aosta, Italy
e-mail: s.occhipinti@mail.scuole.vda.it; susocchip@gmail.com

R. B. Singh et al. (eds.), *Global Geographical Heritage, Geoparks and Geotourism*, Advances in Geographical and Environmental Sciences,
https://doi.org/10.1007/978-981-15-4956-4_12

contents of the discipline are likely to be seen exclusively as elements of danger and vulnerability.

In schools, the theory is too often presented using not involving approaches, a bit boring and not very engaging to promote passion and interest while the environmental and geological peculiarities should represent a useful and significant tool to get in touch with the culture of geosciences.

This research, based on the knowledge of the Italian context, rich in cultural but also natural heritage, tries to enhance, through cross-sectional surveys, sensitivity towards naturalistic and geological values, as geo-sites or geoparks.

In my personal experience, as a teacher and researcher, I have focused the attention on finding more effective ways to create "snappy and smart" teaching tools and paths, to support the work of teachers. I have been working to produce educational tools which can be described as friendly, easy to make and share, and coherent for effective educational approaches, efficient in promoting interest, involvement and skills.

It is a widespread belief, based on tested and shared practices, that Science teaching–learning, should be based primarily on active teaching methodological approaches. These approaches are generally centred on investigative processes, such as PBL that require to identify driving questions, and basically need to develop paths through hands-on laboratory activities.

In the case of the so-called hard disciplines, Physics and Chemistry, the experimental and laboratory approach is a widespread and shared heritage, through the use of machines, tools, objects that develop experimental activities with increasing complexity, using the scientific method of Galilean memory but leaving little room for intuition and autonomous reasoning.

In natural Sciences, the experimental practice is perhaps less widespread, but it allows the development of operative paths which favour investigation, promote curiosity and push the student towards the research, at first guided by the teachers/experts, then gradually let to become more autonomous.

In Earth science, the hands-on approach is naturally part of the teaching of petrography and palaeontology, where the learning object can be manipulated, observed, studied, analysed and compared; while complex phenomena like global tectonics, earthquakes or the observation and study of faults and folds, require a different approach.

12.2 Working Hypothesis

Moving from a classroom level, based on hands-on practices, to an external context, which allows an approach to natural environments, the aim is to promote sensitivity towards geoscience, passing through geo-sites, whether they are isolated sites or inserted in geoparks, in order to acquire the concepts and principles of geo-heritage and geo-conservation, but also simply to discover the not always understandable beauties of geology.

It is not easy, for an "uninitiated", to appreciate the history of a rock, the beauty of a fold, the dynamics of a landslide, the richness of information and connections of a stratigraphic sequence: the task of a teacher should be not only to scientifically illustrate the phenomenon of the fold, or the rock in which it was formed, but also to create/show a link to Physics, temperature and pressure, to Chemistry, composition of materials, to geological history, the event that formed it, but also to its beauty, as if it were a masterpiece of nature.

But, mostly, each of these natural phenomena can become a kind of *learning objects* necessary to promote sensitivity and passion. Whether they can be observed directly in situ, (Figs. 12.1a, 12.2a) or represented through models, (Figs. 12.1b, 12.2b) analysed through apps or virtual images, they allow to enhance the relationship with the territory and its geo-conservation, and they are effective teaching tools for the active teaching of Earth Science.

The use of *learning objects* has been actually experimented in various contexts: the usefulness of the hands-on actions, and the effect of emotions towards the teaching–learning process have been demonstrated by using models and tools.

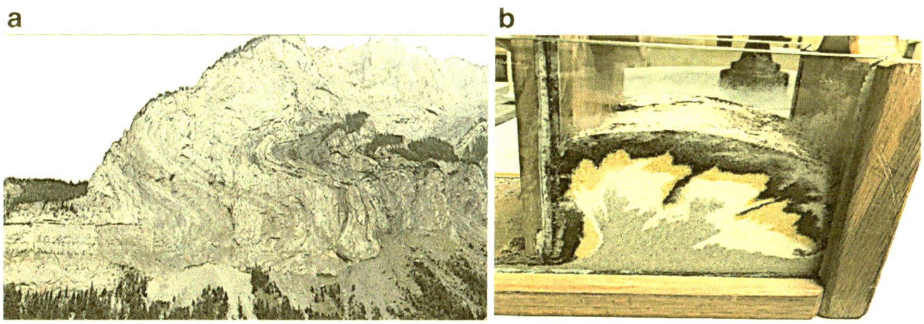

Fig. 12.1 **a** A fold in Lower Jurassic limestone layers of the Doldenhorn nappe, Switzerland. **b** A model of fold, made with cocoa, white and yellow flour and sand—regional scientific laboratory

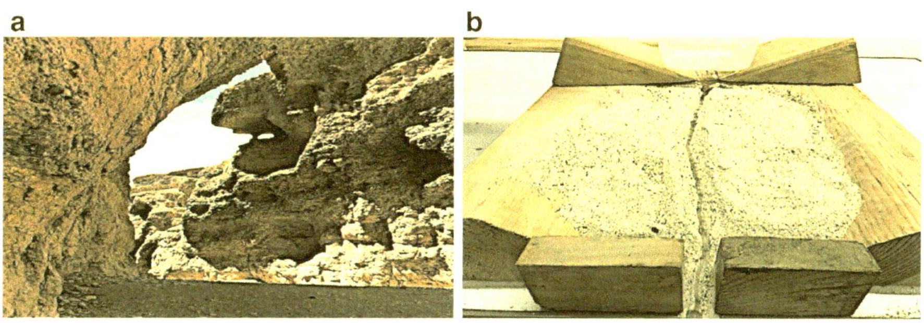

Fig. 12.2 **a** A deep alluvial Ouadi- Namibia. **b** A model of water erosion—regional scientific laboratory

An important factor favouring and implementing the effectiveness of the teaching/learning process is, in addition to the passion that the teacher must put in transmitting curiosity and interest, certainly the choice of the learning object.

The learning object must stimulate curiosity and interest. Jérome Lalande (Fig. 12.3), an astrophysicist who lived in the eighteenth century, in order to attract Parisians along the Seine to observe his telescope, extracted spiders from his snuffbox pretending he wanted to eat them, attracting the attention of people on what he was doing. Anyway, without reaching these extremes, we still need to find our "own spiders" to engage the students. In recent years, my "spider" has been a vibrating plate (Fig. 12.4) built with an old drill and an eccentric way to simulate catastrophic earthquakes. As many of my teaching tools, it was homemade, realized with cheap

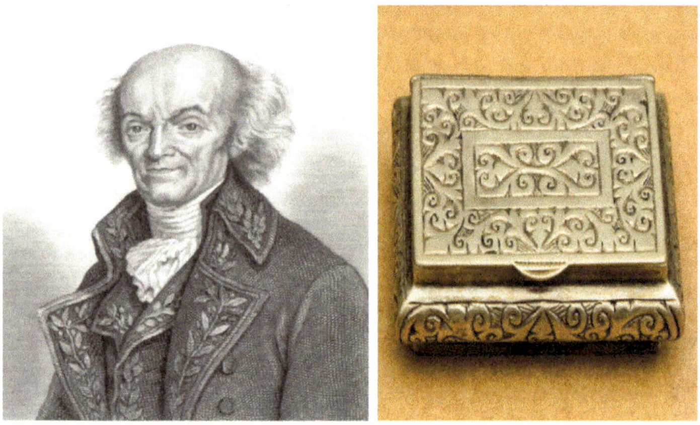

Fig. 12.3 Gerome Lalande and his snuffbox

Fig. 12.4 The vibrating plate

materials mostly coming from my house, but for this reason this vibrating table attracted the attention of other teachers.

They saw the easiness to recreate it at school, maybe as a school project, and also understood the potentials to use it to carry out a number of activities (and discuss interdisciplinary topics) with their students. The knowledge of a "geological object" or of a phenomenon of the evolution of the territory, both for its scientific interest, its beauty or for its dramatic aspect, in the case of a catastrophic event, represents an opportunity for cultural enrichment and awareness.

It is, however, necessary to use effective methodological approaches, able to turn the "geological object" into an opportunity to deepen in the scientific contents, an instrument to promote attention and interest and to develop skills, also through the use of known or unexpected connections: the mediator must be able to change the learning object into a starting point for insights, scientific and historical connections, emotions, all seasoned with a little passion to be transmitted.

It could be useful to remember that a geo-site is a physical element of the landscape that surrounds us for which it is possible to "define geological-geomorphological interest for conservation" (Wimbledon 1995). The landscape emergencies definable as geo-sites are generally characterized by a high scenic landscape value to which representativeness, didactic exemplary, rarity and scientific value are added. Important witnesses of the history of the Earth, geo-sites, are an opportunity to reveal to an audience hidden aspects of the geology of a landscape that still has much to teach.

In the educational context, geo-sites, whether they are environments of great scientific value, or they contain naturalistic or cultural values, or that are simply but significantly representative areas of natural hazards, can, therefore, overcome their natural function, and can be transformed into effective *educational tools*.

12.3 Educational Approaches

To become definitely learning objects, geo-sites need appropriate requirements: accessibility, scientific interest, richness of suggestions, but they require also the teacher's ability to transform an object, often not so captivating, as a rock or a landslide, into an educational tool through effective and intriguing educational paths.

The geological phenomena identified therein, in fact, are relatively interesting from a pure scientific point of view and do not need exceptional characteristics, but, as for more complex phenomena, some elements are, of course, necessary to understand the phenomenon:

- a brief geological framing of the context,
- an analysis of the dynamics, history and evolution, and of characteristics of the phenomenon.

Moreover, they must

- be easily accessible and free of dangers, even by students and tourists,

- or clearly visible, from far;
- be effective from an educational point of view,
- be rich in connections with other scientific aspects or other subjects,
- allow to propose foreseeable or unexpected connections.

From the educational point of view; some methodological educational approaches can be suggested properly depending on the type of pedagogical approach, geo-sites can be used as.

- *demonstrative educational objects*, in the case of a deductive approach, from which then recognize the correctness of what the teacher said: the steep slope can trigger landslides, etc.
- *case analyses*, in the case of the inductive approach, from whose observation and analysis it is possible to derivate principles and laws
- *reasoning cues*, especially when students are leaded by *driving questions* to reasoning, favouring the promotion of skills such as observation and research reports, development of investigative skills, but also modelling and extrapolation of rules and laws.

Finally, to make the activity more effective, less banal and engaging some suggestions could be

- to research and to propose all the elements than can solicit in-depth curiosity, wish to analyse, to research and to find connections;
- to be able to make the observation of an event, past, completed and otherwise unattractive in an effective, exciting educational instrument.

12.4 Analysis of Some Cases

To explain and to demonstrate the effectiveness of this proposal, three geo-sites have been chosen, all located in Italy, Aosta Valley, two of which within the Regional Natural Park of Mont Avic, but many others could have been tested and could effectively represent useful examples.

These three were chosen because the author has studied and has analysed them, with their different geological and historical aspects and they are now included in the National Land Registry of Italian *geo-sites*: https://www.isprambiente.gov.it/it/progetti/suolo-e-territorio-1/tutela-del-patrimonio-geologico-parchi-geominerari-geoparchi-e-geositi/il-censimento-nazionale-dei-geositi.

Finally, the author has tested them as learning objects with students or visitors, on several occasions.

The phenomena that characterize them are in fact well represented and allow numerous connections, predictable and unpredictable, with other phenomena of Earth Sciences, as well as other areas of the natural sciences and with strong relationships with human history.

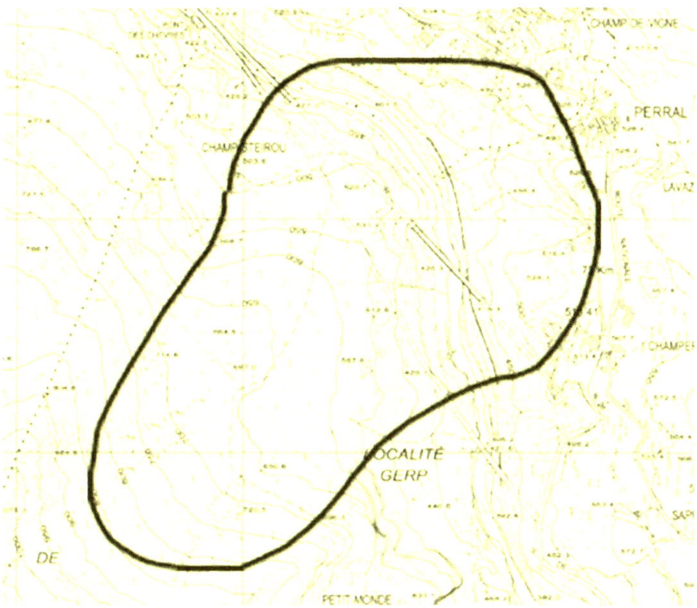

Fig. 12.5 The area of the landslide

12.4.1 The Case of the Mount Avi Landslide

12.4.1.1 Location and Short Description

The geological site of the landslide of the Mont Avi, also called Bellecombe, from the nearby homonymous place, is located on the right side of the central valley of the Dora Baltea in the Aosta Valley, between the towns of Saint-Vincent and Montjovet (Fig. 12.5).

It is a significant accumulation of debris derived from the collapse of a slope, a deep gravitational deformation of the slope (DGPV), which affected the slope of the Bec-Banquettes Mont starting from a height of 1800 m.

12.4.1.2 Geology and Geomorphology

The landslide develops in serpentines and subordinates meta basalts on an area of about 3.5 km^2: two sectors separated by a septum of rock were distinguished, a smallest one in the western part, a biggest one in the south-eastern part.

The glacial tongue in this stretch deviates to the south-east, compared to the initial outflow towards the east, energetically pressing the right bank of the valley.

At the time of the withdrawal, the foot of the slope, free from the thrusts of the glacial mass, slipped towards the valley floor; the body of the landslide shattered into

Fig. 12.6 The body of the landslide

large blocks, it rose on the opposite side and barred the valley of the Dora Baltea with an imposing dam (Fig. 12.6).

12.4.1.3 History

The river, barred by the dam formed by the landslide body, formed a large lake, occupying the entire valley floor at the current village of Saint-Vincent.

In the lake basin, typical lake sediments have been deposited, in particular, sands, which are now found in a place called Sabbie Bianche (White sands).

Later, the pressure of the lake basin broke through the dam, which overturned the mass of water in the narrow valley below, allowing the River to flow back into the valley and digging a deep, suggestive canyon.

12.4.1.4 Impact on Human History

Obviously, it is not easy to date the event: but it is certainly post-glacial, and an ancient Roman tomb on the opposite side of the landslide, indicates that the event is certainly prior to the Roman occupation, about 20 BCE.

We know that the breaking of the dam led to the flooding of the valley below, which remained marshy for a long time, conditioning the route of the Roman road.

Fig. 12.7 The canyon and the ancient road

The valley still has a deep incision that has made the passage of inhabitants and armies very difficult. Communications took place, from the Middle Ages, mainly using light signals, torches and bonfires, which from one castle to another connected peoples and nations.

The road that in the past connected the Low Valley to the Saint-Vincent basin and, therefore, to the rest of the region developed for many centuries on the opposite side.

Only in 1771, it was possible to build the current road, with futuristic engineering works for those years (Fig. 12.7).

12.4.1.5 Predictable Connections

The opening of the road naturally favoured communications between northern Europe and Italy.

In those years the vogue of the Grand Tour spread among young aristocrats, a long journey in continental Europe carried out by young men of the European aristocracy and it usually included many areas of Italy from the North to the South.

In Aosta Valley, as tourism started to develop in the Alps, especially for spas, such as Saint-Vincent, mountaineering and climbing in Courmayeur and other areas, an easier access contributed to the development of culture and the economy of the territory.

12.4.1.6 Unpredictable Connections

The landslide, partly shaped by the action of the river, gravity and covered by dense vegetation, preserves the typical forms of a landslide, despite having affected a deep depth of the slope: landslide body, detachment niche, main escarpment.

The landslide of the Mont Avi is an effective didactic model of how many landslides, ancient and recent, with the body have barred the valley below, forming a dam that has led to the formation of the lake.

12.4.2 The Case of a Glacial Fluvial Stratigraphic Sequence

12.4.2.1 Location and Short Description

The geo-site includes the valley bottom of the Chalamy Torrent, in Champdepraz, on the edge of the Mont Avic Natural Park. It develops on both sides of the valley floor, from q. 650 up to q. 1200 and covers an area of about 1 Kmq with a thickness of about 600 m (Fig. 12.8).

Fig. 12.8 The Chalamy Valley

It consists of deep and widespread fluvio-glacial deposits, with impressive erupting escarpments and cyclopean blocks in a precarious balance.

12.4.2.2 Geology and Geomorphology

The glacial fluvial deposits have a mixed origin and it is not always easy to distinguish them from the morainic, to which they pass upstream, and from the alluvial deposits to which they are progressively intercalated downstream.

They are deposits with heterogeneous lithological and granulometric characters, generally chaotic, consisting of a predominantly fine fraction, sandy and silty, of a grey-blue colour and of a coarse, heterometric fraction, with extremely variable dimensions (Fig. 12.9).

A stratification is recognizable in the outcrop, linked to the different sedimentary origin of the materials: we can distinguish the horizons coming directly from glacial deposition, consisting of blocks of considerable size, several metres in diameter, immersed and incorporated in a silty, non-stratified and chaotic matrix; and the horizons in which the action of melting water prevails, better selected, rich in coarse, rounded, flattened and iso-oriented pebbles and with more or less clear stratification in the sedimentation.

Fig. 12.9 Impressive cyclopean blocks

They are widely subjected to accelerated erosion phenomena, which take on "Badlands" forms, rarely present in the Alps, despite the absence of clay, due to the dissolution of the carbonate matrix, the lack of vegetation cover and the erosive and scaling action at the base of the Chalamy Torrent and its tributaries.

12.4.2.3 History

As a whole, the deposit results from the alternation of flooding events of the Chalamy river with more typically glacial events, which correspond to the numerous oscillations in the period of the glacial Wurmian expansion and the late Wurm withdrawal (e.g. Egesen Phase).

Of particular interest is the thickness of the deposits, as they originated from the dam, caused by the presence of the most powerful body of the Dora Baltea main glacier, which here could reach 800 m height, enriched mainly by the source of the Valle del Chalamy and the contribution of the Monte Rosa glacier, which flowed from the left along the Ayas Valley.

At least a dozen different sedimentary horizons corresponding to the different climate pulsations can be observed.

At present, the continuous passage of water is causing the slopes to rise continuously, leading to the withdrawal of the debris accumulation with a consequent widening of the valley groove.

12.4.2.4 Impact on Human History

On these deposits the erosive action is stabilized, their conoid is still partially active, as evidenced by the importance of the solid contributions that even in the past century (1952) caused catastrophic floods, which destroyed the village of Mure, at the top of the alluvial conoid.

The construction of the road in the 70s–80s, which replaced the old mule track, together with deforestation have triggered a process of instability, with widespread and channelled surface erosion, in correspondence of some side valleys.

12.4.2.5 Predictable Connections

The presence of many sedimentary horizons, derived from the alternation of glacial events, consequent to cold climatic pulsations, and of alluvial supply horizons, consequent to warm climatic pulsations, allow the comprehension of the complex phenomenon of the main ice retreat and of the lateral tributaries (Figs. 12.10 and 12.11).

It was a complex, long-lasting phenomenon, conditioned by many factors; the periodic motions of the Earth, on a large scale, the exposure, the microclines, the seasonal variations that interacted with each other.

Fig. 12.10 The stratigraphic sequence

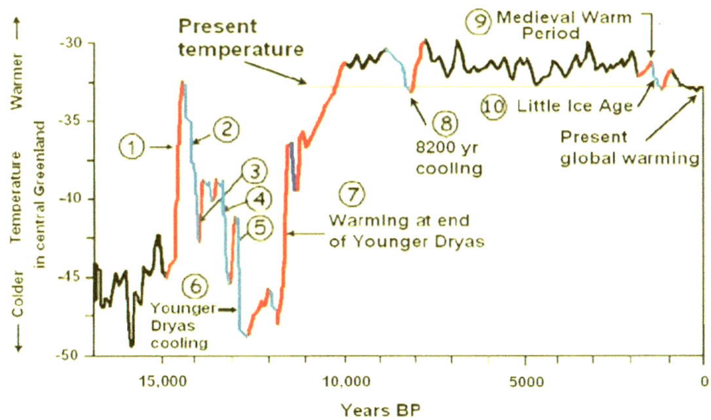

Fig. 12.11 Climatic oscillations

These observations can help to understand the mutual relationships of cause and effect and the dynamics of complexity are fundamental competences in all scientific disciplines, but particularly in Earth Science.

12.4.2.6 Unexpected Connections

The access to the valley of the Chalamy Torrent, due to the presence of landslide deposits, of unstable rocks, was, in the past centuries, very limited. Only a mule track could reach the high mountain pastures.

If, from an economic point of view, it was a limiting factor, which prevented the development of tourism, skiing, or craft and industrial activities, from an environmental and naturalistic point of view, the valley was able to preserve its complete integrity: uncinate pine woods, bogs rich in carnivorous plants, herds of ibex and chamois.

These are unspoiled landscapes of great beauty, of great scientific value, which have led to the establishment of the Mont Avic Regional Natural Park in the 1980s.

12.4.3 The Case of an Ancient Copper Mine

12.4.3.1 Location and Short Description

The mining site is located in the municipality of Champdepraz, on the right bank of the lower Aosta Valley at 1600–1700 m.

12.4.3.2 Geology and Geomorphology

From the geological point of view, the area can be framed in the context of the Zermatt-Saas s.s. unit, part of the lower and eclogitic portion of the Piemontese Ophiolite Zone (Dal Piaz et al. 2010).

The mine of Hérin appears in the form of lenticular bodies of sulphides and minor oxides, from massive to disseminated. It is a typical deposit of calcschists with ophiolites: within the materials of the series, prasinites, chlorites and serpentines, there are stratified or disseminated mineralization, often matching with intense plastic deformations with consequent shortening and enlargement of the lenses (Dal Piaz et al. 2010).

The particular mineralogical composition depends on a profound pre-metamorphic alteration of the original material (hydrothermal manifestations in an expanding oceanic environment). It is precisely in these lithotypes and in the associated quartzite schists with garnet $\pm$ carbonate that the pyrite cupriferous mineralization (impregnations, flattened lenticular masses, structurally) concentrates.

In general, the fundamental minerals (pyrite, chalcopyrite) are also associated with molybdenite and galena (Fig. 12.12).

The observed lithologies are mainly schist with garnets, with muscovite or dominant chlorite, with more or less quartz, carbonate and diffused chloritoid and glaucophane.

Fig. 12.12 The abandoned buildings of Hérin mine

12.4.3.3 History

The earliest evidence of exploitation of the Herin mine dates to 1703, although it is not excluded that it was already known in Roman times, or even pre-Roman times.

Exploited alternately for more than three hundred years, the deposit has played a significant role in the past, contributing largely to the regional production of copper and pyrite until the closure of the mines, which took place in 1957 (Castello 1995).

Various disputes between the families owners and the wrong choice in mining techniques heavily influenced the productivity of the mine and foundry from the seventeenth to nineteenth centuries. Many errors were committed in the exploitation of the mine: the direction of the excavation was wrong and inefficient machinery was used.

12.4.3.4 Impact on Human History

The Hérin mine represents a remarkable evidence/proof of industrial archaeology and a potential landscape and scientific attraction: the abandoned mining structures are inserted in the landscape offering fascinating views; the galleries make large geological sections available for observation and study; the acid drainage of the water causes the precipitation of spectacular metal deposits in the underground and the phenomena related to it are of geo-environmental interest (Fantone and Grieco 2013).

The history of the inhabitants had been strongly influenced by the mine: the men went to work in the mine and stayed the entire week in the houses located at 1800 m,

Fig. 12.13 Copper mineralization

went down on Sunday to go to church, walking each time over 1500 m in altitude. On the contrary, the women walked the same path to bring them food.

As in many other places, an entire community lived thanks to the presence of the mine, a weary, demanding and risky job: inside the mountain the mine developed on 13 levels, facing steep slopes and dangerous paths.

The material had to be transported downstream by hand, after a mechanical and chemical reduction (Figs. 12.13 and 12.14).

12.4.3.5 Predictable Connections

The Hérin mine is a typical lenticular and stratified deposit, associated with ophiolites, then in rocks that bear witness to the oceanic origin of this portion of the Alps, of the ancient Ligurian Piemontese basin, made up of gabbro and basalt altered in the hydrothermal environment and subsequently overthrown during the Alpine orogeny (Castello 1995).

It is not easy to understand the geographic and temporal dimension of the movement that these rocks have undergone, a transport that has taken them from the bottom of an ocean, albeit small but with characteristics of basin in expansion due to the presence of dorsal rocks, to travel a difference in height not less than 4000 m and a distance of not less than 1500 km.

It is a complex dynamic, difficult to understand, but whose study helps students to approach the movements of global tectonics.

Fig. 12.14 The "green river"

We should not forget that this submarine environment, with its hot springs rich in sulphur, saw the appearance of strange organisms, the tubular worms, precursors of life.

12.4.3.6 Unexpected Connections

This territory, whose wilderness justified the creation of a natural park is surprisingly the result of great transformations.

During the 1700s, the Napoleonic inspectors were forced to limit the period of operation of the melting furnaces, particularly in the presence of sulphides, because their reduction produced sulphur fumes which made the air unbearable and damaged the crops of the inhabitants of the area.

For the two centuries that saw the activity of the mine, the furnaces required fuel with a high calorific value to reach the melting temperatures: in the absence of coal mines, the charcoal was produced using the extensive forests present in the area.

The resulting deforestation, the disappearance, for long periods, of the vegetation cover, was the cause of widespread instability, landslides and avalanches, which affected the valley floor; a framework of instability that changed only with the use of blast furnaces and the subsequent closure of the mines.

12.5 Evaluation

Generally, students and visitors appreciate the path that illustrates the different geo-sites.

Each location had, predictably, a different response: the more impressive the site was, like the "green river" in the Hérin mine, or impressive, like the steep slopes of the fluvial glacial deposit, the more the answer was positive.

But when, with the students, we wanted to deepen the contents and the relationships acquired, the predictable and unexpected connections are the ones that left the most sign.

Overall, the objectives that were given to us:

- to understand the geological value and complexity of the phenomenon, from the scientific point of view;
- to recognize the impact that the phenomenon has had on man and his history, or on the natural context,
- to identify the numerous relationships of cause, of complexity that the different contexts contained, and finally.
- to arouse the curiosity and the desire to apportion themes that at first impact are not always grasped; the result, overall, it is more than satisfactory.

12.6 Conclusions

The examples presented, although simple and in some cases trivial, can be a useful example to prove how a geological form, a territory, no matter what the scientific value and the geological meaning, can be effectively used as learning objects.

If the public (students, tourists, visitors) succeed through these "objects" in understanding the cultural, educational, historical significance of a phenomenon, then the object has done its job.

The purpose is always to promote and extend the sensitivity to the Geosciences in the various fields in order to change the usual image that seems mainly devoted to the study of the rocks and to have a voice only during catastrophic events. It wants to promote, with a renewed involvement, a widespread culture of the diverse and complex areas of the discipline, climate, water, security, environmental quality, and living standards.

It wants to seek a different focus on the natural phenomena, responsible for hazards and thus risk, but also complexity and natural resources; it wants to propose the study of the Earth's geological history and its evolution, with an innovative approach that remembers us how the events that have affected the past are the elements that allow us to understand the future (Occhipinti 2014). The numerous catastrophic events which affected the country in recent years show that this sensitivity, as citizens, as users of the natural resource, such as local technical competent in planning, monitoring, safety measures, prevention, demonstrate that such a culture needs to be formed from the base, starting from primary schools, involving all levels of educations, students and teachers, and then families.

We need to pay more attention to the spread of culture and information, changing the common perception of catastrophic phenomena that are instead natural, even if subject to periodic accelerations and not physiological concentrations due to human actions, such as a not always prudent urbanization and planning territorial, or global, such as global warming.

The development of these themes can be a valid support to construct an epistemological model of the Earth Science, that gives strength and consistency to the geological culture, absolutely necessary to enhance a widespread and deep-rooted culture of geo-sites and to a perception of them as examples of the dynamics of the Earth.

References

Castello P (1995) Inventario Dei Minerali E Delle Mineralizzazioni Del Parco Naturale Del Mont-Avic. Monografia, Ed. Parco Regionale Del Monte Avic

Dal Piaz GV, Pennacchioni G, Tartarotti P, Carraro F, Gianotti F, Monopoli B, Schiavo A (2010) Carta Geologica D'italia Alla Scala 1:50.000, Foglio 091 Chatillon

Fantone I, Grieco G (2013) La Miniera Di Herin: Uno Spaccato Storico E Geologico Nel Paesaggio Valdostano. PhD research

Occhipinti S (2014) Models and guidelines for more effective tools and paths in an active teaching-learning in earth sciences: looking for a unifying principle. PhD research

Chapter 13
Geoheritage Sites and Scope of Geotourism in Land of Chhattisgarh, India

Suchita Tripathi

Abstract Geoheritage sites are scientifically, educationally and culturally significant for mankind. These include the areas like rock sites, cave sites or shelters, waterfalls, springs, lakes, minerals or mining sites having Outstanding Universal Value with notable geological features. Besides having great opportunities for tourism in areas of Eco-tourism, Bio-tourism, Tribal tourism, Ethno-Tourism, Village tourism, Folk tourism, Medical tourism and Archaeological tourism, Chhattisgarh state has also a wide scope of geotourism. Chhattisgarh is the abode of a large number of geological sites. The Kutumsar cave and Chitrakot Waterfall in Bastar, Bailadila iron ore mines in Bastar, coal mines in Korba, Raigarh, Raipur, etc., Tatapani Hot Water Lake in Balrampur and many other sites in Chhattisgarh are the good examples of natural geosites. The overexploitation of natural resources for the sake of heavy industrialization and mining activities will lead to the extinction of these sites in Chhattisgarh in the near future. Therefore, it is high time to aware and educate the local people about the scientific, religious and educational importance of these geoheritage sites. The government should take more and more initiative in promoting and popularizing the geotourism in these areas and should make geosites protection law and geoethics for the use of geoheritage sites. Promoting geotourism with proper laws and ethics will not only lead to the conservation and protection of these geosites but also help in economic upliftment of the local people.

Keywords Geoheritage site · Scope of geotourism · Conservation

---

S. Tripathi (✉)
Department of Anthropology and Tribal Development, Guru Ghasidas Vishwavidyalaya, Bilaspur, Chhattisgarh, India
e-mail: suchi.mnt@gmail.com

R. B. Singh et al. (eds.), *Global Geographical Heritage, Geoparks and Geotourism*, Advances in Geographical and Environmental Sciences,
https://doi.org/10.1007/978-981-15-4956-4_13

13.1 Introduction

13.1.1 Geoheritage Sites

Geoheritage studies are very old and have been practised since the eighteenth century but the scientist had rediscovered such studies within a period of 25 years (Reynard and Brilha 2018) for geoconservation and sustainable development. 'Geoheritage' refers to those extra valuable and extraordinarily important elements of the geodiversity or natural heritage of 'Outstanding Universal Value' (Migoń 2018) that are considered for conservation and protection (Brilha 2018a, b; Gray 2018) by 1972 World Heritage Convention of UNESCO. Geodiversity that supports elements of geological importance and their conservation describes characteristic features of the abiotic ecosystem along with its processes like the landscapes formation, minerals, geomorphological sites and geological and hydrological features (Gray 2018).

The places of geological importance having significant sociocultural, historical, educational and scientific value are called as "Geoheritage sites". Such sites with geological and geomorphologic features are meant for geoconservation and geotourism (Colin et al. 2018), for protection of the natural environment and cultural landscapes (Asrat and Zwoliński 2012), at international, national and subnational levels (Gray 2018). "Geoheritage sites" is a term applied to all geological sites like landscapes, distinctive rock or mineral types, cave shelters (Biswas 2010) mountain ranges, lakes or springs, unique fossils or paleontological evidence (Page 2018) that are scientifically or educationally significant. Many unique and distinctive geoheritage and geomorphological sites help in understanding the earth's history and the relationship between cultural and biological heritage and are developed as popular tourist destinations (Coratza and Hobléa 2018). In this way, they not only help in increasing the cultural and aesthetic value of the place but also help in economic benefits for local communities (GSA 2012; Gordon et al. 2018).

Scientifically important geological sites play a very important role in enhancing scientific knowledge among public and school students and also used for recreation and economic support to the locals. They provide knowledge about natural processes, natural hazards, climate, evolution and origin of life, mineral, sources of energy, origin of the earth, etc. Geoheritage sites can be small like a road cut or a rock stone, a mineral, mining site as well as large and extensive like Boiling Springs in Pennsylvania, USA, and Grand Canyon and Yellowstone National Park. Most of the World Heritage sites are mountainous landscapes, volcanoes, while the 'classic' geological sites are very rare, including fossils, stones and stratigraphy. These sites may be public or private property. But regardless of their size and ownership, the need is to protect them from land erosion, urbanization, over-exploitation, etc.

For maintenance and protection of geosites, conservation strategies are essential. In the United States, geological areas like World Heritage Sites, historic places, monuments and national parks are managed by land and forest management agencies to protect them for future generations. Many geomining heritage sites all around the world because of their geological importance are also affirmed as geoheritage sites (Mata-Perelló and Carrión et al. 2018).

Besides the in situ heritage sites, the different valuable objects like fossils, stones, rocks pieces, tools of stone and bone, etc., gathered by museums also provides knowledge to all (Weve and Guiraud 2018). Both culture and geoheritage influence each other. The objects or elements of geological heritage and geomorphological phenomenon sturdily persuade sociocultural institutions, traditions, beliefs, etc., while culture influences the willingness of people to protect geosites. During the last decade, geocultural heritage has been intensively investigated in areas like urban heritage (Chylińska and Kołodziejczyk 2018), literature, arts (Reynard and Giusti 2018).

13.2 Geopark

Geopark is a global designation given to a number of culturally and historically significant geological heritage sites (Brilha 2018a, b). Geoparks include geosites, meant for conservation and protection of geological heritage. In many cases, geoparks are not nationally protected areas. By providing opportunities for geotourism, conservation and education, geoparks help in sustainable rural development (Brilha 2018a, b; Singh and Anand 2013) by transferring knowledge of geosciences to common people, connecting local people to tourism, economic activities and geological research, etc. (NPS 2015; UNESCO 2006). It is opined by many organizations that rural areas are most suitable places for establishing Geoparks (Zouros and Martini 2003) as it reduces the rate of migration to urban areas and promotes sustainable rural development (Farsani et al. 2011).

13.3 Geotourism

Since human origin on earth, man has a great relationship with their environment for existence on earth, without which one cannot imagine their survival. The whole culture of human being is dependent on natural surroundings in one or another way. Out of many institutions which human has created with their interaction with the environment, one is tourism sector that uses cultural heritage and natural surroundings for various activities. Geotourism and geoheritage have close relationships. Geoheritage knowledge is promoted with the help of geotourism and related practices. Geotourism may have either good or bad impacts on geoheritage (Newsome and Dowling 2018). Geo-tourism is a new movement in the tourism market with a special interest in the preservation of sites of geological importance by popularizing geosites as places of cultural, educational and aesthetic significance. It is a new way to make travellers to know about their natural resources and science hidden in them, to make them interact with local communities, their culture and lifestyle and above all a way for sustainable development of communities and area. In this way, geotourism is a branch of tourism with an approach of sustainable development of society. The knowledge of geotourism can be helpful for the students having an educational background in subjects like geography, geology, tourism, etc. (Farsani et al. 2014). The three interrelated aspects of sociocultural, economic and environmental sustainability make geotourism as a way for sustainable development (Cottrell et al. 2007). Geotourism and geoparks help in promoting the economy of an area (McKeever and Zouros 2005) as local goods and manufactured items and products were purchased by the travellers (Frey et al. 2006). Local communities get job opportunities through several mediums like restaurants, local good market, geotour guides (Robinson and Roots 2008), lodging, geosports, museums, bakeries, etc. In one way, it increases the economy of the local people, while in another way, it promotes cultural identity of the people to other places, scientific, geological knowledge and recreation (Alexandrowicz 2006), new experiences (Poon 1993), exchange of emotions, expressions, language and public awareness.

Besides all these positive impacts of Geotourism, there may be negative impacts of it in lack of proper planning. Often due to ineffective management of geological attractions or geosites like caves, springs or other, geotourism led to a great threat to the geosites and the biodiversity of the place that in turn affect the ecological balance of the area (Newsome and Dowling 2018). Usually, due to uncontrolled tourism and anthropogenic activities, purposeful damage by visitors, ignorance of common ethics, illegal collection of cave resources, hunting, mining, pollution, deforestation and urbanization, the fragile ecosystem of caves (Biswas 2016) and other geosites is disturbed or destroyed.

13.3.1 Geological Sites and Scope of Geotourism in Chhattisgarh

India is a land of wonderful landforms and landscapes. There is immense geodiversity with unique geographical structures and variable climate. Outstanding geodiversity and extraordinary geoheritage of Indian subcontinent are very well recognized all over the world. These wonderful landforms should be given attention to conserve our unique heritage for meeting various needs of present and future generations of India (Kale 2014).

Chhattisgarh is the state having large opportunities for tourism in every sector whether it is ecotourism, tribal tourism, village tourism, archaeological tourism and geotourism. The state has many natural and historical sites of geological importance including natural parks, caves, rock shelters, falls, springs, coal and iron ore mines, folded mountains, etc. Some of the sites of geological importance are described here.

13.3.2 Geomorphological Sites

13.3.2.1 Kanger Valley National Park, Bastar

Plateaus, streams, caves, valleys, mountains, with huge biodiversity, make the identity of land of Chhattisgarh. Valleys like Kanger, Bastarnaar, Darbha, Keshkaal are situated in the Bastar plateau. Kanger Valley National Park, a major ecotourist spot for Chhattisgarh, with a lush green scenic beauty is abounded with all types of geological and geographical features at one place. Kanger Valley stretched over a large area being the abode of uncommon species of plants and animals is also home of some of the most amazing natural caves. Due to these interesting and unique features of the park, this park is proposed by the state government to designate it as a heritage site. More than a half dozen of caves are founded by the forest department and local people but tourists can visit only three of them, namely Kutumsar Cave, Dandak Cave and Kailash Cave. A total of 553 species of flora are reported in the park, out of which 43 species are rare and 12 are new of its kind in Chhattisgarh. Kanger Valley is one of the last patches of intact forests remained in the peninsular region. One of the unique features of it is that this is situated in an ecotone region where the southern edge of sal forests and the northern edge of teak forests meet and overlap. Diverse flora includes tall trees with spotted epiphytes, bamboo trees, climbers, medicinal plants and many varieties of herbs and shrubs are found in the park. Varied wild fauna includes different species of mammals, reptiles, amphibians, fishes, insects and birds, etc., like wild boar, chital, jackal, wolf, leopard, wild dog, sloth bear, civet cat, bats, spiders, butterflies, fishes, snakes, crocodiles and many kinds of birds."Binturong", a rare civet giant squirrel, and myna of Bastar are important among them.

13.3.2.2 Kutumsar Cave in Bastar (Earlier Name was GopansarCave Means Hidden Cave)

Since prehistoric times, Caves have been the cultural and religious centres (Amarnath, Kailash and Vaishno Devi temples exist inside caves) in India. Stalagmites formed in caves are worshipped as Lord Shiva by Hindus. Prehistoric human being during Middle Paleolithic and Upper Paleolithic periods were cave dwellers. Art was emerged first time in caves during these prehistoric periods in the form of cave art. This was the first interpretation of human's feeling and imagination in the form of art. Bhimbetka cave is an example of wonderful cave art in India. Caves are an important place for geotourism (Biswas 2016).

Kutumsar or Kotumsar cave (longitude: 18°52′09″ North; Latitude: 81°56′05″ East) is situated at an altitude 560 m (Biswas 1992) near Kutumsar Village. This cave was known to people since British time, but the first serious study and exploration of cave chambers was made by professor Dr. Shankar Tiwari (Geography) during the 1950s, while the first systematic mapping (in 1980s) and complete biodiversity description (2010) was done by Dr. Jayant Biswas.

Entry is followed by a narrow, twisted tubular path. The total length of cave is 150 m (Biswas 1992). Kotumsar cave has two zones; a transition zone and a deep zone (Vermeulen and Whitten 1999; Biswas 2009, 2010). This is not only an interesting place for ecotourism and geotourism but also for biotourism (Biswas 2010). Besides this, the cave has religious or mythological significance among the local people, especially tribals since ancient times (due to formation of some speleothems which look like deity) and travellers.

The main attractive features of these caves are biodiversity (cavernicoles) and speleothems (stalagmites, Stalactites, dripstones) showing wide panoramic views. A big speleothem (stalagmite) in one of its chambers is a major attraction. Many species of cave-dwelling animals like bat (Rhinolophusrouxii, Hipposideroscineraceus), frog (Hylaranamalabarica), fish (Nemacheilusevezardi, crickets and plants were discovered inside Kutumsar cave; of them, the most interesting and unique species are blind and albinic fish and new species of Gryllidae (crickets). Indoreonectesevezardi (then Nemacheilusevezardi), a species of blind fish and Kempiolashankari (named in honour of Dr. Shankar Tiwari), a new species of crickets were discovered by a Biospeleologist, Dr. S.M. Agarwal, with the help of the British Museum (Biswas 2015). Some exciting experiences and fascinating features of Kutumsar caves as told by Professor S.M. Agarwal, a Pioneer Biospeleologist during his first journey in 1956):

"Getting inside the cave was in itself a thrilling experience; the twilight zone was very narrow then. One had to descend with the help of ropes. Inside the cave it was all clay moist soil and one had to move on all fours. It was very slippery. The stalactite and stalagmite rocks and projections with musical notes when touched with stick or some other object were all fascinating. Water dripping from the roof and feeding the small pools inside the cave and suddenly a cluster of those gryllids appearing from nowhere were all very exciting. I enjoyed the experience when I went into the cave for the first time. I went perhaps half a dozen times more. In one of these visits, it started raining and it was a most exciting event that we saw pools of water entering into the mouth of the cave. This suggested that the normal pigmented variety of fish got into the cave through these water pools. Those that got in, could not come out and with very long passage of time began to lose their epigean traits and began to lose pigmentation and some very few even lost all pigmentation and became albinic and their eyes got covered with fold of skin. These we called blind fishes. We studied the eyes thoroughly and concluded that the eyes only regressed. Could the high concentration of calcium in water pools or alternatively, complete darkness affecting melatonin titer in the body be the cause for loss of pigmentation and for the regression of eye, it is an enigma. I deeply regret I could not get the work completed to fruition." (Biswas 2015).

Existence of phenomenon of colour change in troglophilic cavefish Nemacheilusevezardi (Day) along a circadian time scale is found disturbed in natural conditions in a subterranean cave (Pradhan and Biswas 1994). Cave organisms developed adaptation for living in the cavernous atmosphere (Biswas 2010). Later, several other cave species (mostly arthropods) and various bacteria were also discovered by taxonomists. Two species of roosting microchiropteran were found to roost together in Kutumsar (Biswas and Kanoje 1991). Rhinolophusrouxii and Hipposideroscineraceus occupy the dark and light zones according to their adaptational characteristics (Biswas et al. 2011). While in another study only one species of bat, i.e. Hipposideroscineraceus had been reported from Kotumsar cave (Chakraborty 2008). During early days, due to being unexplored or untouched regions, having small disturbances, landscapes, especially cave, were proved good habitat sites for bats.

13.3.2.3 Dandak Caves

Cave ecosystem is unique due to the presence of continuous darkness, high humidity and constant geophysical factors. Dandak cave is very difficult to reach and that is why this cave is not allowed for tourists. The natural stalagmite and stalactite formations and cave-dwelling mammals are unique features. Its total length is about 360 m and has two identical compartments linked by undersized, twisted channels. There is a big entrance followed by a large outer room, a narrow inner chamber with a hole and some small thin cavities. Dandak cave is enlightened by sunlight during the daytime. Due to these features of cave chambers, this cave got name Dandak as one needs to go on knees to go from the first to the second compartment (Biswas et al. 2011). Due to different structure and several geophysical factors, the two cave chambers present distinctive ecologies. The groups of mammals in the outer chamber of cave are completely different from the group living in the inner one. The organisms observed in the outer chamber are Rhinolophusrouxii (Subtroglophile), Viverriculaindica Hydrophylaxmalabaricus (Subtroglophile), Duttaphrynusmelanostictus (Eutroglophile, Rainy season), Heteropodavenatoria (Eutroglophile, round the year), Giant millipede (Trogloxene, Round the year). In inner chamber lives Hipposideroscineraceus (Subtroglophile, Round the year), Hystrixindica (Subtroglophile, round the year) and Kangerosithyriskotumarensis (Troglobite, Rainy Season).

A small colony of Rufous Horseshoe bats (Rhinolophusrouxii, 50–60 number) inhabit cave's outer chamber (Biswas et al. 2011; Chakraborty 2008). Few solitary bats were also visible ashy leaf-nosed bats (Hipposideroscineraceus) are found to be highly sensitive to human disturbance. Small Indian Civet (Viverriculaindica) is a regular visitor. Viverriculaindica as well as Hystrixindica help in maintaining the cave ecology. The cave cricket (Kempiolashankari) is the most fascinating species, abundantly found in both the chamber of the cave. It helps in binding the whole ecosystem of cave together. Indian (Crested) Porcupines, anurans (Fungoid Frog), Guano moth, Giant crab spider or the banana spider etcetera are other animals found in the cave.

13.3.2.4 Kailash Caves, Jagdalpur

Discovered in 1993, Kailash cave lies in one of the hills in Koleng forest range. Inside the cave, there is a large number of conical karst structures. It is named by local peoples as “KailashGufa” because some of the conical karst structures look like “Shivling”. Kailash and Kotumsar Caves are about 2 km deep. the cave is nearly 1000 ft long and 120 ft deep. Budding structures of stalactites and stalagmites can be seen 100 ft down in cave are most interesting features of the cave. These are generally formed in caves or caverns. The cone-shaped structures hanging from the roof are known as stalactites. Similarly, on the floor of cavern, inverted conical pillar-like structures known as stalagmite, develop in the place where limewater drops. A column called “dripstone” is formed when a stalactite and a stalagmite meet. Geode, a deposit is also found in cavities of rocks. The whitish conical structures present a world of fantasy and magic for tourists and are called by name of “Jhumar”, dripstones as “elephant nose” and “Bhim ki Gada”. Kailash Lake, Bhainsa-Darha and Crocodile Lake are other important features of the valley.

13.3.2.5 Other Geological Cave Sites Are

Limestone caves, namely Kailash Gupha Caves and Khuria rani exist near Jashpur in Chhattisgarh.

13.3.2.6 Chitrakote or Chitrakot Waterfalls

The Chitrakot waterfall is a natural waterfall in Bastar District, Chhattisgarh. It is 30 m in height (Chhattisgarh Tourism Board 2015; Kale 2014). It is the widest fall in India (Singh 2010) and especially during the rainy season, it covers a broad area (Puffin Books 2013). During the year, water flow of fall varies, the highest flow in a single expanse forms during the rainy season and makes a horseshoe-shaped ravine (Menon 2014; Puffin books 2013) and due to flood, silt is found in water flow (Jagdish 2013; Terryn 2011). Teerathgarh Falls is another big fall in the Kanger Valley (Menon 2014). Quartzitic sandstone is present when the Indravati River rises and flows downstream the valley but as reaches near Chitrakot Falls, geological formation changes to Archaean granite and gneisses (Sharma 2000). Due to the presence of geological features, it is recognized as geomorphosites. This site is also considered as a prehistoric site as a complex of Mesolithic sites has been found in debitage around Chitrakot in Bastar (Cooper 1983). Many small Shiva Lingas are found (The Economic Times 2012). In the stormy part of this pond, adventure sports are done. In the low flow season, paddle boats, bathe and swimming is permitted. In 2003, this site is found suitable for eco-tourism.

13.3.2.7 Amritdhara Waterfall

Situated on the Hasdeo River near Manendragarh, Amritdhara waterfalls represents an untouched geomorphological wonder in which the river falls from a height of

30–35 m. The morphology and dense vegetation make this waterfall similar in look to that of prehistoric times.

13.3.2.8 Folded Mountains in Hasdeo Valley, Korba District, Chhattisgarh

Folds and faults can be seen in mountains in the Hasdeo River valley in Korba District.

13.3.3 Geo-Paleontological Sites

13.3.3.1 Fossil Park

A fossil park known as Marine Gondwana is situated at Sarguja district of Chhattisgarh. There is a unique exposure of fossiliferous marine Permian (240–280 Ma) rocks of the Talchir Formation belonging to Gondwana Super Group. It covered a length of about 1 km upstream to the confluence of Hasdeo River and Hasianala near village Amakherwa in Manendragarh. The marine fauna is dominated by pelecypods and crinodal stems along with bryozoans are numerous in these rocks (Venkat Ramaiya and Hussain 2015). The park is not properly guarded and the fence erected by the forest department is broken and deteriorated.

13.3.4 Geomining Sites and Minerals

The state of Chhattisgarh has rich mineral resources. There are 28 different types of major minerals are found like tin, iron, coal, bauxite, diamond, dolomite, limestone, etc., Chhattisgarh is a power hub because of a large number of coal deposits. All the tin ore and a large amount of good quality iron ore are found in Chhattisgarh.

13.3.4.1 Iron Ore in Bastar

In Bastar, large deposits of iron ore are present in Bailadila hills. Because of the shape of hills which looks like ox's hump, native inhabitants called it Bailadila. An approximate quantity of 3000 lakh tonnes iron ore deposits is present in 14 iron ore deposits of Bailadila. One of the world's best quality iron ores is exported from Bailadila mines to Japan. In Bailadila area, iron ore is available in significant amount (Info portal of geology 2003).

It was Mr. P. N. Bose who did first geological mapping at this region during 1898–1900. Later for 6 years, another mapping was done between years 1932 and 38. Viewing the mineral potentialities, in December 1958, a circle of Geological

Survey of India was established and during late 1961 NMDC entered in the area. Prof. Euemura of Japan made the commercial discovery of Bailadila during 1955–56, and after signing an agreement in 1960, NMDC in 1964, approved the establishment of mine plant in region in collaboration with Japanese steel mills.

13.3.5 DhalliRajhara

This iron ore site is located in the Durg district. This is an important deposit of hematite ore. The iron ores are found to be deposited in KondeKasa site.

13.3.6 Kalwar-Kauchar Region

Kalwar-Kauchar region covers 16 iron ore deposits which have been discovered by Geological Survey of India. Kauchar Dongri and DondiLohara are the important sites of iron ore deposits.

13.4 Important Minerals in Chhattisgarh

The land of Chhattisgarh ranks second in India as it covers a large amount of mineral resources. Of total mineral produced in India, 10–15% is contributed by the state. The important mineral found in Chhattisgarh is Coal (52,169 million tonnes), Iron ore (2731 million tonnes), Bauxite (148 million tonnes), Limestone (9038 million tonnes), Dolomite (847 million tonnes), Tin ore (32.62 million tonnes), Diamond (1.30 million carats), Corundum, Gold (0.1–0.3 gm/m^3), Quartzite (26 million tonnes), Tin metal (14,449 tonnes). Other minerals found in Chhattisgarh are Aluminium, Uranium, Alexandrite, Graphite, Garnet, Quartzite, Fairclay, etc. (Directorate of geology and mining 2017).

Gold—Gold deposits are found in the Sonakhan area of Chhattisgarh. In Kotri rift zone, gold deposits are found.

Diamond—In Ib–Maini Rivers of North Chhattisgarh, placer diamonds are found. Diamonds were also discovered in Behradih and Payalikhand areas of Central Chhattisgarh.

Coal—The state is the second largest producer of coal in India. Korba, Raigarh, Surguja and Koriya districts are the major producer of coal in the state. Large production of thermal power is done by NTPC Korba and NTPC Sipat Bilaspur.

Bauxite—Mainpat, Samripat, Jamirapat plateau sites in Sarguja, Pandrapat, in Jashpur, Keshkal area in Kanker district, Asna—Tarapur area of Bastar, Bodai Daldali and some other sites of Kabirdham district are the major bauxite-mining sites.

Limestone and dolomite—Throughout Chhattisgarh, there are many sites of limestone and dolomite deposits. Raipur, Bilaspur, Janjgir-Champa, Durg, Raigarh and Jagdalpur are some of these sites.

Base Metal—In Surguja, Koria, Durg and Rajnandgaon districts of the state, Base metals like lead, copper, silica, etc.,are reported.

Tin Ore—Whole Tin ore production in India is made from Chhattisgarh. The state is the only producer of tin ore.

Corundum—It occurs mainly in few places of south Chhattisgarh.

Lepidolite—Lithium bearing pegmatites were identified in few areas of south Chhattisgarh.

13.5 Geothermal Sites

13.5.1 Tatapani Geothermal Reservoir

Geothermal energy is a renewable and clean energy resource and is the largest source of energy today. Geothermal energy is mainly found in structures below the earth and in water or steam saturated hot rocks. This energy can be utilized for electricity production, heating purposes, fish farming, bathing, etc. (Dickson and Fanelli 2003).

Tatapani geothermal reservoir is situated near Kunkuri in Balrampur district of Chhattisgarh. This is an attractive hot water spring well known for its medicinal properties. Tatapani spring has water temperature up to 50–60 Degrees. Water is alkaline in nature (7.9–8.9 pH), containing helium, carbon dioxide, oxygen and nitrogen (Venkat Ramaiya and Hussain 2015). Tatapani would be the first geothermal power project in India (The Hitvada 2016). Various studies had been made to assess the geothermal nature of reservoir (The pioneer 2017). To estimate the power generation potential, a series of surveys at a number of geothermal sites in India were conducted and among them, the Tatapani geothermal field is identified as likely the first site and logistically best location among other identified areas in the country for a geothermal power plant in India. The estimated potential to produce electricity by the Tatapani geothermal field is found as 17 Megawatts by the Geological Survey of India.

13.6 Challenges and Suggested Measures

Geoconservation, that is aimed with the protection of geodiversity and geoheritage for science, education and sustainable development of society and environment has evolved towards a broader discipline. Geosites have wider intrinsic, aesthetic, cultural and ecological values. The functioning of entire ecosystem is maintained by Geoconservation. But, this is not so easy to implement in practice. There are so many challenges. The society or the people do not know much about their geo-heritage and biodiversity. Second thing is that the natural processes, calamities and increasing human activity like urbanization and industrialization, etc., in such areas has led to huge loss and destruction of these resources. Many of these sites give us glimpses of our ancient prehistoric geological formations, earth structure and biodiversity of that time. But by unethical human activity can lead to loss of these sites forever and will never recover (Singh and Anand 2013). These are non-renewable resources (Bruschi and Coratza 2018), therefore, it is high time to pay much attention to protection and conservation of these heritage sites. Firstly, there is needed to make people aware about geoconservation of geosites by improving the scientific basis of geoconservation. The second need is to link or interrelate geoconservation with natural conservation and sustainable development. And third is to integrate geoconservation with protected area planning and management (Gordon et al. 2018). Historical urban cities should be promoted for urban geotourism (Chylińska and Kołodziejczyk 2018) Environmental Impact Assessment (EIA) since the end of 1960s is practised in many countries for geoconservation by integrating EIA with geoconservation but did not get good results (Bruschi and Coratza 2018). Geoconservation strategies often get failure due to the lack of communication with the public. Scientists often do not know how to successfully communicate with the public about geoconservation. But if they do not communicate, there will be a risk to both visitors and geoheritage. Therefore, random information and knowledge should be spread to visitors nor the designer intended as there is a chance of not getting interested in the information by the travellers (Macadam 2018). Due to difference in landforms, size and shape, proper protection law same for all the sites is a challenge. Excessive use of land causes the risk to geoheritage, therefore, proper law and protective measures for geoconservation should be taken and applied by the government with the help of local peoples (Singh and Anand 2013). During recent years, new digital knowledge and techniques are found to be useful for geoconservation. In order to prevent natural hazards, virtual models and digital archive of the natural environment and vulnerable sites can be used. By using digital technologies, geoheritage management is possible (Cayla and Martin 2018). Mapping geographical heritage is important. In the case of India, local initiatives play a key role in building consciousness regarding geoheritage and conservation (Singh and Anand 2013). In the case of Chhattisgarh, in spite of being home of so many rich and wonderful geological sites and natural beauty, the people do not know much about these natural treasures and their geo-bio heritage. Therefore, the tourism department should promote more and more eco, geo and bio tourism in Chhattisgarh. Schools and colleges should send their students for excursion or educational tour in these places. In this way, we can generate science and knowledge related to these geoheritage sites and make them aware of their natural environment and its protection and preservation for sustainable development. By promoting these heritage sites, the economy of the native inhabitants of the place, especially tribals will also grow. By making proper laws or legal framework for geoheritage protection, maintaining geoethics, awaring students and common people about the treasure of knowledge in form of these geosites and by restricting many anthropogenic activities in these sites, by engaging local people as caretakers of the site and by involving them in planning policies for the development of site, we can preserve geoheritage for future generations.

References

Alexandrowicz Z (2006) Geoparki – nowewyzwaniedlaochronydziedzictwageologicznego. Przeglad Geologiczny 54(1):36–41

Asrat A, Zwoliński ZB (2012) Geoheritage: from geoarchaeology to geotourism. Editorial. Quaestiones Geographicae 31(1); Bogucki Wydawnictwo Naukowe, Poznań, 5–6

Biswas J (1992) Kotumsar Cave ecosystem: an interaction between geophysical, chemical and biological characteristics. NSS Bull 54:7–10

Biswas J (2009) The biodiversity of Krem Mawk hyrdop of Meghalaya, India, on the verge of extinction. Curr Sci 96:904–910

Biswas J (2010) Kotumsar Cave biodiversity: a review of cavernicoles and their troglobiotic traits. Biodivers Conserv 19(1):275. https://doi.org/10.1007/s10531-009-9710-7

Biswas J (2015) Feature: Professor S.M. Agarwal, a pioneer biospeleologist from Free India, ambient science, 02(1):53. https://www.caves.res.in/

Biswas J (2016) Caves hold clues to evolution and climate change. Why is India dragging its feet over framing a law to protect the fragile structures? Caving in, down to earth. https://www.downtoearth.org.in/news/environment/caving-in-52339

Biswas J, Kanoje RS (1991) Seasonal and photoperiodic control of the activity and daily exodus of a colony of a cave dwelling bat. J Ravishankar Univ 3–4:7–13

Biswas J, Shrotriya S, Rajput Y, Sasmal S (2011) Impacts of ecotourism on bat habitats in caves of Kanger Valley National Park, India. J: Res J Environ Sci 5(9):752

Books P (2013) The Puffin Book of 1000 fun facts. Penguin Books Limited, UK

Brilha J (2018a) Geoheritage: inventories and evaluation. In: Emmanuel R, José B (ed) Geoheritage assessment, protection, and management. Elsevier, pp 69–85

Brilha J (2018b) Geoheritage and geoparks. In: Emmanuel R, José B (ed) Geoheritage assessment, protection, and management. Elsevier, pp 323–335

Bruschi VM, Coratza P (2018) Geoheritage and Environmental Impact Assessment (EIA). In: Emmanuel R, José B (ed) Geoheritage assessment, protection, and management. Elsevier, pp 251–264

Cayla N, Martin S (2018) Digital geovisualisation technologies applied to geoheritage management. In: Emmanuel R, José B (ed) Geoheritage assessment, protection, and management. Elsevier, pp 289–303

Chakraborty R (2008) Vertebrate fauna of kangerghati, guru ghasidas and sanjay national parks: Madhya Pradesh and Chhatisgarh. Zoological Survey of India, Conservation Area Series, vol 36

Chaudhuri AH (2012) Indravati river's mighty Horseshoe falls of Chitrakoot have earned it the title of 'Little Niagara'. The Economic Times. https://economictimes.indiatimes.com/industry/services/travel/indravati-rivers-mighty-horseshoe-falls-of-chitrakoot-have-earned-it-the-title-of-littleniagara/articleshow/16604310.cms. Accessed 26 Apr 2020

Chitrakote Waterfalls, Bastar (2015) Chhattisgarh Tourism Board. https://cgtourism.choice.gov.in/node/117. Accessed 26 Apr 2020

Chylińska D, Kołodziejczyk K (2018) Geotourism in an urban space? Open Geosci 10(1):297–310

Colin DP, Enrique DM, Jonathan GL (2018) The conservation of geosites: principles and practice. In: Emmanuel R, José B (ed) Geoheritage assessment, protection, and management. Elsevier, pp 193–212

Cooper ZM (1983) Prehistoric habitation patterns around Chitrakot falls_ Bastar District_ Madhya Pradesh. ir.inflibnet.ac.in

Coratza P, Hobléa F (2018) The specificities of geomorphological heritage. In: Emmanuel R, José B (ed) Geoheritage assessment, protection, and management. Elsevier, pp 87–106
Cottrell SP, Vaske JJ, Shen F, Ritter P (2007) Resident perceptions of sustainable tourism in Chongdugou, China. Soc Nat Resour 20:511–525(19) (PDF) Stakeholder Perceptions of Indicators of Tourism Use and Codes of Conduct in a Coastal Protected Area in Alaska. https://www.researchgate.net/publication/262067915_Stakeholder_Perceptions_of_Indicators_of_Tourism_Use_and_Codes_of_Conduct_in_a_Coastal_Protected_Area_in_Alaska. Accessed 27 Apr 2020
De Wever P, Guiraud M (2018) Geoheritage and museums. In: Emmanuel R, José B (ed) Geoheritage assessment, protection, and management. Elsevier, pp 129–145
Dickson MH, Fanelli M (2003) Geothermal energy: utilization and technology, United Nations Educational, Scientific and Cultural Organization (UNESCO), Paris, France
Farsani NT, Coelho C, Costa C (2011) Geotourism and geoparks as novel strategies for socio-economic development in rural areas. Int J Tour Res 13(1):68–81. https://doi.org/10.1002/jtr.800
Farsani NT, Coelho COA, Costa CMM et al (2014) Geoheritage 6:185. https://doi.org/10.1007/s12371-014-0099-7
Final report on 20 year perspective plan for development of sustainable tourism in Chhattisgarh (PDF). Ministry of Culture and Tourism, pp 12–3 of 5
Frey M-L, Schäfer K, Büchel G, Patzak M (2006) Geoparks – a regional, European and global policy. In: Dowling R, Newsome D (eds) Geotourism, pp 95–117. Elsevier
Geoheritage-GSA Position Statement (2012) The geological Society of America. https://www.geosociety.org/documents/gsa/positions/pos20_Geoheritage.pdf. Accessed 27 Apr 2020
Gordon JE, Roger C, Enrique D-M (2018) Geoheritage conservation and environmental policies: retrospect and prospect. In: Emmanuel R, José B (ed) Geoheritage assessment, protection, and management. Elsevier, pp 213–235
Gray M (2018) Geodiversity: the backbone of geoheritage and geoconservation. In: Emmanuel R, José B (ed) Geoheritage assessment, protection, and management. Elsevier, pp 13–25
India: tender process for geothermal project at Tatapani (2009). https://www.credacg.com/nit.pdf, https://www.thinkgeoenergy.com. Accessed 27 Apr 2020
Iron geology data info-info portal of geology with special reference to Rajasthan, India (2003). https://www.geologydata.info/.ironore_deposits.htm. Accessed 27 Apr 2020
Jagdish P (2013) Chitrakoot waterfalls–the Niagara Falls of India. India Study channel. https://www.indiastudychannel.com/resources/163040-Chitrakoot-Waterfalls-The-Niagara-Falls-India.aspx. Accessed 26 Apr 2020
Kailash and Kotumsar Caves. https://www.hamarajagdalpur.com/kailash-and-kotumsar-caves-jagdalpur/. Accessed 27 Apr 2020
Kale PVS (2014) Geomorphosites and geoheritage sites in India. In: Kale V (eds) Landscapes and landforms of India, World Geomorphological Landscapes. Springer, pp 251–253. ISBN 978-94-017-8029-2
Macadam J (2018) Geoheritage: getting the message across. what message and to whom? In: Emmanuel R, José B (ed) Geoheritage assessment, protection, and management. Elsevier, pp 267–287
Mata PJ, Carrión P, Jorge M, Villas-Boas R (2018) Geomining heritage as a tool to promote the social development of rural communities. In: Emmanuel R, José B (ed) Geoheritage assessment, protection, and management. Elsevier, pp 167–177
McKeever P, Zouros N (2005) Geoparks: celebrating earth heritage, sustaining local communities. Episodes 28(4):274–278
Menon G (2014) Evergreen leaves: recollections of my journeys into Wild India. Partridge India, p 103. ISBN 978-1-4828-4170-1
Migoń P (2018) Geoheritage and world heritage sites. In: Emmanuel R, José B (ed) Geoheritage assessment, protection, and management. Elsevier, pp 237–249
National Park Service and American Geosciences Institute (2015) America's geologic heritage: an invitation to leadership. NPS 999/129325. National Park Service, Denver, Colorado

Newsome D, Dowling R (2018) Geoheritage and geotourism. In: Emmanuel R, José B (ed) Geoheritage assessment, protection, and management. Elsevier, pp 305–321

NTPC to establish geothermal potential of Tatapani reservoir (2017) The Pioneer. https://www.dailypioneer.com/2017/state-editions/ntpc-to-establish-geothermal-potential-of-tatapani-reservoir.html. Accessed 27 Apr 2020

Page KN (2018) Fossils, heritage and conservation: managing demands on a precious resource. In: Emmanuel R, José B (ed) Geoheritage assessment, protection, and management. Elsevier, pp 107–128

Poon A (1993) Tourism, technology, and competitive strategies. C.A.B International, Wallingford England

Pradhan RK, Biswas J (1994) Towards regressive evolution: the periodic colour change behaviour of a troglophilic fish Nemacheilusevezardi (Day). Int J Speleol 23:191–201

Reynard E, Brilha J (2018) Geoheritage: a multidisciplinary and applied research topic. In: Emmanuel R, José B (ed) Geoheritage assessment, protection, and management. Elsevier, pp 3–9

Reynard E, Giusti C (2018) The landscape and the cultural value of geoheritage. In: Emmanuel R, José B (ed) Geoheritage assessment, protection, and management. Elsevier, pp 147–166

Robinson AM, Roots D (2008) Marketing tourism sustainably. In: Inaugural global geotourism conference, Fremantle, WA, August 2008, pp 303–317

Sharma AK (2000) Archaeo-anthropology of Chhattīsgaṛh. Sundeep Prakashan, New Delhi, India, p 23

Singh S (2010) Lonely Planet India, Lonely Planet. https://en.wikipedia.org/wiki/Chitrakote_Falls#CITEREFSingh2010. Accessed 27 Apr 2020

Singh RB, Anand S (2013) Geodiversity, geographical heritage and geoparks in India. Int J Geoheritage. -igu-cog-org.igu-geopark.com

Tatapani has great potential for Geothermal power generation: CREDA (2016) The Hitavada. https://www.thehitavada.com/Encyc/2016/10/29/Tatapani-has-great-potential-for-Geothermal-power-generation--CREDA.aspx, https://geothermalresourcescouncil.blogspot.com/2016/10/india-geothermal-power.html Accessed 26Apr 2020

Terryn WC (2011) Chitrakoot falls. Fer Publishing, 72p

UNESCO (2006) Guidelines and Criteria for National Geoparks seeking UNESCO's assistance to join the Global Geoparks, Network, Paris, January 2006. Internal document, 10(18)(PDF) Geotourism and geoparks as novel strategies for socio-economic development in rural areas. https://www.researchgate.net/publication/227697868_Geotourism_and_Geoparks_as_Novel_Strategies_for_Socio-economic_Development_in_Rural_Areas. Accessed 25 Apr 2020

Venkat Ramaiya D, Hussain SM (2015) Geoheritage sites of MP and Chhattisgarh and their preservation and management.In: 5th international symposium 2015 – International Symposium 2015, SEUSL 433. https://ir.lib.seu.ac.lk/handle/123456789/1226

Vermeulen J, Whitten T (1999) Biodiversity and cultural property in the management of limestone resources. World Bank, Washington, DC, p 120

Welcome to Directorate Of Geology and Mining, Chhattisgarh (2017) Mineral Resources Department, Government of Chhattisgarh, India. https://chhattisgarhmines.gov.in/en/welcome-directorate-geology-and-mining-chhattisgarh. Accessed 27 Apr 2020

Zouros N, Martini G (2003) Introduction to the European geoparks network. In: NHM of Lesvos Petrified Forest. Proceedings of the 2nd international symposium of natural monuments and geological heritage. Natural History Museum of the Lesvos Petrified Forest, Greece, pp 17–21

Chapter 14
The Exploration into Evaluation Index System for the Protection Effectiveness of Natural Heritage Protected Areas

D. Wei

Abstract As an important way to protect the natural heritage, a natural heritage protected area has a significant role in the historical, cultural, and scientific values of geological heritage. So the protection should be strengthened. This paper is committed to explore the evaluation index system for the protection effectiveness of natural heritage protected areas, so as to obtain the general understanding of the current situation and problems of the heritage and provide the comprehensive and scientific basis for the protection of geological and paleontological heritage and the rational development of nature reserves.

Keywords Natural heritage protected area · Evaluation of protection effectiveness · Index system · Geological heritage

A nature reserve is an important means of protecting natural resources and biodiversity, as well as an important basis for the human understanding of the history of Earth and biology. Geological heritage is an important way to obtain the information about the evolution of Earth. There is a consensus that the natural heritage protected area could promote the protection and development of relics. Therefore, based on the status and existing protection measures, a comprehensive evaluation index system for protection effectiveness of natural heritage protected areas is theoretically and practically significant to the construction and development of the protected areas in China.

D. Wei (✉)
Beijing Normal University, Beijing, China
e-mail: weidybnu@126.com

R. B. Singh et al. (eds.), *Global Geographical Heritage, Geoparks and Geotourism*,
Advances in Geographical and Environmental Sciences,
https://doi.org/10.1007/978-981-15-4956-4_14

14.1 The Current Situation of the Natural Heritage Protected Area

Establishing a natural heritage protected area, which is crucial to conserve the biological diversity and improve the spiritual civilization, is an effective measure to protect the ecological environment and natural resources, as well as a positive means to accelerate the transformation of economic development and the sustainable development. The development of nature reserves has become an important part of environmental protection in many countries of the world and attracted an extensive attention from the government and society. Over the years, the Chinese government has attached great importance to the development of nature reserves and achieved significant results.

Up to the end of 2011, China (excluding Hong Kong, Macao and Taiwan) has established 2,640 nature reserves of various types and levels, covering a total area of about 149.71 million hectares. The national nature reserve is the essence of nature reserve. Currently, there are 335 national nature reserves, covering an area of 93.15 million hectares and accounting for 10.01 % of the total land area of China.

At present, there is no uniform principle for grades of nature reserves. The nature reserves of China are classified into 3 categories and 9 types according to Principle for Categories and Grades of Nature Reserves (GB/T14 529-93). The three categories are natural ecosystems, wildlife, and natural heritage. The category of natural ecosystem reserve includes five types of nature reserves for protection of forest ecosystem, grassland and meadow ecosystem, desert ecosystem, inland wetland and aquatic ecosystem, and marine and coastal ecosystem. The category of wildlife reserve includes two types of nature reserves for protection of wild animals and wild plants. The category of natural heritage reserve includes two types of reserves for protection of geological heritage and paleontological heritage. Each nature reserve is managed according to the Nature Reserve Regulations of the People's Republic of China (released in 1994) and protected as a strict nature reserve.

A natural heritage protected area is a nature reserve to conserve the geological relics and paleontological relics which are of special significance. China has diversified space–time conditions and complex geological structures which has left a colorful natural heritage in the evolution of millions of years. Some of the natural heritage is typical, representative and rare, and are of great value. The first batch of natural heritage reserves, including the first geological relics protected area and the first paleontological relics protected are, was built in 1980 (24 years later than Guangdong Dinghu Mountain Nature Reserve).

A geological heritage protected area refers to the nature reserve for protecting the special geological structures, geologic sections, peculiar geological landscapes, rare minerals, springs, waterfalls, and relics of geological disasters. A paleontological heritage protected area refers to the nature reserve for protecting the relics of ancient human and paleontological fossils.

Among 2,640 nature reserves of China, there are 123 natural heritage reserves, accounting for 4.66 %, of which there are 91 geological heritage protected areas and 32 paleontological heritage protected areas. Among 123 natural heritage reserves, there are 17 reserves at the national level, including 10 geological heritage protected areas (Table 14.1) and 7 paleontological heritage protected areas (Table 14.2).

Table 14.1 National geological heritage protected areas of China

No.	Name	Administrative division	Area	Protected objects	Establishment time
1	Jixian stratigraphic section of the Middle-upper Proterozoic Erathem	Jixian, Tianjin	900	Stratigraphic section of the middle-upper proterozoic erathem	1984.10.18
2	Liujiang Basin Geological Heritage	Funing, Hebei	1395	Geological heritage	1999.05.01
3	Nihewan	Yangyuan, Hebei	1015	Sedimentary strata of the Cenozoic Era	1997.02.18
4	Chengshantou Coastal Landform	Jinzhou District, Dalian, Liaoning	1350	Geological heritage and coastal karst landform	1989.04.01
5	Yitong Volcanoes	YitongManchu Autonomous County, Jilin	765	Volcanic geological heritage	1983.10.22
6	Dabusu	Qianan, Jilin	11,000	Mud forest, paleontological fossil and wetland ecosystem	1993.01.01
7	Wudalianchi	Wudalianchi, Heilongjiang	100,800	Volcanic geological heritage and spring water resources	1980.03.29
8	Changxing Geological Heritage	Changxing, Fujian	275	Stratigraphic section of limestone of the Permian Period	1980.03.14
9	Mount Ma	Jimo, Shandong	774	Geological heritage such as the columnar jointing stone column and silicified wood	1993.01.13
10	Mount Danxia	Renhua, Guangdong	28,000	Danxia Landform	1995.11.06

Table 14.2 National paleontological heritage protected areas of China

No.	Name	Administrative division	Area	Protected object	Establishment time
1	Paleo-coast and Wetland	Tianjin	35,913	Paleo-coast relics such as the shell bank and oyster beach and coastal wetland	1984.12.01
2	Otog Dinosaur Fossils	Otog Banner	46,410	Dinosaur footprints fossils	1998.01.01
3	Beipiao Bird Fossils	Beipiao, Liaoning	4630	Paleontological fossils including the bird fossils of the late Mesozoic Era	1997.05.18
4	Shenhuwan Underwater Paleo-forest Relics	Jijiang, Fujian	3100	Underwater paleo-forest relics and oyster beach	1991.10.09
5	Shanwang Paleontological Fossils	Linqu, Shandong	120	Paleontological fossils	1980.01.17
6	Nanyang Dinosaur Egg Oryctocoenosis	Nanyang, Henan	78,015	Dinosaur egg fossils	1998.01.01
7	Qinglongshan Dinosaur Egg Oryctocoenosis	Yunxian, Hubei	205	Dinosaur egg fossils	1997.01.13

According to the establishment time of each natural heritage protected area, China started the formal protection of natural heritage in 1980s. In the past 30 years, Chinese government has attached great importance to the protection of natural heritage, but there are still potential threats to the natural heritage reserves, mainly including damaging the original features of the geological heritage due to the construction of unrelated or artificial landscapes in order to attract tourists; damaging the environment and geological heritage due to the construction of artificial buildings or restaurants seeking for benefits instead of the environmental protection; insufficient protection for the geological heritage landscapes which are weak in resistance to the weathered erosion or other natural forces; low quality of tourists who may destroy the geological heritage at will.

Therefore, it is necessary to develop a more specific and effective evaluation system for the protection effectiveness and evaluate the protection results of natural heritage protected areas.

14.2 Domestic and International Related Researches

14.2.1 International Related Researches

At the United Nations Conference on Environment and Development (UNCED) which was held in 1992, governments reached a new agenda for sustainable development. The agenda includes a new Convention on Biological Diversity (CBD), and in particular calls for the establishment of a protected area system by each government to support the sustainable use and the fair and equitable sharing of benefits. In 1997, the World Commission on Protected Areas (WCPA) proposed an evaluation framework for assessing management effectiveness of protected areas (Table 14.3) based on 6 elements of the conservation and management process. The basic criteria that are assessed for each element are listed in the framework (Hockings 2000; Hockings et al. 2002). The meaning of each element is specified as follows: (1) context including the policy environment and the information that may support the decision-making; (2) planning, namely, the evaluation indexes that are selected based on the evaluation purpose, especially on the basis of the protected area system ora single protected area; (3) inputs, namely, the resources needed to carry out management, including the personnel, fund, equipment, and facilities; (4) process including the daily management of the protected areas, and the management of community, natural and cultural resources; (5) outputs including the implementation of management programs and actions; (6) outcomes, namely, the extent to which they achieved objectives. The effectiveness assessment requires long-term monitoring, including the system/regional biological and cultural resources, community economy, and the impact of system/regional management on the community. In the final analysis, the effectiveness assessment truly reflects the management effectiveness.

Table 14.3 WCPA framework for assessing management effectiveness of protected areas

Focus of evaluation	Elements of evaluation	Criteria that are assessed
Design	Context Planning	Significance; threats; vulnerability; national context Protected area legislation and policy; protected area system design; reserve design; management planning
Appropriateness of management systems and processes	Input Process	Resourcing of agency; resourcing of site; partners Suitability of management processes
Delivery of protected area objectives	Outputs Outcomes	Results of management actions; services; and products Impacts; effects of management in relation to objectives

The WCPA framework for assessment is closely related to the concerns of the administrators of protected areas and can be applied to the management effectiveness assessment of the protected areas worldwide (Hockings 2000). This assessment framework is not only the basis for the development of specific management effectiveness evaluation methods for protected areas, but also has implications for the content and standard of the evaluation report on the management effectiveness of protected areas.

The evaluation methods for the management effectiveness of protected areas include the Rapid Assessment and Prioritization of Protected Area Management (RAPPAM). Based on the "rapid assessment questionnaire", the RAPPAM methodology is designed to evaluate the protected areas, analyze the evaluation results, and determine the priority and countermeasures through the participatory symposiums including the administrators of protected areas, the policy-makers and other stakeholders. This methodology has been implemented in more than 1000 protected areas of more than 40 countries, such as Europe, Asia, Africa, Latin America, and the Caribbean. The Management Effectiveness Tracking Tool (METT) is a rapid assessment tool based on the scorecard questionnaire. The scorecard contains six elements defined by IUCN-WCPA, namely context, planning, inputs, process, outputs, and outcomes, but more emphasis is placed on the context, planning, inputs and process which define the needs, restrictions, and priority of activities for the administrators to improve the management of protected areas. In addition, another methodology is Enhancing our Heritage Toolkit (EoH) proposed by the UNESCO. EOH could help administrators and stakeholders to assess the current activities, identify the gaps, and discuss the solutions to problems. The IUCN-WCPA framework combines the topics together, in which the evaluation index and tool could constitute a diagram showing the adequacy and appropriateness of management and the extent to which the objective is achieved.

In the mid-1970s, the Nature Conservation Council (NCC) began the evaluation and record of the scientific research on geology and geomorphology of the most important heritage sites in the UK, and the Geological Conservation Review (GCR) project was officially launched in 1977. GCR is the first project in the world to systematically assess the overall geological heritage of a country.

As for the research on how to protect the geological heritage, the French Guy Martini and the Greek Nickolus Zoulos proposed to establish the geopark to conserve the geological heritage (1996) and the relatively complete evaluation systems for the geological heritage were set up in Britain, Germany, and Australia (2007). Many countries have given attention to the protection of geological heritage. Matthew R. Bennett et al. (1996) thought that the UK has devoted much time to conserve the wild animals and plants and now it was turning its attention to the protection of geological heritage. They put forward a management technique for the geological protection of the abandoned quarry, namely the classification based on the bareness of the quarry and the evaluation of management effectiveness on the hydrology, discharge of harmful substances, and human factors. For example, to evaluate the discharge of harmful substances, the size of the landfill should be assessed.

As for the research on the management effectiveness of protected areas, Jamison Ervin (2003) studied the application of RAPPAM in South Africa, Bhutan, Russia, and China, and analyzed the pressure and threats confronted by the protected areas. The rapid assessment methodology is designed for the scope of the activity, the impact on the natural resources of protected areas, and the renewal cycle of the natural resources with or without human management. This assessment method pays more attention to the relative threats and weakness of a protected area. After determining the most urgent problems confronted by a protected area, such as the poaching, alien plant invasion, damage from the tourism, and deforestation, the management could be improved through the evaluation of management effectiveness. The criteria (such as the biological importance, planning, inputs, and process) of the management effectiveness of a protected area should be scored in four cases, namely "yes", "mostly yes", "mostly no", and "no". The past pressure and future threats of the protected area should also be assessed. The pressure in the past mainly refers to the reduction of biodiversity, limited regeneration capacity, and exhaustion of natural resources. The threat in the future is the potential or upcoming pressure which may lead to or continue the harmful effects.

Davide Geneletti and Iris van Duren used the multiple-criteria decision analysis to divide the Paneveggio-Pale di S. Martino Park of Italy. The park was divided into three areas with different protection levels, ranging from strict nature reserve to tourism and entertainment area. Based on the GIS-based land-use suitability analysis, the land was assigned to the designated protection level. Then the stability of the results was evaluated through the sensitivity analysis. This study provides a management methodology for the administrators and other stakeholders of parks.

Reis and Henriques (2009) showed how geological heritage can be evaluated using an open system of values or contents displayed by the geological objects. Contents are designated as indicial, documental, iconographic, symbolic, scenic, and conceptual. Their qualification depends on both the relevance of the meaning attributed to the objects by scientific communities (defined as relevance grade) and the public understanding of such meanings related to the social use of the objects (defined as abstract perceptiveness). Contents are ranked into three categories of increasing importance. They are rank I—indicial contents; rank II—documental, iconographic, and symbolic contents; and rank III—conceptual and scenic contents. A higher category (rank IV) is available for a feature with universal content, for example, geological heritage in outer space.

Fassoulas et al. (2011) thought that a quantitative methodology for the assessment of geotopes can be used for the management of geological heritage. Their proposed methodology is based on a series of criteria that cover the geological and geographical importance of a geotope, and its scientific, ecological, cultural, aesthetic, and economic significance and the potential for use are evaluated in six aspects. The value indexes for each geotope on a scale range from 1 to 10 which is graded into 5 levels (1, 2.5, 5, 7.5, and 10) (Table 14.4). This methodology was implemented and tested in two areas in the island of Crete, namely the Psiloritis Natural Park and the Lassithi Mountains, producing reliable results, which are in agreement with the geopark's activities and values. The proposed quantitative assessment method serves

Table 14.4 List of criteria used and description of the scoring system

Criteria/Score	1	2.5	5	7.5	10
1. Scientific					
1.1 Geologic history	Single type history	Combination of at least 2 types	Combination of most types	Local story	Tells the whole local story
1.2 Representativeness	No	Low	Moderate	High	Very high
1.3 Geodiversity	<5%	25%	50%	75%	>75%
1.4 Rarity	>7	>5, <7	>3, <4	>1, <2	Unique
1.5 Integrity	Almost destroyed	Strongly deteriorated	Moderately deteriorated	Weakly deteriorated	Intact
2 Ecological					
2.1 Ecological impact	No	Low	Moderate	High	Very high
2.2 Protection status	No protection	Limited	In spots	In large parts	Complete
3 Cultural					
3.1 Ethics	No	Low	Moderate	High	Very high
3.2 History	No	Low	Moderate	High	Very high
3.3 Religious	No	Low	Moderate	High	Very high
3.4 Art and culture	No	Low	Moderate	High	Very high
4 Aesthetic					
4.1 Viewpoints	No	1	2	3	4
4.2 Landscape difference	No	Low	Moderate	High	Very high
5 Economic					
5.1 Visitors	<5000	>5000	>20,000	>50,000	>75,000
5.2 Attraction	No	Local	Regional	National	International
5.3 Official protection	International	State	Regional	Local	No
6 Potential for use					
6.1 Intensity of use	Very intense	Intense	Moderate	Weak	No use
6.2 Impacts	Very high	High	Moderate	Low	No
6.3 Fragility	No	Low	Moderate	High	Very high
6.4 Accessibility	Close to hiking trail	Close to cobble or forest road	Close to local paved road	Close to regional road	Close to highway or town
6.5 Acceptable changes	No	Low	Moderate	High	Very high

the requirements for the adequate management and protection of geoheritage within a territory as it can reveal priorities for sustainable tourism development, including the geotourism and educational tourism activities and the conservation of geotopes.

Robert S. Pomeroy, John E. Parks, and Lani M. Waston introduced 42 indicators including 10 biophysical indicators, 16 socioeconomic indicators, and 16 governance indicators in A Guidebook of Natural and Social Indicators for Evaluating Marine Protected Area Management Effectiveness. They analyzed each indicator and determined the merits and demerits. The selecting of indicators follows the principles of measurableness, accuracy, consistency, sensitivity, and simplicity.

To sum up, the evaluation methodology and the index system for the management effectiveness assessment of the nature reserves are relatively mature internationally, and more attention is paid to the assessment of the ecological effect and management effectiveness of the protected areas. However, the researches on the evaluation for geological heritage protected areas are insufficient. Researchers have paid much attention to the value of geological heritage, and the study on protection measures and effectiveness is starting.

14.2.2 Domestic Related Researches

Domestic researches on geological heritage protection pay more attention on the existing problems in protection process and the corresponding measures and rational development model. Qian (2001) took Hebei Province as a study area to study the protection of geological heritage. He believed that the establishment of geological heritage reserves is the most effective protection measure. Zhang (2005) analyzed the significance of the geological heritage reserves and the inadequate legislation during the development of geological heritage protected areas and she proposed the legislative proposals. Hang (2006) proposed 6 protection steps including the protection classification, protection form, protection mode, protection grade, protection sequence, and protection zoning. Hu and Wu (2007) pointed out the problems confronted by the rapid development of national geoparks, mainly including the relatively excessive growth, unbalanced distribution and the little tourism brand effect of national geoparks. In addition, Wu et al. (2004), Li et al. (2004), Fang (2004), and Li and Lu (2007), respectively, investigated and gave advice to the protection of geological heritage in Cuihua National Geopark, Sichuan Longmenshan National Geopark, Anhui Huangshan World Gopark, and Sichuan Janmenguan Geopark.

The contents and methods of evaluation for geological heritage are varied. The evaluation indicators in the relevant documents issued by the Ministry of Natural Resources of the People's Republic of China focus on the nature, the protectability and the management basis of geological heritage resources. The Technical Requirements and Instructions of National Geopark Construction of China (Trial) proposes five indicators including the typicality, rareness, naturalness, systematicness, and integrity for assessment of the natural attribute, three indicators including the suitability of protected area, economic and social value, and scientific value for assessment of the protectability, and five indicators including the staff of institutions, regulations of borders, land ownership, foundation work, and management for assessment of the management basis. Compared with other contents of evaluation, the management effectiveness assessment of heritage is relatively deficient, and the evaluation system is imperfect.

Hao et al. took Luochuan loess geological heritage resources as an example, and proposed to assess the protection in three aspects: the current situation of geological heritage, the conservation and management, and the environmental protection (2004). Then he graded the protection of Luochuan loess geological heritage through the fuzzy comprehensive evaluation (2007). An and Chen (2010) also graded the protection of Danxia landform geological heritage in Fuyun County, Xinjiang. On this basis, Peng and Wu (2006) selected seven indicators including the protection and management agency, conservation status, complexity of protection, protection planning, implementation of protection measures, impact of development, and impact of engineering, and then assessed the geological heritage through the expert-based evaluation. Shen et al. (2009) proposed four indicators: conservation status, protection and management, complexity of protection, and impact of engineering, each of which had three grades of evaluation.

Overall, these evaluation systems have difficulties in fully reflecting the protection effectiveness of natural heritage, therefore, it is significant to further study and develop a more scientific evaluation system for the protection effectiveness.

14.3 Selecting the Indicators for Protection Effectiveness Assessment of Natural Heritage Protected Areas

On the basis of the analysis on the protection measures and effectiveness of natural heritage protected areas, the research background is analyzed and the basic indicators that affect the protection effectiveness evaluation of geological heritage are selected to provide the theoretical support for the index system. At the same time, the relevant researches on the assessment of protection of natural heritage and the evaluation index systems both at home and abroad are reviewed to establish the evaluation index system for the protection effectiveness of natural heritage protected areas in China.

14.3.1 Selecting the Indicators

It is an important step to establish an index system in the protection effectiveness assessment of natural heritage protected areas. In order to better reflect the achievements of protection, this study selects the key elements in the index system of protection effectiveness evaluation which are of scientificity, representability, measurability, comparability, and operability. With the protection effectiveness assessment of natural heritage protected area as the general objective, the goals include the evaluation of natural heritage and the evaluation of management. The evaluation criteria include the integrity of natural heritage, the threats confronted by natural heritage, management basis, management mechanism, management behavior, and management effectiveness. The indicators are shown in Table 14.5.

14.3.2 Establishing the Evaluation System

An indicator for the evaluation which shows the characteristic and the quantity of the assessed object plays a role in both the qualitative analysis and the quantitative analysis. A comprehensive evaluation of the natural heritage and the management measures of the protected area is necessary to understand the protection effectiveness of natural heritage and the current situation and problems of the heritage, which would provide a scientific and comprehensive basis for protecting the geological and paleontological heritage and developing the nature reserve.

14.3.2.1 Evaluation of Natural Heritage

The protection of natural heritage is assessed from the integrity and the potential threats.

The area proportion (%) of the natural heritage reflects the spatial distribution of natural heritage in a reserve. The larger the proportion is, the higher attention the natural heritage receives and the greater value the natural heritage has. As an important indicator for the protection effectiveness of the natural heritage, the change in the area proportion of the natural heritage shows the persistent existence of the protected object and whether it has been destroyed by human or nature. The integrity of information about the natural heritage is an important basis for the conservation and restoration of the heritage, so the detailed and accurate data of monitoring and scientific research reflects the good protection of the natural heritage to a large extent. In addition to the establishment of protected areas, some important natural heritage in a protected area requires the engineering isolation, restoration and conservation, and dynamic monitoring. Therefore, the area proportion of natural heritage protected area (%) reflects the protection of the natural heritage. In addition to the overall protection of natural heritage, the internal diversity and changes can also show the

Table 14.5 Criteria used for evaluating the protection effectiveness of natural heritage protected areas

Goal	Criterion	Indicator	Parameter
Evaluation of natural heritage	Integrity of natural heritage	Stability of natural heritage	Area proportion of natural heritage (%) Changes of area proportion of natural heritage Integrity of information about natural heritage Area proportion of natural heritage protected area (%)
		Diversity and utilization of natural heritage	Changes of diversity of natural heritage Availability of important natural heritage
	Threats confronted by natural heritage	Human disturbance	Density of road network Impact of mining Volume of visitors Protection measures Theft and loss Commercial activity
		Natural factor	Impact of geological conditions Impact of extreme climate Pollution of solid waste Water environment quality Air environment quality
Evaluation of management	Management basis	Agency and personnel	Whether the agency and personnel could satisfy the management
		Border and functional zoning	Scope and ownership Functional zoning
		Infrastructure	Furnishing of equipments
		Fund guarantee	Whether the fiscal appropriation could satisfy the wage and management

(continued)

Table 14.5 (continued)

Goal	Criterion	Indicator	Parameter
		Support of professional technique	Training of professional technical personnel
	Management mechanism	The institutional guarantee for management	Implementation of the management system
		Community participation	Opinions of local residents and their impact on the policy-making
		Monitoring and evaluation	The presence or absence of monitoring and evaluation of the management effectiveness
		Personnel management (incentive mechanism)	The management system and incentive mechanism
		Law enforcement power	The presence or absence of law enforcement power
		Management system	Competent authority of the protected area
	Management behavior	Protection and development planning	Implementation of the protection and development plans
		Background survey of resources	Implementation of background survey of resources
		Scientific research	Cooperation in the scientific research
		Dynamic monitoring	Dynamic monitoring of the protected object
		Training of the staff	The number of employees being trained and the frequency of training
		Publicity and education	Activities of publicity and education

(continued)

Table 14.5 (continued)

Goal	Criterion	Indicator	Parameter
		Community co-management	The presence or absence of corresponding community management agency
		Tourism management	Control and management of the tourism
		Patrol	Frequency and record of the field patrol
		Fund management	Whether the fund is managed and allocated reasonably
	Management effectiveness	Control of the protected area	Control of the core area and buffer area to the external persons
		Harmonious development with the community	Coordination between the management of protected area and the local community

protection effectiveness. As for the nature reserve which has the stable biodiversity, the protection measures are effective. Reasonable utilization and protection are complementary to each other. Therefore, the rational utilization of important natural heritage is important to reflect the protection effectiveness.

Road construction, commercial activities, and mining are the main threats confronted by many geological heritage protected areas. Unreasonable activities may have a fatal effect on the heritage. In the scenic area, excessive tourists or improper behavior will greatly destroy the natural heritage. Nowadays, many scenic areas are overly pursuing the economic benefits, and the protection of the heritage is deficient, which cause irreparable loss. The protection measures and the frequency and loss of thievery are also important indicators. In addition to the natural geological conditions and extreme climatic conditions, the man-made pollution to the water and air also damages the natural heritage.

14.3.2.2 Evaluation of Management

The methods and indicators for assessing the management are diverse, and many scholars have done empirical researches and achieved good results. Internationally, the more mature concepts and methods have been used in the assessment of protection effectiveness of nature reserves, but the scope of application is limited, and the selection of appropriate evaluation methods and evaluation projects is indefinite.

On the basis of the RAPPAM methodology, this paper adjusts the index system, and selects 24 indicators from 4 aspects including the management basis, management mechanism, management behavior, and management effectiveness.

The management basis shows the conditions of the agency, staff, fund, equipment, and facilities of the protected area. The management mechanism refers to the regulations, plans, objectives, allocation mechanism of the charge, coordination of the local residents, surrounding residents, local community and other related institutions, departments, and enterprises. Management behavior is the activity that the protected area takes to achieve the protection objective and the functions of protected area. The management effectiveness is assessed by the protected objects, the management and protection of resources in the protected areas, and the impact on the surrounding communities and regional economic development. The four criteria evaluate the management progressively. On the basis of the foreign evaluation index systems, the localized investigation has been carried out to make evaluation system closely related to the actual situation of the management of nature reserves in China, so as to improve the evaluation and reflect the management effectiveness of the protected area in an all-round way.

14.4 Conclusion and Discussion

After reviewing the researches on conservation of natural heritage, this paper analyzes the influence factors and characteristics of the protection effectiveness of heritage, and establishes the index system for the protection effectiveness of natural heritage protected areas. The index system consists of four parts: the goal, the criterion, the indicator, and the parameters.

The above indicators are different in the status and role in the protection effectiveness assessment of natural heritage, so different weights are given to different indicators, and a comprehensive evaluation model is established for the quantitative evaluation. In order to further explain the specific application of the evaluation model, it is necessary to test the scientificity and practicability of the model through the case study, which is the direction for further research.

References

An GR, Chen Q (2010) Assessment of Danxia landform geological heritage in Fuyun County, Xinjiang based on fuzzy synthesis judgment. Explor Eng 9:142–144

Fang W (2004) Conservation and planning of world geoheritage in Mount Huangshan. Res Soil Water Conserv 14(5):118–120

Fassoulas C, Mouriki D, Dimitriou-Nikolakis P, Iliopoulos G (2011) Quantitative assessment of geotopes as an effective tool for geoheritage management. Geoheritage 2011(11):46–63

Hang S (2006) Optimal selection mode and step of geological heritage protection: A case study of Xinjiang. J Guilin Univ Technol 26(1):148–152

Hao JQ, Wu CJ, Tao YK (2004) Protection and utilization assessment of geological remains-a case study of the Loess remains of Luochuan. J Mt Sci 22(1):7–11
Hao JQ, Li YQ, Wen M (2007) Assessment of Luo-Chuan Loess geological vestige based on fuzzy synthesis judgment. Geol Shanxi 25(1):7–11, 45–50
Hu WX, Wu CJ (2007) The Characteristics of national geopark in China, some problems and a study on some measures for its rapid construction. Geol Rev 53(1):98–103
Li XQ, Lu ZM (2007) Study on geological relics protection of Jianmenguan Geopark in Sichuan. Res Soil Water Conserv 14(5):123–125
Li XQ, Qin JX, Yin JC (2004) Protection of geological relics in Longmen Mountains National Geopark. Mt Res 22(1):12–16
Peng YX, Wu CJ (2006) Evaluation on the Relationship between conservation and utilization of geological relics: a case from Shaanxi Province. Resour Sci 28(1):192–197
Qian JP (2001) The protection recommendations for geological heritage diversity in Hebei. China Popul Resour Environ 11(1):136–137
Reis RPD, Henriques MH (2009) Approaching an integrated qualification and evaluation system for geological heritage. Geoheritage 2009(1):1–10
Shen J, Hao CY, Wang HF (2009) Accordance between utilization and protection of geoheritage in Mentougou, Beijing. Resour Ind 11(4):37–41
Wu CJ, Han LY, Tao YK (2004) The coordinative operation of national geoparks based on the protection and utilization of geological remains–a case study of the national geopark of the landslide landscape, Cuihua Mountain. Mt Res 22(1):17–21
Zhang YF (2005) The legal protection of nature reserve of geological trace. Sci Technol Prog Policy 22(8):80–82

Chapter 15
Geoheritage and Potential Geotourism in Geoparks—Indian Perspective

Sudesh Kumar Wadhawan

Abstract Indian subcontinent has a variety of geological domains ranging from Precambrian/Achaean to active Neogene and Anthropocene with several very interesting world-class geoheritage sites that display unique geological features, rock types, fossils, geological boundaries or tectonic discontinuities, processes and landscapes. Geoheritage sites or the Geosites are natural sites of rare and unique geological and geomorphologic significance. Studying, protection and development of Geoheritage sites achieve its goals through a three-pronged approach: conservation, education and geotourism, respectively. These Earth heritage sites must be suitably delineated and protected initially by Geological Survey of India (GSI) as are part of an integrated concept of protection, geosciences education and sustainable development of the area around with active support from the Tourism/Forest Departments of the State governments. Ever since the International Union for Conservation of Nature (IUCN) has adopted a Resolution (2015) that affirmed Geodiversity and Geoheritage as integral parts of Natural Diversity and Natural Heritage; it is imperative to treat geodiversity and geoconservation as inseparable from biodiversity and nature conservation. Presently development of the Geoparks is the major international theme implying geosciences application for inclusive growth of society and protection/conservation of unique Geoheritage. Geoparks are places (delineated and designated area in contrast to a singular specific Geosite or the Geological Monument) where rare geoheritage sites with geomorphic landscapes and geological phenomena are preserved intact and where sustainable economic development plans are pursued through low-impact recreational, scientific, and educational activities. So far, although a few National Fossil Parks have been delineated by GSI as the National Geological Monuments or the Geotourism Hotspots in Indian subcontinent, yet there is no formally recognised Geopark in India, neither at the State level or at the National level, nor at the UNESCO Global level. It is time now to consider

---

S. K. Wadhawan (✉)
Geological Survey of India, Jaipur, India
e-mail: wadhawansk.leo@gmail.com

Present Address:

Adjunct Professor, Amrita Vishwa Vidyapeetham, Kerala, India

R. B. Singh et al. (eds.), *Global Geographical Heritage, Geoparks and Geotourism*, Advances in Geographical and Environmental Sciences,
https://doi.org/10.1007/978-981-15-4956-4_15

giving the needed impetus and importance to the delineation and recognition of Geoparks in India. Since GSI has the mandate as per the listed Charter of Functions, it has to take lead in this matter and achieve tangible results before the IGC-2020 mega geosciences event in New Delhi. Geotourism is generally dependent on earth's geoheritage, educative through geointerpretation, geoguides and increased awareness, locally beneficial through sustainable economic viability community involvement and foster geoconservation. Present contribution enumerates on inventory of geoheritage sites and recommends approaches for geoconservation with two case studies on optimally utilising potential of Jaisalmer Geopark that constitutes part of the declared Desert National Park as a unique and fragile ecosystem in dryland environment and the Jodhpur Rocks Geopark being developed by a Trust managed by erstwhile royal family, regarding protection, conservation and development of geodiversity and biodiversity thus promoting geotourism for tourist satisfaction in unique rocky dryland environment of the Thar Desert in western India.

Keywords Geoheritage · Geodiversity · Geoparks · Geotourism · Geological Survey of India · India

15.1 Introduction

Earth science studies about geodiversity and geoheritage have been a recently incorporated subject into Geosciences and are yet to gain momentum and wider acceptance amongst policymakers and administration in India, whereas better informed public needs to be involved for their geoconservation. Spectacular geodiversity in Indian terrain has been very well recognised all over the world despite lack of a detailed inventory of geoheritage sites. Nevertheless, the public awareness is growing for practising geoconservation and exploring geotourism potential of the listed geoheritage and other picturesque geomorphosites in India. The lofty young mountain ranges of the Himalaya, dissected rugged continental volcanic eruptions of the Deccan Traps, the Precambrian Indian peninsula, the dryland environment of the Thar Desert and the rain-soaked North-eastern India have universal appeal and important position in Earth's geological history.

In order to increase awareness about geological heritage (together with biological heritage) that forms part of natural heritage and must be considered in nature management and land-use planning, it is imperative to elaborate on the basic concepts on the five relatively new 'geos': geoheritage, geodiversity, geopark, geoconservation and geotourism (Unesco.org; Anantharamu et al. 2001; Bhargava et al. 2010; Brocx and Semeniuk 2007; Geological Survey of India 2012). **Geodiversity** is expressed in geological formations as a variety of rock formations, minerals, fossils, tectonic structures; geohydrological features soils and geological processes that created distinctive rock assemblages and landscapes. It is the major abiotic component of lithosphere that supports geomorphic processes and landforms, biodiversity and ecosystems. **Geoheritage** is the abbreviated version of the term geological heritage. It is part of the natural heritage of a certain area constituted by geodiversity elements with particular geological value and hence worthy of safeguard for the benefit of present and future generations. Geoheritage sites or the geosites are natural sites of rare and unique geological and geomorphologic significance. Geoheritage can include both in situ elements (geosites) or ex situ elements (collections of geological specimens) having paleontological, geomorphological, mineralogical, petrological or stratigraphic significance, etc. Studying, protection and development of geoheritage sites achieve its goals through a three-pronged approach: conservation, education and geotourism, respectively.

A **Geopark** is a unified area that advances the protection and use of geological heritage in a sustainable way, and promotes the economic well-being of the people who live there. Therefore, geoparks are places (delineated and designated area in contrast to a singular specific Geosite or the Geological Monument) where rare geoheritage sites with geomorphic landscapes and geological phenomena are preserved intact with local community involvement and where sustainable economic development plans are pursued through low-impact recreational, scientific, and educational activities. Geoparks can be of National and/or Global significance. **Geoconservation** deals with effective mechanisms for preservation of significant earth science features—the geoheritage sites, for purposes of education, science and heritage and value-added recreation (Geotourism Hotspots of Indian Subcontinent 2016; Gray 2004). However, as awareness about the unique heritage and geological component of ecosystem services is raised, its sustainable advantages for the society are increasingly realised for protection and conservation of these geosites. Well-managed geological sites can support different types of sustainable use with clear benefits for the society, such as educational, scientific and economic uses. Geoconservation needs to promote geotourism circuits for sustainable growth of the locally involved communities. This is already happening in many territories around the world through the establishment of Global Geoparks which have recently been fully recognised by UNESCO. Ever since the International Union for Conservation of Nature (IUCN) has adopted a Resolution in 2015, that affirmed Geodiversity and Geoheritage as integral parts of Natural Diversity and Natural Heritage, it is imperative to treat geodiversity and geoconservation as inseparable from biodiversity and nature conservation. Presently development of the Geoparks has become a major international theme implying geosciences application for inclusive growth of society and protection/conservation of unique Geoheritage. As on July 2020, there are 161 Global Geoparks in 44 countries.

As per charter of functions, Geological Survey of India (GSI) has been mandated to identify, preserve and make initial and sustained efforts in conservation and protection of the unique Geoheritage sites in the country. Consequently, GSI has been making programmed attempts in studying and protecting natural sites of rare and unique geological and geomorphologic significance (Hose 2012; Kale 2009). Such endeavours were initiated way back in 1951, and presently 36 Geoheritage sites have been part of an integrated concept of protection, education and sustainable development. However, in Indian context, lot more needs to be accomplished for systematic promotion and upkeep of geoheritage sites/geoparks beyond identification and declaration of a geological monuments or geosites. Geotourism is based on appreciation of natural geological features of the terrain and aesthetic values of landscape. In addition to the fauna and flora, geotourism thrives on abiotic dimension of the environment and is treated as an integral part of the UNESCO's Geopark programmes (Kale 2010; Lazzari and Aloia 2014; Mazumdar 2007). Besides, geotourism and recreational activities based on geodiversity elements are completely integrated in the aims of the International Year of Sustainable Tourism, as was proclaimed by the United Nations for 2017.

Indian subcontinent has a variety of geological domains ranging from Precambrian/Achaean to active Neogene and Anthropocene with several very interesting geodiversity and world-class geoheritage sites that display unique geological features, rock types, fossils, geological stratigraphic boundaries or tectonic discontinuities, processes and landscapes (Mathur et al. 2017; National Heritage Division of Indian National Trust for Art and Cultural Heritage (INTACH) 2016; Panizza 2001; Mc Keever and Zouros 2005). India generally lags the world trend in practicing geoconservation. Given that the geoheritage is a non-renewable natural resource and gets adversely affected by natural factors such as weathering, erosion, climate change and human interventions for changed land uses that can result in a partial or total loss of geological sites. The risk of its damage is triggered also by urban development, vandalism, smuggling of geological samples owing to absence of a proper legal protection and international agreements, or lack of expertise and poor understanding by local and national authorities. Therefore, proper identification of geosites and landscapes, their characterisation and suitable management of these localities will allow their preservation and for reaping the benefits of promoting geotourism, thereby building mutually beneficial relationships between geoscientists, site owners and/or local and regional authorities (Reddy 2013; Reynard and Panizza 2005; Dowling and Newsome 2010; Dowling 2011). Present contribution enumerates with case studies on opportunities and challenges faced for optimally utilising potential of geoheritage and geoparks in India and proposes new approaches and reviews the procedures used in the development and management of potential Geoparks. It is hoped that emerging economies including Indian society, gear up to fully recognise the importance of geoheritage and envision aspirations for its protection, particularly in conjunction with or extension of the existing many national and international policies and strategies for the protection of biodiversity: wildlife sanctuaries and other protected habitats.

15.1.1 Inventory of Geoheritage Sites in India

As the National custodian, vide its Charter of Functions, it is the bounden duty of Geological Survey of India to delineate and maintain the Geoheritage sites in India either on its own or in collaboration with the State Government authorities who have been convinced for its intrinsic value and geotourism potential. GSI has been making concerted efforts in studying and protecting natural sites of rare and unique geological and geomorphologic significance. Such endeavours were formally initiated way back in 1951 when two major geosites: Fossil wood occurrences near Tiruvakkarai, South Arcot district and Sattanur area near Tiruchirapalli district in Tamilnadu in southern India were declared National Geological Monuments by GSI. Thereafter, between 1974 and 1981, over 20 geosites were declared as geological monuments by GSI. A comprehensive account of 26 geological monuments was compiled by GSI in a Special Publication No.61 in 2001 (Hose 2012). Several other prominent geoheritage features and landscapes have also been added to make the list go up to 40 geosites landscapes that need to be preserved as part of geoconservation and promoted as geotourism hotspots (Reddy 2013; Reynard and Panizza 2005; Dowling and Newsome 2010; Dowling 2011). GSI got published a book on the **40** nos. of Geotourism Hotspots in India under the aegis of 36th IGC Secretariat and released it during the 35th IGC at Cape Town, South Africa in September 2016 (Dowling 2011). Out of these 40 listed Geotourism Hotspots in India, four are large geoecological zones/regions namely: Kerala Backwaters, Kerala; Sundarban Mangrove Forests, West Bengal; Thar Desert, Rajasthan and Shyok-Nubra Valley, Ladakh that also preserve the pristine natural beauty. Although there are many more potential geosites, such as the Bellum Caves in Kurnool district, Andhra Pradesh, Ajanta and Ellora Caves in Maharashtra, Deccan Trap landscapes of distinctive volcanic lava flows in Maharashtra, that are already well-known Geotourists hotspots and maintained by the concerned State Governments, the other 36 officially recognised National Geological Monuments/Geoheritage sites in India are categorised here in the following genetic, geoscientific and geomorphic groups. A very brief description of these 36 Geoheritage sites is enumerated below.

15.2 Large and Unique Geological Deposits and Rare Minerals and Rock Assemblages

15.2.1 Peninsular Gneiss, Lalbagh, Bangaluru (Banglore), Karnataka

The site was declared as National Geological Monument in 1975. The Peninsular gneisses are extensively used as building stones mined form several quarries around Banglore (Bangaluru). One such quarry adjacent to Kempagowda Tower is declared as National Geological Monument. The Peninsular gneiss bears witness to several geological events spanning thousands of millions of years. The major rock type exposed in the quarries is dark biotite gneiss of granitic to granodioritic composition. Vestiges of older mostly amphibolites rocks are seen in the form of enclaves or inclusions within this gneiss. The monument is maintained by the Karnataka State Govt.

15.2.2 Charnockite, St. Thomas Mount, Chennai (Madras), Tamilnadu

This geosite is preserved as the type area of the rock type 'Charnockite'—a unique Archaen greasy looking hypersthenes bearing granulitic textured, high grade metamorphic granitic rock type and commonly associated with another rare basic meta-igneous rock called norite. The term Charnockite was coined in honour of Job Charnock the founder of Kolkata city (formerly Calcutta) and occurs at the St. Thomas Mount in Chennai. It was declared as National Geological Monument in 1975. The charnockites originally identified in St. Thomas Mount have been recognised and studied in many Precambrian shield areas of the world, enabling geologists to understand the complexity of charnockite problem related to signifying the process of primordial crustal evolution. That really is the geological significance of Charnokite geohritage site in India.

15.2.3 Laterite Near Angadipuram PWD Rest House Premises, Malapuram District, Kerala

This site was declared as National Geological Monument in 1979 on the occasion of International Seminar on lateritisation processes. Laterite is a secondary residual product resulting from chemical weathering of a variety of parent rocks. Laterite is used commonly as a dimension stone in local building blocks. The geoheritage site is maintained by State Govt. of Kerala.

15.2.4 Volcanogenic Bedded Barytes, Mangampeta, Cuddapah District, Andhra Pradesh

Mangampeta bedded Barytes deposits are largest amongst the known occurrences in the world. Barytes occurs in two well-developed massive blocks within the Pullampet Formation of Nallamalai Group of rocks in Cuddapah Supergroup. Barytes deposits form distinct layers separated successively by shale/phyllites and represent precipitation from vapours of volcanic origin under submarine environment. With huge proven reserves of chemically inert, high specific gravity and stable barium-sulphate occurrences in the Mangampeta bedded Barytes deposits India has attained global top rank in the production of Barytes. The site was declared as National Geological Monument in 1982.

15.2.5 Gossan in Rajpura-Dariba Mineralised Belt, Rajsamand District (Formerly Part of Udaipur District), Rajasthan

This site was declared as National Geological Monument in 1977. The importance of the gossans lies in its diagnostic characteristics which help in the recognition and location of sulphide minerals beneath the surface. This area is preserved considering the unique nature and educative value. The gossans' capping extends over a strike length of 4.5 km with width ranging from a few metres to 40 m.

15.2.6 Kishangarh Nepheline Syenite, Ajmer District, Rajasthan

This site was declared as National Geological Monument in 1976. The Nepheline Syenite bodies occur within a group of Precambrian metasedimentary rocks comprising predominantly of kyanite-staurolite-biotite schist, quartzite and amphibole-biotite schist which are considered basal members of the Delhi Supergroup. The nepheline syenite bodies comprise nepheline, theralite and camptonite the theralite occurring as inclusions and the camptonite occurring as dykes cutting across the nepheline syenites.

15.2.7 Naga Hill Ophiolite Suite, Near Pungro, Nagaland

The Naga Hill Ophiolite suite of rocks represent unique evidences defining the eastern part of India–Eurasia collision zone that preserves the rare segments of oceanic lithosphere that once was the Neo-Tethyn ocean in late Jurassic period. Ophiolite sequence represents section of oceanic lithosphere emplaced on continental crust or within the accretionary prism of sediments of subduction/supra subduction zones. At Pungro, Naga Hills, ophiolite outcrops occur as disconnected lenses enclosed between the Nimi Formation on the east and the Disang Flysch sediments on the west, as emplaced bodies in the Indo-Myanmar Range in Nagaland and Manipur, the North-eastern states of India. Naga Hill Ophiolite belt is exposed near Pungro, near Kohima. It is characterised by the presence of discontinuous thrusted sheets of highly tectonised ultramafic mantle sourced peridotite-altered to serpentinite, with dominance of harzburgite at the base followed upwards by ultramafic–mafic cumulates, gabbros and pillow basalt, and pods of dunite and anorthosite, to intermediate volcanic, mostly with pyroclastic components, oceanic pelagic sediments and high-grade metamorphic rocks such as glaucophane bearing metabasalt, metapelites and cherts. The site was declared as National Geoheritage site on July 24, 2014.

15.2.8 Panjal Volcanics in Srinagar District, Jammu & Kashmir

Panjal Volcanics in Srinagar district in Jammu & Kashmir state represent about 289 million years old Permian volcanic eruptions in the Himalayan Mountain building orogeny and occupy vast areas around the picturesque valley of Kashmir. These volcanic rocks are andesitic, dacite and rhyolites as the basal flows and grade to basalt in the upper younger flows. Besides being amygdaloidal in nature and vesicles being filled with secondary mineral of calcite, chlorite and epidote imparting green colour to the Panjal Volcanics, a typical glomeroporphyritic texture showing clustering of plagioclase feldspar crystals in mafic groundmass is seen at several places. The Panjal traps are best preserved at the area around famous and oldest shrine dedicated to the Lord Shiva–the Shankaracharya Temple overlooking the Dal Lake in Srinagar Kashmir.

15.3 Geological Boundaries and Contacts of Stratigraphic Significance

15.3.1 Eparchaean Unconformity, Tirumala–Tirupati Road, Chittor District, Andhra Pradesh

This geosite shows a boundary between two geologically contrasting groups of rocks and has great stratigraphic significance. The site was declared as National Geological Monument in 1976. Here the Nagari Quartzite belonging to lower part of the Nallamalai Group of the Proterozoic Cuddapah Supergroup occurs overlying unconformably the Achaean granite-gneiss basement rocks. The Nagari Quartzite is dated around 1600 million years old, whereas this prominent unconformity represents a hiatus or a break in geological deposition in Cuddapah sedimentary basin in the Eastern Dharwar Craton.

15.3.2 Barr Conglomerate, Pali District, Rajasthan

The site was declared as National Geological Monument in 1976. The Barr conglomerate rests unconformably upon the basement gneiss in the vicinity of Barr and is composed of pebbles of quartzite and rarely granite gneiss set up in a fine-grained pelitic matrix. The pebbles are stretched to extraordinary extent such that greatest dimension is as much as 20–30 times their least dimension.

Fig. 15.1 **a** and **b** Precambrian Malani Igneous Suite of rocks (volcanic tuff and rhyolites) underlie unconformably the horizontally bedded sedimentary rocks of the Jodhpur Group

15.3.3 Malani Igneous Suite of Rocks Contact with Jodhpur Group of Sedimentary Rocks, Jodhpur District, Rajasthan

The unconformable contact between the Malani Igneous Suite of the Precambrian age (700+ _ 50 My) and overlying aranaceous sedimentary rocks—the sandstones belonging to the basal part Jodhpur Group. The Malani Igneous Suite represents the last major phase of Precambrian igneous activity in the area. It is represented by 40–50 m thick welded tuff followed by flows of rhyolites indicating violent and fissure type acidic volcanic eruptions besides their plutonic equivalents. The Malani–Jodhpur contact is erosional and best exposed at base of the Mehrangarh Fort in Jodhpur city (Fig. 15.1).

15.4 Tectonic and Volcanic Structures/ Features of Regional Significance

15.4.1 Great Boundary Fault at Satur in Bundi District, Rajasthan

The western boundary of the Vindhyan Basin is delineated by a regional fault zone named as the Great Boundary Fault (GBF). It extends for over 400 km along NE/NNE–SW/SSW direction from south of Mandalgarh to south of Barodia in Rajasthan. GBF marks a significant tectonic unconformity and stratigraphic boundary between the Gwaliors (Pre-Aravalli group of rocks) in the NNW and the upper Vindhyan group of sedimentary rocks towards the SSE separated a by a over 20 km wide fault zone of parallel and oblique pervasive tectonic discontinuities. The geosite is located at Satur, about 10 km NW of the Bundi on Jaipur highway. It is unique and has immense educational value for geoscientists.

15.4.2 Meteorite Impact Structures

15.4.2.1 Lonar Lake Buldana District, Maharashtra

Impact crater is produced when a moderately large meteorite hit the earth's surface. Lonar impact crater in Buldana district, Maharashtra is one such unique feature and is reported to be a singular example of impact craters in the basalt volcanic provinces of the world. Lonar impact crater in Deccan Basalt rocks of Cretaceous-Tertiary age is circular depression, 137 m deep and 1830 m in diameter and supports a saline lake. The present lake is rimmed by 20 m high circular ridge comprising the 'shock ejecta', black glassy material characteristic of intensely shocked basalt due to high velocity impact of extraterrestrial object. Drilling investigations by GSI yielded highly fractured and brecciated basalt even at a depth of 335 m with typical *maskelnite* mineral formed under the shock impact.

15.4.2.2 Dhala Impact Structure—A Geosite in Shivpuri District, Madhya Pradesh

Dhala Impact Structure is located in the eastern fringe of Dhala village in Shivpuri district, Madhya Pradesh. It covers an area of 65 km^2. This rare geoheritage site is another confirmed terrestrial meteorite impact structure on earth. It is recorded as the largest, oldest and the only complex impact structures in South East Asia. It occurs entirely confined within Achaean Crystalline Basement of Bundelkhand Craton. Dhala Impact Structure geoheritage area is close to historic and cultural heritage cities of Jhansi and Khajuraho, and shows diagnostic evidences of impact shock metamorphism with multiple sets of planar deformation features in minerals clasts of 'ballen' quartz and checkerboard feldspar with granular zircon, etc. The impact event has been estimated to have taken place between 1.6 and 1.7 Ga.

15.4.3 Rare Volcanic Features of Past Geologic Events of Eruption

15.4.3.1 Pyroclastic and Pillow Lavas, Kolar Gold Fields, Kolar District, Karnataka

These were declared as National Geological Monument in 1974. These indicate the KGF as a centre of major volcanic activity. The pillow lavas are of two types, one of them is characterised by chilled margin and the other contains not only a chilled margin and the other contains not only a chilled margin but also an inner amygdular border in some places. The pyroclastics consist of basaltic rock fragments some of which might be distorted volcanic bombs granite, granite gneiss, etc. The rock fragments are of various shapes, in some cases, they are fused with one another, their boundaries getting blurred at their contacts.

15.4.3.2 Columnar Lava, St. Mary Island, Karnataka

Columnar Lava, St. Mary Island in Karnataka was declared a National Geological Monument in 1979. The columnar structures/joints are considered to be the result of contraction during the slow cooling of large volume of hot mobile lava. These rocks formed part of the major geological event which witnessed stupendous outpouring of basaltic lava in the vast Deccan Plateau during the Cretaceous-Eocene times.

15.4.3.3 Pillow Lavas Near Mardihalli, Chitradurga District, Karnataka

The site was declared as National Geological Monument in 1979. In cross section, the pillows show a thin chilled glassy margin and radial cracks and vesicles. At the base of the flow, the pillows are bun shaped with a flat bottom and rounded top and have downward-pointing projections that fit into the V-shaped space between two lower pillows. The age of the pillow lavas is estimated to be 2500 million years.

15.4.3.4 Pillow Lava in Iron Ore Belt at Nomira, Keonjhar District, Odisha

This site was declared a National Geological Monument in 1976. The pillow lavas are reported from basaltic and andesitic volcanic rocks associated with Bonai–Keonjhar Iron Ore formation in the Gorumahisani Belt of Odisha. Ellipsoidal pillow structures generally exhibit chilled margins and concentric compositional layering. These volcanic primary structures and associated tuffs at Nomira are underlain by quartzite towards east and overlain by shale-cherty-shale and Banded Hematite Jasper on the west implying subaqueous nature of eruptions. This geosite of rare Precambrian pillow structures was declared as National Geological Monument in December 1976.

15.4.3.5 Welded Tuff, Malani Igneous Suite, Jodhpur District, Rajasthan

Welded Tuff, belonging to the Malani Igneous Suite, is exposed near the Mehrangarh Fort within Jodhpur city. Welded tuffs are silica-rich glassy pyroclastic rocks that of rhyolites composition are indurated by welding of glass shards and other violent volcanic ejects by combined action of heat and superimposed overlying ash material and hot gases. The columnar joints developed in the rhyolites of the area show hexagonal and rectangular spectacular columns attaining a length of over 30 m at places (Fig. 15.2).

Fig. 15.2 Welded Tuff, belonging to the Malani Igneous Suite, is exposed near the Mehrangarh Fort within Jodhpur city

15.5 Rare Sedimentary Structures and Fossiliferous Stratigraphic Markers

15.5.1 Sedimentary Structures—Eddy Markings, Kadana Dam, Panch Mahals District, Gujarat

The eddy currents are preserved in the Upper Aravalli Lunavada Group of rocks, as spirals of 8–16 cm diameter on top of bedding plane of a fine-grained granite bed. In the centre of these spirals, whirl balls of 1–2 cm diameter are seen. The petrified record of eddies around the whirl balls form spiral-like structure. The outcrop showing the eddy markings was preserved and declared as geological monument by GSI in 1976.

15.5.2 Fossils and Fossiliferous Stratigraphic Markers

15.5.2.1 Fossil Wood Park near Tiruvakkarai, South Arcot District, Tamilnadu

This site was declared as National Fossil Wood Park in 1951. The fossil trees generally ranging in length between 3 and 15 m are found in coarse and friable Cuddalore Sandstones of Mio–Pliocene age. The trees belong to several rare species of conifers and pals and their fine woody structure and the annular rings as also the pit structure, the knots are preserved with astonishing fidelity. This park is maintained by GSI.

15.5.2.2 National Fossil Wood Park, Sattanur, Perambalur District, Tamilnadu

This geosite situated in Cauvery basin in Perambalur district, Tamilnadu is known for its shallow marine sedimentary rock sequence with rare fossil faunal assemblage of Upper Cretaceous age. The area also hosts the drifted petrified fossil wood logs of the non-flowering trees within the Trichinapoly Group of rocks, about 100 million years old, which grew in the deltaic-marshy environments. This site is declared as National Fossil Wood Park in 1951. The fossil wood occurs in the Upper Cretaceous rocks and maintained by GSI.

15.5.2.3 Siwalik Fossil Park, Saketi, Sirmur District, Himachal Pradesh

The park was initially inaugurated in 1974 with the erection of life-size models of Sabretooth tiger, Hippopotamus and Tortoise. Later models of giant elephant, four-horned giraffe and *gharial* were added. Inside the park upper Siwaliks with rich collection of fossil vertebrates which thrived during Plio-Pliestocene period are exposed. A museum is set up displaying large collection of vertebrate fossils collected from Saketi and adjacent areas. Saketi Fossil park was renovated and rare catalogued fossils shifted to a new and large building hosting a museum in May 2014. The park was developed by GSI in collaboration with the State Government and is presently maintained by the Himachal Pradesh State Govt. However, the museum and models are being maintained by the GSI.

15.5.2.4 Stromatolite Fossil Park, Jharmarkotra Rock Phosphate Deposit, Udaipur District, Rajasthan

This site was declared as National Geological Fossil Park in 1978. This area displays a wide variety of stromalities in dolomitic limestone of the Aravalli Supergroup. The stromatolites are mostly of branching type, but domal, club shaped and long columnar forms are also present. The stromatolites are identified as *Collenia Coloumanasis, Fenton and Fenton Collenia Kusiensis* and *Collenia Baicalcio Maseov*. The park is maintained by the Govt. of Rajasthan.

15.5.2.5 Stromatolite Park Near Bhojunda, Chittaurgarh District, Rajasthan

This site was declared as National Geological Fossil Park in 1976. This area shows spectacular development of stromatolite (conophyton, collenia and weddelia assemblages of algal structures). The stromatolite structures are associated with the massive Bhagwanpura Limestone Horizon of Lower Vindhyan age. This park is maintained by the State Government of Rajasthan.

15.5.2.6 Akal Fossil Wood Park, Jaisalmer District, Rajasthan

Jaisalmer sedimentary basin in western Rajasthan has preserved fossil evidences dating back to the Jurassic Period indicating hot and humid climate characterised by dense forests. 180 million years old fossils of animals and plants are preserved at and around Wood Fossil Park at Akal, situated 18 km away from Jaisalmer city. The Jaisalmer Basin formed part of the southern shelf of the Neotethys ocean during the Jurassic times. The area is well known for its rich geodiversity both in terms of landscapes and outcrops of rock types and the variety of fossils that these rocks have preserved. Present Thar Desert area once hosted luxuriant forests in a tropical climate, bordering the sea some 180 m.y. ago. The fossil wood logs considered to represent gymnosperms are found in the sandstone belonging to Lathi Formation of Lower Jurassic age. This fossil wood park was declared in 1977. This park has been maintained by the State Forest Department, Government of Rajasthan (Fig. 15.3).

Fig. 15.3 Akal Fossil Wood Park, Jaisalmer District, Rajasthan

15.5.2.7 Lower Permian Marine Beds at Mahendragarh, Surguja District, Chhattisgarh (Formerly Part of Madhya Pradesh)

The site was declared as National Geological Monument in 1982 to preserve the Lower Permian marine fossiliferous and indicating faunal affinity of Lower Gondwana age. The fauna reported include various species of brachiopods, gastropods, crinoids and foraminifers. The Fossil Park is located on the right bank of the Hasdeo River, near the railway bridge.

15.5.2.8 Stromatolite Bearing Dolomite Limestone of Buxa Formation at Mamley, Near Namchi, South Sikkim District, Sikkim

The geoheritage site at Mamley, 6 km NNE of Namchi, Sikkim, exposes rare lithounits of Buxa Formation of the Daling Group of Proterozoic age. There are purple calcareous slates, minor quartzite and dominant grey-purple limestone/dolomite that are stromatolite bearing—the Precambrian blue-green algal structure. This stromatolite occurrence is created as layered biochemically accreted sequence of biostructures that were formed under shallow marine water conditions. These algal structures are indicative of the earliest life evolving on earth and presently preserved in south

Sikkim. The site was declared as National Geoheritage by GSI site in June 2014 and is being protected and developed into geotourism site by Government of Sikkim.

15.5.2.9 Salkhan Stromatolite Fossil Park, Sonbhadra District, Uttar Pradesh

Stromatolite Fossil in Salkhan area of Sonbhadra district, Uttar Pradesh shows a rare occurrence of columnar/cylindrical forms of algal Conophytons colonies, attaining up to 30 cm diameter and 110 cm height, of Precambrian microbial growth. The host rock Salkahan limestone belongs to the Semri Group of Lower Vidhyan sequences in the areas. Stromatolite is the layered biochemical accretionary structures formed in shallow marine water environment by trapping and cementing fine grains by cyanobacteris—a unicellular photo-synthesising microbe.

15.5.2.10 Plant Fossil Bearing Inter-Trappean Beds of Rajmahal Formation, Upper Gondwana Sequence Around Mandro, Sahibganj District, Jharkhand

Plant fossil bearing inter-trappean continental sedimentary beds of Rajmahal Formation, Upper Gondwana sequence around Mandro, Sahibganj district in Jharkhand, occur enclosed within a series of volcanic lava flows of basalt. These less than one to 11 m thick sedimentary deposits preserve a rich collection of Ptilophyllum flora fossils and petrified wood belonging to the Upper Gondwana sequence of Permo-Carboniferous age (Fig. 15.6). The rare geosite is well preserved as plant fossil bearing inter-trappean beds in the area (Fig. 15.4).

Fig. 15.4 **a** and **b** Petrified tree trunk lying vertically and showing annular ring in transverse section

Fig. 15.5 Dinosaur Fossil Park at Raiyoli—Balasinor, Gujarat (Photo courtesy R. Arya)

15.5.2.11 Raiyoli—Balasinor Dinosaur Fossil Park, Kheda District, Gujarat

Several dinosaur fossil skeletal remains and best-preserved fossilised nests of dinosaur eggs were discovered by GSI in 1981 at the Raiyoli—Balasinor Dinosaur Fossil Park in Kheda district of Gujarat. These fossils are preserved within calcretised sandstone and limestone beds of the Lameta Formation belonging to the Upper Cretaceous (Maestritian) age. This geosite spread over an area of 70 acres, has yielded over 13 different species of dinosaur that thronged the area for more than 100 million years until their extinction some 66 million years ago. Presently the privately owned area having dinosaur fossil occurrences is protected and conserved as an enterprising local family who have developed geotourism hotspot in the area (Fig. 15.5).

15.5.2.12 Karai Badlands-Kulakkalnattam Geological Section, Perambalur District, Tamilnadu

Marine sedimentary rock sequences in the present-day Cauvery basin show excellent evidences of marine transgression of Mid-Late Cretaceous age and support a rich collection of well-preserved invertebrate fossils. The Karai-Kulakkalnattam geological section in Perambalur district of Tamilnadu hosts a variety of marine Cretaceous fossils of ammonites, nautiloids, belemnites, pelecypods, gastropods, oysters, etc. and has been recognised as the best section of marine Cretaceous rocks exposed in the world. The site shows highly dissected terrain as 'badlands' and fossiliferous horizons are best exposed for over 4 km on either side of the Allatur-Ariyalur road [NH-45].

15.5.3 Geomorphic Sculptures and Landscapes

15.5.3.1 Sendra Granite, Pali District, Rajasthan

Declared as National Geological Monument in 1976, this granite suite occurs as isolated granite exposures emplaced within calc-schist, calc-gneisses and migmatites of Khumbhalgarh Group of Delhi Supergroup. The granite suite comprises granite, granite gneiss and associated mignatite and syenite. The granite shows grain size variations from medium to coarse grained and hypidiomophic texture and crude schistose and gneissic foliation is conspicuous at places.

15.5.3.2 Natural Geological Arch, Tirumala Hills, Chittor District, Andhra Pradesh

This natural arch is a unique natural marvel a bridge carved out of jointed quartzite (Nagri Quartzite about 1400 million years old, belonging to the Cuddapah Supergroup) by collective erosional actions of winds and water operative over geological time. Natural arches of such scenic beauty and dimension are rare in the world and only two similar well-known examples have been geoconserved and developed as geotourism sites: (1) The Rainbow Arch in Utah region of USA and (2) Natural Arch in the Dalradian Quartzite in United Kingdom. The Natural Geological Arch at Tirumala Hills was declared as National Geological Monument and is being maintained by the Tirumala Tirupati Devasthanam Authorities since 1981, as an important geomorphosite or the heavenly creation of Nature.

15.5.3.3 Varkala Cliff, Thiruvananthapuram District, Kerala

Varkala Cliff has been established as a major Geotourists destination, as it provides a great scenic view of high aesthetic value. It is a major cliff in the state of Kerala facing the Arabian Sea, extending from Edava in the north to Vettoor in the south. It exposes a geological type section of the Tertiary sedimentary rocks (Mio–Pliocene age of 15–23 million years) of the Warkalli Formation, showcasing shoreline littoral deposits of shallow marine—estuarine environment and Neogene tectonism. The site was declared as National Geoheritage site by GSI in May 2014 and is preserved by the State Government of Kerala (Fig. 15.6).

a b

Fig. 15.6 **a** and **b** Varkala Cliff sections, Thiruvananthapuram district, Kerala

15.5.3.4 Erra Matti Dibbalu, Vishakhapatnam District, Andhra Pradesh

Erra Matti Dibbalu (EMD), the dissected stabilised coastal red sand dunes and poorly consolidated sediment mounds carved into 'badlands' are located between Vishakhapatnam and Bheemunipatnam. EMD covers 1083 km^2 area amongst the Khondalite suite of rocks exposed in the vicinity as variously eroded ridges imparting spectacular scenic beauty along the recreational beaches. The geoheritage area preserves rare geological and geomorphic phenomena of marine, winds and fluvial agencies that have been active for evolution of east coastal plain. The site was declared as National Geoheritage site in early 2014 and with active involvement of INTACH and support from the State Government, area is geoconserved and promoted as geotourism resource.

15.5.3.5 Arwah Lumshynna cave, East Khasi Hills District, Meghalaya

Arwah Lumshynna cave is formed within fossiliferous limestone of Shella Formation of Jaintia Group belonging to the Palaeocene epoch of lower Tertiary period in Meghalaya. The Arwah Lumshynna cave in limestone horizon is developed for about a kilometre in length as a karstic dissolution feature and is set in spectacular natural scenic surroundings. It has already been promoted as geotourism site in Meghalaya and attracts ever-growing footfall of nature lovers and adventure tourists.

15.5.4 Challenges for Geoconservation and Geotourism and Ways Forward

It is remarkable that while initiating an outreach programme on Indian Geoheritage sites and launching a movement for conservation of the formerly listed 26 National Geological Monuments by the GSI, the Natural Heritage Division of Indian National Trust for Art and Cultural Heritage (INTACH) took up a survey in 2016 on their status of geoconservation, outlining the threats and recommending ways to protect and popularise the invaluable geodiversity and geosites for lasting value and sustainable development. The results were compiled and published as a monograph that updated on current scenario with aims to influence the policymakers for urgent need for a comprehensive central legislation for geoconservation and associated systematic promotion of geotourism (Wadhawan 2013; Singh and Anand 2013). Several new potential geosites in Indian terrain were also indicated with captioned field photos. A draft Geoheritage Law was also prepared that has been pending due diligence with the Government of India.

The Geoheritage and its sustainable management for the education and well-being of present and future generations are rapidly developing in many different settings from resource extraction to cultural and environmental protection. Using geoheritage for education can also raise awareness about earth processes and the historical interactions between people and their planet. Educational geotourism based on an exceptional geoheritage in national, provincial and municipal protected areas, as well as in UNESCO Global Geoparks provides opportunities for recreation, education and tourism, and can be a new focus for sustainable economic development. As an added bonus, geoheritage also provides opportunities for presenting and exploring indigenous interpretations of landscapes. Therefore, geoheritage in many settings, such as UNESCO World Heritage Sites, UNESCO Global Geoparks, areas of historical resource extraction or other protected sites of all kinds, including local, national level geosites, may be used for educational and geotourism purposes.

In spite of assertions made in the High Power Committee Report 2009 by the Ministry of Mines, the cadre controlling parent authority for GSI, 'advancement of the cause of Geosciences by documentation, propagation, archiving and education, including creation and management of Geological Monuments and Parks for use of the public, students, and researchers, and future generations. The purpose is to popularise geosciences for public…', much remains to be achieved. However, way back in 1990, under directions from the then Board of Management, Ministry of Mines and GSI had attempted to draft a Bill for conservation, protection and maintenance of such declared national geological monuments. Article 49 of the Indian Constitution provides for protection of monuments, places and objects of national importance. It is under this Article that the Central Government will have to enact a law to define 'a natural heritage resource and place or object of historic interest to be of national importance', put in place a system to identify and protect all such places or objects and notify the same for protection. As the issue involves collateral intervention of different Departments/Ministries such as Ministry of Tourism, Culture, Roads and

Transport, Environment and Climate Change, Earth Sciences and Mines, this could not fructify and remained unattended for lack of political will and administrative apathy. This attempt may now be revived and the Act can be called the National Geoheritage Sites and Geoparks (Conservation, Protection and Maintenance) Act, 2020. Besides, we need to evolve the relevant Rules and Regulations against destruction or mining of such rare sites and check smuggling or selling of rare fossils from different geosites (Wadhawan 2020).

Opportunities for Earth scientists to influence important policies and decisions have been constrained with difficulties more so in a democratic and diverse setup like ours in India. This is amply ratified by the proactive functioning of some States in sharp contrast to several others who have remained dormant on the subject in India. Although some State Governments in India have been convinced to take proactive measures on the need to protect the geodiversity and to geoconserve and promote geotourism, such as in States of Tamil Nadu, Sikkim, H.P., yet hardly any dent is made on others. As national custodian and Nodal Agency, GSI can certainly play a leading role for deliberating with and inviting participants from all stakeholders, government and non-government and coordinating the result-oriented activities. GSI needs to ensure its national commitment to preserve the National Geoheritage sites/Geological Monuments. In any case, I think the electorate of the region is supreme and most political or bureaucratic decisions are made that are considered economically viable and socially responsible. Hence more proactive approaches are needed to project the importance of geodiversity and make geosciences contributions clearly visible to the public. However, the role and geoscientific ownership of Geological Survey of India (mandated by Charter of Functions, para-9, Government of India 2009) must remain intact and with added responsibility of supervising the protection and conservation efforts.

In view of general apathy and inordinately long-pending legislation, need to realise geotourism potential has been gaining momentum, outlined below are the two approaches that were successfully adopted with gainful results to safeguard geoheritage, promote geoconservation and convert these unique geodiversity areas into geotourism hotpots. These are enumerated below.

Case Study-1: Jaisalmer Jurassic Park, Rajasthan

Jaisalmer sedimentary basin in western Rajasthan has preserved fossil evidences dating back to the Jurassic Period indicating hot and humid climate characterised by dense forests. 180 million years old fossils of animals and plants are preserved at and around Wood Fossil Park at Akal, situated 18 km away from Jaisalmer city. The Jaisalmer Basin formed part of the southern shelf of the Neotethys ocean during the Jurassic times. The area is well known for its rich geodiversity both in terms of landscapes and outcrops of rock types and the variety of fossils that these rocks have preserved. Therefore, the Jaisalmer Basin in the Thar Desert has cultural, historical, biodiversity and geoheritage sites and landscapes——all the necessary ingredients and built-in infrastructure to utilise potential for developing Geopark in west Rajasthan, India (Wadhawan 2016). The city of Jaisalmer in western Rajasthan was founded in 1156, located within western part of the Thar Desert in India and has been on a historical trade route between Delhi and West Asia. It is presently a famous tourist destination very well connected by rail, road and air. Entire city with over 60,000 inhabitants is built with yellow coloured siliceous limestone of Jaisalmer Formation of Jurassic age - a global heritage stone resource. It has world famous over five centuries old Golden Fort built on a 100 m high Trikuta Hill surrounded by golden aeolian sands that glow majestically with the evening sun rays. Several historical buildings called *Havellis* or the private mansions built by merchants have intricately carved yellow sandstone facades. Other tourist attractions include the Jain and Hindu temples built during twelfth and fifteenth centuries. In addition, owing to the presence of a diverse range of geological phenomena including, amongst many others, unique stratigraphic units, structures, minerals and rare dinosaur fossils, foot-prints and petrified wood logs, the Jurassic Jaisalmer Basin and geomorphological landforms forming part of the present-day dryland environment in western India is a potential candidate for becoming a Geopark (Wadhawan 2016).

Presently a notified territory of about 3162 km^2 constitutes a Desert National Park (DNP) as a unique and fragile ecosystem in dryland environment. DNP harbours a wide array of flora and fauna species and naturally supports the symbolic and protected Rajasthan State bird (Great Indian Bustard), animal (Camel) tree (Khejri) and flower (Rohida). It also has fossil evidences dating back to the Jurassic Period indicating hot and humid climate characterised by dense forests. Lithostratigraphy of the mapped formations namely the Lathis, Baisakhi and Bhadasar formations is well established that displays an array of Jurassic siliciclastic, mixed carbonate-siliciclastic and carbonate rocks that range in age from Early Jurassic to Tithonian. A variety of depositional environments ranging from continental fluvial to near-shore and off-shore deep marine are well established and documented. The desert landscape includes a variety of stable and active dune fields including parabolic, linear transverse and barchans of different size and generations, desiccated rocky plains and pebble-boulder spreads as deflation lag deposits and salt lakes. Besides the prevalent traditional water harvesting practices, application of modern geotechnical knowledge in dryland setting of western India has facilitated inter-basin transfer of perennial Himalayan fresh-water of the Sutlej River through *Indira Gandhi Nahar Priyojna*—a system of canals and distributaries—thereby addressing the scarcity and availability of potable quality water for the human settlements within the Thar Desert. Various distinctive geological and geomorphological features preserved within the delineated territory showcase ideal palaeo-environmental settings in time and space and offer primary tools for geoscientific research, education, training and Geotourism for socioeconomic development of the region. Therefore, efforts were made to involve a well informed and enlightened participation of the Rajasthan Forest Department for promoting geosites, at least within the notified, protected and delineated Desert National Park which has the assigned responsibility of preserving and promoting the rich biodiversity of the desert landscape within the demarcated territory of Jaisalmer Basin.

This case study emphasises on identifying and elaborating on unique geoheritage sites and landscapes as available within the delineated and notified territories of the National Parks or Wildlife Sanctuaries or Botanical Reserves, etc., listed by the Forest departments of the States or the Central Government authorities. As it involves no extra burden on the given logistics and infrastructure and provides value addition and the needed third dimension was strengthened through conjunctive utilisation of geodiversity together with biodiversity and the cultural heritage prevalent within the allocated resources. Easy acceptability of such joint action plans needs to be supported with expert technical guidance and geoscientific explanatory notes by Geoscientists of the GSI, research organisations and academia. Colourful attractive brochures, published literature and detailing on the geotourism aspects will have to prepared and distributed through digital media and websites. Next logical step is to get the area declared as a National Geopark with active support from stakeholders, an NGO like INTACH and the Rajasthan State government.

Case Study-2: Rao Jodha Desert Rock Geopark Around Jodhpur Fort Mehrangarh Ridge and City Area

The sprawling rocky tracts in Jodhpur host a variety of national geological features and monuments, cultural and historic heritage forts and temples, and traditional water harvesting structures and conservation of rich lithophytes—the sturdy desert flora that can thrive on the rocky boulder landscapes in the dryland environment are preserved and developed as a geopark (Wadhawan 2013, 2016; Mathur et al. 2017). It is now creatively maintained and promoted as geotourism attraction by the Mehrangarh Museum Trust managed by the erstwhile royal family of Jodhpur. The Jodhpur city that served as a capital of Marwar region under the Rathore dynastic clan was founded by the Rajput king Rao Jodha in 1459. Spread over 200 acre plot of rock garden, the two Geoheritage sites of 1) Welded Tuff belonging to the Malani Igneous Suite, exposed near the Mehrangarh Fort within Jodhpur city and 2) The unconformable contact between the Malani Igneous Suite of the Precambrian age (700 + _ 50 My) and overlying arenaceous sedimentary rocks—the sandstones belonging to the basal part Jodhpur Group (Hose 2012; Dowling 2011; Singh and Anand 2013) are now geoconserved and developed into geotourism sites through guided tours or well-marked walking trails with information signage. Over 300 native desert species have been planted to create a colourful sanctuary-cum-nursery of desert flora and fauna at the Rao Jodha Desert Rock Geopark. This success story depends entirely on the private enterprise and ownership of the land that happens to be hosting the geoheritage sites/georesources together with other significant historic or cultural heritage and traditional societal aspects. The managers of the site or museum needed to be enthused and motivated to see the long-term benefits of initial investment for developing the intrinsic geoscientific themes with collateral value additions and basic amenities for promotion of geotourism, sale of geoproducts handicrafts and mementos, etc. A similar approach adopted at the Raiyoli—Balasinor Dinosaur Fossil Park in Kheda district of Gujarat has been yielding rich dividends on utilising the inherent potential of geotourism, thereby conserving the rare fossil sites for future generations and generating added source of income and employment in services and catering industry.

15.6 Concluding Remarks

1. The geoheritage sites must be suitably delineated, notified and protected initially by Geological Survey of India (GSI) as are part of an integrated concept of protection, geosciences education and sustainable development of the area around with active support from the State governments and stakeholders. The mandated responsibility of GSI to maintain or protect the identified Geoheritage sites/National Geological Monuments and Fossil Parks must be strengthened on a continuing basis. The Siwalik Fossil Park at Saketi, Sirmaur district in Himachal Pradesh, amongst others, is an outstanding example of such approach.
2. GSI also has to identify several more Geoheritage sites (many of these are known or reported and are waiting to be officially recognised), and to elaborate on their unique geoscientific characters and delineate the area (with definite Lat./Long. coordinates), protect and maintain these through pillar-fencing and labelling with large signages.
3. The moot question is how best to convert these georesources to realise the geotourism potential of these inventoried geoheritage sites and promote conjunctive utilisation and geoconservation of available geodiversity, biodiversity and cultural heritage through incentivised involvement of public–private sector as have been elaborated in the aforesaid two case studies.
4. However, to develop a recognised Geopark of National and International standing, India still has to go a long way and encourage formulation of comprehensive management plans designed to foster socioeconomic development that is sustainable and based on geotourism preferably in Private–Public-Partnership mode that would take adequate care for conserving and enhancing geological heritage and provide means for teaching geoscientific disciplines and broader environmental issues directed towards protection of notified geodiversity and optimum utilisation of georesources.
5. To facilitate this objective the need to accord top priority to enact comprehensive legislation for delineation, notification and conservation of geoheritage sites and development of Geoparks to promote sustainable geotourism with active participation of private sector and as per international good practices as delay has been leading to neglect and destruction of vulnerable geodiversity and geoheritage sites.

References

Anantharamu TR, Bellur D, Bhasker AA (2001) National geological monuments. Geological Survey of India, Special Publication No. 61, vol vi, 98p

Bhargava ON, Kumbkarni S, Ahluwalia AD (2010) Geomorphology and landscapes—illustrations from Himalayas. Technology Publications, Dehradun 211p

Brocx M, Semeniuk V (2007) Geoheritage and geoconservation—history, definition, scope and scale. J R Soc West Aust 90:53–87

Dowling RK (2011) Geotourism's global growth. Geoheritage 3(1):1–13
Dowling RK, Newsome D (eds) (2010) Global geotourism perspectives. Goodfellow Publishers, Oxford
Geological Survey of India (2012) Travel brochure on National Geological Monuments of India: Region South. Geological Survey of India, Hyderabad, 12p
Geotourism Hotspots of Indian Subcontinent (2016) 36th International Geological Congress (IGC), C/o Geological Survey of India, New Delhi, 101p
Government of India (2009) Report on the functioning of Geological Survey of India, Ministry of Mines. New Delhi. 369p
Gray M (2004) Geodiversity: valuing and conserving abiotic nature. Wiley, Chichester 434p
Hose TA (2012) 3G's for modern geotourism. Geoheritage 4:7–24
Kale VS (2009) DEM-based studies of some outstanding fluvial landforms in India. In: Sharma HS, Kale VS (eds) Geomorphology of India. Prayag Pustak Bhavan, Allahabad, pp 59–68
Kale VS (2010) The Western Ghat: the great escarpment of India. In: Migon P (ed) Geomorphological landscapes of the world. Springer, New York, pp 257–264
Lazzari M, Aloia A (2014) Geoparks, geoheritage and geotourism: opportunities and tools in sustainable development of the territory. GeoJ Tour Geosites (GTG) 13(1):110–123
Mazumdar MK (2007) The Geopark Initiative. Curr Sci 92(1):13
Mathur SC, Pradip K, Rathore SS, Mathur S (2017) Mehrangarh ridge of Jodpur City, Western Rajasthan, India: its geoheritage and geotourism sites. In: Proceedings of International Geopark Conference in Taiwan, pp 45–51
Mc Keever PJ, Zouros N (2005): Geoparks: Celebrating earth heritage, sustaining local communities. EPISODES, December 2005, pp 274–278
Natural Heritage Division of Indian National Trust for Art and Cultural Heritage (INTACH) (2016) A monograph on National Geoheritage Monuments of India. Natural Heritage Division, New Delhi, 146p
Panizza M (2001) Geomorphosites: concepts, methods and examples of geomorphological survey. Chin Sci Bull 46(suppl):4–6
Reddy DR (ed) (2013) Proceedings of National Workshop and brainstorming session on geoheritage: need for an Indian activism. Department of Geology, Andhra University, Visakhapatnam, 30–31 May 2013, 94p
Reynard E, Panizza M (2005) Geomorphosites: definition, assessment and mapping an Introduction. Géomorphol Relief Process Environ 3:177–180
Singh RB, Anand S (2013) Geodiversity, geographical heritage and geoparks in India. Int J Geoheritage 1:10–26
UNESCO (1999) What is a global geopark?|United Nations Educational, Scientific and Cultural Organization. Unesco.org. Accessed 16 Aug 2018
Wadhawan SK (2013) Conservation, protection and maintenance of national geoheritage sites – contributions from Geological Survey of India. In: Proceedings of the National Workshop brainstorming session on Geoheritage Sites in India, Andhra University & INTACH, Vishakhapatnam, 30–31 May 2013, 23–45p
Wadhawan SK (2016) Jaisalmer basin in Thar desert - potential for developing Geopark in Rajasthan, West India. In: Proceedings of the 35th international geological congress, IUGS, Cape Town South Africa. Association of Geoscience Societies of America, 11p
Wadhawan SK (2020) Responsible Mining and Conservation of Geodiversity through Geotourism in India – a perspective. Mining Engineers' J 21(10): 22–26

Part III
Geotourism

Chapter 16
New Routes of Geotourism for the La Campana–Peñuelas Biosphere Reserve, Chile

Rodrigo Figueroa-Sterquel, Jorge Negrete-Sepúlveda, Andrés Silva-Marambio, and Fernando Cosio

Abstract The Mediterranean ecosystem of Central Chile is considered a global "*hotspot*" of biodiversity due to its high level of endemism and rapid rate of destruction of its habitats. This is why the conservation of biodiversity and geodiversity has recently gained greater importance, including the potential of Biosphere Reserves to new international management agreements. The concept of Biosphere Reserves (associated to the UNESCO Man and Biosphere Program) and its foundations of land and territory management gives the possibility of a Participatory Action Research with the different actors of the territory. For the characterization La Campana–Peñuelas Biosphere Reserve as a Geotouristic destiny, an innovative methodological procedure was developed with diverse theoretical, conceptual, and technical approaches. This includes the assessment of geodiversity sites and also the cadastral registers and inventories of attractions and tourist services in the countryside, with the most suitable places were determined to locate activities/services for nature and rural tourism. In addition, we connected the character of the attractions with the trends of international demand in natural and cultural tourism, which allowed us to formulate and predesign 9 routes. The criteria for the final selection of tourist routes were (a) Territorial aptitude including traceability of agro-sustainable products at community level; (b) Tourist equipment; (c) Complementary services; (d) Investments; and (e) Associativity. Applying these criteria, four tourist routes were designed (a) Casablanca Valley Wine Route; (b) Charles Darwin Route (La Campana and Palmas de Ocoa); (c) Royal road Route "La Dormida"; and (d) Railroad route Valparaíso—Santiago. Finally, and responding to the need to measure and quantify the perception of the quality of services and the environment of the new routes, a survey was applied to evaluate the satisfaction of visitors to the La Campana–Peñuelas Biosphere Reserve. In addition, national, regional, and local tour operators were invited to evaluate critical points and gaps in tourist services and the management of the territories.

R. Figueroa-Sterquel (✉) · J. Negrete-Sepúlveda · A. Silva-Marambio · F. Cosio
Pontificia Universidad Católica de Valparaíso, Valparaíso, Chile
e-mail: rodrigo.figueroa@pucv.cl

R. B. Singh et al. (eds.), *Global Geographical Heritage, Geoparks and Geotourism*, Advances in Geographical and Environmental Sciences,
https://doi.org/10.1007/978-981-15-4956-4_16

Keywords Biodiversity hotspot · Mediterranean Chile · Sustainable tourism · Geotouristic routes · Tourism governance

16.1 Introduction

The La Campana–Peñuelas Biosphere Reserve (BR LC-P) is an iconic BR in Central Mediterranean Chile, a territory recognized by its global importance as a center of plant diversity and a biodiversity hotspot, recognized by its high levels of plant endemism (2014), related to the geodiversity and landscape heterogeneity in tune with longstanding evolutionary and biogeographic processes (2011). In spite of this ample recognition, Central Chile, that ranges from the southern Atacama Desert at 30°S toward the temperate forests around 40°S, is considered as one of the most threatened ecoregions nationwide. The most significant threats are currently urban expansion, wildfires, electric projects, pollution, and a general depletion of geobiological values and environmental services. This is a common issue for Mediterranean ecoregions, encompassing huge challenges for conservation and restoration, related to population growth and land-use pressures of industrial agricultural expansion, urban sprawl and also increased mass-tourism (Naveh 2007).

Central Mediterranean Chile encompasses three Biosphere Reserves: Fray Jorge BR, La Campana–Peñuelas BR and Nevados de Chillán BR (Moreira-Muñoz and Borsdorf 2014). The pitfalls of the Chilean BRs have recently been recognized, including wide environmental menaces and a low societal recognition, associated to the (too) slow implementation of the MaB National and Regional Committees; and a weak (if any) strategy to the achievement of the Lima Action Plan goals (Carvajal-Mascaró et al. 2019).

In the proposed expansion of the La Campana–Peñuelas Biosphere Reserve (BR LC-P), the Pontifical Catholic University of Valparaíso (PUCV) played a very important role. In 2008, the Geography Institute made available to the forestry agency that administers the protected areas in Chile (CONAF), 30 students of a fifth-year geography course, led by professors Jorge Negrete, Rodrigo Figueroa, and Sebastien Velut, for the realization of a diagnosis of the territorial context of the RB LC-P. The request of the Ministry of Foreign Affairs of Chile, which included the results of the diagnosis, allowed to approve its expansion as a World Biosphere Reserve by UNESCO in 2009. Since then, the PUCV and its Institute of Geography are part of the Management Committee as the scientific representative of the reserve. The tourism potential of the La Campana–Peñuelas BR in particular has been long recognized) (Figueroa et al. 2012a, b; Fuenzalida et al. 2013), but the pitfalls and challenges are still there. Impacts of urban sprawl and wildfires have been assessed by (Salazar et al. 2015); also the presence of mesocarnivores within the Biosphere Reserve but outside the limits of the core areas has been evaluated by Muñoz et al. (2017).

The relation between geomorphology, geoconservation, and geotourism has been remarked by (Thomas 2012). Several aspects of geoethics come nowadays into play (Acevedo and Martínez Frías 2018).

The critical interdependency between Geoheritage and Geotourism has been discussed by Newsome and Dowling (2018).

For the case of Valparaiso region, the relation of biodiversity values, geoheritage, and cultural values has been remarked recently by Manríquez et al. (2019) and the value for the children's education for Arenas-Martija et al. (2019) in this book.

In 2010 the Government of Chile, through its productive development agency (CORFO) launched an ideas competition for the development of public goods for tourism innovation, this instrument was seen as an opportunity by PUCV scholars. They applied for and won the contest with the proposal "New Destinations and Products of Nature and Special Interests for the Valparaíso Region", which was intended to help boost the central territory of the Valparaíso region through the diversification of destinations linked to Nature Tourism and Special Interests, with focus on the World Biosphere Reserve "La Campana–Peñuelas" and the corridor between Santiago and Valparaíso.

This contribution seeks to explicitly develop a proposal of geotourism routes that enhance the potential of the La Campana–Peñuelas BR as a territory in transition toward sustainability. Together with the selection of the routes, the coarse aspects of a governance strategy for the Reserve are outlined.

16.2 Methods

The most suitable places to locate activities and services for nature and rural tourism were assessed associated to the presence of geodiversity sites and also the cadastral registers and inventories of attractions and tourist services in the countryside.

For the inventory of potential and current tourism resources, the methodology proposed by the Organization of American States and its Tourism Training Center (OEA-CICATUR 1978) was used, adapted based on the experience of the research team, also addressing the services and tourist facilities in the territory (accommodation, food, travel agencies, transport, night clubs, among the main ones), generating descriptive sheets of the attractive and obtaining their absolute location through georeferencing by GPS, in order to generate both a geographic information system, and a means of representation and distribution of the open information, using the Google Earth(c) software.

In addition, once selected the definitive routes, specialized professionals made a characterization of the natural ecosystems and cultural landscapes, focusing on geotourism heritage and archeology, history, and rural culture, to give an integrated foundation to the proposed geotourism routes.

The methodology used for the identification and evaluation of natural ecosystems suitable for nature tourism and special interests derives from the clinical ecosystem methodology (Gastó 1979). It is used to address complex problems related to ecosystems. It consists of a series of logical steps for the collection and subsequent analysis of the phenomena under study. The adaptation to the present study consists of the following stages: (1) Characterization, (2) Diagnosis, (3) Design, and (4) Monitoring.

In relation to the objective of determining the most suitable places for the location of leisure and recreation activities designed to meet the needs of the tourist visiting BR, five criteria were used to determine the different degrees of territorial aptitude: (1) spatial accessibility, (2) distance to populated localities, (3) distance to priority sites for conservation, (4) distance to tourist attractions, and (5) distance to watercourses. Of these, the first four were considered as factors and the last as a restriction. Finally, as a decision rule, "expert judgments" were used on the relative importance of each criterion, resulting from the application of the reciprocal ranking method proposed by Malczewski (1999).

The character of the attractions with the trends of international demand in natural and cultural tourism was also assessed, allowing the preselection of nine tourism routes. The criteria for the selection of touristic routes were (a) territorial aptitude including traceability of agro-sustainable products at community level; (b) tourism equipment and infrastructure; (c) complementary services, (d) investments, and (e) associativity.

A survey was applied to evaluate the satisfaction of visitors to the La Campana–Peñuelas Biosphere Reserve. The survey encompasses these aspects:

- Demographic and socioeconomic characteristics of the visitor.
- Reasons, time of stay, forms of mobility, and spaces traveled on trips.
- Evaluation of positive and negative perceptions about the tourist destination and tourist services.
- Proposals for improvements to the offer of the tourist destination and services.

In addition, national, regional, and local tour operators were invited to evaluate critical points and gaps in tourist services and the management of the territories. The objective of the visit of the tour operators is to give an account of their perception with respect to the proposed geotourism products. Tour Operators of the Metropolitan Region and the Valparaíso Region participate, local operators and a representative of the SERNATUR Quality Unit. A survey was developed where Fam Tour participants carry out their evaluations. These surveys were delivered and explained to tour operators at the beginning of their trip. The evaluations were anonymous.

For the adequate governance of the project and the BR's routes, a protocol was designed in a participatory manner, covering these aspects:

- At the beginning and to ensure compliance with the project's administrative requirements, a top-down model was used.
- Form a public–private directory, committing the national and regional public administration from the beginning, to the communal administration and to the actors of the private tourism sector.
- Obtain the representation of the national tourism agency, the national productive development agency, the agency that administers the protected areas, the national agency that administers the national assets, a grouping of municipalities and a grouping of small and medium-sized businesses.
- The head of the regional government, Intendente, acted nominally as president of the Board of Directors.

- Subsequently, once the financing was obtained and the project started, the representatives of the municipal government were gradually included individually, and the representatives of business organizations of the local scale were involved.
- Strategic decisions were made with the consent of the national tourism agency and the national agency for productive development. These agencies would be the ones who would manage the geotouristic routes in the future once the project is finished.

16.3 Results

One of the first results was the creation of a Board of Directors chaired by the regional Intendant with representatives of the national and regional governments, representatives of local governments (municipalities), and small and medium-sized entrepreneurs at the regional level (Fig. 16.1). He held sessions at least twice a year and approved the proposals of the project's scientific team.

The summary matrix is presented, where a series of variables was evaluated to calculate the index of landscape importance, scientific interest, implementation of tourism projects and choice. Table 16.1 shows the calculated indices for each of the ecosystems ordered descending by the choice index.

First, there is the ecosystem "Bosque húmedo de roble de Santiago" (*Nothofagus macrocarpa*) with an index of 21.34. The presence of this ecosystem in a latitude so far north of the country and at an altitude between 900 and 1000 m above sea level is extremely important, possibly due to the nearby presence of El Roble and El

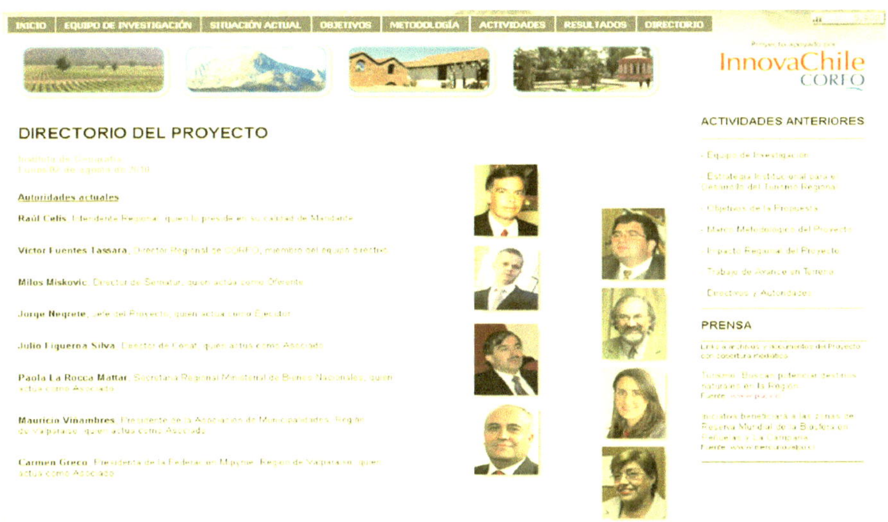

Fig. 16.1 Board of directors

Table 16.1 Summary matrix with the calculated and ordered indexes according to the choice index

Ranking	Identification		Diagnosis			
	Name	Route	Landscape importance index	Scientific interest index	Tourism projects implementation index	Choice index
1	Bosque Húmedo de Roble de Santiago	Darwin	21.5	10.5	7.1	21.3
2	Bosque Latifoliado de umbría	Darwin	20.5	10.5	7.1	20.8
3	Bosque Umbría de altura	Darwin	20.0	10.5	7.1	20.6
4	Bosque Camino la Virgen	Gold	19.5	8.3	4.5	17.3
5	Portezuelo Ocoa y panorámica	Darwin	8.0	10.5	7.1	14.6
6	Fundo Santa Laura	Royal route	15.8	8.3	3.3	14.4
7	Bosque latifoliado costero (Ocoa)	Darwin	7.5	10.5	7.1	14.3
8	Mina Nueva y panorámica	Darwin	6.5	9.5	7.1	13.3
9	Bosque Solana Cimarrón	Darwin	5.5	10.5	7.1	13.3
10	Playa Laguna Verde	Water	11.3	7.3	5.3	13.2

Iman hills, which would change a microclimate, from the summer dry domain to the wet domain, and consequently the predominance of this species. This ecosystem, in addition, presents a great landscape beauty, due to its dynamics of colors regarding the foliage, standing out the change of color from intense green in summer to red in the autumn season. You can see gorges with pure water that slips under the native vegetation, where certain birds in the middle fly or jump over the water. Place of recollection and propitious for the biophilia or affinity with nature.

They follow very closely, in 2nd and 3rd place, with 20.84 and 20.59, the ecosystems "Bosque latifoliado de umbría" and "Bosque Umbría de altura", respectively, corresponding to the summer dry domain, also high importance, for its dense native forest and its shelter. It should be noted that these three ecosystems correspond to the Darwin route and are within the La Campana National Park, as well as belonging to the core area of the Biosphere Reserve. Located at 484 m above sea level, with a steep slope, inside the Umbria broadleaved forest and where the native species of peumo (*Criptocarya alba*), maitén (*Maytenus boaria*) and boldo (*Peumus boldus*), expelling at sunset its penetrating aroma, together with an intense smell of humidity, while listening as the birds sing that seek their roost at nightfall. It is a harmonious cultural landscape of high scenic beauty and high eco and biodiversity.

In 5th place, with 14.6, the ecosystem «Bosque latifoliado costero» within La Campana National Park, through the entrance of Ocoa, this is perhaps, one of the most scenic beauty, for its high bio and ecodiversity (Fig. 16.2). This means a high presence of species, both plants and birds and other animals, forming a dense coastal broadleaved forest, with endemic characteristics, that is to say that there are unique species for the country and other areas. Among these we can highlight the Chilean palm (*Jubaea chilensis*), presenting itself as a forest, unique in the country, associated with other native trees and shrubs of the central coast. A series of wildlife is found there. At the end of this trail, after a journey of about two hours is La Cascada, also of a high beauty.

Within La Campana National Park, through the entrance of Ocoa, this trail is, perhaps, one of the most scenic beauty, for its high bio and ecodiversity. This means a high presence of species, both plants and birds and other animals, forming a dense coastal broadleaved forest, with endemic characteristics, that is to say that there are unique species for the country and other areas. Among these we can highlight the Chilean palm (*Jubaea chilensis*), presenting itself as a forest, unique in the country, associated with other native trees and shrubs of the central coast. A series of wildlife

Fig. 16.2 Coastal broadleaved forest in Ocoa sector of BR

is found there. At the end of this trail, after a journey of about two hours is La Cascada, also of a high beauty.

In this exercise, the Darwin route is the one that presents the greatest attractions, either because of its landscape quality as of scientific interest and the feasibility of implementing tourism projects. Among the outstanding attractions, is the La Campana National Park, for being also a core area of the BR.

16.3.1 The Tourist Demand of Geotourism

In relation to tourism demand, data from SERNATUR and CONAF show that the BR has a high potential for international tourist visitation when it is located one hour away from the main points of arrival of tourists to the country and the region. These points reached 3,009,815 tourists a year and correspond to the international airport of Santiago, with 34.3% of the country's total, and the Los Libertadores mountain pass, which receives 19.1% of the country's total tourists (SERNATUR 2017) and receives tourists from Argentina and the South American cone from the Atlantic slope. In the core areas of the RB, in 2016, the entrance of 102,292 visitors was registered, of which only 3.3% were foreigners. In addition, the potential demand for national tourists corresponds to 8,903,074, corresponding to travelers in high season (SERNATUR 2017).

In order to know the demand and the characteristics of the national and international market, a visitor satisfaction questionnaire was applied to the BR LC and its environment, raised in high tourist season to 117 groups of visitors, around La Campana NP.

The visitors mostly lodge in the surroundings (90%), therefore, they enter the category of tourists. About 30% know the place for the first time, giving an account of a greater diffusion of the destination. The majority origin corresponds to the Valparaíso Region (51%) from its metropolitan area distant 75 km to the coast and from the surroundings, forming a demand for proximity, in the vicinity of 100 km.

They are mainly adults with young children who preferably lodge in cabins or hotels and a minority, in the house of relatives or friends. The main reason for the trip corresponds to vacations, rest, and recreation, configuring a tourist destination of traditional interior.

They give very high importance for your tourist experience to the cleaning of tourist sites, accommodation, hospitality, protection of the environment, the quality/price ratio of services, and gastronomy (food and drink). The results of the evaluation of the experience of the tourist in the destination show a dissatisfaction with the quality/price of the offers, with the protection of the environment, with the cleaning of the tourist sites and a relative dissatisfaction with the gastronomy and the accommodation. A very high dissatisfaction was obtained from the offer of local products in meals and information and tourism signage. It highlights a very high satisfaction with hospitality, constituting a fortress of destiny. The overall satisfaction level of your stay in the destination exceeds 97% of tourists.

Among the activities that seek to make highlights the enjoyment of gastronomy (food and drink) and photography, also, secondarily, observe flora, ride, visit museums and monuments and make purchases. The main place visited corresponds to the La Campana National Park and the offers of the localities of the surroundings to the 30 km.

The aspects described as positive surprises correspond to attributes that the visitor gives to the destination, the most mentioned were hospitality, tranquility, weather, customer service, programmed activities, nature protection, native fauna, gastronomy, pollution-free place, and offer of crafts. We can observe that a few escape from the control of the destination's administrators (tourism services and public administration) such as the climate and the native fauna, the rest corresponds to a broad sphere of coordinated public–private action. In turn, the aspects mentioned as negative surprises correspond to the state of the roads, the carelessness of the environment, the lack of drinking water, the presence of insects, the lack of parking, the presence of stray dogs, careless grooming, lighting of the streets, the supply of restaurants after 10 pm, and tourist signage. A set of elements of shared responsibility, both public and private. Even so, there is a high level of loyalty, given that 96% would return to visit the destination and 93% would recommend the destination to another person.

16.3.2 Preliminary Routes Proposal

As a result of a first stage of obtaining territorial information from documentary and primary sources (fieldwork), a preliminary proposal from the scientific team identified nine geotouristic routes (Figs. 16.3 and 16.4), which corresponded to

1. Peñuelas–Laguna Verde Water Route, an area close to the Lago Peñuelas National Reserve core area, which was valued by society at the end of the nineteenth century by the presence of freshwater, which was a key resource for urban development and the electrification of the port city of Valparaíso.
2. Waddington Canal Route, next to the core area of the La Campana National Park, which highlights an industrial irrigation project that allowed the growth of export agriculture in the late nineteenth and early twentieth centuries with de influence of UK technology.
3. Margamarga Gold Route, in the buffer zone, accounts for the process of original occupation of the territory, and valorization of space by the Inca Empire, in prehistory, and the Spanish Empire, in modern times, mining gold alluvial by washing.
4. Darwin's route, which reconstructs the milestones of Charles Darwin's two trips to Chile and the Valparaíso region in his global journey to ground his natural history studies and his theory of evolution.

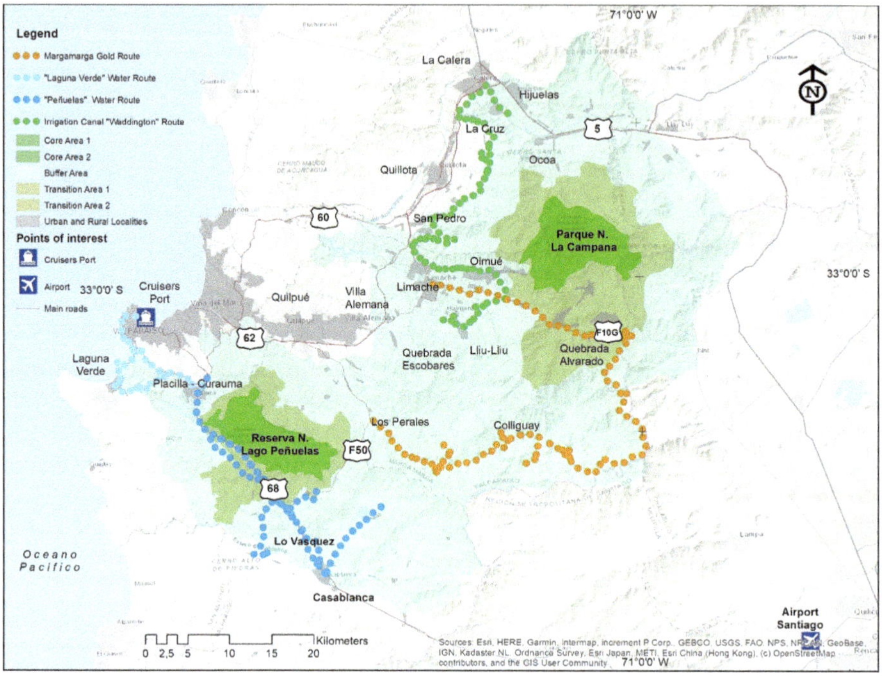

Fig. 16.3 Preliminary proposal geotouristic routes and not developed in second stage

5. La Campana–Peñuelas: Focused on the core areas of the Reserve and its linkage axis through the routes of the Inca Trail and the subsequent royal road of the Spanish conquest and colony.
6. Palmas de Ocoa, close to the core area of the Reserve, which highlights one of the areas with the highest concentration of palms (*Jubaea chilensis*) in good condition in the central zone of Chile.
7. La Dormida Royal Route, in the buffer zone of the La Campana National Park, gives an account of the process of Spanish conquest and colonization and an axis of connection between the valleys of Aconcagua and Mapocho whit Peru and the metropolis.
8. Route of the Valparaíso–Santiago Railroad, which puts value in the area of buffer and transition, the remnants and the railroad environment and passenger service that ran between the port and the capital between 1860 and 1980.
9. Casablanca Wine Route, in the buffer area of the Lago Peñuelas National Reserve core area, which highlights a valley of new white wine production, influenced by contemporary Californian production techniques.

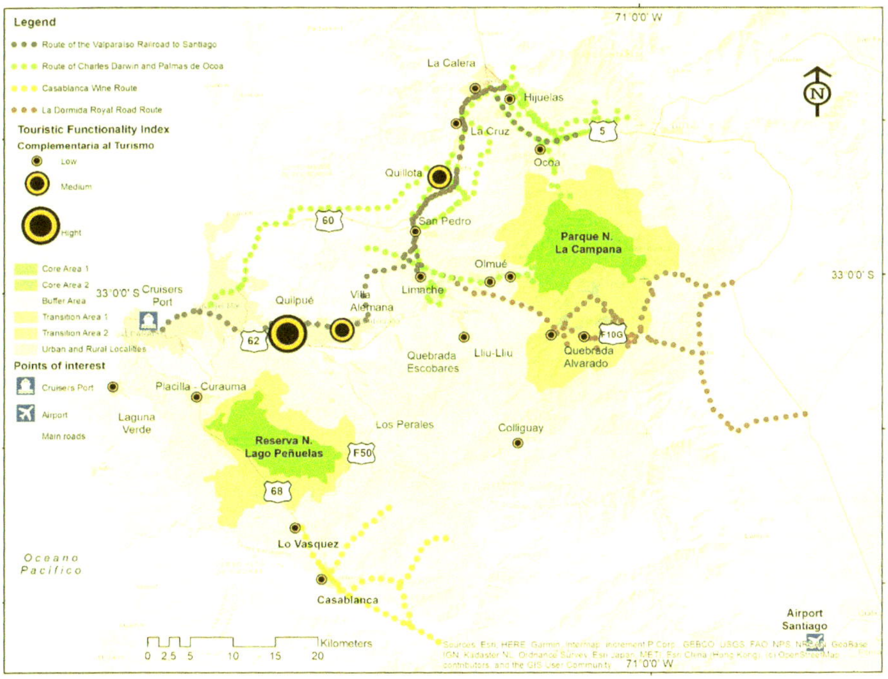

Fig. 16.4 Proposal geo touristic routes for developed in second stage

16.3.3 Workshops with the Territorial Actors of Tourism

The Directory Board together with the scientific team designed participation workshops, inviting tourism entrepreneurs and complementary services, as well as public officials at the national, regional, and local level linked to the regulation and promotion of tourism activities. The workshops were held in locations close to their workplaces. The dynamics of the workshops consisted in a presentation of the intermediate results of the project presented by the scientific team and then they met in small groups to interact with instruments such as maps, photographs, and tables that they had to complete, discuss, and validate. Finally, through brief presentations each group made a summary of what was worked on. Once completed, the scientific team performs an integration and synthesis on the desk of the results of the workshops.

Among the relevant results of these workshops were asked to identify the attributes that the BR had. The results of the workshops identified (i) the cultural heritage of the area, (ii) considering it an articulated territory, (iii) the Biosphere Reserve, (iv) multiple economic and social uses, (v) the diversity of landscapes, and (vi) diversification of services tourist. With these first signs of the territory, we proceeded to request their hierarchy, resulting in an order of the following attributes: 1° the Biosphere Reserve area, 2° the cultural heritage, 3° the diversity of landscapes, and 4° the climate. Thus, it includes the geographical aspects that underlie the occupation of the territory, its uses by people, the results of the millenary relationship between culture and space and the conservation objectives of its attributes.

The criteria for the selection of the final products (tourist routes) were the result the Directory Board reunions and discussions. The scientific committee proposed decision criteria based on the information analysis made and with the Directory Board the following were selected (i) territorial aptitude for geotourism, (ii) habitability of the territory (financial services, security, among others), (iii) the existence of public and private investments supporting for tourism, and (iv) the evidence of the association between territorial actors of tourism. Applying this criteria, four tourist routes were proposed (a) Casablanca Valley Wine Route; (b) Charles Darwin Route (La Campana and Palmas de Ocoa); (c) Royal road Route "La Dormida", and (d) Railroad route Valparaíso—Santiago. Each of these routes was characterized and mapped (Fig. 16.4).

The Directory Board was agreed to define the concept of "Special Interest Tourism Routes" to carry out the creative work of designing and launching a graphic campaign to disseminate and publicize.

16.3.4 Route of Charles Darwin and Palmas De Ocoa

This product is based on the voyage of scientific exploration that Charles Darwin, young English naturalist, made from 1831 on board the HMS Beagle, arriving in Valparaiso in the winter of 1834. For 9 months he made observations on flora, fauna, and geology of the various landscapes that he could visit in the region and especially on the route that followed from the city of Valparaiso to the hill La Campana, twice, described in his Travel Journal.

This Route offers a series of varied and beautiful ecosystems, cultural attractions and tourist services in places, such as La Campana National Park, the Rabuco Valley and Ocoa.

The National Park La Campana, a core area of the Biosphere Reserve is of high ecological and scientific value, allowing a series of nature tourism activities. In its environment, there are a series of productive systems, where agro-sustainable tourism activities can be projected (Figs. 16.5 and 16.6).

There are three types of resources and tourist attractions on the route: (1) the relict of the natural represented by the National Park La Campana (PN LC); (2) the multiple agricultural use of the Valley of Hijuelas; and (3) the traditional festivities of Olmué and Hijuelas that reflect the culture and traditions of the Chilean countryside.

16.3.5 La Dormida Royal Road Route

This product is based on the old route of the "Camino Real de La Dormida", which was part of a road network that the Spaniards built to communicate the conquered

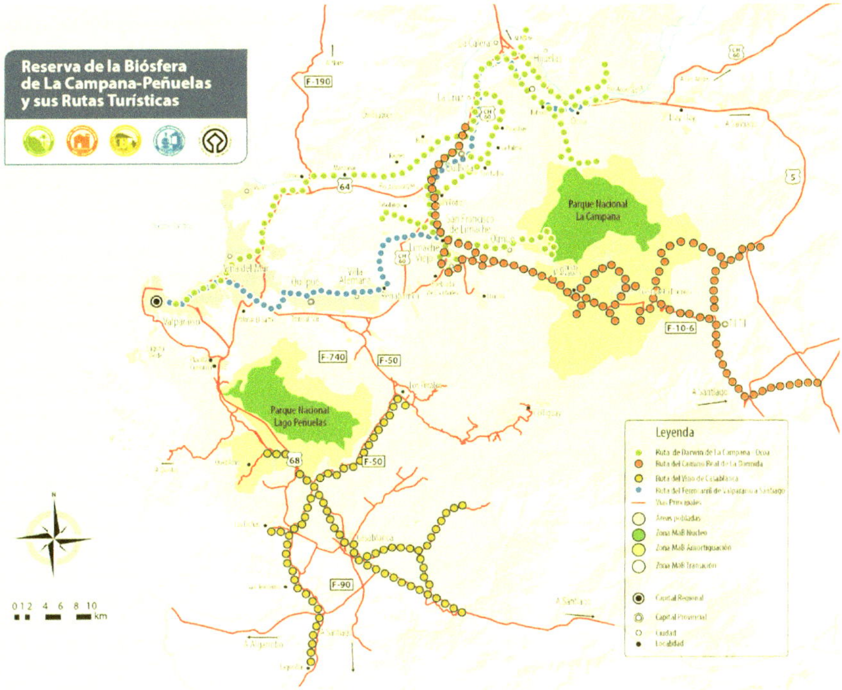

Fig. 16.5 Map of four geo tourism routes La Campana–Peñuelas Biosphere Reserve

territory and allow, in turn, to control and extract the economic resources they needed for the subsistence of their colonies. La Dormida communicated Santiago, in the first centuries of the conquest, with the port of Valparaíso on the coast, a vital space for its supply. It also allowed to maintain control of the Marga Marga's washing alluvial gold, from its administrative center in Quillota.

This Route is highlighted by a series of natural landscapes, cultural attractions, and tourist services in unmissable places. Due to its high ecodiversity it allows to carry out various activities of nature tourism and mountain life at different times of the year, hiking, horseback riding, among others, and establishing high mountain boutiques with naturalistic base products (Figs. 16.5 and 16.7).

The observation of flora in different stations, viewpoints, and small distribution centers stands out. As a general conception, the route already presents some points of commercialization of processed products in which passers-by stop to buy.

Fig. 16.6 Guide of diffusion Charles Darwin La Campana Ocoa Route

Fig. 16.7 Guide of diffusion La Dormida Royal Road Route

16.3.6 Route of the Valparaíso Railroad to Santiago

It presents a high Eco Biodiversity, due to the presence of different basins, geomorphology, and microclimates, in addition, a diverse rurality. This route allows a strong connectivity with others, meeting points, and views to different mountains of the area (Figs. 16.5 and 16.8).

Regarding its biodiversity, in general, Acacia caven spinal fields are found. Along with this, the corridors of Broadleaved Forest in the Quebradas and some high peaks on the southern exposure slopes stand out, and in the northern exposures there are shrubs and xerophytic species of quisco and chaguales. Finally, in the bottoms of ravine you can see fertile plains of a high diversity of aquatic plants.

From the geodiversity point of view, the diversity of the geomorphology in the area of the Las Cucharas bridge is remarkable for the depth and width of the river bottom and the deep cliffs that occur there, as well as rocks of different colors and origins. The rest of the geoform has smooth to flat hills.

It is considered the sub-route San Pedro to Quintero, where the Ritoque wetlands stand out, highlighting the dunes, grasslands, fertile plains, and Mediterranean meadows, where there is an important wildlife and high rurality.

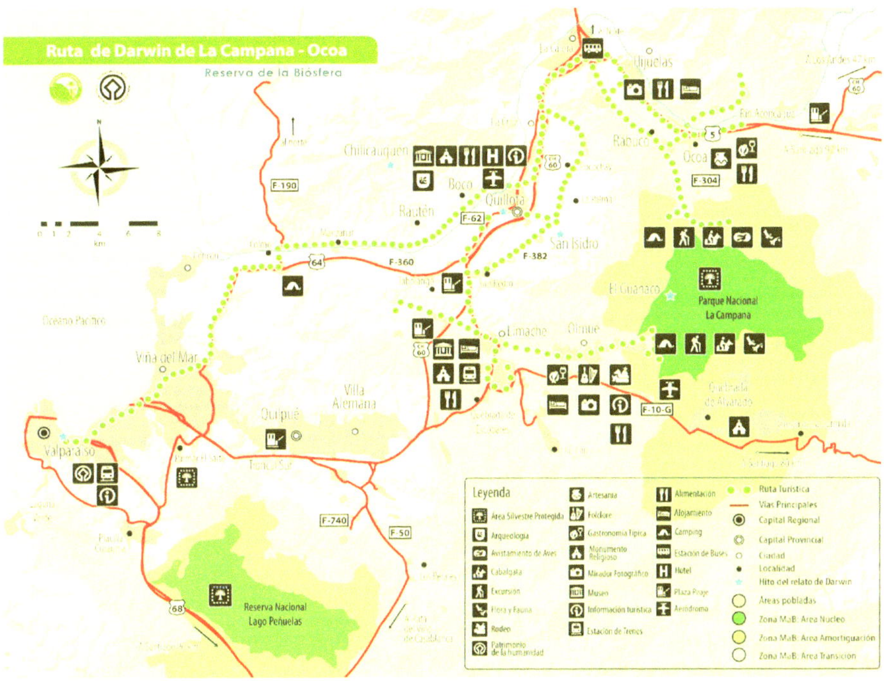

Fig. 16.8 Map of Charles Darwin La Campana Ocoa Route, like model routes

Fig. 16.9 Guide of diffusion Route Valparaíso Railroad to Santiago

16.3.7 Casablanca Wine Route

This route is based on the contemporary vineyards present in the sector that are the culmination of a long process of vine cultivation in Chile, and especially in the central zone of the country and in the valleys of Casablanca, Marga Marga, Limache, and Aconcagua. The production of wines in Chile dates back to the early days of the colony. After a long process, viticulture in Chile is renewed with the introduction of French stocks in the first half of the nineteenth century, being one of the outstanding vineyards Hacienda Los Perales de Quilpué.

This Route offers beautiful ecosystems, cultural attractions and tourist services in places you can not leave visit, in a tour that also includes many attractions that exist throughout the area of the BR.

The model proposed for the Casablanca Wine Route relates the flow of traceability of agro-sustainable products toward communal tourism (Figs. 16.5 and 16.10). From the Municipality of Casablanca, who supports the creation of a tourist route, according to the characteristics of natural resources. An experience of "traceability" is valued, that is, a process that goes from the beginning of the product or system to its consumption. This requires a process of reconversion of conventional agriculture and livestock to agro-sustainable ecological systems, supported by specialized public services and developed by producers of the association of local producers,

Fig. 16.10 Guide of diffusion of Casablanca Wine Route

who generate healthy products for people and the environment and sustainable and that have a market preference.

On the other hand, under the rules of the Biosphere Reserve, the wine industry produces wines of excellent organoleptic quality, both for export and for local consumption, in a strategic alliance with gourmet restaurants, who, by using sustainable agro-ecological products from the area raise their prestige. In addition, agro-ecological products are obtained at local fairs by residents and by passing tourists.

Finally, the evaluation of tourism service operators measure and quantify their perception of geotouristic routes, the quality of the services that the companies belonging to the routes have, and also to evaluate the environment, keeping in mind the tourist demand to the which is directed, to identify its critical points and guide the improvement of the quality of the offer and have the marketing channels represented by the tour operators.

In relation to the qualification of the offer, the experts stated that both the equipment and the facilities do not meet the standards demanded by tourists. The main problems detected were (1) There is a manifest difference in quality between the services observed, (2) Demand exceeds supply and is intensive in one season, (3) lack of awareness among residents, local capacities must be created to support to the routes, (4) Lack of legal tools to order, monitor, and address problems that create sustainable services and products (the recycling of waste, as an example). In relation to human resources, it exists but without professionalizing. In relation to the demand, it is indicated that it is demanding, informed and growing for this type of tourism. Regarding marketing, they consider that (1) there is an absence of marketing techniques, the dissemination material is inadequate, (2) The offer does not have an objective image addressed with the product proposal, and (3) The offer does not identify the target market and does not know the needs and demands of that market.

16.4 Discussion

When we conceived the design of geotouristic routes, and the process of building them, we put ourselves on the stage that this objective and its concretion would be the center of interaction and summon the different actors that inhabit, think, and build the territory in the BR LC-LP, including the tourism sector. Then we see tourism and the products proposed as intermediate instruments and management situations (Girin 1990; Vinck 1999) for the identification of common objectives and to begin to build what is a Biosphere Reserve. We rely on the theory of territorial governance proposed by different authors (Berry 1983; Aggeri and Labatut 2010; Chia et al. 2008, Ubilla-Bravo 2016), especially in touristification as an instrument of territorial governance (Figueroa-Sterquel et al. 2016, 2018). The University is part of the management system of the BR and has a legitimate interest and a duty to promote the objectives pursued by the program Man and the Biosphere of UNESCO.

In relation to the system of governance of the project, the design of a top-down management model is positively evaluated, since it allowed to grant an umbrella to the idea of building a new instrument for spatial planning (BR) and to summon different actors for a new geotourism destination and new tourism products for the region. In addition, the original design of the unpublished public innovation policy in Chile was responded to. So with this design, the initial administrative obstacles were overcome to obtain the financing of the idea.

We agree with Bonin and Velut (2008) that the governance of a complex space such as the La Campana–Peñuelas Biosphere Reserve can not only be an exercise of the political authorities or only a call for consultation with other actors, but must rely on the authority local policy, in the territorial and sectoral organizations of the residents, also of the entrepreneurs and entrepreneurs of the territory in order to collectively and jointly design the future objectives for that territory. Thus, the initial design was complement during the execution stage of the project, with a bottom-up management model. Some municipalities, into the territory of the reserve, and the tourist and territorial organizations of those communes where the new geotourist products were proposed are incorporated through agreements.

In this perspective, the construction of materials by the scientific team should be thought of as intermediate objects (Vink 1999) to provoke the necessary conversation between the different actors involved in the process of common construction of a new reality that arises with the expansion of the BR (residents, visitors, public officials, farmers, entrepreneurs, territorial organizations, scientists, students, tourists, among others). In this sense, the various instruments such as workshops, interviews, surveys, inventory cards, questionnaires, maps, photographs, and digital and online technological systems were designed and thought to disseminate, communicate, exchange opinions, propose, think about the territory lived and dream it, while the pretext was to build a shared geotouristic product, learning to conceive, design, build, and reproduce what was learned about the BR generating cumulative loops in the management process (Argyris and Schön 1978).

The methodology of multicriteria spatial analysis, which provides the geography, was a starting point to develop dialogues between the actors in relation to the territorial aptitudes to founding geotourist routes and, as an intermediate object, as a result of the interaction with others, allowed to incorporate other variables of a political and social nature among the decision criteria, in particular the continuity of public policies and the support of the routes in existing social organizations.

Our research revealed the existence of an extraordinary biodiversity, geodiversity, and cultural heritage in the mountain areas and inland valleys within walking distance of the metropolitan peripheries that run the risk of urbanization without including ecosystem criteria for their value, enjoyment and use, reproducing forms of relationship with nature associated with tourist coasts (Figueroa et al. 2016). The dissemination of this knowledge among the actors allows integrating new perspectives to systems of governance of conservation (Manríquez et al. 2019).

The visitors of this territory are directly related to the proximity tourism that allows the accessibility to a different landscape to the origin: either from the dry Pampas Argentina that escapes toward the nearest coast; Whether from the Santiago basin, in search of escape to a coastal landscape, through natural and open spaces in a rural context with no environmental pollution and extensive urbanizations. The frequentation is directly motivated by the visit and the enjoyment of the coast and the coastal urbanization and among the explanatory factors are the socioeconomic conditions and the alternation in the time of the vacations with different environments of the origin (coast versus interior, temperate versus warm). In this context, the protected natural spaces are presented as circulation spaces and in practice they are scarcely frequented and used.

One of the relevant proposals of the scientific team, agreed with the Board, was to design products that have a global or international significance, understanding that tourism is one of the most globalized phenomena and the new geotouristic routes should consider the representations of tourists and the global imaginary built in their travels and places of origin. Then the routes were thought of: of the royal road, of Charles Darwin, of the wine and of the railroad, that have their presence in diverse latitudes of the planet. This is how it is thought globally and is done locally.

These geotourist products have the advantage of having been built in a participatory manner with the political establishment, farmers, tourism entrepreneurs, and social organizations, accompanied by scientists. In particular, its scientific foundations and integrated information of the territory are the basic aspects to support the proposed Geotourism routes and, therefore, the best-achieved aspect of the project.

The greatest challenges or weaknesses are the political, human, and social aspects, in relation to ensuring the continuity of the public policies of conservation of bio (geo) diversity and the development of the leadership of the territorial actors to tackle on their own the management of these spaces, based on the learning achieved in this work, to develop their shared development goals.

This challenge continues today in the management of these geotourism routes and the BR, since the formal Biosphere Reserve management system has been paralyzed with the changes of political cycles and the common actions of civil society have lost strength and coherence in exchange for processes of fragmentation of the actors and social movements that individually value the routes and the BR.

In this co-construction of geotourism routes has been the first time that the actors have met, they have summoned others and they have thought about the territory of the BR LC-LP, driven by the leadership of the knowledge of the scientists making use of public policy of innovation. A second effort has been led by another scientific center, mobilizing the networks around the concept of territorial ordering of the BR (Lebuy et al. 2018), however, social and political leadership is a pending aspect of the current implementation process of the geotourism and management routes of the BR.

16.5 Conclusions

This contribution seeks to explicitly develop a proposal of geotourism routes that enhance the potential of the La Campana–Peñuelas BR as a territory in transition toward sustainability. Together with the selection of the routes, the coarse aspects of a governance strategy for the Reserve are outlined.

The challenge of co-innovation is part of the new role of the academic stakeholders to promote and be a facilitating partner of shared coordination, innovation and leadership networks, for the human security of resident populations and tourists, in the contexts of globalization, adaptation to climate change and more sustainable development.

The design of routes must be understood as articulating processes of actors in the territory, which promote the governance of the new destination and the products, facilitating a process of mutual learning.

The execution of the study and the implementation of the routes allowed to include the MaB Reserve as an official mark of geotourism planning in the regional context, which represents the acceptance of the public and private actors of the proposal.

The greatest challenges or weaknesses are the political, human, and social aspects, in relation to ensuring the continuity of the public policies of conservation of bio (geo) diversity and the development of the leadership of the territorial actors to tackle on their own the management of these spaces, based on the learning achieved in this work, to develop their shared development goals.

In this co-construction of geotourism routes has been the first time that the actors have met, they have summoned others and they have thought about the territory of the BR LC-LP, driven by the leadership of the knowledge of the scientists making use of public policy of innovation.

References

Acevedo RD, Martínez Frías J (2018) Geoethics in Latin America. Springer.

Aggeri F, Labatut J (2010). La gestion au prisme de ses instruments. Une analyse généalogique des approches théoriques fondées sur les instruments de gestion. Financ Contrôle Stratég 13(3):5–37

Arenas-Martija A, Orrego-Araya J, Ortiz-Pavéz C, Gajardo-Tapia C, Zanetta-Colombo N (2020) Collaborative elaboration of geo educational routes: strategy for teaching-learning of sciences and geography. In press in this book, Puchuncaví, Chile

Argyris C, Schön DA (1978) Organizational Learning. Addison, Readings

Berry M (1983) Une technologie invisible. L'impact des instruments de gestion sur l' évolution des systèmes humains CRG-1133. Cahier du laboratoire numéro

Bonnin M, Velut S (2008) La contribución del concepto de Reserva de la Biosfera al desarrollo sustentable. Un enfoque comparado Francia-Chile. In: Duran V, Montenegro S, Moraga P (eds) IV jornadas de derecho ambiental, June 2008, Santiago, Chile, pp 167–184. halshs-00432202

Carvajal-Mascaró F, Moreira-Muñoz A, Salazar-Burrows A, Leguia-Cruz M, & Jorquera-Guajardo F (2019). Divergencias y contradicciones en la planificación sustentable del periurbano rural metropolitano de Valparaíso. Caso Reserva de la Biosfera La Campana-Peñuelas, Chile central/Divergences and contradictions in the sustainable planning of the. Urbano 22(39):64–87. https://doi.org/10.22320/07183607.2019.22.39.04

Chia E, Torre A, Rey-valette H (2008). Vers une « technologie » de la gouvernance territoriale! Plaidoyer pour un programme de recherche sur les instruments et dispositivos de la gouvernance des territoires. Norois 209(4):167–177

Figueroa-Sterquel R, Negrete J, Velut S (2012a) Valorización Turística De La Reserva De La Biosfera La Campana Peñuelas En El Contexto De La Metropolización De Chile Central. Turismo Rural Experiencias Y Desafíos En Iberoamérica, Mexico, pp 191–218. ISBN: 978-607-7533-92-4

Figueroa-Sterquel R, Negrete J, Fuenzalida M, Allesch R, Inostroza J, Cosio F (2012b) Desarrollo de nuevas rutas de turismo para la Reserva de la Biosfera La Campana–Peñuelas. Un nuevo destino de turismo para la región de Valparaíso, Chile. Revista Geográfica de Valparaíso 46:16–25

Figueroa R, De Kartzow P, Arenas A, Flores L, Inostroza J, Portal M, Larraguibel C (2016) Los conflictos entre el desarrollo inmobiliario y la conservación de la biodiversidad. Estudio de caso en los Acantilados de Quirilluca. In Hidalgo R, Santana D, Alvarado V, Arenas F, Salazar A, Valdebenito C, Álvarez L (organizers) Las costas del neoliberalismo. Naturaleza, urbanización y producción inmobiliaria: experiencias en Chile y Argentina. Impresión Gráfica LOM, Santiago de Chile, mayo de 2016. ISBN 978-956-14-1801-1

Figueroa-Sterquel R, Chia E, Tapia L, Andrade J (2016) Efectos de la certificación turística en la gobernanza territorial: el caso del Sello "Biosfera", Olmué (Chile). PASOS. Revista de Turismo y Patrimonio Cult 14(3, 2016 Special Issue):675–690

Figueroa-Sterquel R, Szkutnicki A, Szmulewicz P, Martínez P, Chia E (2018) Innovación territorial desde los instrumentos de gestión: estudio del aprendizaje y la gobernanza territorial en destinos turísticos de Chile. In: Muñoz A, Osorio M, Guijarro G (eds) Tendencias de investigación en turismo en América Latina. Estudios de caso. Pontificia Universidad Católica del Ecuador

Fuenzalida M, Figueroa-Sterquel R, Negrete Sepúlveda J (2013) Evaluación de la aptitud territorial para el turismo de naturaleza y rural. Reserva de la Biosfera La Campana - Lago Peñuelas, Chile. Estudios y Perspectivas en Turismo 22:120–137

Gastó J (1979) Ecología, el hombre y la transformación de la naturaleza. Editorial Universitaria, Santiago, Chile

Girin J (1990) L'analyse empirique des situations de gestion: éléments de théorie et de méthode. Epistémol Sci De Gestion 1990:141–182

Lebuy R, González B, Huerta P, Cáceres T (2018) Propuesta de Ordenamiento Territorial Participativo para el Territorio de la Reserva de la Biosfera La Campana- Peñuelas: Un análisis desde lo local. In: Bosque Mediterráneo y Humedales: paisaje, evolución y conservación. Aportaciones desde la Biogeografía. Almud Ediciones

Malczewski J (1999) GIS and multicriteria decisión analysis. John Wiley & Sons, New York.

Manríquez H, Mansilla P, Figueroa-Sterquel R, Moreira-Muñoz A (2019) Geodiversity meets biodiversity: a landscape approach for biogeocultural conservation and governance in Mediterranean central Chile. Eco.mont – J Prot Mt Areas Res Manag 11:43–48.

Moreira-Muñoz A (2011) Plant geography of Chile. Springer, Dordrecht

Moreira-Muñoz A (2014) Central Chile Ecoregion. In: Hobohm C (ed) Endemism in vascular plants. Springer, Dordrecht, pp 221–233

Moreira-Muñoz A, Borsdorf A (2014) Reservas de La Biosfera de Chile: Laboratorios para la Sustentabilidad. Academia de Ciencias de Austria, Instituto de Geografía, Pontificia Universidad Católica de Chile. Santiago (In Spanish)

Muñoz AE, Beltrami E, Osorio C (2017) Evaluación de presencia de mamíferos en predios frutícolas del norte chico y su variación a escala intra-predial. Sustentabilidad(es) 8(16): 46–60

Naveh Z (2007) Transdisciplinary challenges in landscape ecology and restoration ecology, 1st edn. Springer, Dordrecht

Newsome D, Dowling R (2018) Geoheritage and geotourism. In: Reynard E, Brilha J (eds) Geoheritage. Elsevier, Netherlands

OEA-CICATUR (1978) Metodología de inventario turístico en Quito. Curso Interamericano de Planificación del Desarrollo Turístico. Organización de Estados Americanos, Centro Interamericano de Capacitación Turística. México D F.

Salazar A, Moreira-Muñoz A, del Rio C (2015) La Campana-Peñuelas biosphere reserve in Central Chile: Threats and challenges in a peri-urban transition zone. Eco.mont – J Prot Mt Areas Res Manag **7**:66–71

SERNATUR (2017) Anuario de turismo 2016. Subsecretaría de Turismo—Servicio Nacional de Turismo, pp 27. Santiago, Chile

Thomas MF (2012) A geomorphological approach to geodiversity – its applications to geoconservation and geotourism. Quaestiones Geographicae 31:81–89

Ubilla-Bravo G (2016) Gobernanza territorial: bases, características y la necesidad de su estudio en Chile. Seminario Sustentabilidad y Gobernanza, September 2016, Santiago, Chile. https://doi.org/10.5281/zenodo.164823, halshs-01389912

Vinck D (1999) Ingénieurs au quotidien- Ethnographie de l'activité de conception et d' innovation. Grenoble, PUG

Chapter 17
Modern Geotourism's UK Antecedents

Thomas A. Hose

Abstract This chapter provides an overview of geotourism's development from the late seventeenth century to the present day. It mainly draws upon examples from the United Kingdom (UK), in which the geotourism concept was first formally recognised and defined. Specifically, it examines, as the birthplace of early geotourism, the English Lake District and Peak District. It also examines central southern England as the location of the beginnings of early modern geotourism. It explores, as geohistory, how the rise of countryside-focused travel and the emergence of aesthetic landscape movements supported the development of early geotourism. Modern geotourism, a twentieth paradigm, is then examined, especially in relation to its role in geoconservation and the provision of geo-interpretation; it is also set within the framework of modern tourism. Finally, possible future geotourism developments in the UK and elsewhere in Europe are explored.

Keywords Geoconservation · Geohistory · Geo-interpretation · Geotourism

17.1 Introduction

Geotourism's emergence required a fundamental shift in how landscapes were perceived, by travellers and/or tourists, and then exploited for tourism; the recognition that wild landscapes were worthy places to visit was essential. These early leisure travellers were preceded by pilgrims journeying to religiously significant sites. This shift, especially in Europe and later in North America and Australasia is apparent in the changing preferences of wealthy well-educated elite travellers, especially during the eighteenth century; such changes were eventually adopted by the wealthier middle classes in the nineteenth century, before their widespread acceptance by the wider population in the twentieth century. In other parts of the world, especially China, such changes have probably followed a similar but presently uncharted pattern over a different timeframe. A Ming Dynasty account, *The Travel Diaries of Xu Xiake*, by

T. A. Hose (✉)
School of Earth Sciences, University of Bristol, Queens Road, Clifton, Bristol BS8 1RJ, UK
e-mail: gltah@bristol.ac.uk

R. B. Singh et al. (eds.), *Global Geographical Heritage, Geoparks and Geotourism*, Advances in Geographical and Environmental Sciences,
https://doi.org/10.1007/978-981-15-4956-4_17

Xu Xiake (1587–1641) is probably one, if not the earliest, of China's early geotourism writings. Xiake documented his journeys over some thirty years throughout south-west China. His English near-contemporaries are Thomas Hobbes (1588–1679) and Charles Cotton (1630–1687). Much later than in the United Kingdom (UK), the despicable Opium Wars (1839–1842 and 1856–60) allowed European, especially British, geological ideas to enter a China previously closed to foreign influences; much earlier than these pleasure travellers were both Christian and Muslim pilgrims who journeyed to the major religious sites in Europe and North Africa. For example, Ibn Jubayr (1145–1217), an Arab traveller and poet, on his 1183-1185 Mecca pilgrimage returned to his home in Moorish Spain via Sicily where he recorded volcanic eruptions by Stromboli and Mount Etna, which former "At the close of night a red flame appeared, throwing up tongues into the air…As for the great mountain [Mount Etna] in the island…it also presents a singular feature in that some years a fire pours from it in the manner of the 'bursting of the dam'. It passes nothing it does not burn until, coming to the sea, it rides out on its surface and then subsides beneath it." (Broadhurst 1952, pp. 343–344).

Curiosity and aesthetic worth were significant motives for the early landscape leisure travellers such as Hobbes and Xiake. Later travellers, such as Arthur Young (1741–1820) and John Farey (1766–1826), were particularly motivated by adventure and scientific exploration. There was some later overlap with the aesthetic motives of visual artists and poets with geological travellers' motivations; for example, the poets Samuel Taylor Coleridge (1772–1834) and William Wordsworth (1770–1850) were acquainted with geologists' ideas (Dean 2004; Wyatt 1995). Eighteenth- and nineteenth-century visual artists' and later photographers' visualisations and travelers' writings on landscapes still influence modern geotourism provision. Modern geotourists continue the historic preference for supposed 'wild' or 'natural' landscapes, rather than agricultural landscapes and 'brutalized' heavy industry and mining places. Modern dedicated geotourists would probably refer to themselves as 'amateur geologists' (in the UK), 'fossickers' (in Australia), 'rockhounds' (in North America) and 'recreational geologists' (in Europe). They are largely unadventurous in their choice of places to pursue geotourism. A seminal Australia-based geotourism study on 'fossicking' (coined in that country for participants in the 1850s 'gold rush') noted the peace and quiet of rural landscapes was just about as important to its participants as the activity itself (Jenkins 1992, p. 134). The physical basis of the appeal of such landscapes can only be maintained by legally defined and managed preservation measures; their geosites require pro-active geoconservation. Unlike most other forms of essentially countryside recreation, geotourism is not necessarily seasonally constrained (Hose 1996). It could extend the vacation period in some resort, and underpin economic regeneration strategies in former mining areas.

Before the seminal mid-1990s definition of geotourism as "The provision of interpretive and service facilities to enable tourists to acquire knowledge and understanding of the geology and geomorphology of a site (including its contribution to the development of the Earth sciences) beyond the level of mere aesthetic appreciation." (Hose 1995, p. 17) various authors had referred to tourism with some geological experience (e.g. Anze and Weixing 1985; De Bastion 1994; Jenkins 1992; Maini and Carlisle 1974; Martini 1994; Page 1998; Spiteri 1994). 'Geotourism' (Komoo and Deas 1993; Komoo 1997) and 'tourism geology' were employed in Malaysia, with the latter intended to place geoconservation at the same level of importance as wildlife conservation (Komoo 1997, p. 2973), but they were not defined. In China, 'tourism earth science geology' was defined as a marginal discipline "...to find, evaluate, plan and protect natural landscapes and cultural relics with tourism value, and discuss their formation causes and evolution history on the basis of earth scientific theories and methods...with a view to promoting the development of tourism." (Anze and Weixing 1985); with no mention of geosites (or rocks, minerals and fossils) it is not geotourism *sensustricto*. Much later, it was suggested that tourism earth-science incorporates 'tourism geology' which studies the "... distributions, types, characteristics, causes of formation and changes of varied scenic spots by geological theories, methods, technologies and results. The comprehensive survey and evaluation of basic geology, karst geology, dynamic geology and environmental geology of scenic areas and spots are conducted to organise targeted earthscientific travels, reasonably select travel routes and supporting facilities, maximally display the aesthetic, cultural and scientific values of scenic areas and spots..." (Anze et al. 2015, p. 4); as such, it is a form of landscape tourism.

A relatively recent geotourism redefinition as "The provision of interpretative and service facilities for geosites and geomorphosites and their encompassing topography, together with their associated in situ and ex situ 'artefacts', to constituency-build for their conservation by generating appreciation, learning and research by and for current and future generations." (Hose 2012, p. 11) overtly reinforces the geoconservation need, but admittedly places less overt emphasis on tourism—that being left to the moniker itself. In Europe, several authors have employed broader definitions; for example, "travelling in order to experience, learn from and enjoy our Earth heritage." (Larwood and Prosser 1998, p. 98) but appositely they indicated its provision is partly "...a consequence of successful Earth heritage conservation." (Larwood and Prosser 1998, p. 98). A major reproach to this and similar definitions is that, without any overt geoconservation content, they could encourage 'geo-exploitation' (Hose 2008) to the detriment of the geotourism resource. Martini (2000) proposed geotourism, due to most governments' inability or probable unwillingness, could help fund geoconservation. True geotourism incorporates an examination and understanding of the physical basis of geosites and geomorphosites together with their interpretative media and marketing, as well as geoscientists' lives, work, collections, publications, artworks, field-notes, personal papers, workplaces, residences and even final resting places. It necessarily includes 'geo-history', the study, evaluation and application of a systematic narrative of geological and geomorphological discoveries, events, personages and institutions contextualised within contemporary socio-economic and cultural trends (Hose 2012, p. 12).

Geotourism acquired a newer geographical usage in the United States of America when geology-focused studies elsewhere were unfortunately missed by National Geographic when it belatedly claimed to have coined the term itself for "...a destination's geographic character—the entire combination of natural and human attributes that make one place distinct from another..." (Stueve et al. 2002, p. 1); as such it is a repackaged element of sustainable landscape tourism. Because National Geographic's geotourists travel to see specific scenery and/or wildlife "...or experience a particular local culture, climb a particular mountain or kayak a particular river..." (Buckley 2003, p. 79) it has simply rebranded cultural and adventure tourism. Whilst its supporters dismiss geology-focused geotourism as a somewhat small specialist subsector (Buckley 2003, p. 79), the seminal Australian study noted it is actually one of the world's largest dedicated leisure pursuit groups (Jenkins 1992, p. 129). National Geographic erroneously suggested its approach had led to geotourism's rapid acceptance, but it had already gained widespread recognition within the European geoscience community by the 1990s and was then also beginning to be recognised by the tourism industry in Europe and Australasia; indeed, the world's first dedicated national geotourism conference was at the Ulster Museum, Belfast in Northern Ireland in 1998 (Robinson 1998).

17.2 Geotourism Set Within a Tourism Framework

Modern geotourism is a late twentieth century paradigm (Hose 2016) built upon the eighteenth- and nineteenth-century improvements in physical and intellectual access that facilitated elite travellers' ventures into previously inaccessible places. It can be considered an explicit form of 'special interest' tourism, in which a particular interest mainly determines traveller's motivation and decision-making, that implies 'active' or 'experiential' travel (Hall and Weiler 1992, p. 5). 'Special interest travel', when people goto a particular region or destination because it is only there it can be pursued (Read 1980, p. 195), encapsulates the basic geotourist experience. Special interest tourism developed from the early 1980s as a field of tourism studies when academics recognised significant near-contemporary changes in tourism product development and consumption. It is an expanding tourism segment potentially overlapping with other tourism forms such as 'eco-tourism', 'sustainable' and 'alternative' tourism and potentially 'educational travel', 'environmental', 'nature-based' and 'heritage' tourism. Heywood (1990, p. 46) suggested that tourism which incorporates active components inclined towards conservation, scholarship, science and environmental awareness depends upon scarce well-educated wealthy tourists, broadly corresponding to Plog's (1974) 'allocentrics'(Hall and Weiler 1992, p. 4); probably only pertinent to 'dedicated geotourists' (Hose 2000, p. 136), seeking intellectual improvement. 'Casual geotourists' (Hose 2000, p. 136), primarily motivated by pleasure-seeking and social engagement, are considerably commoner; research (Hose 1997, pp. 2956–2957, 2000, pp. 137–138, 2005, p. 55) identified their general characteristics, noting their own, and any accompanying (grand)children's, desire for informal educational experiences.

17.3 The Necessary Pre-cursors for Geotourism's Growth

Modern geotourism's growth is undeniably linked to potential geotourists' geology and physical geography experiences during schooling and the mass-media promotion of geological concepts and attractions. However, the role of populist science volumes, especially on the history of geological ideas and about individual geologists, is also a factor. Of course, there is also the continuing fascination, with or without some tie-up with a blockbuster movie or television series, children have with dinosaurs and other spectacular fossils. Until the mid-twentieth century geology was little studied formally in schools. It was usually taught by geographers because it met their needs and methodology to understand and explain geographical issues; thus geology had a service function for geography (Fisher 1994, p. 477). Few geology graduates generally entered teaching, partly due to the limited numbers graduating in the discipline. Whilst in the 1930s UK universities annually graduated only about 30 geologists (Hamilton 1976, p. 105), by the 1960s this had substantially increased to several hundred. To encourage geology education in UK schools the Association of Teachers of Geology was founded in 1967. Renamed in 1988 as the Earth Science Teachers' Association (ESTA), it organises an annual conference, publishes its own journal and develops learning support materials.

The burgeoning list of geology field-guides (Hose 2006), academic and populist, is a good measure of geology interest within schools and universities, and by the public. Despite the increasing numbers involved in it today, geology is not as enthusiastically and proportionately pursued by the public as it was in the late nineteenth century; then it was at the forefront of scientific enterprise and innovation. Concomitantly, there were immensely popular local societies and numerous gifted amateur field naturalists and geologists (Allen 1978); the latter were drawn to examine geo-attractions such as coastal cliffs, quarries, and major private geology collections and public museums.

17.4 Early UK Geo-Attractions

Early UK geotourism provision (see Fig. 17.1 for the locations herein mentioned), an aspect of geohistory, originated in central England's Peak District in the late seventeenth century when its limestone caves became popular with travellers. Charles Cotton, in *The Wonders of the Peake* of 1681, noted the entrance to one of these, Poole's Cavern (Fig. 17.2), was already gated and an old woman made a living as the keyholder. Almost a century later, Oatlands Park grotto (Barton and Delair 1982), near Weybridge in Surrey was constructed between 1760 and 1778 as an architectural folly—a solely aesthetic structure. Intended to be a limestone cave recreation it walls were encrusted with fossils from the Bath area, modern shells and natural cave features such as stalagmites. Naturally formed rock features developed as geo-attractions first emerged in north-west England in the Lake District at the beginning of the nineteenth century. There, the Bowder Stone (Fig. 17.3), a perilously

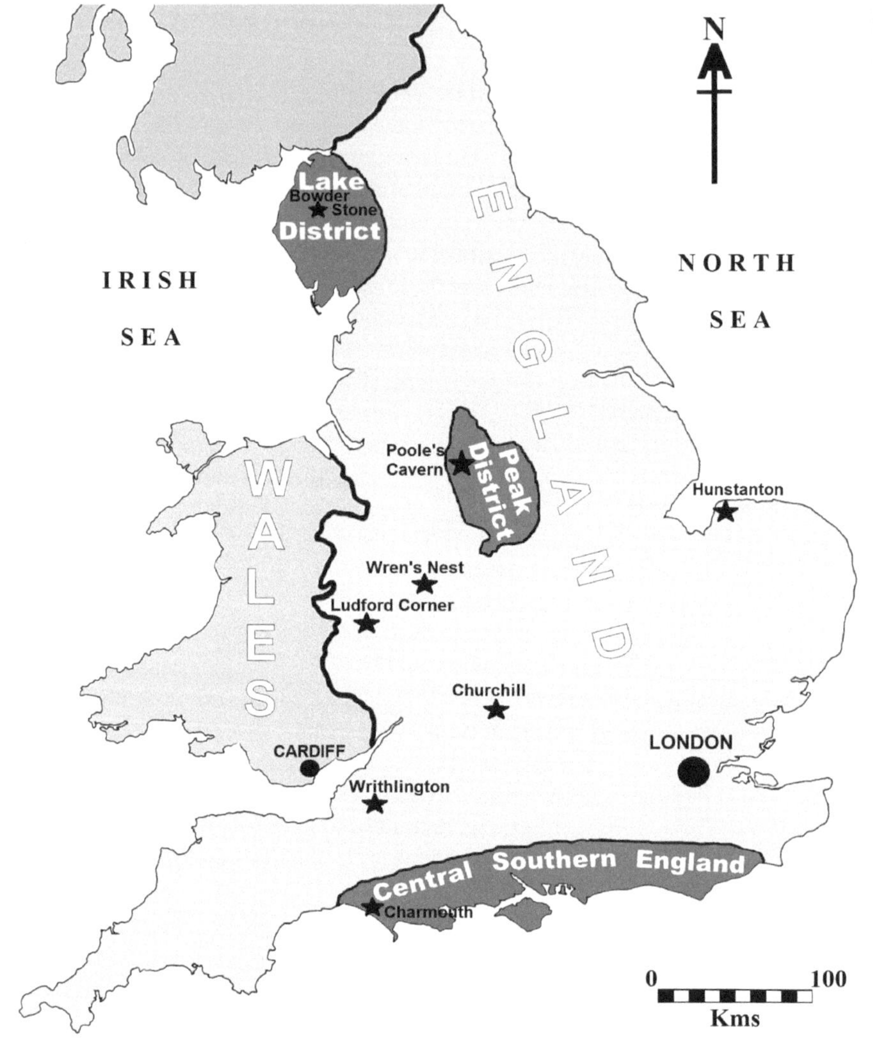

Fig. 17.1 Map of the UK's Historic Geotourism Regions and Geosites—showing the three geotourism regions and some of the historically important geosites mentioned in the text

balanced boulder in Borrowdale, had a hermitage folly, new druidical stone, and a small house for its guardian by 1807; William Green's *Tourist's New Guide* of 1819 recorded the latter began, expecting a tip to follow, and on travellers' arrival "…an exordium preparatory to the presentation of a written paper, specifying the weight and dimensions of the stone…" (Green 1819, vol. 2, p. 134). The painting *Duria antiquior* of 1830 by Henry De la Beche (1796–1855) was the first-ever visualisation of fossils, based on Dorset examples, as they might have appeared in life. At the Crystal Palace

Fig. 17.2 Poole's Cavern, Buxton, Derbyshire—is named after a fifteenth-century robber reputed to have used the cave as his lair and has been a show cave since 1854; carved out of Lower Carboniferous limestone, it was occupied by people in Neolithic, Bronze Age, Iron Age and Roman times

Fig. 17.3 The Bowder Stone, Rosthwaite, Cumbria—this 12-metre high perched 2000- ton block, as illustrated by this late nineteenth-century hand-coloured postcard, composed of Upper Ordovician andesite is still accessed by ladder-steps to a viewing platform; it has been a tourist attraction since 1798 when Joseph Pocklington (1736–1817) set a ladder upon it

in Greater London the world's first geological educational theme park, with several three-dimensional reconstructions of prehistoric animals and plants on accurately rendered geological sections (Doyle 1993, 1995; Doyle and Robinson 1993; Hawkins 1854; McCarthy and Gilbert 1994), was established in 1858. The *in situ* preservation of spectacular fossils began in the late nineteenth century when two 'fossil forests' were discovered. One, unearthed in 1873 at the South Yorkshire County Lunatic Asylum (Cleal and Thomas 1995a, pp. 208–210; Sorby 1875, p. 458) was protected by two viewing sheds; another, discovered in Edinburgh in 1887, 'Fossil Grove' (Cleal and Thomas 1995b, pp. 189–191), is the UK's longest continuously open geo-attraction. The first-ever urban geotrail was established in Rochdale, near Manchester by 1881; it still has 30 upstanding stone pillars, engraved with biblical quotes in accordance with the moral educational role than usually ascribed to geological study, setup in a churchyard (Baldwin and Alderson 1996, p. 227).

After these auspicious beginnings little further was achieved in the UK until the 1950s. In 1956, the first purely geological National Nature Reserve was established at the Wren's Nest (Fig. 17.4), Dudley in the West Midlands. By the early 1960s its management had become a challenge to the then statutory nature conservation body, the Nature Conservancy Council; the danger of over-collecting its superbly preserved fossils meant a site management and interpretative approach of 'hammer-and-take' necessarily had to be replaced with one of 'look-and-see' (Robinson 1996a, p. 211). Several versions of its field-guide have been published since the late 1960s, supplemented in 1996 by an on-site display (Robinson 1996a). In 1973, the first

Fig. 17.4 The Wren's Nest National Nature Reserve, Dudley, West Midlands—this former limestone quarry, with both exposed and extensive underground workings of the Wenlock Group (Upper Silurian) limestones, shows numerous original sedimentary features—including the ripple bed of a former shallow tropical seafloor, seen on the right of the view from an interpretative platform; over 700 fossil species, 86 unique to the site, have been recorded from its rocks

Fig. 17.5 Ludford Corner, Ludlow, Shropshire—is on the route of the Mortimer Forest Geology Trail and the world-famous Ludlow Bone Bed Member (Lower Devonian) can be seen in the low cliff as the notch running from the commemorative plaque, behind the bench which obscures it, to the "No hammering" sign on the right; discovered in 1835 it was once thought to mark the first appearance of fishes, and hence vertebrates, in the fossil record

purposefully established educational geotrail, *The Mortimer Forest Geology Trail* (Lawson 1977), was established near Ludlow in Shopshire; it incorporated Ludford Corner (Fig. 17.5) geo-historical site along with some small roadside quarries. It was a joint venture between the national forestry agency, the Forestry Commission and the Nature Conservancy Council.

In the late 1980s, two geology-focused visitor centres were established, the National Stone Centre (Thomas and Hughes 1993) at Wirksworth, Derbyshire in the Peak District and the Heritage Coast Centre (Edmonds 1996) at Charmouth (Fig. 17.6), Dorset on the Jurassic Coast World Heritage Site.

Elsewhere, visitor centres were established by Local Authorities at sites with considerable geoarchaeological interest; for example, at the Cresswell Crags Ice Age site, near Worksop in Nottinghamshire. In Dorset, a private estate opened the Lulworth Visitor Centre, on the Jurassic Coast World Heritage Site, in the early 1990s; its populist guidebook noted "...all are struck by the beautifully shaped coast and the contrasting rocks that form it. This booklet has been written for both geography students and visitors who would like to understand a little more about this unique landscape." (Pfaff and Simcox 1998). All the centres provided a range of activities such as talks, identification services and guided walks. In the 1990s, numerous interpretative panels were sited at geosites popular with tourists; for example, by English Nature at Hunstanton (Fig. 17.7) in Norfolk and Scarborough in Yorkshire in 1993.

Fig. 17.6 Charmouth Heritage Coast Centre, Dorset—the Centre, opened in 1985, is located on the top floor—of a two-storey former cement factory—above a cafe, beach souvenirs and fossil shops; it gets more than 30,000 visitors every year, to view its displays and go on organised Jurassic fossil walks, over the six months it is open

A UK national geoconservation initiative, the Regionally Important Geological and Geomorphological Sites (RIGS) scheme (Harley and Robinson 1991; Harley 1996), sponsored by English Nature, led in the early 1990s to the formation of numerous local (usually county based) and regional (particularly in Scotland and Wales) voluntary geoconservation RIGS groups and Earth Heritage Trusts. Most of these developed interpretative schemes, usually involving outdoor panels and leaflets, often funded by the Curry Fund (Green 2008, p. 100) of the Geologists' Association (GA); for example, in 1998 at Cleeve Common near Cheltenham in Gloucestershire and in 2001 at the Moorfield and Wellfield Quarries near Huddersfield in West Yorkshire. The North-East Wales (NEWRIGS) RIGS Group began in the mid-1990s their ongoing series of geotrail leaflets, such as the innovative *Steaming Through the Past* (Burek and France 1998; Green 2008, pp. 99; NEWRIGS 1997) based on the route of a heritage railway at Llangollen in Clwyd. Concomitantly, the voluntary wildlife conservation bodies, the Wildlife Trusts, produced trails with some geo-interest; for example, at Brown End Quarry (Green 2008, pp. 97–98) in Staffordshire (originally in 1991 and renewed in 2004). Some geotrails included on-site collecting facilities, from rock face clearances and spoil-tip material, with limited or no on-site interpretation; for example, at Writhlington (Fig. 17.8) Geological Nature Reserve (Duff et al. 1985, pp. 61–64; Jarzembowski 1989, p. 219; Robinson 1993) in the 1980s.

Fig. 17.7 Hunstanton, Norfolk—the interpretative panel, placed in 1993 by English Nature, is as the northernmost end of the resort's promenade and overlooks the 18-metre high red (in the lower part of the cliff) and white (Lower Cretaceous) Chalk cliffs; fossils found in the rocks include common shrimp burrow trace fossils and bivalves

Selling minerals and fossils to (geo)tourists was widespread in the nineteenth century. It continues in localities such as the Jurassic Coast and the Isle of Wight (Taylor 1992). It was noted at the turn of the last century that "Fossils had been collected at sundry localities for sale to visitors in the latter part of the eighteenth century, especially at Lyme Regis and Charmouth." (Woodward 1907, p. 115). Today, there is very occasionally limited collaboration between commercial collectors and public agencies, such as museums, to provide interpretation during fossil collecting and preparation; the best-publicised was at a Glasgow housing estate in the 1980s where during a fossil excavation (Wood 1983) souvenirs were sold and site-tours given. Commemorative plaques and monuments have been sited since the late nineteenth century. For example, at Churchill in Oxfordshire a monument to William Smith (Fig. 17.9), the 'Father of English geology', was erected in 1891.

At Ludford Corner near Ludlow in Shropshire, the Ludlow Bone Bed's type-locality has a plaque; its text includes "Sir Roderick Murchison in 1839 placed the Fish Bed near the upper limit of his Silurian System.", citing the geology and the famous geologist. Geological 'roadshows' and 'fairs' (Reid 1993), such as the Dudley Rock and Fossil Fair, are a recent and sometimes commercial interpretative development. The chief source for the past 150 years of popular geological interpretation and information, apart from the mass media, has been museums.

Fig. 17.8 Writhlington Geological Nature Reserve, Radstock, Somerset—this is on the landscaped remains of former coal-mine spoil heaps and has a collecting tip, the mound in the middle distance with a collector 'at work', of dumped shales—particularly rich in obvious Upper Carboniferous plant fossils and obscure insects—established in the early 1990s for amateur collectors and children

The UK's first purpose-built geology museum was the Rotunda Museum at Scarborough opened in 1829. The Natural History Museum, with some geology displays, opened in London in 1857; it specifically aimed to attract a broad, including working-class, audience (Bennett 1996). The Geological Survey's Museum of Economic Geology, in a building in London adjacent to Scotland Yard, opened in 1841; it displayed useful rocks and minerals, and manufactured items (such as gun-barrels!) and had a laboratory which could analyse visitors' rocks and soils (Bailey 1952). The Museum of Economic Geology re-opened in 1851 in premises near Piccadilly, together with a Government School of Mines and of Science Applied to the Arts, with exhibits supporting lecture programmes. It closed in 1923 and reopened, with traditional and particularly regional (supporting its handbooks) geology displays, in 1935 in purpose-built South Kensington, London premises. These displays remained virtually unaltered into the 1960s. These began to be replaced from the 1970s by thematic exhibitions; for example, the 1973 ground-breaking multimedia 'The Story of the Earth' (Tresise 1973; Dunning 1974, 1975) was the first permanent exhibition anywhere to cover plate tectonics. Other innovative galleries followed in the late 1970s and 1980s; of these, 'British Fossils' adopted a modified traditional specimen-rich approach. For its 'Dinosaurs' gallery the museum departed from its usual populist soft-backed geological booklets, opting for an A3 hardback publication (Gardom and Milner 1993)—undoubtedly aided by the Reader's Digest sponsorship.

Fig. 17.9 William Smith memorial, Churchill, Oxfordshire—erected in 1891 from locally quarried limestone to record the local birthplace of William Smith (1769–1839) creator of the first geological map of England and Wales; the metal plaque on the 2-metre monolith records him as the "Father of British Geology"

In the 1990s, a £12-million programme, with sponsorship from mining and power companies duly acknowledged on prominent plaques, to replace and renew individual displays (Clarke 1991) and whole galleries (Dagnall 1995; Robinson 1996b; Smith 1996); the first two of the new galleries, 'The Power Within' (geology-based) and 'Restless Earth'(geomorphologically-based), opened in 1996 (Sharpe et al. 1998). Elsewhere in the UK, multi-media galleries were opened at the National Museum of Wales and the National Galleries of Scotland in the mid-1990s. Most provincial museums, due to financial and staffing constraints, did not adopt the approach (Knell 1993, p. 20). Concern was expressed about their displays' quality because "Poor displays serve only to reinforce negative attitudes to geology…Textbook in style and jargon-riddled—even the keen amateur would have difficulty in understanding some geological displays." (Knell and Taylor 1991, p. 24).

The presumed innovative approaches in museum gallery and visitor centre exhibition design were mainly those already pioneered by the mass-media. The communicative potential of the mass-media, such as radio, to promote science was recognised in the USA; probably the world's first, WGBH (Boston)'s, radio broadcast of a twenty-hour classroom geology course (Lyons et al. 1993) was in 1954. It was not until 1988 that a six-part school's radio geology programme was broadcast in the UK. Also, in the UK, the first television programmes to explore emerging geological concepts were two single BBC programmes; *The Restless Earth* (on plate tectonics) in 1972 and *The Weather Machine* (on ice ages) in 1974, both accompanied by books (Calder 1972, 1974). The BBC broadcast *On the Rocks,* a further-education production, in 1984 and the six-part radio *Rock Solid* series, both accompanied by books (Wood 1978; Grayson 1988a, b). The BBC broadcast *Earth Story* in 1988, narrated by an eminent biologist, accompanied by a coffee-table style book (Lamb and Singleton 1998). The only similar UK commercial television offering was the 1990 *Landshapes* series.

17.5 Four Landscape Movements

'Tourist' first appeared as a synonym for 'traveller' in the late eighteenth century. It was incorporated within the title of Mavor's (1798–1800) *The British Tourists; or Traveller's Pocket Companion, through England, Wales, Scotland, and Ireland...*, probably the UK's first national guide-book. Seemingly, "The traveller exhibits boldness and gritty endurance under all conditions (being true to the etymology of 'travel' in the word 'travail'); the tourist is the cautious, pampered unit of a leisure industry. Where tourists go, they go *en masse*, remaking whole regions in their homogeneous image." (Buzzard 1993, p. 2); this underscores the major issue, from the earliest appearance of tourism, about tourists' potential to destroy the places they visit. Several less pejorative labels have been employed (such as 'Lakers', 'excursionists' and 'eco-tourists') over the past 200 years to describe persons who chiefly visit landscapes for leisure purposes. Landscapes are socio-cultural constructs as much as they are physical entities. In Europe, from Renaissance times, leisure travellers sought out landscapes that matched their quest for the authentic, imagined, novel and exotic. Their expectations of the landscapes they visited were nurtured and subjugated by images in visual art, poetry and prose, especially when included within guide-books. The perceptions of such idealised landscapes and the values attributed to them by travellers and tourists are a combination of their processing of direct observation and subsequent interpretation, influenced by their education, knowledge, interests and expectations. Travellers (and later tourists), artists, writers, geographers, geologists and earth scientists delineate, describe and illustrate landscapes from their specific mind-sets.

Until the middle of the eighteenth century, travellers' "...preferred rural landscape was generally a humanised scene of cultivation, evidence of the successful control of nature." (Towner 1996, p. 138). Subsequently, three major aesthetic movements underpinned the changing public perception of, and subsequent relationship with British and European, landscapes. The first, the 'sublime', overlapped with and was superseded by the 'picturesque'. The 'romantic' was a later over-arching movement from about 1780 to 1850. In the middle of the eighteenth century, Edmund Burke, in *A Philosophical Enquiry into the Origin of Our Ideas of the Sublime and Beautiful* of 1757, equated the sublime with astonishment, fear, pain, roughness and obscurity. The Romantics associated the sublime with the tumultuous chaos of mountains lying beyond rolling foothills, deep valleys and dangerous rocky precipices. The poet William Wordsworth in *A Guide to the Lakes* (Fig. 17.10), first published in 1810, noted in its most popular fifth edition of 1835 that two landscape formation phases contributed separate but linked elements, for "Sublimity is the result of Nature's first great dealings with the superficies of the earth; but the general tendency of her subsequent operations is towards the production of beauty; by a multiplicity of symmetrical parts uniting in a consistent whole." (Wordsworth 1835, p. 35); the masses of rock, hill and lake, due to their wildness and ruggedness, solicited awe and wonder from their observer; such was the case with the earliest, late seventeenth century, travellers into the UK's countryside. Conversely, the 'picturesque' thrilled observers with the softer

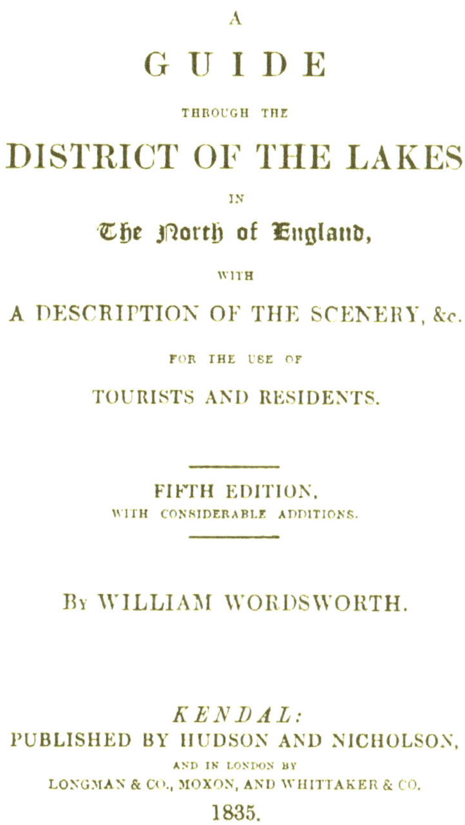

A

GUIDE

THROUGH THE

DISTRICT OF THE LAKES

IN

The North of England,

WITH

A DESCRIPTION OF THE SCENERY, &c.

FOR THE USE OF

TOURISTS AND RESIDENTS.

FIFTH EDITION,

WITH CONSIDERABLE ADDITIONS.

By WILLIAM WORDSWORTH.

KENDAL:

PUBLISHED BY HUDSON AND NICHOLSON,

AND IN LONDON BY

LONGMAN & CO., MOXON, AND WHITTAKER & CO.

1835.

Fig. 17.10 *A Guide to the District of the Lakes…*—facsimile of the title-page of the 1835 fifth edition of William Wordsworth's acclaimed Lake District guide-book; this is the most republished version of the volume and is still in print today effects due to Nature's subsequent operations producing variegation and harmony as seen, for example, in a river meander's or lake shore's curve, the grouping of rocks and trees flanking them, and the interplay of light and shade over them with the subtle colour-gradations seemingly melding the scene. It was an approach adopted, from the late eighteenth century, by the pioneering travellers into the UK's countryside. Essentially the landscape is framed as for a picture, hence the significance of scenic 'stations' or viewpoints. Waterfalls were especially favoured by those interested in the sublime and the picturesque; the moniker 'cataractist', for persons choosing to visit waterfalls, briefly passed into contemporary if uncommon usage.

The two main aesthetic movements overlapped, and this is obvious when the 'picturesque' was defined as "What pleases the eye" in the 1801 Supplement to Dr Johnson's Dictionary. By the nineteenth century's opening it was in popular enough usage to overlap with the 'Sublime'. The 'Romantic' saw the expression of feeling for landscape and its evocation in art and literature, especially poetry. The late-twentieth century emergence of 'ruralism' (Hose 2012, p. 14) saw the countryside as a location for leisured pursuits and an outdoor adventure-sports' backdrop. It is a reworking of earlier themes within the framework of post-modern consumerism, exemplified by the popularity (it remained in *The Sunday Times* newspaper's best-seller list for over three years, and number one on the same list for longer than any other book) of *The Country Diary of an Edwardian Lady* (Holden 1977)—a bit of a misnomer because the author had resided in a rapidly urbanising town!

Each of the major (Fig. 17.11), and to some extent the modern minor, movements is a reflection of three entwined threads, the

1. Characteristics of the travellers or visitors;
2. Meanings ascribed to, and understanding of, natural phenomena witnessed by them;
3. Fundamental UK shift from rural (land-based economy) to industrial (manufacturing-based) society and concomitant middle-class rise in numbers and influence.

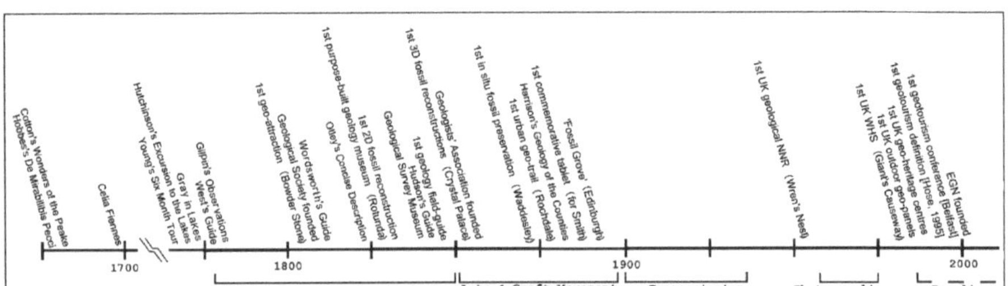

Fig. 17.11 Geotourism Timeline—showing the major UK events and influences, with some significant publications, in the development of modern geotourism

The movements were recorded and promoted by artists and writers especially those, now considered quintessentially English: John Constable (1776–1837), John Keats (1795–1821), J.M.W. Turner (1775–1851) and William Wordsworth (1770–1850). Similarly, there are those generally unknown today such as William Gilpin (1724–1804) and Samuel Palmer (1805–1881). In the twentieth century their place, although their landscape images are generally less idealised and more stylised (see Mullins 1985), was taken by the likes of Paul Nash (1889–1946), Ben Nicholson (1894–1982), John Piper (1903–1992) and Sir Stanley Spencer (1891–1959). Into the twenty-first century artists such as David Gentleman (b 1930) and David Hockney (b 1937) have developed these traditions. The aesthetic movements' impact on travellers, the forerunners of modern geotourists, can be most readily traced in the landscapes in which they were first perceived and promoted.

17.6 The English Peak and Lake Districts' Role in Early Geotourism

The Peak District was the first UK region to have its countryside explored, from the late seventeenth century onwards, by travellers. These visited caves and mines, readily accessible from the industrialising provincial centres, such as Derby and Sheffield, and broke their journeys to visit potteries and textile mills—much as modern tourists now visit industrial archaeology sites. Hence, it can be considered geotourism's birthplace. By the time railways had opened the Peak District, between the 1840s and the 1870s. to travellers from across the UK its major contribution to landscape-based tourism development had passed. The railway reached Derby in 1839 but it was not until 1863 that its passengers could alight at Buxton. Surprisingly the GA did not organise a major excursion to the region until 1877; this was over five days with much time spent in the 'White Peak'. The National Stone Centre (Thomas and Prentice 1984) near Wirksworth, the UK's first such visitor centre opened in 1989, at the junction of the 'White Peak' and the 'Dark Peak'. The Peak District National Park, the first such established in the UK 1951, is the country's most visited National Park.

Before the nineteenth century only the social elite travelled afar, at least in any style and that was literally horse-powered. Hence, there was a very limited market for publications to inform such travellers. One of the earliest was *Britannia* (Ogilby 1675), a strip map of the roads of England and Wales, that was far too bulky until the 1721 pocket edition to carry on a journey. By the late seventeenth century, the region's major sights and sites were organised and promoted in the books of Thomas Hobbes (Fig. 17.12) and Charles Cotton, of 1678 and 1681, respectively; both recognised seven accessible 'wonders' of 'two fonts' (the ebbing and flowing wells at Tides well and St Ann's Well), 'two caves' (Poole's Cavern and Peak Cavern), 'one palace' (Chatsworth House), 'one mount' (Mam Tor) and 'a pit' (Eldon Hole pothole). The

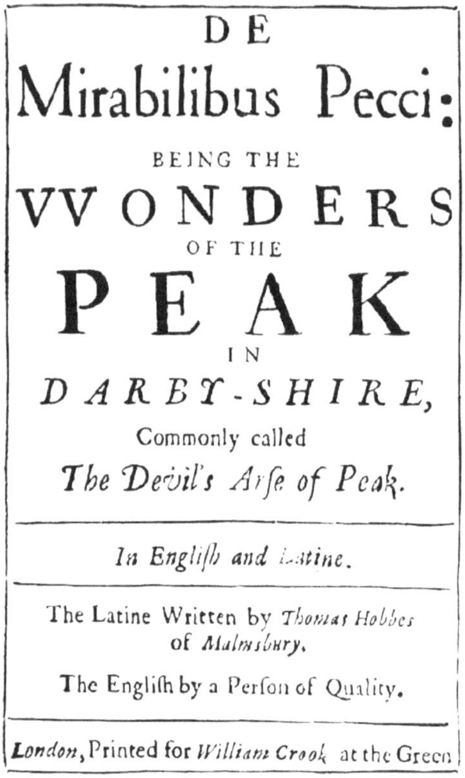
DE
Mirabilibus Pecci:
BEING THE
VV ONDERS
OF THE
PEAK
IN
DARBY-SHIRE,
Commonly called
The Devil's Arſe of Peak.

In Engliſh and Latine.

The Latine Written by *Thomas Hobbes* of *Malmsbury.*

The Engliſh by a Perſon of Quality.

London, Printed for *William Crook* at the Green

Fig. 17.12 *De Mirabilibus peci: Being the Wonders of the Peak…*—facsimile of the title-page of the volume published, as the first attempt at a tourist guide-book for the Peak District, in Latin and English in 1678; it was a reprint of the original Latin poem published in 1636

caves of the 'White Peak' (underlain by Lower Carboniferous limestones) were the most popular natural (geo)attractions.

Celia Fiennes (1662–1741), a privileged traveller through both the Peak District and Lake District in the late seventeenth century (Fiennes 1949), rode to Buxton and explored Poole's Cavern in 1697; afterwards she rode to Ashbourne and visited its copper mines, perhaps becoming the earliest recorded true geotourist. She also rode through the Lake District in 1698, recording potted char and bread recipes but, presumably because she saw it as an unprofitable barren wilderness, not the scenery. Daniel Defoe (1660–1731) in the 1720s also considered it a wild, barren and frightful place; likewise, he concomitantly described the Peak District as a "houling wilderness" (Defoe, nd, vol. 1, p. 160) and its 'wonders' overrated. In the late eighteenth century, the agricultural writer Arthur Young (1741–1820) explored Dove Dale and recorded in 1771 that "It is bounded in a very romantic manner by hills, rocks and hanging woods…forming a wide assemblage of really romantic objects…" (Trench 1990, p. 158). Earlier, in 1768, he had toured the Lake District and found the view of Derwent water elegant but the mountains surrounding the entrance to the nearby Borrowdale valley were too wild and split with 'dreadful chasms'.

By the 1750s the entrance to Poole's Cavern was surrounded by modest cottages whose inhabitants made a living from travellers. Defoe specifically commented, with redolence to modern geotourists' expectations raised by hyperbolic promotion, that "The wit that has been spent…had been well enough to raise the expectation of strangers, and bring fools a great way…is ill bestowed upon all those that come…with a just curiosity…they generally go away, acknowledging that they have seen nothing suitable to their great expectation, or to the fame of the place." (Defoe, nd, vol. 1, p. 168). The antiquary William Bray (1736–1832), recollecting his 1773 visit to Peak Cavern, drew attention to the cave experience commercialization for "…you come to the first stream…to be crossed by lying down in a boat, which is pushed forward by the guide…After crossing the water a second time, on the guide's back, you enter Roger Rain's House…Here you are entertained by a company of singers, who…are stationed in a place, called the Chancel." (Bray in Mavor, vol. 2, 1783, pp. 338–339). Thomas West (1720–1779), in an 'Appendix' to his 1778 Lake District guide-book, provided an itinerary of the region's major caves and limestone crags; his sub-title *Some Philosophical Conjectures on the Deluge…* indicates how observations of natural phenomena inspired geological speculation. Almost concomitantly with these general accounts, the first account of the Peak District's geology, *An inquiry into the original state and formation of the Earth*, was published in 1778 by John Walcott (d 1813), but despite two further editions in 1786 and 1792 it had a very limited readership amongst travellers. There is also no evidence that the near contemporaneous *Petrificata Derbiensia; or fig.s and descriptions of petrifactions collected in Derbyshire* of 1809 by William Martin'(1767–1810) and *General View of the agriculture and minerals of Derbyshire* of 1811 by John Farey (1728–1798) had a greater, if any, readership amongst the early Peak District travellers. Likewise, the ground-breaking geological cross-sections by White Watson (1760–1835) in his *A delineation of the Strata of Derbyshire…* of 1811 and *A section of the strata in the vicinity of Matlock Bath…* of 1813; despite travellers visiting his Bakewell museum-shop, the contemporary readership was probably limited to a few geologists and natural scientists. At least both the travellers and geologists were keen collectors of minerals and fossils. Blue John and other coloured flourpsars and limestones (in their native or carved forms such as vases) were particularly in demand. Indeed, the novelist Jane Austen (1775–1817) had the character of Elizabeth Bennet in *Pride and Prejudice* of 1813 noting the opportunity to collect 'a few petrified spars' as one of the attractions of a Peak District visit. They were purchased, at the mines and caves as well as from shops in Bakewell, Matlock and even Derby cathedral, to fill the 'cabinets of curiosities' of the adventurous travellers. The Peak District's overt commercialism, with rising prices and increasing accommodation difficulties, eventually prompted travellers to seek curious natural phenomena elsewhere, such as the Lake District, in quieter wild landscapes.

Numerous historians and writers of literary and art criticisms have charted the stages in the Lake District's rise to its present popularity with tourists, but most have generally failed to mention its many industrial, especially mining, landscapes. These were deliberately ignored in promoting its wild rural idyll, especially by the 'Lake School' of poets; these were named as Samuel Taylor Coleridge, Robert Southy

(1774–1843) and William Wordsworth in the *Edinburgh Review* of August 1817 (Daiches and Flower 1979, p. 115). From the 1750s onwards travellers to the region, or 'Lakers', visited because of the alleged quality of its scenery and antiquities; the poet Thomas Gray (1716–1771) visited in his later years in 1767 and 1769, the farmer Arthur Young in 1768, the artist William Gilpin in 1772 and the antiquary William Hutchinson (1732–1840) in 1773. Gray's letters describing his second visit were published in William Mason's posthumous edition of his work in 1775. In 1778 Thomas West, did for the Lake District what Charles Cotton had done a century earlier for the Peak District; he organised its chief sights and sites into a guide-book, the first devoted to the region that established the pattern for later tours and their guide-books. Near-contemporary illustrated guides to its antiquities were also available, such as that by Hutchinson of 1794.

Thomas West's 1778 guide-book is mainly accounts of numbered 'stations' from which travellers can view scenic wonders whilst reading the accompanying description. The accounts guided tourists with the textual precision necessary before detailed topographic maps were available; so, the location of Coniston's first station was "A little above the village of Nibthwaite the lake opens in full view. From the rock, on the left of the road, you have a general view of the lake upward. This station is found by observing an ash tree on the west side of the road, and passing that till you are in a line with the peninsula, the rock is then at your feet." (West 1778, pp. 50–51). West's stations, along with some of his own, were indicated on the maps issued from 1783 by Peter Crosthwaite (1735–1808), who styled himself as a geographer and hydrographer. He particularly added a couple for Derwentwater, especially popular with early tourists. For that halfway up Latrigg he had a flight of steps cut and a marked cross laid out on the ground. The other, conveniently located near his museum, was advertised by beating a gong and publishing customers' names in the local newspaper; in 1793 it recorded 1,540 travellers (Ousby 1990, p. 172). West's guide-book noted the view of Derwentwater from Cockshott Hill on its north-eastern shore almost fulfilled the ideal requirements of the 'picturesque' since "On the floor of a spacious amphitheatre, of the most picturesque mountains imaginable, an elegant sheet of water is spread out before you, shining like a mirror, and transparent as chrystal; variegated with islands, that rise in the most pleasing forms above the watery plane, dressed in wood, or clothed with forest verdure, the water shining round them." (West 1778, pp. 89–90). Arthur Young, after his 1768 Lake District tour, declared the stations should be made readily accessible, and resting places provided, to a greater range of travellers than thitherto possible due to the precipitous nature of many paths; because woodland obscured some views, he recommended their trees be pruned or felled. The stations often had structures for travellers but they were not universally liked; in 1799 James Plumtree thought the first Windermere station "too finished and artificial" (Ousby 1990, p. 158). The 1802 account of *A Tour through the Northern Counties...* by Richard Warner (1763–1857) included route maps that still could lead modern tourists to popular places. West and his contemporaries, by reducing the region's scenery to stations with erudite explanations, established tourist behaviours seen almost anywhere today; for example, car-parks at 'viewpoints' or 'beauty spots' where snapshots can be taken.

A real impetus to the Lake District's exploration, especially by artists and writers, was given by Continental Europe's closure to the British, from 1789 to 1815, during the French Revolution and the succeeding Napoleonic wars. Its mountainous landscape was then promoted as mimicking the Alps, which the likes of Wordsworth had already visited. It was well promoted to the elite educated and discerning tourist by William Green (1760–1823) who in 1819 published the illustrated two-volume *Tourist's New Guide.* He was a commercially successful landscape painter, originally encouraged by Thomas West, with exhibitions at his salerooms in Ambleside for which travellers were charged a shilling's viewing fee. Wordsworth was well acquainted with Green and fulsomely complimented, as a "complete Magazine of minute and accurate information." (Wordsworth 1835, p. 6), the *Tourist's New Guide* in his own guide-book. Whilst Green popularised the region in paintings and prints, Wordsworth did so in poetry and prose. Wordsworth's innovative guide-book coupled descriptions of the tourist sites with accounts of the area's history, characters, and natural history—including geology. It had developed from the anonymous text accompanying an 1810 volume of the engravings by the Rev. Joseph Wilkinson (1764–1831). Eventually, in 1820 it evolved into an appendix to his River Duddon sonnets (Wordsworth 1835) and then into a guide-book in its own right by 1822; continuously revised and expanded, it ran into several editions upto the 1840s; the later ones included three geology letters by the renowned geologist Adam Sedgwick. Later guide-books are noteworthy for their strong literary content drawing the reader's attention to site-focused poetry and other literary allusions; for example, *The Scenery and Poetry...* of 1846 by Charles Mackay (1814–1889) used Shelley's brief stay in the area as an excuse to quote the description of the waterfall from 'Alastor' for the unremarkable Stock Ghyll Force. There was a crossover with the more conventional guide-books which began to include similar literary allusions. For modern geotourism the creation of these literary landscapes is probably their most enduring achievement.

By the first decades of the nineteenth century most of the necessary work to make the Lake District's landscape readily accessible to elite visitors had been completed. The downside was the high number of visitors was then threatening the survival of the tranquil landscape they had come to enjoy. Wordsworth in *A Guide to Lakes*, in two 1844 public letters and his sonnet opposing the Kendal and Windermere Railway, espoused this concern. However, he was seeking to protect the landscape, for the educated elite for whom West's and his own guide-book had been written, from the lower middle and working classes from the northern industrial towns such as Manchester. He recognised the railways, then encircling the Lake District, would from cheap let trains hundreds of tourists at a time visit; many of these would then probably spend much of their visit in ale-houses. The railway reached Carlisle in 1846 and Windermere in 1847, but Keswick (for Derwentwater) was not reached until 1865. Carlisle was eventually on the London to Edinburgh via Birmingham main line. Although much of the limited track was laid to export the region's rich mineral resources, as the associated revenues declined, it was increasingly focused on tourism. The Kendal to Windermere line had been specifically laid to promote tourism to revive the local economy. Wordworth's spiritual successor, John Ruskin, echoed his disapproval of the railway 30 years later when an extension was proposed. Such condemnation was not universal; for example, Harriet Martineau's 1855 *A Complete Guide to the English Lakes* encouraged mass-tourists, and suggested they spent a day in the mountains, possibly with the services of a guide.

Prior to the aesthetic tourist guide-books, travellers recorded their private impressions which were sometimes published as journals. Numerous amateur and professional artists visited and recorded the sights in various media; these were sometimes published as sets of engravings either on their own or accompanying prose or poetry. For example, the young J.W.M. Turner made his living as a topographical artist executing numerous watercolours and a few oil paintings; two of the latter Lake District landscapes were exhibited at the Royal Academy in 1798, the most noteworthy being *Morning Amongst the Coniston Fells, Cumberland*. He contributed two plates to Samuel Rodger's *Poems* of 1834. Gray's journey from the southern end of Derwentwater into Borrowdale reminded him of Alpine passes where travellers were threatened by avalanches! Gilpin on the same valley wrote "As we proceeded in our route along the lake, the road grew wilder, and more romantic. There is not an idea more tremendous, than that of riding along the edge of a precipice, unguarded by any parapet, under impending rocks, which threaten above; while the surges of a flood, or the whirlpools of a rapid river, terrify below." (Gilpin 1786, vol. 1, p. 187). Meanwhile, the first populist account of Lakeland geology was by Johnathan Otley (1766–1856), a clockmaker by profession and amateur geologist—renowned as the 'Father of Lakeland Geology'—by choice, in *A concise description of the English Lakes...and observations on the mineralogy and geology of the district* (Fig. 17.13) of 1823; this was popular enough to go to eight editions by 1850.

Adam Sedgwick (1785–1873) added, as he had for Wordsworth's guide-book, some geological observations to *A Complete Guide to the Lakes* of 1842 by John Hudson which went to at least five editions by 1860 and is still in print today.

17.7 Central Southern England and the Birth of the Modern Geotourist

Just as for the Peak District and Lake District, railways opened central southern England (Hampshire, the Isle of Wight, Dorset and Sussex) to tourists and geological exploration from the 1850s. In the east, Brighton was linked to London in 1841, Portsmouth (for the Isle of Wight) in 1859 and Bournemouth in 1870. In the west, Bridport was linked to the railway network in 1857, Weymouth (for the Isle of Portland) in 1865, but Lyme Regis had to wait until 1903. The far western part of the region is promoted as 'Wessex' (Daiches and Flower 1979, pp. 158–171) due to novels of Thomas Hardy (1840–1928). Sussex, undoubtedly due to its proximity to London and its active geological community, centred on The Geological Society established in 1808, was its first county subjected to detailed geological examination. The whole region much featured in the field excursion programme of the GA, founded

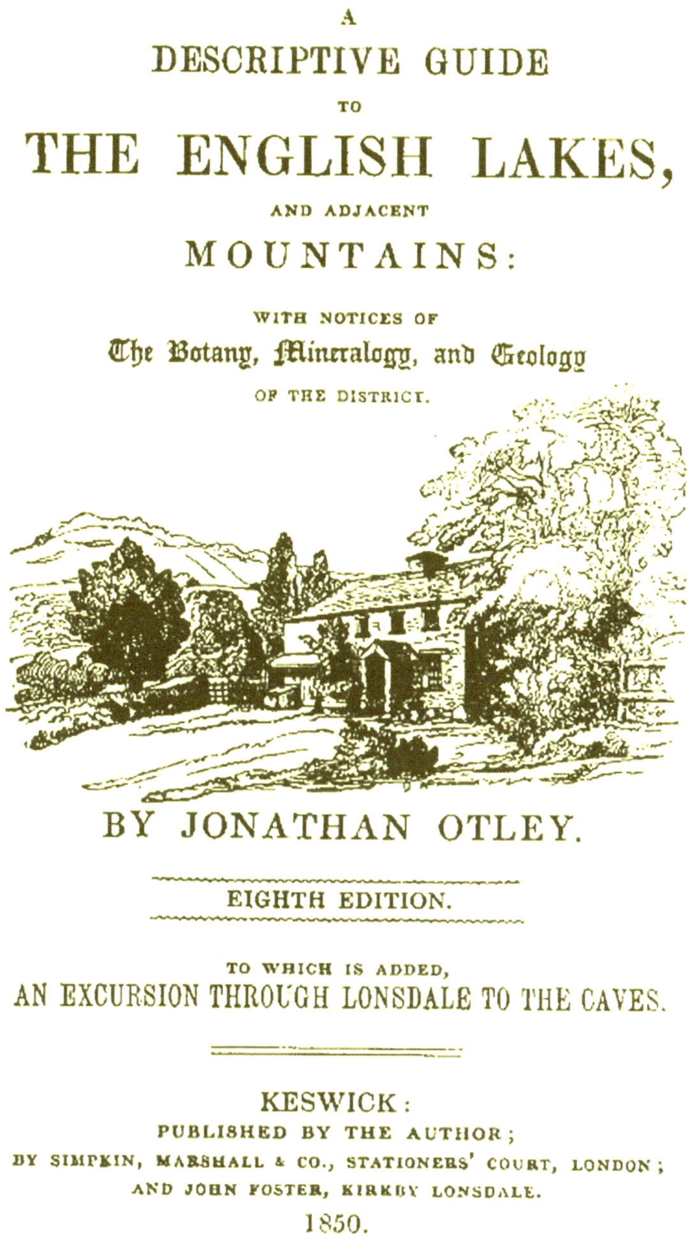

A

DESCRIPTIVE GUIDE

TO

THE ENGLISH LAKES,

AND ADJACENT

MOUNTAINS:

WITH NOTICES OF

The Botany, Mineralogy, and Geology

OF THE DISTRICT.

BY JONATHAN OTLEY.

EIGHTH EDITION.

TO WHICH IS ADDED,

AN EXCURSION THROUGH LONSDALE TO THE CAVES.

KESWICK:

PUBLISHED BY THE AUTHOR;

BY SIMPKIN, MARSHALL & CO., STATIONERS' COURT, LONDON;

AND JOHN FOSTER, KIRKBY LONSDALE.

1850.

Fig. 17.13 *A Descriptive Guide to the English Lakes*...—facsimile of the title-page of the eighth (1850) edition of Jonathan Otley's Lake District guide-book, the first populist account of the region's geology and mines in 1858. There is a noticeable relationship between its excursions to the area and the expanding railway network. Excursions were announced in its *Circular* and written up as reports in its *Proceedings*; for example, Lyme Regis in 1906 (Lang 1906).

The first summaries of the area's geology, *The Fossils of the South Downs* (Mantell 1822) and *Illustrations of the Geology of Sussex* (Mantell 1827) by Gideon Algernon Mantell (1790–1852) were large and costly library volumes. Both were well illustrated with locality-specific information on where fossils could be found—useful in excursion planning. Dixon's *The Geology and Fossils of the Cretaceous and Tertiary Formations of Sussex* of 1850 (with a second edition in 1878) reproduced the plates from Mantell's *Fossils of the South Downs*.... Mantell's *Geology of the South-East of England* of 1833 is a pocketable volume; its Sussex section includes geotourism information such as "The beach near this place [Rottingdean] contains semi-translucent pebbles of agate, and chalcedony, of a bluish grey colour. These are collected by visitors and when cut and polished are used for bracelets and other ornamental purposes..." (Mantell 1833, pp. 40–41). His 1847 *Geological Excursion Round the Isle of Wight and the Adjacent Coast of Dorsetshire* (Fig. 17.14) was cheap enough to be taken into the field, possibly explaining the poor condition of many surviving copies.

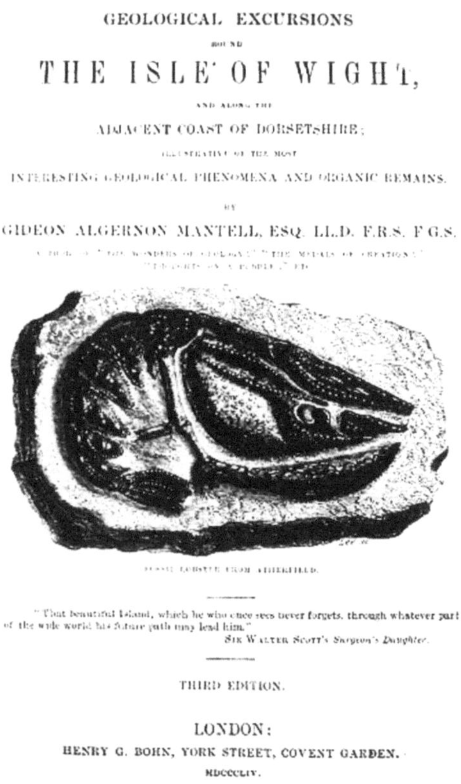
GEOLOGICAL EXCURSIONS

ROUND

THE ISLE OF WIGHT,

AND ALONG THE

ADJACENT COAST OF DORSETSHIRE;

ILLUSTRATIVE OF THE MOST

INTERESTING GEOLOGICAL PHENOMENA AND ORGANIC REMAINS.

BY

GIDEON ALGERNON MANTELL, ESQ. LL.D. F.R.S. F.G.S.

FOSSIL LOBSTER FROM ATHERFIELD.

"That beautiful Island, which he who once sees never forgets, through whatever part of the wide world his future path may lead him."

Sir Walter Scott's *Surgeon's Daughter*.

THIRD EDITION.

LONDON:

HENRY G. BOHN, YORK STREET, COVENT GARDEN.

MDCCCLIV.

Fig. 17.14 *Geological Excursions around the Isle of Wight...*—facsimile of the title-page of the third and final (1854) edition of Gideon Mantell's illustrated geological pocket-guide to the geology of the Isle of Wight and Dorset coast

Its sold well enough to warrant two further editions in 1851 and 1854; the first edition's Preface justified its publication in a manner common to modern field-guides for "...the Geology of the Island is but little known or regarded by the majority of the intelligent persons who every season flock by thousands to its shores, and, rapidly traversing the accustomed routes, visit the picturesque localities noted in the numerous handbooks, and take their departure without suspecting that they have been travelling over a country rich with the spoils of nature...of the highest interest to the instructed observer." (Mantell 1847). Although Dorset was significant in scientific geology's development from the nineteenth century's beginning, it had to wait until mid-century for a dedicated populist field-guide, the *Handbook of the geology of Weymouth and the Isle of Portland* published in 1860, with a second expanded edition almost a quarter of a century later in 1884; its Preface indicated the need for a geology guide "...written in a somewhat scientific yet elementary and popular form... pointing out...where the various formations can be best examined, and by rendering the latter more easily identified by means of views, sections and other illustrations." (Damon 1860). Towards the nineteenth century's close it was noted, in the compendium volume *Geology of the Counties of England and of North and South Wales* (Harrison 1882) that "Some portions of Dorset—more especially the coast—offer such unrivalled opportunities for geological study, that they early attracted the attentions of writers on that science." (Harrison 1882, p. 72). The same could be said of the Isle of Wight.

Despite difficult access, until the late-nineteenth century advent of steam-powered ferries, the Isle attracted numerous geological publications because "In the splendid sections exposed along the line of coast we are enabled to examine the strata and trace their relationships to each other...while the beauty of the scenery...lends an additional charm to the investigation" (Harrison 1882, p. 103); such aesthetic landscape emphasis underpins modern populist field-guides. Its earliest guide, *A Description of the Principal Picturesque Beauties, Antiquities, and Geological Phenomena, of the Isle of Wight* of 1816 was a large illustrated library volume; it had much detail on woodlands and archaeological sites and is noteworthy for the twelve letters by Thomas Webster containing detailed observations upon coastal geology. Its Preface recorded that it was result of observations made between 1799 and 1801 and prepared for publication in 1802 when "Circumstances...occasioned me to lay aside the whole for ten years...I had expected that my work would be superseded by the labours of later travellers...I was induced to re-examine my own work, and at length prepare it for publication." (Englefield 1816, pp. i–iii). The Geological Survey published a mid-nineteenth century (Bristow 1862) account of the Isle's geology. *A Concise Exposition of the Geology, Antiquities, and topography of the Isle of Wight* (Wilkins 1861) attempted to address the niche tourist market but was curiously a subscription volume. The later *A Popular Guide to the Geology of the Isle of Wight* (Norman 1887) had a map and 22 plates of fossils and topographic views—many more than today's equivalents; its essentially stratigraphical accounts focused on fossil collecting localities were accompanied by appendices on Newport Museum and the Ventnor Collection. Its Preface noted that the author had already published in a local newspaper a series of letters on the Isle's geology and "At the solicitation of many friends and readers, and seeing that Dr. Mantell's—the only popular work on the subject—was published nearly forty years ago, I have decided with much diffidence to re-publish these letters in the form of a Popular Guide…I have endeavoured to give my descriptions in such a manner that the work may be useful to the geologist as well as entertaining to the general reader…" (Norman 1887, pp. iii–iv). A populist field-guide, with the usual fossil collecting emphasis, from the twentieth century's first quarter (Hughes 1922) was prefaced "The Isle of Wight is classic ground of Geology. From the early days of the science it has been made famous by the work of great students of Nature…" (Hughes 1922). The Geological Survey's sole twentieth century summary of the Isle's geology, in its traditional 'memoir format, *A Short Account of the Geology of the Isle of Wight* (White 1920), was issued for 80 years without any changes, excepting a new bibliography, until 1990. However, the various historic publications' emphases on collecting, especially in situ fossils, was by the late twentieth century at odds with geoconservation and modern geotourism practice. Unsurprisingly, the first regional geoconservation publication, *Guidelines for Collecting Fossils on the Isle of Wight* (Anon, nd) was because "In recent years, increasing pressure has been imposed on the eroding coastline by growing numbers of collectors and visitors."

The area's ready accessibility by railway from London allowed excursions to its major places of geological interest to be included in several London-based guidebooks; for example, *Geological Excursions Round London* (Davies 1914) the Preface of which indicated its intention to be "…useful to many who are interested in the natural sciences and desire a field-acquaintance with the geological formations which occur in their district. To those who take pleasure in country walks the route here described will prove attractive, while a knowledge of the structure of the country will add interest to future excursions." (Davies 1914). The same author later published *The Dorset coast: a geological guide*, in two editions in 1935 and 1956; the latter noted the shift in transport and impact on the visitor's experience, but its Introduction still promoted in situ collecting because Dorset's geology "…is well known, at least from books… a poor substitute for hammering the actual rocks and seeing how they fit into the structural fabric of the country…" (Davies 1956, p. 1). Somewhat unusually, but quite logical to non-geologists, his *Geology of London and South-East England* (Davies 1939), reversed the usual practice of beginning with the older rocks (instigated by Charles Lyell), so that readers began with rocks containing fossils similar to modern forms before progressing onto less familiar older forms; it also promoted GA membership and the benefits of utilising museums and public libraries. The GA itself published from 1958 onwards several centennial field-guides for the area; for example, *Geology of the Central Weald…* (Allen 1960). The GA has also published more recent guides to the region, especially for the coastal sections (Ager and Smith 1973; Allison 1992; House 1989; Kirkaldy 1976). Their descriptive itineraries are well-detailed routes, with some attempt at interpretation, primarily for the geologically knowledgeable rather than the casual visitor.

In 2001 the region's western coastal section became England's first natural UNESCO World Heritage Site, popularly promoted as the 'Jurassic Coast' because of its significant contribution to the development of geology and its spectacular coastal sections. However, the first such UK location was the Giant's Causeway and Causeway Coast in Northern Ireland in 1986. This designation has resulted in a steady stream of purportedly populist field-guides, such as *Geology and Scenery of Dorset* (Bird 1995), an example of those guides forging a link between what the tourist sees and the underlying geology. Other guides, such as *Finding Fossils in Lyme Bay: A Guide to Lyme Regis, Charmouth and Surrounding Areas* (Coram 1989) and *Lyme Bay Fossils Beach Guide* (Clarke 1998), continue the historic focus on fossil collecting.

17.8 Summations and Additions

Modern geotourism's emergence required the recognition and promotion of geological phenomena concomitant with tourists' willingness to engage with seemingly untamed landscapes into which reliable transport was developed. However, tourism's beginnings and geotourism's antecedents were in the late seventeenth century when transport was road based. Indeed, "Travel had been a rarity for all but a minority throughout much of the eighteenth century. Getting around was difficult, and expensive. It was only with the arrival of maps and guidebooks, the creation of the turnpikes and the improvements to the technology of the stagecoach…that even many of the elite began to travel." (Flanders 2006, p. 211). The first metalled turnpike toll road was opened in 1706. By 1800 there were some 1600 (Briggs 1983, p. 207) linking major cities and towns. Any journey away from turnpikes was on gravel and mud tracks that tested, especially in the winter months, their literal horse-power and coach technology. Long-distance road journey times were slow. The London to Manchester coaching route took four days in 1754, but by 1790 this had halved. A swifter and more reliable means of transport was clearly required for mass-tourism.

This was offered by the railways, as demonstrated in *Bradshaw's* compendium timetables from 1839 onwards. Mass-tourism's genesis is generally incorrectly credited to Thomas Cook's organised rail trips that began in 1841 with a temperance excursion. However, these were predated by other excursions from 1836 (Brendon 1991, pp. 5–9). The first recorded special railway train, taking excursionists to a church bazaar, was in 1839 on the Whitby and Pickering Railway. The railways from the 1850s onwards chiefly provided the transport to field areas for most nineteenth-century geotourists. By then railways and steamships had opened both coast and countryside; some 8,000 miles (12,870 kms) of railway track (Briggs 1983, p. 210) linked England's cities and major towns to the newly emerging excursion areas. Indeed, "The excursion fever which seized the working and lower classes in the 1850s was the result of…the desire of the railway companies to fill their trains, and the desire of the public itself to take to the rails… the private railway companies fought to gain passengers, much as package holiday companies do today…" (Swinglehurst 1974, p. 22). As the nineteenth century progressed the social elite also succumbed to railway conveyance (Faith 1990, p. 17) admittedly in first-, unlike their servants stuck in third-, class railway carriages. The GA made considerable use of railways often with either special excursion trains or reduced-price excursion tickets. Its first published excursion in 1860, to Folkestone in Kent, was by a specially arranged South Eastern Railway train. The railway's unprecedented passenger numbers created a need for purpose-built railway hotels; the first was Bridge House, at the southern end of London Bridge, opened in 1837. The first at a station was at London Euston, opened in 1839. Derby's Midland (Railway) Hotel, ideally located for excursions into the nearby Peak District, opened in 1840.

In 1907 the GA employed for the first time, following their appearance in the twentieth century's opening decade, a motor omnibus for its Tonbridge, Kent excursion. After the Great War (1914–1918) it was their usual excursion transport mode, as it was for many day-trippers and tourists. The twentieth century's rise in private motor vehicle ownership opened the countryside to the social elite before the Second World War and to the masses by the 1960s (Lavery 1971, pp. 97–111). By the 1960s, two-thirds of tourists travelled by car. Motorists' maps, guide-books and educational posters were published by the major petrol suppliers; one company, Shell (1964), especially promoted nature and geology education. The 1960s witnessed the decline of the branch-line railway network—that began in the 1930s—and the rationalisation of the main-line railway network; around a third was closed to passenger traffic (Simmons and Biddle 1997).

By the twentieth century's beginning the guide-book was an established travellers' tool given "…in the nineteenth century [it] was, initially, a British and a German invention, people from those countries being the first to have the money and the intellectual curiosity to travel, at least in any numbers." (Sillitoe 1995, p. 221). The first tourist guide-books were for a discerning educated wealthy elite readership. By the mid-nineteenth century the burgeoning middle-classes with more modest means and education—the market into which Karl Baedecker tapped—was the main readership for guide-books. Baedecker's first England guide, but in German, was published in 1862. He was the first to employ asterisks as commendation marks for hotels and restaurants, sights and sites of outstanding natural beauty, architecture and artworks in guide-books; his intention was to familiarise readers with such as were most likely to be seen by travellers. Baedeker's guide-books were always cheaper (in price and production quality) and sold in greater numbers than those of his major rivals, such as John Murray whose guide-books were especially for wealthy travellers. Murray also published 40 volumes of populist county and cathedral guides. Numerous rival publications, especially as regional series, were produced by smaller publishing houses; for example, Black's of Edinburgh published 44 volumes from 1826 onwards. Ward Lock's near 90-volume *Red Shilling Guides* were published from the 1870s, Baddeley's 19-volume *Thorough Guides* with maps and plans by Bartholomew were published from the 1880s. Macmillan's 24-volume *Highways and Byways* series were published from the late 1890sand Methuen's 50-volume*Little Guides* from the mid-1900s. Hence, towards the close of the nineteenth century the UK was well covered by good-quality affordable guide-books, many of which incorporated some geological information.

However, dedicated geotourists, from the mid-nineteenth century, could avail themselves, finances permitting because of expensive short print runs, of a burgeoning range of geology field-guides and textbooks. Jonathan Otley's 1823 Lake District guide is perhaps the first to be tourist-focused, whilst Mantell's *Geological Excursion Round the Isle of Wight...* of 1847 was the first modern pocketable geotourists' guide-book. Two major works summarising southern Britain's geology, Phillips's *A Selection of Facts from the Best Authorities, Arranged so as to Form an Outline of the Geology of England and Wales* of 1818 and Conybeare's and Phillips's *Outlines of the Geology of England and Wales* of 1822 were also available. The former was the first compilation of the UK's known stratigraphy. The latter was the standard introduction to geology which, through various revisions, represented the evolving compromise between biblical theology and geology.

Concomitant with these guides, the first geological field-manuals appeared; principal amongst these were De La Beche's *A Geological Manual* of 1831 (available in French, German and USA editions), *How to Observe* of 1835 and *The Geological Observer* of 1851. Some of them were aimed at the public and children; for example, Mantell's *Thoughts on a Pebble* of 1849. In the late nineteenth century, county-based geology accounts were incorporated within trade directories. One set of such entries, Harrison's *Geology of the Counties of England and of North and South Wales* of 1882, was bound into a single volume; its modern counterpart is Anderson's *Field Geology in the British Isle* of 1983. The GA's excursions were twice published, as *Geologists' Association: A Record of Excursions Made Between 1860 and 1890* (Holmes and Sherborn 1891) and Geology*in the Field... (1858–1908)* (Monckton and Herries 1909), as discrete volumes; they provide invaluable accounts of its excursions and contemporary field practices. The UK's most influential specialist's field-guides appeared in the late 1950s, the GA's *Centennial Guides*. The non-specialist's *Geology Explained* series, published by David and Charles, appeared in the late1960s and recently have been reprinted.

17.9 Looking to the Future

The locations and activities established by the late nineteenth century are embedded within much modern geotourism provision. The major changes have been the mode of transport and the type of accommodation utilised by geotourists, as well as their relationship with the landscape. Private cars have replaced mass rail transport as the preferred mode of travel. There is a greater tourist emphasis on pleasure and leisure than intellectual endeavour compared with the nineteenth century. A new UK impetus was given, from the mid-twentieth century, to landscape tourism by the establishment of National Parks (NPs) and Areas of Outstanding Natural Beauty (AONBs) following the 1949 National Parks and Access to the Countryside Act. Whilst in the NPs measures and funding were available to facilitate and promote access for tourists, the AONBs generally lacked such resources, although progress was painfully slow (Green 1996, pp. 102–105). The 1968 Countryside Act further promoted landscape-based tourism, albeit on the urban fringe, with the establishment of Country Parks; as areas of open country for informal recreation they were intended to relieve visitor pressure on the NPs and AONBs. Some of the Country Parks, such as Park Hall in Staffordshire, were established on old mining sites, but any geological interest was generally ignored in their interpretative provision.

UNESCO's global recognition for landscapes has been available since 1972; the most significant citation is the World Heritage List containing around 630 sites, 7% of which are inscribed primarily for their geological interest. Because not all significant geosites met the 'outstanding universal value' criterion required by the World Heritage Convention an alternative designation was sought. In 1999 the geoparks programme envisaged they would recognise the relationships between people and geology, together with the potential for economic development; they would promote landscape elements, rather than discrete geological outcrops and be managed holistically to protect and enhance their natural characteristics. The geopark programme's major intended benefit was to focus attention on geoconservation and sustainable development. Geosites within geoparks must be scientifically significant, have educational potential and some aesthetic appeal. Ideally the geological interest should be allied to some archaeological, historical, cultural or ecological interest. Geoparks must develop and maintain educational provision to retain their status. Geoparks must also prohibit the sale of local and imported, geological specimens within their confines. The creation of European Geoparks as a small network of sites with significant geology was supported by UNESCO.

The European Geoparks Network (EGN) was established in June 2000, to widely share information and expertise, by four geoparks, the: Reserve Geologique de Haute-Provence in France; Lesvos Petrified Forest in Greece; Geopark Gerolstein/Vulkaneifel in Germany and Maestrazgo Cultural Park in Spain. The EGN signed a collaboration agreement, in April 2001, with UNESCO's Division of Earth Sciences. The EGN signed the Madonie declaration in October 2004 to become the official European branch of the UNESCO Global Geoparks Network. Today, it has 25 Geoparks in 10 countries. Each organises regular meetings and conferences, facilitating the exchange of information and providing promotional opportunities. The first UK geopark, created in 2001, was the Marble Arch Caves in Northern Ireland. The first in England, in 2003, was the North Pennines AONB. The creation of Fforest Fawr Geopark, in 2006, encompassing the Brecon Beacons NP suggests that the two regions, now the UK's most popular NPs, significant in early geotourism development might be similarly recognised. However, the only outstanding, and at different stages of development, UK geopark proposals are for The Black Country, Shropshire's Welsh Borderland and the South Wales coalfield'; the first of these achieved formal EGN designation in April 2020.

Europe's major geotourism enterprise is also new geopark designations. Primarily focused on economic and social regeneration they continue and develop the interpretative provision trends established by the recognition of industrial heritage in the 1970s. The Internet will increasingly host geopark information and virtual geotrails. Geo-interpretation will become holistically integrated within other discrete heritage industry provision, especially at industrial archaeological and heritage transport sites. Geopark's increasing popularity will create new demands and pressures on protected landscapes and geosites, fueling the inevitable debate on the necessary balance between geoconservation and geo-exploitation. Consequently, the use, at least from nodal park-and-ride schemes, of rail and bus services, perhaps integrated with bicycle usage, to reduce private care usage will be encouraged. Detailed geotourism and modern scientific geological research will be required to underpin new geopark selection and aid their sustainable management. Finally, to rewrite the great Scottish geologist James Hutton's 'uniformitarian principle' the past really is the key to present geotourism provision.

References

Anon (nd) Guidelines for Collecting Fossils on the Isle of Wight. Isle of Wight County Council, Newport

Ager DV, Smith WE (1973) Guide no. 23: the coast of South Devon and Dorset between Branscombe and Burton Bradstock. Geologists' Association, London

Allen P (1960) Guide no. 24: geology of the Central Weald: The Hastings Beds. Geologists' Association, London

Allen DE (1978) The naturalist in Britain: a social history. Allen Lane/Penguin Books, London

Allison RJ (1992) Guide no. 47: the coastal landforms of Dorset. Geologists' Association, London

Anderson JGC (1983) Field geology in the British Isles: a guide to regional excursions. Pergamon Press, Oxford

Anze C, Weixing L (1985) Status quo and prospects of tourism earth-science. The Earth 4–16

Anze C, Yunting L, Ng YCY (2015) The principles of geotourism. Springer, New York

Bailey E (1952) Geological survey of Great Britain. Thomas Murbyand Company, London

Baldwin A, Alderson DM (1996) A remarkable survivor: a nineteenth century geological trail in Rochdale, England. Geol Curator 6(6):227–231

Barton ME, Delair JB (1982) Oatlands Park Grotto and its ammonite fossils. Geol Curator 3(6):375–387

Bennett T (1996) How the other half visit. Mus J 96(2):37

Bird E (1995) Geology and scenery of Dorset. Ex Libris, Bradford on Avon

Bray W (1783) Tour through some of the Midland Counties, into Derbyshire and Yorkshire by William Bray, F.A.S. Performed in 1777 paraphrased in volume 2, Mavor, W. (1798–1800) The British Tourists; or Traveller's Pocket Companion, through England, Wales, Scotland, and Ireland. Comprehending the Most Celebrated Tours in the British Islands (6 volumes), London

Brendon P (1991) Thomas Cook: 150 years of popular tourism. Secker and Wurburg, London

Briggs A (1983) A social history of England. Wiedenfeld and Nicholson, London

Bristow HW (1862) The Geology of the isle of wight. Geological survey of Great Britain, London

Broadhurst RJC (1952) The Travels of Ibn Jubayr: being the chronicle of a mediaeval Spanish Moor concerning his journey to the Egypt of Saladin, the holy cities of Arabia, Baghdad the city of the Caliphs, the Latin kingdom of Jerusalem, and the Norman kingdom of Sicily. Cape, London

Buckley R (2003) Environmental inputs and outputs in ecotourism: geotourism with a positive triple bottom line? J Ecotourism 2(1):76–82

Burek CV, France DE (1998) NEWRIGS uses a steam train and town geological trail to raise public awareness in Llangollen, North Wales. Geoscientist 8(9):8–10

Buzzard J (1993) The beaten track: European tourism, literature, and the ways to culture 1800–1918. Oxford University Press, Oxford

Calder N (1972) The restless earth. BBC Books, London

Calder N (1974) The weather machine. BBC Books, London

Clarke G (1991) Geology and the public at the Natural History Museum. Geol Today 7(6):217–220

Clarke NJ (1998) Lyme Bay fossils beach guide. Nigel J. Clarke Publications, Lyme Regis

Cleal CJ, Thomas BA (1995a) Wadesley fossil forest, geological conservation review. Palaeozoic palaeobotany of Great Britain, vol 12. Chapman and Hall, London, pp 208–210
Cleal CJ, Thomas BA (1995b) Victoria park, geological conservation review. Palaeozoic palaeobotany of Great Britain, vol 12. Chapman and Hall, London, pp 188–191
Conybeare WD, Phillips W (1822) Outlines of the geology of England and Wales. Longman Hurst, London
Coram R (1989) Finding fossils in Lyme Bay: a guide to Lyme Regis Charmouth and surrounding areas. British Fossils, Wimbourne
Cotton C (1681) The wonders of the Peake. Nottingham
Daiches D, Flower J (1979) Literary landscapes of the British Isles: a narrative atlas. Paddington Press, New York
Dagnall P (1995) A new era for the earth galleries. Environ Interpret 10(2):4–5
Damon R (1860) Handbook of the geology of Weymouth and the Isle of Portland. Edward Stanton, London
Davies GM (1914) Geological excursions round London. Thomas Murby, London
Davies GM (1935) The Dorset coast: a geological guide. A and C Black, London
Davies GM (1939) Geology of London and South-East England. Thomas Murby, London
Davies GM (1956) The Dorset coast: a geological guide, 2nd edn. A and C Black, London
Dean R (ed) (2004) Coleridge and geology. Scholars' facsimiles & reprints, Ann Arbor
De Bastion R (1994) The private sector - threat or opportunity? In: O'Halloran D, Green C, Harley M, Stanley M, Knill J (eds) Geological and landscape conservation. The Geological Society, Bath, pp 391–395
De la Beche HT (1831) A geological manual. Treuttel and Würtz, London
De la Beche HT (1851) The geological observer. Longman, Brown Green, and Longmans, London
Defoe D (nd) A tour through England and Wales divided into circuits, vol 2. J.M. Dent, London
Dixon F (1850) The geology and fossils of the cretaceous and tertiary formations of Sussex. Longman, Brown Green, and Longmans, London
Dixon F (revised by Jones TR) (1878) The geology of Susses or the geology and fossils of the cretaceous and tertiary formations of Sussex, 2nd edn. William J. Smith, Brighton
Doyle P (1993) The lessons of crystal palace. Geol Today 9(3):107–109
Doyle P (1995) Crystal palace park geological time trail: Paxton's heritage trail number two. London Borough of Bromley, Croydon
Doyle P, Robinson JE (1993) The Victorian geological illustrations of Crystal Palace Park. Proc Geol Assoc 104(4):181–194
Duff KL, McKirdy AP, Harley MJ (eds) (1985) New sites for old: a student's guide to the geology of the east Mendips. Nature Conservancy Council, Peterborough
Dunning FW (1974) The story of the earth exhibition at the geological museum, London Museum 26(2):99–100
Dunning FW (1975) The story of the earth in the geological museum: a view from the inside. Geol–J Assoc Teach Geol 6:12–16
Edmonds RPH (1996) The potential of visitor centres: a case history. In: Page KN, Keene P, Edmonds RPH, Hose TA (eds) English nature research reports no. 176 - earth heritage site interpretation in England: a review of principle techniques with case studies. English Nature, Peterborough, pp 16–23
Englefield, H.C.:A description of the principal picturesque beauties, antiquities and geological phenomena of the Isle of Wight. With additional observations on the strata of the Island, and their continuation in the adjacent parts of Dorsetshire by Thomas Webster. T. Webster, London (1816)
Faith N (1990) The World the Railways Made. Pimlico, London
Farey J (1811) General View of the agriculture and minerals of Derbyshire. Sherwood, Neely and Jones, London
Fiennes, C.:The Journeys of Celia Fiennes (edited by Morris, C.). Cresset Press, London (1949)
Fisher, J.A.: The changing nature of Earth science fieldwork within the UK school curriculum and the implications for conservation policy and site development.In: O'Halloran, D., Green, C.,

Harley, M., Stanley, M., Knill, J. (eds.) Geological and Landscape Conservation, pp 477–481. The Geological Society, Bath (1994)
Flanders J (2006) Consuming Passions: Leisure and Pleasure in Victorian Britain. Hodder and Stoughton, London
Gardom T, Milner A (1993) The Natural History Museum Book of Dinosaurs. The Natural History Museum, London
Gilpin, W.:Observations, Relative Chiefly to Picturesque Beauty, Made in the Year 1772, on Several Parts of England: Particularly the Mountains and Lakes of Cumberland, and Westmorland(in 2 volumes). R. Blamire, London (1786)
Grayson A (1988a) Radio geology. Geol Today 4(2):62–63
Grayson A (1988b) Rock Solid: Britain's Most Ancient Heritage. Natural History Museum, London
Green, B.: Countryside Conservation (3rdedn.). E and FN Spon, London (1996)
Green CP (2008) The Geologists' Association and geoconservation history and achievements. In: Burek CV, Prosser CD (eds) The History of Geoconservation, Special Publications 300. The Geological Society, London, pp 91–102
Green, W.:The Tourist's New Guide, Containinga Description of the Lakes, Mountains, and Scenery in Cumberland, Westmorland, and Lancashire, with Some Account of Their Bordering Towns and Villages. Being the Result of Observations Made During a Residence of Eighteen Years in Ambleside and Keswick(2 volumes). R. Lough and Co., Kendal (1819)
Hall CM, Weiler B (1992) What's Special about Special Interest Tourism? In: Weiler B, Hall CM (eds) Special Interest Tourism. Belhaven, London, pp 1–14
Hamilton BM (1976) The Changing Place of Geology in Science Education in England. J Geol Educ 24:105–110
Harley MJ (1996) Involving a wider public in conserving their geological heritage: a major challenge and recipe for success. In: Stow DAV, McCall GJH (eds) Geoscience education and training in schools and universities, for industry and public awareness. Balkema, Rotterdam, pp 725–729
Harley MJ, Robinson E (1991) RIGS - a local Earth-science conservation initiative. Geol Today 7(2):47–50
Harrison WJ (1882) Geology of the counties of England and of North and South Wales. Kelly and Co/Simpkin Marshall and Company, London
Hawkins BW (1854) On visual education as applied to geology illustrated by diagrams and models of the geological restorations at the crystal palace. J Soc Arts (republished as a leaflet by James Tennant, London)
Heywood P (1990) Truth and beauty in landscape - trends in landscape and leisure. Landsc Aus 12(1):43–47
Hobbes T (1678) De Mirabilibus Pecci: being the wonders of the peak in Darby-shire, commonly called the Devil's Arse Peak. In English and Latine. The Latine written by Thomas Hobbes of Malmsbury. The English by a person of quality. William Crook, London
Holden E (1977) The country diary of an Edwardian Lady. Michael Joseph, London
Holmes TV, Sherborn CD (1891) Geologists' association: a record of excursions made between 1860 and 1890. Stanford, London
Hose TA (1995) Selling the story of Britain's stone. Environ Interpret 10(2):16–17
Hose A (1996) Geotourism, or can tourists become casual rockhounds? In: Bennett MR (ed) Geology on your doorstep. The Geological Society, London, pp 207–228
Hose TA (1997) Geotourism - selling the earth to Europe. In: Marinos PG, Koukis GC, Tsiamaos GC, Stournass GC (eds) Engineering geology and the environment. A.A. Balkema, Rotterdam, pp 2955–2960
Hose TA (2000) European geotourism – geological interpretation and geoconservation promotion for tourists. In: Barretino D, Wimbledon WP, Gallego E (eds) Geological heritage: its conservation and management. Instituto Tecnologico Geominero de Espana, Madrid, pp 127–146
Hose TA (2005) Writ in stone: a critique of geoconservation panels and publications in Wales and the Welsh borders. In: Coulson MR (ed) Stone in Wales. Cadw, Cardiff, pp 54–60

Hose TA (2006) Leading the field: a contextual analysis of the field-guide and the field-excursion in England. In: Wickens E, Hose T, Humberstone B (eds) Current trends and research in pedagogy. Leisure and Tourism Education Research Centre, Buckinghamshire Chilterns University College, High Wycombe, pp 115–132
Hose TA (2008) The genesis of geotourism and its management implications. In: Abstracts volume, 4th international conference, geotour 2008, geotourism and mining heritage. AGH University of Science and Technology, Krakow, pp 24–25
Hose TA (2012) 3G's for modern geotourism. Geoheritage 4(1–2):7–24
Hose TA (2016) Introduction: Geoheritage and Geotourism. In: Hose TA (ed) Geoheritage and geotourism: a European perspective. The Boydell Press, Woodbridge, pp 1–13
House M (1989) Geology of the Dorset coast. Geologists' Association, London
Hudson J (1842) Complete guide to the lakes. Hudson, Kendal
Hughes JC (1922) The geological story of the Isle of Wight. Stanford, London
Hutchinson W (1774) An excursion to the lakes, in Westmoreland and Cumberland, August 1773. J. Wilkie/W. Goldsmith, London
Hutchinson W (1794) The history of the county of Cumberland, and some places adjacent from the earliest accounts to the present time (in two volumes). Carlisle
Jarzembowski E (1989) Writhlington geological nature reserve. Proc Geol' Assoc 100(2):219–234
Jenkins JM (1992) Fossickers and rockhounds in Northern New South Wales. In: Weiler B, Hall CM (eds) Special interest tourism. Belhaven, London, pp 129–140
Knell S (1993) No stone unturned. Mus J 93(3):20
Knell S, Taylor MA (1991) Museums on the rocks. Mus J 91(1):23–25
Kirkaldy JF (rev. Middlemiss FA, Allchin LJ, Owen HG) (1976) Guide no. 29: the weald. Geol' Assoc London
Komoo I (1997) Conservation geology: a case for the ecotourism industry of Malaysia. In: Marinos PG, Koukis GC, Tsiambaos GC, Stournas GC (eds) Engineering geology and the environment. Balkema, Rotterdam, pp 2969–2973
Komoo I, Deas KM (1993) Geotourism: an effective approach towards conservation of geological heritage. Unpublished paper. In: Symposium of Indonesia - Malaysia Culture, Badung
Lamb S, Singleton D (1998) Earth story: the shaping of our world. BBC Books, London
Lang WD (1906) Excursion to Lyme Regis, easter, 1906: from april 12th to april 17th. Proc Geol' Assoc 19(9):323–340
Larwood J, Prosser C (1998) Geotourism, conservation and tourism. Geol Balacania 28(3–4):97–100
Lavery P (1971) Recreational geography. David and Charles, Newton Abbot
Lawson JD (1977) Mortimer forest geological trail. Nature Conservancy Council, London
Lyons PC, Robertson EC, Milton L (1993) C. Wroe Wolfe's geology course on radio-station WGBH (Boston) in 1954. J Geol Educ 41:170
McCarthy S, Gilbert M (1994) The crystal palace dinosaurs: the story of the world's first prehistoric sculptures. Crystal Palace Foundation, London
Mackay C (1846) The scenery and poetry of the english lakes: a summer ramble. Longman, Brown, Green, and Longmans, London
Maini JS, Carlisle A (1974) Conservation in Canada: a conspectus - publication no. 1340. Department of the Environment/Canadian Forestry Service, Ottawa
Mantell GA (1822) The fossils of the south downs. Lupton Relfe, London
Mantell GA (1827) Illustrations of the geology of Sussex. Lupton Relfe, London
Mantell GA (1833) The geology of the south-east of England. Longman, Rees Orme, & etc., London
Mantell GA (1847) Geological excursion round the isle of wight and along the adjacent coast of Dorsetshire. Henry G. Bohn, London
Mantell GA (1849) Thoughts on a pebble. Reeves, Benham and Reeve, London
Martin W (1809) Petrificata Derbiensia; or fig.s and descriptions of petrifactions collected in Derbyshire. Wigan

Martini G (1994) The protection of geological heritage and economic development: the saga of the Digne ammonite slab in Japan. In: O'Halloran D, Green C, Harley M, Stanley M, Knill J (eds) Geological and Landscape Conservation. The Geological Society, Bath, pp 383–386

Martini G (2000) Geological heritage and geo-tourism. In: Barretino D, Wimbledon WP, Gallego E (eds) Geological heritage: its conservation and management. Instituto Tecnologico Geominero de Espana, Madrid, pp 147–156

Martineau H (1855) A complete guide to the english Lakes. Whittaker and Co., London

Mavor W (1798–1800) The British tourists; or traveller's pocket companion, through England, Wales, Scotland, and Ireland. Comprehending the most celebrated tours in the British Islands (in six volumes). E. Newberry, London

Monckton HW, Herries RS (eds) (1909) Geology in the field: the jubilee volume of the Geologists' Association. Stanford, London, pp 1858–1908

Mullins E (1985) A love affair with nature. Phaidon, Oxford

NEWRIGS (1997) Steaming through the past: a geological trail for Llangollen Valley. NEWRIGS/Chester College, Chester

Norman MW (1887) A popular guide to the geology of the Isle of Wight, with a note on its relation to that of the Isle of Purbeck. Knight's Library, Ventnor

Ogilby J (1675) Britannia, volume the first: or an illustration of the Kingdom of England and the dominion of wales: by a geographical and historical description of the principal roads thereof. Actually admeasured and delineated in a century of whole-sheet copper-sculps. Accommodated with the ichonography of the several cities and capital towns; and completed by an accurate account of the more remarkable of passages of antiquity, together with a novel discourse of the present state, London

Otley J (1823) A concise description of the English Lakes, and adjacent mountains, with general directions to tourists; and observations on the mineralogy and geology of the district. Keswick, L. Otley, Keswick

Ousby I (1990) The Englishman's England: taste travel and the rise of tourism. Cambridge University Press, Cambridge

Page KN (1998) England's earth heritage resource - an asset for everyone. In: Hooke J (ed) Coastal defence and earth science conservation. The Geological Society, London, pp 196–209

Phillips W (1818) A selection of facts from the best authorities, arranged so as to form an outline of the geology of England and Wales. William Phillips, London

Pfaff M, Simcox D (1998) Lulworth rocks. Weld Estate, Lulworth

Plog SC (1974) Why destination areas rise and fall in popularity. Cornell Hotel Restaur Adm Q 15:55–58

Read SE (1980) A prime force in the expansion of tourism in the next decade: special interest travel. In: Hawkins DE, Shafer EL, Rovelstad JM (eds) Tourism marketing and management issues. George Washington University Press, Washington D.C., pp 193–202

Reid CGR (1993) The local geologist 12: The roadshow. Geol Today 9(4):147–153

Robinson E (1993) Who would buy a coal tip? Earth Sci Conserv 32:20–21

Robinson E (1996a) Geology reserve: Wren's nest revised. Geol Today 12(6):211–213

Robinson E (1996b) Earth galleries open: impression 1. Geol Today 12(4):128–130

Robinson E (1998) Tourism in geological landscapes. Geol Today 14(4):151–153

Rodger S (1834) Poems. T. Cadelland E. Moxon, London

Sharpe T, Howe S, Howells C (1998) Gallery Review: setting the standard? The earth galleries at the natural history museum, London. Geol Curator 6(10):395–403

Shell (1964) The shell book of nature. Phoenix House, London

Sillitoe A (1995) Leading the blind: a century of guidebook travel 1815–1911. Macmillan, London

Simmons J, Biddle G (eds) (1997) The Oxford companion to British Railway history. Oxford University Press, Oxford

Smith PJ (1996) Earth galleries open: impression 2. Geol Today 12(4):130–132

Sorby HC (1875) On the remains of a fossil forest in the coal-measures at Wadesley, near Sheffield. Q J Geol Soc 31:458–459

Spiteri A (1994) Malta: a model for the conservation of limestone regions. In: O'Halloran D, Green C, Harley M, Stanley M, Knill J (eds) Geological and landscape conservation. The Geological Society, Bath, pp 205–208
Stueve AM, Cock SD, Drew D (2002) The geotourism study: phase 1 executive summary. www.tia.org/pubs/geotourismphasefinal.pdf
Swinglehurst E (1974) The romantic journey: the story of Thomas Cook and Victorian travel. Harpurand Row, New York
Taylor MA (1992) The Local Geologist 1: Exporting your geoheritage? Geol Today 7:32–36
Thomas IA, Prentice JE (1984) A consensus approach at the National Stone Centre, UK. In: O'Halloran D, Green C, Harley M, Stanley M, Knill J (eds) Geological and landscape conservation, pp 423–427. The Geological Society, Bath
Thomas IA, Hughes K (1993) Reconstructing ancient environments. Teaching. Earth Sci 18:17–19
Towner J (1996) An historical geography of recreation and tourism in the western world 1540–1940. Wiley, London
Trench R (1990) Travellers in Britain: three centuries of discovery. Arum Press, London
Tresise G (1973) The story of the earth at the geological museum. Mus J 73(2):71–72
Warner R (1802) A tour through the Northern Counties of England, and the Borders of Scotland. G. and J. Robinson, London
Watson W (1811) A delineation of the strata of Derbyshire, forming the surface from Bolsover in the east to Buxton in the west. Sheffield
Watson W (1813) A section of the strata in the vicinity of Matlock Bath. Derbyshire, Chesterfield
West T (1778) A guide to the lakes: dedicated to the lovers of landscape studies, and to all who have visited, or intend to visit the lakes in Cumberland, Westmorland and Lancashire. B. Law, etc., London
White HJO (1920) A short account of the geology of the Isle of Wight (1990 reprint with a new bibliography). HMSO/British Geological Survey, London
Wilkins EP (1861) A concise exposition of the geology, antiquities, and topography of the Isle of Wight. T. Kentfield, Newport
Wilkinson J (1810) Select views in Cumberland, Westmoreland and Lancashire by the Rev. Joseph Wilkinson, rector of East and West Wretham in the County of Norfolk, and Chaplain to the Marquis of Huntley. R. Ackerman, London
Wood RM (1978) On the rocks: a geology of Britain. BBC Books, London
Wood SP (1983) The bearsden project or quarrying for fossils on a housing estate. Geol Curator 3(7):423–434
Wordsworth W (1820) A guide through the district of the lakes of Northern England with a description of the scenery, & c. for the use of tourists and residents, 5th edn. Longman, Hurst, Rees, Orme and Brown, London
Wordsworth W (1835) The River Duddon, a series of sonnets; Vaudracourand Julia: and other poems. to which is annexed, a topographical description of the country of the lakes, in the North of England. Hudson and Nicholson, Kendal
Wyatt J (1995) Wordsworth and the geologists. Cambridge University Press, Cambridge
Young A (1771) The farmer's tour through the East of England being the register of a journey through various Counties of this Kingdom, to enquire into the state of agriculture, & c. Containing I. The particular methods of cultivating the soil. II. The conduct of live stock and the modern system of breeding. III. The state of the population, the poor, labour, provisions, & c. IV. The rental value of the soil, and its division into farms, with various circumstances attending their size and state. V. The minutes of above five hundred original experiments, communicated by several of the nobility, gentry, & c. With other subjects that tend to explain the present state of english husbandry. W. Strahaan and W. Nicoll, London

Chapter 18
Geomorphological Resources for Geoeducation and Geotourism

Lucie Kubalíková, Karel Kirchner, and Aleš Bajer

Abstract Geodiversity (geological, geomorphological, soil and hydrological features) are considered the main resource for geotourist and geoeducational activities which are (or should be) closely related and should support each other. Geoeducation can help to increase recognition of geodiversity and geoheritage at all the levels, can have a positive effect on the behaviour of visitors to geotouristic attractions, it can help maintain the geotourism activities in a rational scale and it helps to avoid the overexploitation of geoheritage for geotourism purposes. In opposite, sustainable geotourism development can make the geoeducational resources more accessible and available. A specific position within said resources is occupied by geomorphological features. While the geotourist and geoeducational importance of landforms is indisputable, especially thanks to their scientific, aesthetical or cultural values, the geomorphological processes are sometimes considered hazards and not resources for such activities. The example from Kokomeren valley in Kyrgyzstan shows that even an active process can serve geotourist and geoeducationl purposes. Another specific issue of geomorphological resources is represented by anthropogenic landforms: although their position within heritage concept is not clear, their potential for geotourist and geoeducational activities is undeniable which is supported by several examples from all over the world.

Keywords Geotourism · Geoeducation · Geomorphological processes · Anthropogenic landforms · Heritage concepts

L. Kubalíková (✉) · K. Kirchner
Department of Environmental Geography, Institute of Geonics of the Czech Academy of Sciences, Drobného 28, 602 00 Brno, Czech Republic
e-mail: Lucie.Kubalikova@ugn.cas.cz

A. Bajer
Department of Geology and Pedology, Faculty of Forestry and Wood Technology, Mendel University in Brno, Zemědělská 3, 613 00 Brno, Czech Republic

R. B. Singh et al. (eds.), *Global Geographical Heritage, Geoparks and Geotourism*, Advances in Geographical and Environmental Sciences,
https://doi.org/10.1007/978-981-15-4956-4_18

18.1 Introduction: Geotourism and Education

Abiotic nature (or geodiversity) is undoubtedly one of the most important resources for human society, it has numerous functions and offers various services and benefits (Gray 2013; Gordon and Barron 2012). Today, the importance of geodiversity is already accepted (Gray 2018; Brilha et al. 2018) and abiotic ecosystem services are included in the classification of ecosystem services (European Environment Agency 2018). Besides the provisioning, supporting and regulating services, cultural and so-called knowledge services of geodiversity are also recognized. The last two include geoeducational and geotourist use of geodiversity (Gray 2013, 2018; Gordon 2018).

Geotourism and education have been always closely related. Environmental education is one of the pillars of the geotourism and it also plays an important role within geoconservation. Since the early 1990s when the concept of geotourism originated, the education and interpretation were emphasized and accepted as an important tool that can raise the awareness of the geodiversity and geoheritage conservation and contribute to the sustainable development of geotourism.

The educational aspect is integrated or reflected in numerous definitions and approaches to geotourism, beginning from the early ones (which define geotourism as niche tourism within ecotourism) up to the present holistic approaches. Hose (1995) says that geotourism means "the provision of interpretive and service facilities to enable tourists to acquire knowledge and understanding of the geology and geomorphology of a site (including its contribution to the development of the Earth sciences) beyond the level of mere aesthetic appreciation". Slomka and Kicinska-Swiderska (2004) offer another definition with emphasis on the educational aspect: geotourism as "an offshoot of cognitive tourism and/or adventure tourism based upon visits to geological objects (geosites) and recognition of geological processes integrated with aesthetic experiences gained by the contact with a geosite". Joyce (2006) also includes education and learning into his brief definition: "People going to a place to look at and learn about one or more aspects of geology and geomorphology".

Likewise, an approach to geotourism as a geographical tourism introduced by National Geographic Society (2005) include the educational (or interpretation) aspect: the key features of geotourism are represented by integrity of place, international codes, market selectivity and diversity, tourist satisfaction, community involvement and benefit, protection and enhancement of destination appeal, land use and planning, conservation of resources, interactive interpretation and evaluation. In addition, National Geographic has adopted the term geoeducation to describe education about our world; a well-rounded geoeducation provides young people with a fundamental understanding of how the human and natural worlds work at local, regional and global scales (National Geographic Society 2018). This approach is wide and it includes both natural features and anthropogenic impact on them (and vice versa).

Dowling and Newsome (2010) state that "geotourism is a form of natural area tourism that specifically focuses on geology and landscape. It promotes tourism to geosites and the conservation of geodiversity and an understanding of Earth sciences through appreciation and learning". According to the authors, geotourism should be geologically based, environmentally educative, generating tourist satisfaction, sustainable and being locally beneficial. Here, geoeducation can be considered a part of the environmental education which is focused especially on Earth sciences and which seeks to create interlinks among geology, pedology and geomorphology within the landscape. The importance of geoeducation and interpretation is also emphasized by Hose (2012) who presents three key interrelated aspects of modern geotourism: geoconservation, geohistory and geo-interpretation.

Later, Dowling and Newsome (2018) say that geotourism promotes geoconservation and foster geoeducation through geo-interpretation among others. The promotion and communication of knowledge about geology and geomorphology (education and interpretation) is vital for the protection of geoheritage and geotourism development.

Farsani et al. (2018) consider the geotourism a form of educational tourism and state that one of the main tasks of geotourism is the transfer and communication of geoscientific knowledge and ideas. The geoeducation is, of course, an important tool for increasing public geoliteracy (Clary 2018). The importance of education and interpretation is also emphasized in Arouca Declaration (2011).

Numerous geoconservation approaches and projects also count on education. For example, Digne Declaration (1991) mentions the importance of education and learning within geoconservation in the article 7: "We have always been aware of the need to preserve our memories—our cultural heritage. Now the time has come to protect our natural heritage. The past of the Earth is no less important than that of Man. It is time for us to learn to protect this Earth heritage, and by doing so learn about the past of the Earth, to learn to read this 'book', the record in the rocks and the landscape, which was mostly written before our advent". Andersen et al. (1990) emphasize the relationships between education and Earth science conservation and state that if Earth-science sites are conserved, they can be used for teaching or research. This form of general education is vital if geological conservation is to become better understood and more widely supported. The authors also stress the importance of links between education and management.

The UNESCO Framework for geological conservation also mentions the vital importance of education and defines the principles of geoparks and heritage sites (Dingwall 2005). Prosser et al. (2013) and Prosser (2019) mention close relationships between geoconservation, appropriate management of geological, geomorphological and soil features and processes and education or research.

The education relevance was emphasized within the Geosite project (IUGS): one of the objectives was to provide a factual basis to support national and international initiatives to protect geological resources for research and education. Likewise, the national projects of inventorying geosites include sites that are primarily used for education (e.g. Czech Geological Survey 2018; MNHN 2018).

This brief overview of selected approaches, definitions and project brings evidence that geoeducation, geotourism and geoconservation are really closely linked and that geoeducation has numerous functions, for example, (1) it helps to increase recognition of geodiversity and geoheritage in international, national, regional and local levels which contribute to the geoconservation activities (inventorying, assessing), (2) it makes geodiversity relevant to where the people live and the places they visit,

(3) it helps to interpret, utilize and widen understanding of geodiversity and geoheritage for numerous purposes (including geoconservation, geotourism and other forms of sustainable tourism), (4) it helps to create and foster the sense of place and regional identity, (5) it contributes to discover the links between abiotic, biotic and cultural components of the landscape by public. These selected aspects make the geoeducation really fundamental for geoconservation and geotourism purposes.

18.2 Geomorphological Resources for Geotourism and Geoeducation

Geodiversity (according to definitions presented by Dixon 1996; Australian Heritage Commission 2002; Gray 2004, 2013 or Brocx and Semeniuk 2007) includes geological, geomorphological, hydrological and soil features, their systems, assemblages and contribution to the landscapes.

In this section, emphasis is given on the geomorphological resources for geotourism and geoeducation. While the geotourist and geoeducational importance of natural landforms is indisputable, especially thanks to their scientific, aesthetical or cultural values (Pralong 2005; Panizza and Piacente 2005; Gordon 2012, 2018), the geomorphological processes are sometimes considered hazards and not resources for such activities. In some cases, ongoing geomorphological processes stand against the geotourism development or they represent a threat to the geoheritage (Smith 2005; Alcántara-Ayala 2017; Cesaro et al. 2017). However, rational geotourist and geoeducational use of these processes can help better explanation of the origin and evolution of the landforms and the correct interpretation can help to know the complex relationships between process and resulting landform. The knowledge and explication of the processes can help the understanding of possible geohazards (e.g. rock fall, landslides) and thus make the planning of geoeducational, geotourist and other locally beneficial activities more effective. It is therefore obvious that geomorphological processes (including the dangerous ones) should be also considered important resources for geotourism and should be also taken into account when planning and managing geotourist and geoeducational activities which are (or should be) closely linked.

Another aim of this chapter is to present the geoeducational and geotourist potential of anthropogenic processes and consequent anthropogenic landforms because they can be also viewed as an important resource for the abovementioned activities. The position of the anthropogenic landforms within natural and cultural heritage is discussed, some specifics of anthropogenic geomorphological heritage are outlined and particular examples of geotourist and geoeducational use of anthropogenic landforms from all around the world are presented.

18.2.1 Active Geomorphological Processes as a Resource for Geotourist and Geoeducational Activities: A Case Study from Kyrgystan (The Ak-Kiol Rockslide Dam)

18.2.1.1 The Geological and Geomorphological Settings of the Study Area

The Ak-Kiol Lake (rockslide dam) is located on the Unkursay river which is the local name of left-hand inflow of the Kokomeren river (Fig. 18.1a) belonging to the Tien Shan Basin. It is a typical basin and mountain range system that has been formed mainly in Neogene and Quarternary (orogenesis has been still running) most likely due to north–south compression. This neotectonic deformation started after a long period of planation which took place during the Mesozoic era. Older intensive tectonic deformations come from Caledonian and Variscan orogenic stages that formed the complex structure of the basement. The basement rocks are formed mainly from Paleozoic granites and Late Precambrian metasediments and granites. The north-east part of the study area is formed by Devonian sandstone and Ordovician, Devonian and Carboniferous sediments. In depressions and valleys, the Neogene deposits represented by red beds (layers of conglomerate, sandstone, siltstone and mudstone) are situated. The Pliocene and Pleistocene sedimentary deposits which show intensive neotectonic orogenesis are also present here. In higher altitudes,

Fig. 18.1 Rockslide dam in Kokomeren valley: **a**, **b** overall views, **c** former lake covered with vegetation, **d** detail of the rockslide with head scarp. Photos: Aleš Bajer

Quaternary deposits represented by glacial moraine and alluvial and fluvial-deluvial deposits are situated (Chedia 1986; Sadybakasov 1972). The extreme neotectonic features occur together with rockslides, rock avalanches and rockfalls which resulted in dammed lakes. Some lakes have persisted to this day, some of them are present only in the form of lacustrine sediments. The most evident tectonic activities are associated with neotectonic faults along which uplifts and depressions of rupturing surface occur (Sadybakasov 1972; Strom and Korp 2006; Evans et al. 2006).

18.2.1.2 The Ak-Kiol Rockslide Dam

The upper rockslide forms a dam with a beautiful lake behind (Fig. 18.1b). Another large rockslide dammed the same valley 2.5 km downstream, but this dam has breached and today, there is a plain covered by vegetation (Fig. 18.1c) with several meters of lake deposits. On the side slopes, the former coastline can be observed. The rockslide originated in Paleozoic brown-red conglomerate and sandstone with thin gypsum interbeds. Figure 18.1d illustrates the rockslide with a head scarp. The thickness of the debris flow is up to 200 m which is evident from 150 m deep gullies below the dam. The rockslide is of Holocene period but the exact age is not known (Strom 2010, 2013, 2014; Evans et al. 2006).

Local people recognize these processes and consider them a geohazard; no buildings are constructed under the rocks and inside the valley, asphalt roads are neither constructed because of the moving terrain. However, locals are able to use these processes and consequent landforms (small depressions filled with water—ponds or swamps) for pasture and agriculture. In nearby Suusamyr, there is a little tourist agency offering horse tours. Local people offer guiding services and their own horses and mules for transporting tourists and their baggage. Accommodation is possible within family houses or it is allowed to camp on the private lands for a fee.

This area serves as an open-air laboratory for studying landslides and rockslides, the annual International Summer School on Rockslides and Related Phenomena has been organized here since 2006 (Strom and Abdrakhmatov 2009; Strom 2014; The International Programme on Landslides 2018).

The whole area is aesthetically attractive which can be considered one of the most important prerequisites for geotourist development. Thanks to its scientific importance and representativeness the area has a high potential for research activities (which are already taking place here) with outreach to public environmental education. The basic tourist infrastructure is also present and it depends mainly on the local people whether they want to continue in developing sustainable geotourist and geoeducational activities.

18.2.2 Anthropogenic Landforms: A Bridge Between Geoheritage and Cultural Heritage

18.2.2.1 Importance of the Anthropogenic Landforms

Already in prehistoric time, people used the geodiversity in different ways: they exploited mineral resources (stone, gems, metals), various landforms served as shelters, communication paths or suitable places for the construction of important buildings, e.g. castles, forts or sacral objects (Fig. 18.2). All these activities have been accompanied by modifications of landscape and terrain and nowadays, the human agent is equal to natural factors in the shaping of landforms (Szabó 2010). New landforms are created and new processes even surpass the effectiveness of natural exogenic processes (Szabó et al. 2010; Goudie 2006a, b). The impact of these activities (respectively, anthropogenic processes) is often very destructive and in some cases, "humans are often victims of an environment created or modified by themselves" (Szabó 2010).

The anthropogenic processes result in anthropogenic landforms. An anthropogenic landform is created by human activity, especially by construction, excavation, hydrological interference and farming (Goudie 2006a, b). Anthropogenic landforms can be sorted by the character of the impact: direct or indirect, respectively, intentional or unintentional (or according to Szabó et al. (2010): primary anthropogenic landforms and secondary anthropogenic landforms), but for the purposes of geotourism and geoconservation, the genetic classification of the landforms is probably the most suitable.

According to the processes which formed the landform, several groups of landforms can be defined: mining, industrial, agricultural, urban/residential, communication/traffic, water management, military, funeral and others (Szabó et al. 2010; Kirchner and Smolová 2010). These landforms often change the original appearance of the landscape, create new dominants or influence the original natural environment and conditions. The creation of such landforms is accompanied by processes which would not normally exist at a place (e.g. superficial subsidence depressions in the

Fig. 18.2 Geomorphological conditions of an area have always influenced the situation of important buildings, e.g. significant elevations have been always suitable for castles, forts or monasteries: **a** Mehrangarh Fort in Jodhpur in Rajasthan, India, **b** Trosky castle ruins in the Czech Republic, **c** Aït Benhaddou—fortified village (ksar) in Morocco. Photos: Lucie Kubalíková

areas of underground mining, landslides and other slope movements on the artificial slopes or abrasion on the shores of artificial lakes and dams).

On the other hand, some anthropogenic landforms (e.g. quarries, pits, communication cuttings or underground landforms) can be considered important from the from scientific, educational, cultural, historical, environmental and tourist point of view (Prosser 1992, 2019; Dávid 2008; Parkes and Gatley 2018): (1) they can be seen as elements that increase the overall landscape diversity and influence the biodiversity (e.g. old quarries, flooded pits), (2) they provide information about the landscape changes or modification in the past which can be an important resource for understanding the cultural and technical level of the society, (3) they allow to trace the use of geodiversity in the past and interpret cultural heritage in relation to abiotic nature, (4) some specific anthropogenic landforms form an inseparable part of cultural heritage objects, e.g. fortification earthen ramparts or irrigation channels, (5) they can serve as an important resource for geotourist activities as some of the landforms are visually attractive or allow to interpret the technical aspects of using the geodiversity resources (e.g. mining tourism), (6) they allow observing stratigraphical, tectonic, palaeopedological and other Earth-science features that would normally remain hidden and unrecorded in the literature or on geological maps (Osborne 2000; Petersen 2002) which can be used in both formal and informal geoeducation.

18.2.2.2 Position of Anthropogenic Landforms: Which Heritage?

The importance of anthropogenic landforms is indisputable, however, their position within the geoheritage/natural heritage/cultural heritage concepts still remains a subject of discussion.

The concept of geoheritage is based on the definition of natural heritage, which was presented in 1972 (UNESCO 1972). The term geoheritage was defined as those components of natural geodiversity of significant value to humans, including scientific research, education, aesthetics and inspiration, cultural development, and a sense of place experienced by communities (Dixon 1996; Dingwall 2005). Sharples (1995) says that "geoheritage comprises those aspects of natural geodiversity which are of significant value to humans for purposes which do not decrease their intrinsic or ecological values; such purposes may include scientific research, education, aesthetics and inspiration, recreation, cultural development and contribution to the 'sense of place' experienced by human communities". ProGEO (2011) states that geoheritage is "part of the natural heritage of a certain area constituted by geodiversity elements with particular geological value and hence worthy of safeguard for the benefit of present and future generations".

In the abovementioned definitions, there appears the word "natural" (natural geodiversity or primary geodiversity that means the features formed without the human impact or activity), so in theory, the anthropogenic landforms should not be included into geoheritage in general.

Some authors (e.g. Coratza and Hobléa 2018) include anthropogenic landforms into the concept of "geomorphological heritage". The special situation can be found in urbanized areas that are usually heavily affected by anthropogenic transformations of relief and thus the abundance of anthropogenic landforms is high there: in these cases, the anthropogenic landforms are also respected as a component of geomorphological heritage (Reynard et al. 2017; Kubalíková et al. 2017, 2019, 2020) or "complex urban geoheritage" (Habibi et al. 2018).

While the position of anthropogenic landforms within geoheritage is still not clear, it is obvious that specific anthropogenic landforms can be respected as a full-value part of mining heritage (Ahmad and Jones 2013; Conlin and Jolliffe 2014; Pearson and McGowan 2000). This type of heritage is considered a subset of cultural heritage, however, the natural aspects of mining are also included (geological settings, type of material extracted) and anthropogenic modifications, landforms and processes are reflected as well. These include mine working and operational areas (open cuts, pits, shafts, adits), infrastructure to support the mine, such as water supply (dams, races, pipelines) and landscape modification due to mining such as deforestation, pollution-induced barren areas, silted dams, open cuts, embankments and mounds, tailings dumps, dredged streams or modified vegetation (Pearson and McGowan 2000).

Other approaches consider some specific examples of anthropogenic landforms as a part of cultural heritage: the following section will present some examples. The convention concerning the protection of the world cultural and natural heritage (UNESCO 1972) defines the term "sites" within the cultural heritage: "works of man or the combined works of nature and man, and areas including archaeological sites which are of outstanding universal value from the historical, aesthetic, ethnological or anthropological point of view". Guidelines on the inscription of specific types of properties on the World Heritage List (UNESCO 2008) specifies also Cultural landscapes and Heritage canals.

Cultural landscapes represent the "combined works of nature and of man" and embrace a diversity of manifestations of the interaction between humankind and its natural environment. Cultural landscapes often reflect specific techniques of sustainable land use, considering the characteristics and limits of the natural environment they are established in, and a specific spiritual relationship to nature. The same document presents the concept of Heritage canals: "A canal is a human-engineered waterway. It may be of outstanding universal value from the point of view of history or technology, either intrinsically or as an exceptional example representative of this category of cultural property. The canal may be a monumental work, the defining feature of a linear cultural landscape, or an integral component of a complex cultural landscape" (UNESCO 2008). These subtypes of cultural heritage were recognized and established in the early 1990s.

Anthropogenic modifications of relief and resulting landforms are an inseparable part of the landscape in general (Szabó et al. 2010; Goudie 2006a, b). European Landscape Convention (Council of Europe 2000) describes the landscape as an area, as perceived by people, whose character is the result of the action and interaction of natural and/or human factors: although it is not mentioned to the letter, it can be assumed that anthropogenic landforms are also a part of the landscape that should be protected.

Specific anthropogenic landforms are included in the archaeological heritage: according to the European Convention on the Protection of the Archaeological Heritage, Revised (Council of Europe 1992), some anthropogenic landforms can be considered "elements of the archaeological heritage ", which include all remains and objects and any other traces of mankind from past epochs. The archaeological heritage shall include structures, constructions, groups of buildings, developed sites, moveable objects, monuments of other kinds as well as their context, whether situated on land or under water.

It is evident that anthropogenic landforms and anthropogenic modifications of the relief are tightly related to the culture and civilization (Szabó et al. 2010) and these links are often presented by spiritual, religious, historical or cultural value of these specific landforms (Gordon 2018). In addition, some anthropogenic landforms have also the scientific, environmental, research and educational value. Based on this, it can be said that anthropogenic landforms are an indisputable part of the heritage, but thanks to the fact that they stand somewhere on the border of cultural and natural heritage and it is not easy to sort them clearly, the position of anthropogenic landforms within the heritage concepts remains a subject of discussions.

18.2.2.3 Examples of Geotourist and Geoeducational Use of Anthropogenic Landforms from All Around the World

Regardless the ambiguities of the position of anthropogenic landforms within heritage concepts, it can be stated that these specific geomorphological features have a high potential for geotourism and geoeducation which have been already widely recognized and supported by numerous papers. The geotourist and geoeducational importance and potential of anthropogenic landforms is discussed, e.g. by Lóczy (2010), who stresses the role of anthropogenic landforms in geoconservation and geotourism, Mata-Perelló (2018), who analyses the relationships between geomining heritage and local/regional development, or Petersen (2002), Powel et al. (2013), who point at the potential of road cuttings or temporary exposures for geoeducation and scientific research. Besides it, there are numerous case studies that underpin the significance of anthropogenic landforms: Hose (2017) introduces mining geoheritage in Peak District in the UK, Lopéz-García et al. (2011) present an example of mines in SE Spain, Margiotta and Sansò (2017) focus on the potential of abandoned quarries for local/regional tourism development in Italy, Pica el al. (2017) and Kubalíková et al. (2017, 2019) outline the importance of anthropogenic transformations of the relief within urban areas, Evans et al. (2018) presents the geotourism within industrial settings with examples of black coal mining in the UK, Carrión Mero et al. (2018) provides an example of using the mining sites for geotourism development in Zaruma-Portovelo mining district in Ecuador, Boukhchim et al. (2018) analyse the geoconservation and geotourism aspects of cave dwellings in Southeast Tunisia, Rybár and Štrba (2015) present the mining heritage at BáňskáŠtiavnica UNESCO WHS in Slovakia, and many others.

Fig. 18.3 UNESCO World Heritage Sites which include anthropogenic landforms: **a** Las Médulas in Spain (the most important open gold pit in the Roman Empire), **b** Mountain Railways of India – Kalka Shimla Railway (railway construction is accompanied by numerous cuttings, ramparts or underground structures—tunnels), **c** Cerro Rico in Potosí, Bolivia (silver ore was extracted here using a series of hydraulic mills). Although these sites are inscribed primarily as cultural sites, the landforms that were created are also very significant and attractive (UNESCO 2018). Photos: Lucie Kubalíková

Some UNESCO Global Geoparks operates with mining history and present the anthropogenic mining landforms as their attractivenesses, e.g. Copper Coast Geopark or Tuscan Mining Geopark (Copper Coast Geopark 2018; Tuscan Mining Geopark 2018) and some World Heritage Sites are former mines or include the anthropogenic landforms too (https://whc.unesco.org/en/list/), for particular examples see Fig. 18.3.

Besides it, there are numerous examples of local tourist activities that are related to the active or recent mines, e.g. organized trips to the copper quarry of Chuquicamata near Antofagasta (Chile), "coal safari" in the former brown-coal open mines in the north of the Czech Republic or excursions to the active sulphur mine of Kawah Ijen in eastern Java (Indonesia). Figure 18.4 presents the abovementioned and other examples of geotourist and geoeducational use of anthropogenic landforms from all around the world.

18.3 Conclusions

The resources for geotourism and geoeducation are very diverse and they include not only natural geodiversity features (rocks, landforms, processes, soils, etc.) but also their links to the civilization and culture. The geomorphological resources for geotourism and geoeducation encompass both landforms and processes. Geomorphological processes can be considered hazards, however, the case study from Kyrgystan shows that active processes possess a high potential for geoeducational activities. In addition, the resulting aesthetically valuable landforms can serve as an important resource for geotourism development. A specific position within geomorphological resources is occupied by anthropogenic landforms. Although their place within the heritage concepts is questionable, their importance for geotourism and geoeducation

Fig. 18.4 Examples of geotourist and geoeducational use of anthropogenic landforms: **a** and **b** Agrarian terraces near Ubud, Bali, Indonesia (these terraces form an inseparable part of the landscape appearance and they have a high aesthetic value which can be considered a basis for tourist use), **c** and **d** Copper mine of Chuquicamata near Calama, Chile (the mine represents one of the popular tourist destinations within the region, organized tours are provided here with a possibility of collecting the samples of the copper ore, **e** and **f** Abandoned limestone quarry of Hády in Brno, Czech Republic: numerous Earth-science aspects (stratigraphic, tectonic, geomorphologic, hydrogeological, palaeontological) can be observed and used both for formal education (pupils of local schools, university students) and informal learning (the site is equipped with educational path and possesses basic tourist infrastructure). The nearby road cutting in granodiorites on Jedovnická Street **g** is also used for educational purposes accompanied by collecting the samples of biotite crystals. Photos: Diego Delso—under the License CC-BY-SA (**c**), Lucie Kubalíková (other photos)

is indisputable: numerous anthropogenic landforms form a part of top tourist destinations (including the UNESCO WHS), some of them are used both for formal and informal education as presented on the examples from all around the world.

Geotourism and geoeducation are closely related and should support each other. The rational geotourist development (including the appropriate development of tourist infrastructure or safety measures) can make the geoeducational resources more accessible and usable by more visitors. Geoeducation is one of the pillars of geotourism and it helps to appreciate scientific, cultural and other values of geodiversity and geoheritage. The correct interpretation of geodiversity and setting the links between it and particular components of the cultural heritage can bring Earth-sciences closer to the public, can help to avoid unsustainable use of the geoheritage for tourism purposes and last but not least, it can contribute to better acceptance of geoconservation measures.

References

Ahmad S, Jones D (2013) Investigating the mining heritage significance for Kinta district, the industrial heritage legacy of Malaysia. Proc-Soc Behav Sci 105:445–457

Alcántara-Ayala I (2017) Landslides and society - a foreword. In: Sassa K et al. (eds.) Advancing culture of living with landslides, conference: workshop on world landslide forum, ISDR-ICL sendai partnerships 2015–2025, vol 1. https://doi.org/10.1007/978-3-319-59469-9_42

Andersen S et al (1990) Earth-science conservation - an absolute need for science and education. Jb Geol B-A 133(4):653–669

Arouca Declaration (2011) International congress of geotourism – arouca 2011, Associação Geoparque Arouca. https://www.europeangeoparks.org/?p=223. Accessed 30 Nov 2018

Australian Heritage Commission (2002) Australian natural heritage charter for the conservation of the places of natural heritage significance, 2nd edn. Australian Heritage Commission and Australian Committee for the International Union for the Conservation of Nature (ACIUCN), Canberra

Boukhchim N, Fraj TB, Reynard E (2018) Lateral and "Vertico-Lateral" Cave Dwellings in Haddej and Guermessa: Characteristic Geocultural Heritage of Southeast Tunisia. Geoheritage 10(4):575–590. https://doi.org/10.1007/s12371-017-0251-2

Brilha J, Gray M, Pereira DI, Pereira P (2018) Geodiversity: an integrative review as a contribution to the sustainable management of the whole of nature. Environ Sci Policy 86:19–28

Brocx M, Semeniuk V (2007) Geoheritage and geoconservation - history, definition, scope and scale. J R Soc West Aust 90:53–87

Carrión Mero P, Herrera Franco G, Briones J (2018) Geotourism and local development based on geological and mining sites utilization, Zaruma-Portovelo, Ecuador. Geosciences 8(6):205

Cesaro G, Delmonaco G, Khrisat B, Salis S (2017) Geological conservation through risk mitigation and public awareness at the Siq of Petra, Jordan. In: Sassa K, Mikos M, Yin Y (eds) ISDR-ICL sendai partnerships 2015–2025, vol 1

Chedia OK (1986) Morphostructures and neotectonic of the Tien Shan. Frunze, Ilim Publishing House, 324p (in Russian)

Clary RM (2018) Geoheritage and public geoliteracy: opportunities for effective geoscience education within US parks. In Dowling RK, Newsome D eds (2018) Handbook of Geotourism. Edward Elgar Publishing, pp 244–253

Conlin MV, Jolliffe L (2014) Mining heritage and tourism: a global synthesis. Routledge, London, p 280

Copper Coast Geopark (2018) Mining heritage. https://coppercoastgeopark.com/mining-heritage/. Accessed 26 Oct 2018
Coratza P, Hobléa F (2018) Specifities of geomorphologic heritage. In: Brilha J, Reynard E (eds.) Geoheritatge: assessment, protection and management. Elsevier, pp 87–106
Council of Europe (1992) European convention on the protection of the archaeological heritage (revised) – archeological heritage. European Treaty Series No. 143. https://www.coe.int/en/web/conventions/full-list/-/conventions/rms/090000168007bd25. Accessed 30 Dec 2018
Council of Europe (2000) European landscape convention. https://www.coe.int/en/web/conventions/full-list/-/conventions/treaty/176. Accessed 30 Dec 2018
Czech Geological Survey (2018). Significant geological localities of the Czech Republic. Available at http://lokality.geology.cz. Accessed 26th October 2018
Dávid L (2008) Quarrying: an anthropogenic geomorphological approach. Acta Montan Slovaca 13(1):66–74
Dingwall P et al. (2005) Geological world heritage: a global framework. A Contribution to the Global Theme Study of World Heritage Natural Sites. IUCN, WCPA, UNESCO.
Dixon G (1996) Geoconservation: an international review and strategy for Tasmania. Occasional Paper 35, Parks & Wildlife Service, Tasmania, 101p
Dowling R, Newsome D (eds) (2010) Geotourism. The tourism of Geology and Landscape. Goodfellow Publishers Ltd., 246p
Dowling RK, Newsome D (eds) (2018) Handbook of Geotourism. Edward Elgar Publishing, 520p
Evans SG, Scarascia Mugnozza G, Strom A, Hermanns RL (eds) (2006) Landslides from massive rock slope failure. NATO science series: IV: earth and environmental sciences, vol 49. Springer
Evans BG, Cleal CJ Thomas BA (2018) Geotourism in an industrial setting: the South Wales coalfield geoheritage network. Geoheritage 10(1):93–107
European Environment Agency (2018) CICES V 5.1 - guidance and spreadsheet. https://cices.eu/resources/. Accessed 30 Dec 2018
Farsani NT, Neto de Carvalho C, Kejianet X (2018) Education as a key tenet of geotourism. In: Dowling RK, Newsome D eds (2018) Handbook of geotourism. Edward Elgar Publishing, pp 234–243
Gordon JE (2012) Rediscovering a sense of wonder: geoheritage, geotourism and cultural landscape experiences. Geoheritage 4:65–77
Gordon JE (2018) Geoheritage, geotourism and the cultural landscape: enhancing the visitor experience and promoting geoconservation. Geosciences 8(4):136–160
Gordon JE, Barron HF (2012) Valuing geodiversity and geoconservation: developing a more strategic ecosystem approach. Scott Geogr J 128:278–297
Goudie A (2006a) Anthropogenic landforms. In: Goudie A (ed) Encyclopedia of geomorphology, vol 1. Taylor & Francis
Goudie A (2006b) The human impact on the natural environment, 6th edn. Blackwell Publishing, 357p
Gray M (2004) Geodiversity: valuing and conserving abiotic nature. Wiley, Chichester, 434p
Gray M (2013) Geodiversity: valuing and conserving abiotic nature, 2nd edn. Wiley Blackwell, 495p
Gray (2018) The confused position of the geosciences within the "natural capital" and "ecosystem services" approaches. Ecosyst Serv 34:106–112. https://doi.org/10.1016/j.ecoser.2018.10.010
Habibi T, Ponedelnik AA, Yashalova NN, Ruban DA (2018) Urban geoheritage complexity: evidence of a unique natural resource from Shiraz city in Iran. Resour Policy 59:85–94. https://doi.org/10.1016/j.resourpol.2018.06.002
Hose TA (1995) Selling the story of Britain's stone. Environ Interpret 10(2):16–17
Hose TA (2012) 3G's for modern geotourism. Geoheritage 4(1–2):7–24
Hose TA (2017) The English Peak District (as a potential geopark): mining geoheritage and historical geotourism. Acta Geoturistica 8(2):32–49
Joyce EB (2006) Geomorphological sites and the new geotourism in Australia. Geological Society of Australia, Melbourne

Kirchner K, Smolová I (2010) Základy antropogenní geomorfologie. Univerzita Palackého v Olomouci, 288p

Kubalíková L, Kirchner K, Bajer A (2017) Secondary geodiversity and its potential for urban geotourism: a case study from Brno city, Czech Republic. Quaestions Geographicae 36(3):63–73

Kubalikova L, Kirchner K, Kuda F, Machar I (2019) The role of anthropogenic landforms in sustainable landscape management. Sustainability 2019, 11, 4331. https://doi.org/10.3390/su11164331

Kubalíková L, Kirhcner K, Kuda F, Bajer A (2020) Assessment of Urban Geotourism Resources: An example of two geocultural sites in Brno, Czech Republic. Geoheritage 12(7). https://doi.org/10.1007/s12371-020-00434-x

Lóczy D (2010) Anthropogenic geomorphology in environmental management. In: Szabó J, Dávid L, Loczy D (eds) Anthropogenic Geomorphology. A guide to man-made landforms. Springer, pp 25–38

Lopéz-García JA et al (2011) Scientific, educational, and environmental considerations regarding mine sites and geoheritage: a perspective from SE Spain. Geoheritage 3(4):267–275

Margiotta S, Sansò P (2017) Abandoned Quarries and geotourism: an opportunity for the Salento Quarry District (Apulia, Southern Italy). Geoheritage 9(4):463–477

Mata-Perelló J (2018) Geomining heritage as a tool to promote the social development of rural communities. In: Reynard E, Brilha J (eds) Geoheritage: assessment, protection and management. Elsevier, pp 167–177

MNHN (Museum National d'Histoire Naturelle, Origines et Évolution, France) INPN (2018) L'inventaire du patrimoine géologique. https://inpn.mnhn.fr/programme/patrimoine-geologique/presentation. Accessed 13 Dec 2018

National Geographic Society (2005) Geotourism Charter. https://travel.nationalgeographic.com/travel/sustainable/pdf/geotourism_charter_template.pdf. Accessed 13 Oct 2017

National Geographic Society (2018) Geo-education: essential preparation for an interconnected world. https://www.nationalgeographic.org/education/geo-education-essential-preparation-interconnected-world/. Accessed 20 Dec 2018

Osborne RAL (2000) Presidential address for 1999–2000. Geodiversity: "green" geology in action. Proc Linn Soc NSW 122:149–173

Panizza M, Piacente S (2005) Geomorphosites: a bridge betwenn scientific research, cultural integration and artistic suggestion. Il Quaternario Ital J Q Sci 18(1):3–10

Parkes M, Gatley S (2018) Quarrying and geoconservation in the Republic of Ireland the effectiveness of guidelines for operators. Geoheritage 10:169–177

Pearson M, McGowan B (2000) Mining heritage places assessment manual. Australian Council of National Trusts: Australian Heritage Commission, Canberra

Petersen J (2002) The role of roadcuts, quarries, and other artificial exposures in geomorphology education. Geomorphology 47:289–301

Pica A, Luberti GM, Vergari F, Fredi P, Del Monte M (2017) Contribution for an urban geomorphoheritage assessment method: proposal from three geomorphosites in Rome (Italy). Quaestiones Geographicae 36(3):21–36

Pralong JP (2005) A method for assessing tourist potential and use of geomorphological sites. Géomorphologie: relief, processus, environnement 1(3):189–196

ProGEO (1991) Digne-les-Bains declaration - declaration of the rights of the memory of the earth. www.progeo.ngo/downloads/DIGNE_DECLARATION.pdf. Accessed 30 Dec 2018

ProGEO (2011) Conserving our shared geoheritage – a protocol on geoconservation principles, sustainable site use, management, fieldwork, fossil and mineral collecting. https://www.progeo.se/progeo-protocol-definitions-20110915.pdf. Accessed 23 Apr 2013

Prosser CD (1992) Active quarrying and conservation. Earth Sci Conserv 31:22–24

Prosser CD et al (2013) Geoconservation for science and society – an agenda for the future. Proc Geol Assoc 124(4):561–567

Prosser CD (2019) Communities, quarries and geoheritage—making the connections. Geoheritage 11, 1277–1289. https://doi.org/10.1007/s12371-019-00355-4

Powell JH, Nash G, Bell P (2013) GeoExposures: documenting temporary geological exposures in Great Britain through a citizen-science web site. Proc Geol Assoc 124(4):638–647
Reynard E, Pica A, Coratza P (2017) Urban geomorphological heritage. an overview. Quaestiones Geographicae 36(3):7–20
Rybár P, Štrba L (2015) How to present unique mining heritage historical mining town of Smolnik. In: 15th international multidisciplinary scientific geoconference (SGEM) Albena, Bulgaria - Ecology, economics, education and legislation, vol I, pp 409–414
Sadybakasov I (1972) Neotectonics of the Central part of the Tien Shan. Frunze, Ilim, 116p
Sharples C (1995) Geoconservation in forest management – principles and procedures. Tasforests 7(12):37–50
Slomka T, Kicinska-Swiderska A (2004) Geotourism – the basic concepts. Geoturystyka 1:2–5
Smith BJ (2005) Management challenges at a complex geosite: the Giant's Causeway World Heritage Site, Northern Ireland. Revue de Géomorphologie 3:219–226
Strom AL (2010) Landslide dams in central Asia region. Landslides – J Jpn Landslide Soc 47(6):1–16
Strom A (2013) Geoogical prerequisities for landslide dams, disaster assessment and mitigation in Central Asia. In: Wang F, Miyajima M, Li T, Fathani TF (eds) Progress of geo-disaster mitigation technology in Asia. Springer, Berlin
Strom A (2014) Rockslides and rock avalanches in the Kokomeren River Valley (Kyrgyz Tien Shan). In Arbanas SM, Arbanas Ž (eds.) Landslide and flood hazard assessment, Faculty of Mining, Geology and Petroleum Engineering, University of Zagreb and Faculty of Civil Engineering, University of Rijeka, pp 245–250
Strom AL, Korp O (2006) Extremely large rockslides and rock avalanches in the Tien Shan Kyrgyzstan. Landslides 3:125–136
Strom A, Abdrakhmatov K (2009) International Summer School on rockslides and related phenomena in the Kokomeren River valley, Tien Shan, Kyrgyzstan. In: Sassa K, Canuti P (eds) Landslides – disaster risk reduction. Springer, Berlin, pp 223–227
Szabó J (2010) Anthropogenic geomorphology: subject and system. In: Szabó J, Dávid L, Loszy D (eds) Anthropogenic geomorphology. A guide to man-made landforms. Springer, Berlin, pp 3–12
Szabó J, Dávid L, Loczy D (eds) (2010) Anthropogenic geomorphology. A guide to man-made landforms. Springer, 250p
The International Programme on Landslides (2018) 2019 International Summer School on Rockslides and Related Phenomena in the Kokomeren River Valley (Kyrgyzstan). https://iplhq.org/icl/wp-content/uploads/2018/09/Summer_School_announcement_2019.pdf. Accessed 20 Dec 2018
Tuscan Mining Geopark (2018). Tuscan Mining Geopark. https://www.coopcollinemetallifere.it/en/parco-colline-metallifere. Accessed 26 Nov 2018
UNESCO (1972). Convention concerning the protection of the world cultural and natural heritage. https://whc.unesco.org/archive/convention-en.pdf. Accessed 22 Oct 2018
UNESCO (2008) Operational guidelines for the implementation of the world heritage convention. https://whc.unesco.org/en/guidelines/. Accessed 30 Dec 2018
UNESCO (2018) The world heritage list. https://whc.unesco.org/en/list/. Accessed 30 Nov 2018

Chapter 19
Dinosaur Geotourism in Europe, a Booming Tourism Niche

Nathalie Cayla and Heidi Elisabeth Megerle

Abstract Dinosaurs are not only a very popular science topic, throughout the world, paleontological tourism highlighting dinosaur paleontological sites is booming. Dinosaur outcrops, "hot spots" of paleontological discoveries, are valorized into geotourism destinations contributing to local economic development based on a specific territorial resource: the paleontological resource. Even though the first discoveries of dinosaurs and the beginning of academic paleontological studies already had their origin in Europe in the nineteenth century, dinosaur geotourism is more recent and has developed significantly in the past decades. The following chapter will present an overview of the diversity of dinosaurian geotouristic sites in Europe. To highlight the geotouristic trajectories of chosen paleontological sites, several case studies will be presented. A few of them have joined the UNESCO Global Geopark network in order to gain a greater notoriety, others are outside of geopark's territory or have chosen out of different reasons to manage their development by private companies or a public–private partnership. However, whatever the scientific interest of the sites, the success of these tourism projects must be based on a structured network of actors, gathering scientists, tourism stakeholders supported by local policy-makers, and of course local population.

Keywords Geotourism · Dinosaur geosites · UNESCO Global Geopark · Dinopolis · Münchehagen

---

N. Cayla (✉) · H. E. Megerle
PRAG, Docteur en Sciences de La Terre, de l'Univers et de l'Environnement, EDYTEM, Université Savoie Mont-Blanc, CNRS Pôle Montagne, 73376 Le Bourget du Lac, France
e-mail: Nathalie.cayla@univ-smb.fr

H. E. Megerle
e-mail: Megerle@hs-rottenburg.de

R. B. Singh et al. (eds.), *Global Geographical Heritage, Geoparks and Geotourism*, Advances in Geographical and Environmental Sciences,
https://doi.org/10.1007/978-981-15-4956-4_19

19.1 Introduction—Historical Development

The academic studies of dinosaurs have their origin in Europe. In the early nineteenth century William Buckland described Megalosaurus, the first species of dinosaur, even at that time, the term did not yet exist. It was only in 1842, when Richard Owen introduced the term "dinosaur" in the scientific community. The first dinosaur ever found on German soil was Plateosaurus engelhardti, which had already been recovered in 1834 by Johann Friedrich Engelhardt near Nuremberg. Plateosaurus was the fifth described dinosaur in the world. The first dinosaur eggs also had been discovered in Europe by the engineer Philippe Matheron in 1846 in the South of France (Taquet 2001).

Even though the first discoveries of dinosaurs and the beginning of academic paleontological studies have their origin in Europe, findings and scientific works globalized meanwhile. One of the oldest known dinosaurs has been unearthed in the Moon Valley (Argentina), today a UNESCO World Heritage Site. Tyrannosaurus Rex, probably the most famous dinosaur, has been found within the Morrison Formation which covers a large part of the United States from New Mexico to Canada, and some of the strangest dinosaurs, such as Deinocheirus, a giant Ornithomimosaurus, has been discovered in Mongolia.

As dinosaurs always fascinated not only scientists, but also the general public, numerous dinosaur remains, "hot spots" of paleontological discoveries, have been valorized as geotourism destinations (Cayla and Duval 2013; Carpenter 2018; Cayla 2018). Based on their specific territorial values, the paleontological resources, these destinations thus contribute to local economic development (Megerle 2008; Landel and Senil 2009; Peyrache-Gadeau et al. 2016). While paleontological research on dinosaurs has a long history in Europe, dinosaur geotourism is more recent and has developed significantly in the past decades. One of the main reasons for this time lag is the role played by museums in ensuring the conservation of these natural treasures. Often, after being described, dinosaur remains were kept for museum display. Gradually, it became equally important to promote the tourism development of remote areas, where the discoveries had been made (Laws and Scott 2003).

The following chapter will present an overview of the diversity of dinosaurian geotouristic sites in Europe. To highlight the geotouristic trajectories of chosen paleontological sites, several case studies will be presented. A few of them have integrated the UNESCO Global Geopark network in order to gain a greater notoriety (Megerle and Pietsch 2017), others are outside of geopark's territory or have chosen out of different reasons to manage their development by private companies or a public–private partnership. Since not only geotouristic offers in general, but also especially dinosaur offers are currently characterized by a high degree of dynamism, the overview cannot claim to be exhaustive.

19.2 Dinosaur Geotourism in Europe, a Wide Diversified Offer

Dinosaur geotourism is a niche tourism, which developed rapidly, mostly during the past decades. The main infrastructures are field or open-air museums, interpretive centers, local museums, or palaeontological parks, valorizing fossiliferous sites (Lipps 2009). Numerous tourist destinations in Europe try to attract visitors with dinosaur offers (Fig. 19.1). Hereby a fundamental distinction has to be made between in situ and ex situ offers. While the former are located at an original find site, the second display objects, which were recovered from other places.

Nearly all natural history museums showing dinosaur fossils are part of the second category. This includes the world-famous Senckenberg Museum in Frankfurt am Main (Germany), which features more than a thousand dinosaur exhibits. Some of them even can be brought to life with virtual-reality glasses (Senckenberg Gesellschaftfür Naturforschung 2018). In combination with new didactical approaches, an additional audience can be attracted, which could not be reached by a "classical" natural history museum. A wide-ranging knowledge transfer might

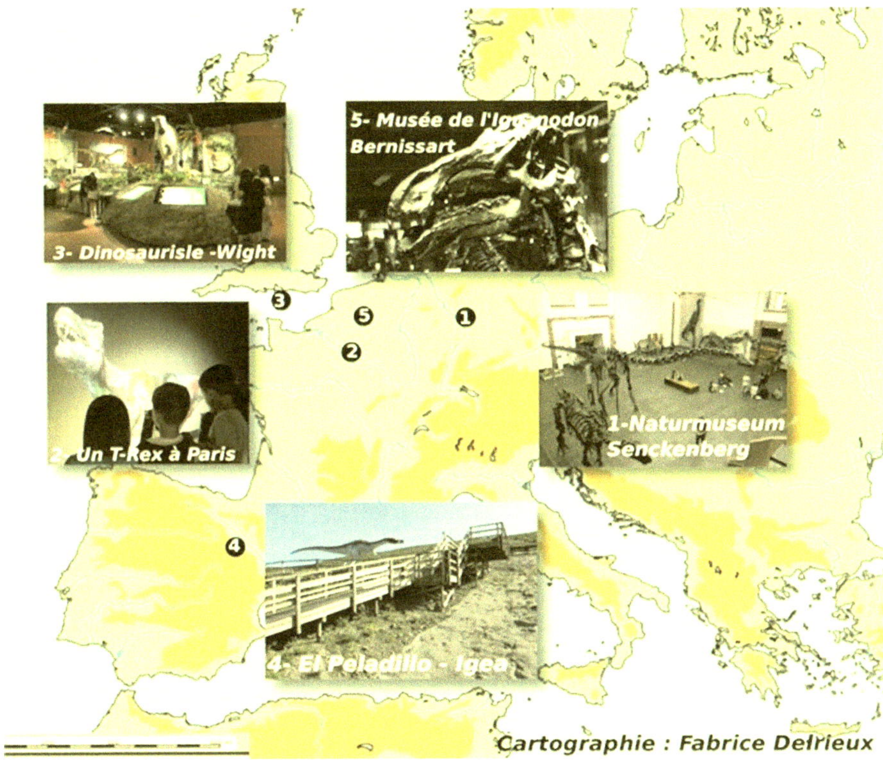

Fig. 19.1 Samples of dinosaur destinations illustrated in the text

be a desirable side effect. Special exhibitions of dinosaurs, like “Un T-Rex à Paris” (6/8/2018–4/11/2019) can attract record numbers of visitors and serve as the key event for the concerned museum. Within 4 months, 340,000 persons came to the Natural History Museum of Paris (David 2018). Also, a considerable financial profit can be gained. This special exhibition resulted in a net profit of 1.5 Mio. €, even taking into consideration the expenses for building up the exhibition and the rental fee for the exhibited objects (David 2018). Nevertheless, the geographical reference is missing due to the ex situ presentation, which poses a serious disadvantage for an authentic geotourism. Therefore, the following chapter focuses on in situ offers that connect original sites with tourist offers and a sound didactic concept.

Only a very small number of the natural history museums lie in close vicinity to the paleontological sites. Usually, these museums had been built up after the paleontological excavations and mostly show only findings from the neighboring sites. One of the most remarkable examples of this type of museum is Dinosaur Isle in the Isle of Wight, South England. Most of the fossil bones discovered in the island since the middle of the nineteenth century have been dispersed in museum collections in the mainland. So the Wight Island Council began its own collection, which in 1914 became the first geological museum of the island located in Sandown. It is currently home to 40,000 specimens of which nearly 200 are types (Munt 2008) and welcomes around 60,000 visitors a year (Simpson 2018). Esparaza (Aude-France) which opened in 1992 and welcomes 40,000 visitors per year or the “Museo del Jurasico” (La Muja-Spain) which opened in 2004 and welcomes 150,000 visitors per years have also been built near the outcrops of important discoveries.

In addition to the two types of tourist destinations mentioned above, open-air museums or field museums are facilities that provide an interpretation of heritage directly on the geosite. This is the choice that is made most of the time in order to highlight dinosaur footprints because the size of the sites and their vulnerability make it difficult to envisage removing it. La Rioja province, in Spain, owns numerous dinosaur footprints geosites (Fuertes-Gutierez et al. 2016). The project “Ruta de yacimientos de la Rioja” allowed the valorization of some of them. Era del Peladillo, near Igea, with more than 1,700 traces is one of the first sites in the world concerning the quantity of dinosaurs footprints with among them those of Baryonix, the fishing dinosaur. In 2013, the interpretive valorization of the geosite has been realized by Paleoymas. The action carried out on the site, located three kilometers from the town Centre and had involved an investment of 168,856€ financed by the Tourism Competitiveness Plan in which the Ministry of Industry, Energy and Tourism, the Government of La Rioja and the Mancomunidad de Desarrollo Turístico La Senda Termal collaborate. Since 20 years, at the French Switzerland border, a lot of dinosaur’s track sites have been discovered. Loulle (Jura-France) presents on an area of 500 m^2 numerous tracks discovered in 2006 and valorized in 2014 for a cost of around 150,000€. There are no visitor numbers for these two cases, as no entrance fees are charged.

A special category must be made (Perini and Calvo 2008) for active excavation sites, where the visitor observes live work of paleontologists and may even participate in it. This can be called experiential tourism since the visitor will be able to interact directly with the researchers. In 2008, an exceptional fossiliferous deposit, dated from −130 to −125 million years ago, was discovered in Angeac-Charente (France). Faced with the scientific importance of the deposit, the authorities of Charente finally acquired on June 22, 2012 a part of the quarry, in order to be able to carry out meticulous excavations there for several years. Every summer, the site welcomes the public to discover the work of paleontologists.

The tourist's fascination evoked by dinosaurs, e.g., dinosaur models, is also used to attract visitors to places, where never any fossils or traces had been discovered. Furthermore, no original objects are integrated. A typical example is the 2017 newly opened Dino Adventure Park in Weiterstadt near Darmstadt (Germany). This adventure theme park, addressing families as the target group, is designed as an indoor play park with life-size dinosaurs. Using the latest animation technology, the visitors should be able to relive the time of the dinosaurs by encountering moving and roaring dinosaurs (dino-world 2018). Even though adventure parks such as Dino-World can sometimes reach a high number of visitors and, given a good and scientifically correct design, can also provide some information about dinosaurs and their habitat, the fun factor and commercial background are clearly dominating (Wilson 2018). Due to the fascination of dinosaurs, there are also links between geotouristic highlights and dinosaurs, although no actual relationships exist. One example is the National Geotope Externsteine in the Teutoburg Forest (North Rhine-Westphalia). The sandstone sediments were deposited in the Lower Cretaceous. Today's form is the result of specific weathering and erosion processes. Nevertheless, dinosaurs are blamed for the actual situation by a cartoon (Fig. 19.2).

Fig. 19.2 Cartoon showing the would-be construction of the Externsteine by dinosaurs

In addition dinosaur sites, which are not valorized as tourist destinations are also excluded. The later include sites, which are no longer accessible today (e.g., backfilled quarries) as well as sites, which are kept confidential to protect them against looters. Similar cases are sites, where only a small panel indicates the high geoscientific importance, and which are mostly known only to a limited circle of experts. This is the case for one of the most important find sites of Plateosaurus in the South of Germany.

Dinosaur geotourism, in general, is based on three types of paleontological remains: bones, eggs or hatcheries and footprints or trackways. Ensuring the in situ geoconservation in order to develop a local geotourism offer is a key challenge, which has to take into account poaches, erosion and further risk factors (Guyomar 2013; Page 2018). This explains why in the past, remains that were studied in museums have been preserved there, sometimes far from the discovery outcrop. The case study of Bernissart is a good example of this:

In 1878, 29 Iguanodon skeletons were discovered in the coal mine of Bernissart, a small village not far from the French–Belgian border. After their study by paleontologists from the Royal Belgian Institute of Natural Sciences, a first mounted skeleton was exposed in Brussels in 1882. Then, in 1902, a group of 11 specimens was presented to the public, a world first and the main attraction of the museum's display. It was not until 2002 that one of the skeletons was returned to the village of Bernissart, where an interpretation center: «Le musée de l'Iguanodon» has been created. The discovery outcrop, 322 m below surface, is no longer accessible and the interpretation center tries to recreate «the sense of place» developing contextual events of this exceptional discovery of one of the most famous dinosaurs of early Cretaceous. This project has been completed thanks to financial support from a European INTERREG IV program with an amount of 645,000€. The aim is to improve tourism offers in order to develop local economy linking to natural and cultural heritage. The center welcomes today around 7,000 visitors a year, in a village of 11,000 inhabitants.

The following overview (Fig. 19.3) lists geotouristic offers, which are located on original find sites of dinosaur fossils in Europe a survey shows the diversity of projects, initiated by public, private, or mixed stakeholders. Some destinations require entrance fees, others offer free access.

19.3 UNESCO Global Geopark, an International Renowned Label

In Europe, only 5 of the 73 UNESCO Global Geoparks have geoheritage linked to dinosaurs. Hateg Country Dinosaurs Geopark (Romania) is a hot spot of paleontology and reveals a unique fauna of dinosaurs. Other geoparks as Conca de Tremp Montsec (Spain), Terra Vita (Germany) or Geopark des Monts d' Ardèche (France), and the aspiring Thuringia geopark (Germany) own unique dinosaur paleontological sites.

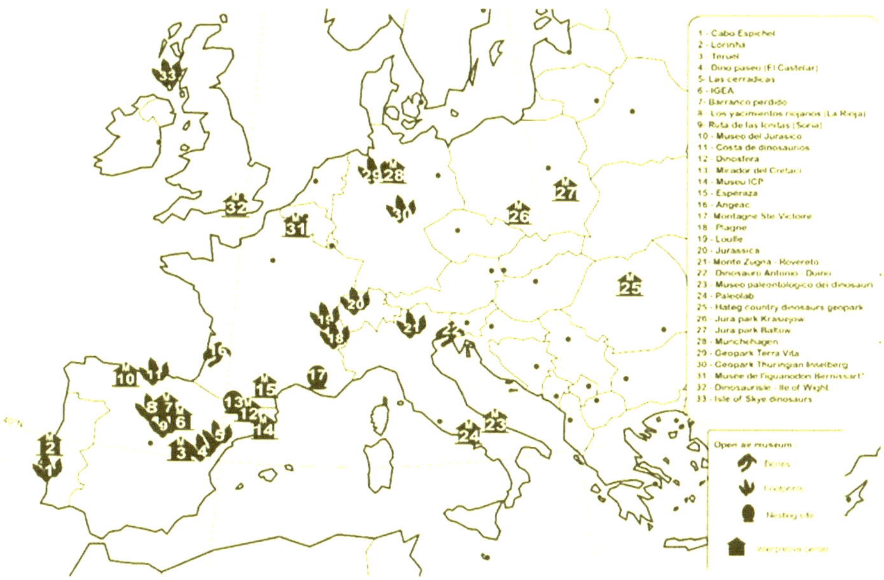

Fig. 19.3 Overview of main European dinosaurian geotourism destinations

19.3.1 Hateg Country Dinosaurs Geopark (Romania)

Peculiar dwarf dinosaurs have been unearthed in the region of Hateg, Romania. The first discoveries of dinosaur's bones were made by Franz Nopcsa and published in 1897 (Grigorescu 2005). The Sanpetru and Densus-Ciula formations of the Hateg Basin, are considered to be Maastrichtian, the latest age of Upper Cretaceous (Van Itterbeeck 2005). At that time, the Hateg region was an island of about 80,000 square kilometers about the size of Tasmania today, located near the equatorial belt. Hateg dinosaurs, such as the theropod *Balaur Bondoc* or the titanosaur *Magyarosaurus Dacus*, are of a spectacular scientific interest because of their smaller size than other contemporary dinosaurs of their genus living elsewhere on the planet. This reflects the particular evolution process of insular dwarfism (Benton et al. 2010). The scientific research that resumed in the 1970s allowed in 1988, the discovery of nesting sites similar to those observed in the south of France. The presence of newborn's bones associated with these eggs made it possible to attribute the hatchery to *Telmatosaurus transylvanicus*, a hadrosaur (Grigorescu 2010) (Fig. 19.4).

In 2004, the Hateg Country Dinosaurs Geopark joined UNESCO Global Geopark network, in order to promote this remote region of Carpathian Mountains (Andrasanu and Grigorescu 2012) . In the Romanian National Spatial Plan, geoparks are considered as territories that must sustain local identity for a sustainable development and to achieve these goals two million euros were invested (Stoleriu 2014). The management of the geopark is carried out by the University of Bucharest, but a staff of five

Fig. 19.4 Telmatosaurus nest—Hateg Country Dinosaurs Geopark

scientist's works in Hateg. Firstly, a lot of projects focused on tourism infrastructures improvement. The main goal of the geotouristic experience is to emphasize the authenticity of the "Land of Hateg" enhancing both natural and cultural heritage as already promoted by the Retezat National Park, which borders the geopark, created in 1935. Seven interpretation centers and a few thematic trails disseminated around Hateg provide knowledge about local geoheritage. The first visitor center opened in the main town in 2010, the "house of the geopark" has welcomed 5321 visitors in 2017, nearly 25% more visitors than in 2016. The central theme of the exhibition is myths and beliefs about dragons and the key element of the display is a cast of *Balaur Bondoc* done by the Canadian artist Brian Cooley in 2014 (Fig. 19.5). A few kilometers away, the house of science and art in the village of General Berthelot, only visited by 2616 tourists in 2017, presents full-scale models of *Magyarosaurus dacus* and *Zalmoxes robustus* and also dinosaur's remains. The journey of Magyarosaurus model, also done by Brian Cooley in Canada from Rotterdam harbour to Hateg, financially sustained by a kickstarted lottery and the Transylvania Dinosaurs Museum project, allowed a cross-cultural collaboration between art and science (Seghedi et al. 2017). With the UNESCO Global Geopark label, the Hateg region was integrated in an international network. Nevertheless one of the main difficulties is still the small number of international tourists, representing at the national level only 22.6% of the total tourist arrivals registered in 2016 and only 18% of the total overnight stays (Iatu et al. 2018). Tourism incomes in Hateg region will gradually progress, supported by numerous local stakeholders involved in the project and a very active community of local volunteers developing an effective communication on social networks.

Fig. 19.5 Life-like model of Balaur Bondoc—House of the geopark–Hateg

19.3.2 UNESCO Global Geopark Conca De Tremp Montsec (Spain)

The Coll de Nargó area, Lleida Province in Spain (Catalonia), which corresponds to the uppermost Cretaceous continental sediments, has yielded thousands of dinosaur's eggs and clutches. First discoveries had been made in the 1970s by German researchers, later they had been investigated by volunteers of the association "Amics dels dinosaures de l'Alt Urgell" (ADAU). In 2007, the largest dinosaur nest in Europe had been discovered near the village. This part of the Pyrenees is a key point of the last time of the dinosaur's era. The remains are included in a 400 m layer of sediment corresponding to about 4 million years of a total deposition between 72 and 68 million years (Selles et al. 2013).

An ambitious project, «Dinosaures dels pirineus», supported by the Catalonia Sustainable development strategy with the horizon 2026, has been developed in order to enhance paleontological geoheritage in this part of the Pyrenees. The project is managed by the Institut Catala de Paleontologia (ICP). Part of «Dinosaures dels pirineus» is included in the Geopark Conca de Tremp Montsec which joined the UNESCO GGN in 2018. But the two territories are not completely overlapped. The brand "Dinosaurs del pirineus" gathered interpretive centers spreading into the Pyrenees as closely as possible to the discovery. The Museu de la Conca Dellà mainly dedicated to hadrosaurs is a registered museum by the Dept. of Culture of Generalitat de Catalunya, used as a new repository of the discovered

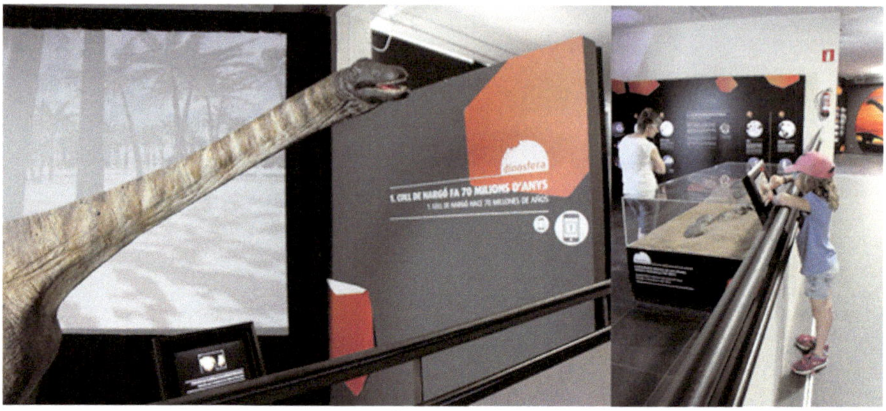

Fig. 19.6 Dinosfera- Geopark Conca de Tremp Montsec

materials. In the visitor center of the geopark in Epicenter, the extinction of dinosaurs linked with the rise of the mammals is presented. Other centers are in Vilanova de Meià, Tartareu, and Berguedà or Fumanya.

In order to valorize the discovery of the largest European dinosaur's nest, near the village of Coll de Nargó, the assigned interpretive center Dinosfera (Fig. 19.6) opened in 2011 welcoming about 7,500 visitors per year. The main theme addressed in the exhibition is the reproduction of dinosaurs: history of oviparity, structure and adaptations of eggs, and above all diversity of dinosaur's eggs and nest. The key element of the display is the cast of the nest discovered not far away.

A few kilometers away the "mirador del Cretacico" is the in situ museum that allows to discover the layers from which the nest had been extracted. It was created in 2008 for a total spending of 24,000€ financed by provincial supports. In 2017, a part of the fossils of the site was destroyed with a hammer in a criminal act.

19.3.3 Other UNESCO Global Geoparks with Dinosaur Sites in Europe

Two other UNESCO Global Geoparks and one aspiring geopark integrate among other geosites also dinosaur sites. In all three cases, Dinosaur's tracks were found. In the little French village of Payzac, inside the Geopark des Monts d'Ardèche, labeled in 2014, dinosaurs tracks led by an ichnospecies Coelurosaurichnus grancieri, a little carnivorous dinosaur of upper Triassic have been discovered by a local amateur and named in 2000 (Courtel and Demathieu 2000). This discovery is of great scientific importance because these footprints, which are about 237 million years old, are among the oldest discovered in France. A private exhibition is proposed by the amateur paleontologist who discovered them.

Within the UNESCO Global Geopark Terra Vita (Northern Germany), 150 million years old traces of two-legged predatory dinosaurs (Megalosaurus) and a whole herd of great vegetarians (Camarasaurus) are exposed. Already in 1921, these traces were discovered by a geologist in the small quarry of Barkhausen. Thereupon, a larger area was uncovered, which released further traces. These dinosaur tracks are protected against weather conditions since 2000 by the construction (280,000€) seen on Fig. 19.7. In order to give visitors a more vivid impression of the animals, which left behind these traces, life-like models of the respective dinosaur species were set up. The dinosaur traces were also integrated into the exhibition "Terra vision" of the natural history museum am Schoenberg in Osnabruck. Today this site is visited the fourth most in the Geopark with around one hundred visitors a day in summer (Härtling and Meyer 2010).

Within the aspiring Geopark Thuringia Inselberg—Drei Gleichen (Eastern Germany) traces and skeletons of the Tambach native dinosaurs were found in sandstone layers of the Rotliegendes (Permian), especially in the world-famous site of Bromacker. A 4.5-km-long dinosaur discovery path connects various sites and uses dinosaur models to transmit knowledge about the various dinosaur fossils and traces found here. In 2017, a new section of the Georoute was completed, showing at six stations the work of paleontologists in the discovery and salvage of corresponding finds (Brauer et al. 2016).

Fig. 19.7 Dinosaur tracks at Barkhausen, Geopark Terra Vita

In all cases joining the UNESCO, GGN had been an opportunity to increase public visibility and awareness. Furthermore, the Global Geopark seal awards the geotourism destination a high value as well as high-quality standards, attracting thereby an increased number of tourists.

19.4 Geotourism Between Private and Public Management—Dinosaur's Sites Outside of Geoparks

Geoheritage like other commons are often under public protection and management. Valorizing these sites and building up a tourism destination is another step which sometimes needs the help of private funds. Reconciling expectations of public and private institutions and establishing a geotourism destination, where protection, environmental education and attractiveness for tourists (activity, fun-factor) are successfully combined, are often considerable challenges.

19.4.1 Dinopolis (Spain)

Spain is home to palaeontological sites of great importance. Most of them are located in economically underdeveloped rural areas far from the main tourist circuits. The province of Teruel, located in the south of Aragon, focused its tourism development on the creation of the "Teruel-Dinopolis Palaeontological Site", a thematic park located in the provincial capital, a city of 30,000 inhabitants. Territorio Dinopolis is a public scientific foundation created in 1998 and a private and public management company employing around 100 people year-round (Fig. 19.8). The project was financed by the Instituto Aragonese de Fomento, which invested 15 million euros to support the economic development of the province, one of the least populated in Spain (Bosquet 2005).

Fig. 19.8 Dinopolis (Spain)—a leisure park and a true museum

The theme park opened in 2001 and has grown ever since (Alcalà 2011a, b). A study conducted in 2011 by the Fundación de Economía Aragonesa shows that from 2001 to 2011, more than 1,750,000 visitors discovered Dinopolis. The incomes generated by all this tourist activity amount to more than 1,560 million euros including more than 184 million euros of direct incomes (Perez y Perez and Gomez Loscos 2011). In ten years, the number of tourist beds around Teruel has increased from 700 to 2000. Dinopolis is above all a family leisure park that offers many recreational activities, from 4D cinema to TRex discovery (a species that is only known in North America, not in Europe). It also offers a museographic section dedicated to the history of life, where visitors can improve their knowledge particularly about the dinosaurs discovered in the province like *Turiasaurus riodevensis*, a titanosaur dated to the Jurassic-Cretaceous boundary, around 145 million years old. In 2018, Territorio Dinópolis received the PAC Awards for the best Nature Park of the Year in Spain.

One of the main objectives of the project was, to increase the length of tourists' stay, in order to promote the development of the entire province. Therefore, six thematic interpretive centers, each being dedicated to a specific palaeontological theme, have been created in the heart of hamlets, all around Teruel, with the aim of distributing tourist flows and economic incomes. For that, the Fondo de Inversiones de Teruel has invested more than 3 million euros. The last one focused on *Turiasaurus riodevensis,* opened in 2012 in Riodeva, a little village of two hundred inhabitants.

Throughout the year, numerous events support the attendance of Dinopolis (Alcalà 2011a, b). Educative programs for children but also students or amateurs, Geolodià, a day for geology, Geodrome an annual meeting on geosciences… At the same time, the paleontological excavations still carried out by the foundation's researchers have led to new discoveries. Since 2002, many dinosaur tracks have been unearthed around the village of El Castellar. There are, in particular, two new ichnogenres: *Deltapodus ibericus*, footprints left by a stegosaurus and *Iberosauripus grandis* made by a theropod. Here again, in order to attract tourists to this remote village, a geotrail was inaugurated in 2015. It shows a great number of footprints and leads to the deposit of El Castellar which presents tracks from the end of the Jurassic period between 150 and 145 million years ago. The creation of the geotrail and the development of the deposit cost 130,000€, financed by the government of Aragon (Cobos and Alcàlà 2018).

The strategy chosen by the provincial authorities is a success. The analysis of the network graph of actors involved in this project shows the balance between the two main pillars of "Territorio Dinopolis" (Fig. 19.9). On the one hand, Dinopolis manages the leisure park and its development with the objective to attract visitors in a concern for legitimate economic profitability. On the other hand, the Dinopolis Teruel foundation guarantees active research, preservation of the provincial paleontological heritage and the quality of the scientific mediation offered to visitors. The current project responds well to these two expectations, thus demonstrating the importance of a thoughtful offer at several levels in order to reach the widest possible audience.

Fig. 19.9 Dinopolis actors graph network

19.4.2 Dinosaurier-Park Münchehagen (Germany)

For several centuries, brownish quartz sandstones of the Berriasian Bückeberg Formation (German Wealden) have been extracted in the district of Münchehagen in Rehburg-Loccum (Lower Saxony, Germany) in smaller quarries and used as versatile building blocks (Wings et al. 2012: 113). In 1979, bowl-shaped depressions were discovered, which turned out to be dinosaur tracks. Further rock quarrying resulted in the uncovering of more than 300 dinosaur tracks, making the site in Münchehagen unique in the world (Richter et al. 2012: 4). Here, the only German traces of sauropod from the Lower Cretaceous were detected (Wings 2012: 59). Consequently, the site of Münchehagen was honored as National Geotope in 2006 (Akademiefür Geowissenschaften and Geotechnologien e. V. 2018). Detailed information on the geological strata and the dinosaur tracks as well as the exploration history can be found in Wings et al. (2012).

Since the discovery of the traces, persons interested in dinosaurs visited the quarry. Therefore, one year after the discovery, a shelter was built to safeguard the traces. In 1986, the entire site was fenced in and placed under protection as a natural monument. To promote geotourism in the region, the natural monument was embedded in an educational theme park about Earth history (Wings et al. 2012: 118). In 1992, the

Fig. 19.10 Dinosaur tracks and shelter in Münchehagen

Wesling Company (quarry operator) recognized the tourism potential and founded today's dinosaur open-air museum, whose centerpiece is the hall with the traces of the dinosaurs (see Fig. 19.10) (Richter et al. 2012: 4). Till today, the Dinopark is operated and financed privately without any public funds, only the shelter, protecting the tracks was financed by the state of Lower Saxony (Knötschke 2019).

With sandstone quarrying continuing adjacent to the shelter, dinosaur tracks continue to be uncovered. These are salvaged, scientifically documented and edited. A part of them is subsequently exhibited in the Dinosaur Open-Air Museum (Richter et al. 2012: 52). Still every year new dinosaur tracks are found in Münchehagen. A further expansion of the vertebrate track record at Münchehagen could even launch new activities of the Dinosaurs Open-Air Museum, as well as public excavation campaigns for park visitors. Finally, the southern parts of the quarry should be united with the park area and form a larger, common national geotope or a future monument zone (Wings et al. 2012: 139).

The integration of the tracks into today's dinosaur park principally entailed the risk that an adventure park could have developed, in which financial aspects and the fun factor would have been in the foreground. This had been prevented with an appropriate conception in Münchehagen from the beginning. Therefore, there are no special offers typically for adventure parks, such as rides, but the edutainment concept focuses exclusively on dinosaurs. The result is Europe's largest scientific theme park, which focuses on communicating scientific findings. With the help of a total of 230 life-like models (see Fig. 19.11) and different activity offers (see Fig. 19.12), various target groups can be reached. Learning about dinosaurs and

Fig. 19.11 Life-like model of Dinosaur in Münchehagen

Fig. 19.12 Activities for visitors in Dinopark Münchehagen

the Earth's history is made interesting for target groups of all ages by experiences and adventure elements (Richter et al. 2012: 5). The 2.5 km long circular route through the open-air museum follows the geological eras, from Devon to Stone Age, focusing on the geological formations of dinosaur's lifespan. The museum also serves as a place of learning, with modular hands-on experience offers and guided tours and is one of the most visited off-campus learning sites in Lower Saxony. For the park, a scientifically correct mediation and a high scientific standard are self-evident (Dinosaur Park Münchehageno.J.).

Beyond that, the Dinoparkis officially recognized as a museum and houses an internationally recognized research and competence center for fossil preparation, in which since 1999 fossils have been partially prepared for the visitors to watch. This center works in close cooperation with the Steinmann Institute of the University of Bonn and the Landes museum (State Museum) Hannover (Dinosaur Park Münchehageno. J.).

In the 25 years of its existence, several million tourists have visited the dinosaur park, which covers a tourism niche (Süddeutsche Zeitung 2018). The park is in a rural region but within a surrounding catchment area with a high population density and major cities like Dortmund, Bremen, or Hannover. So 95% of the visitors come from a radius of a maximum of 200 km around the park. The fascination with dinosaurs is the main motive for visiting the park. The Dinoparkis is a family destination, which offers intergenerational activities and has a high rate of regulars. Therefore, the family season ticket had been introduced. Some of today's family heads had themselves got to know the park as children and now come back with their own children.

Many visitors expect a classic theme park and do not know beforehand that the park is located in a location with original finds. Many of them would not realize that the tracks in the track hall are originals, but take them for well-done reproductions. It is, therefore, difficult at first to "sell" the original find site. By panels and explanations, this has to be made understandable. Only then the visitor's realize the particularity of the place. For the Dinopark, the tracks and the ongoing excavations are a great opportunity, not only to inspire its visitors, but to create a deeper connection to this particular form of geoheritage through the "fascination of the real". This is further enhanced by special offers, such as visits to the original excavation site or children's discovery tours, which allow them to observe the uncovering of the fossils by a taxidermist and sometimes even help with the preparation. Some of the enthusiastic children now study geology or biology or are trained as taxidermists (Knötschke 2019).

So far, Dinopark does not work with new media. This was discussed intensively, but the decision was made against an expansion of digitization. Imagination and active experiences are encouraged. For children, a multitude of activities are offered (see Fig. 19.11), where tablet touchscreens or scanning QR-codes might be detrimental and restrict imagination and creativity. Results of visitor's satisfaction survey showed that guests did not miss digital animation like moving and roaring dinosaurs (Knötschke 2019).

With approximately 100,000 visitors per year, current entrance fees of €12.50 for adolescents and adults over the age of 13 and €10.50 for children over the age of 4 lead to considerable added value for the economically underdeveloped region. Hence the home town of the Dinopark looks at "the dinosaurs of prehistoric times" as a touristic flagship, which has made the city known far beyond its borders (Rehberg-Loccum 2018). Therefore the city offers a 30 km cycling tour, combining dinosaur find sites and natural and cultural heritage of the township. Recently several hotels in the surroundings started to offer packages including a visit to the Dinopark and the Dinopark encourages visitors to stay for more than one day. On the other hand, the nearby tourism destination Steinhuder Meer (largest lake in Lower Saxony) began to advertise the Dinopark as excursion destination for its guests.

As the general concept of the Dinopark proved to be very successful, the Wesling Company opened in February 2018 a second Dinopark following the same basic principle of combining real discoveries, life-like models, and knowledge transfer from scientific research to the visitors. The chosen location is Lourinha (Portugal), the so-called "capital of dinosaurs", where outstanding fossil finds had been made. Already in the first year, about 300,000 visitors came to the Lourinha Dinopark (Knötschke 2018). The subsidiary company Dinosaur Park International offers consulting for comparable projects.

Even if the balance between education and fun is difficult to maintain, the approach for dinosaur theme parks seems to be a success. The Baltow Jurassic park which opened in Poland in 2004 only one year after the discovery of dinosaur footprints of late Jurassic had an attendance of 320,000 visitors in 2013 (Kruczek 2015).

19.5 Conclusion

Meeting the worldwide growth of tourism, there is an increasing specialization of individual practices. Palaeontological tourism is an interesting way of diversifying geotourism, as it is not DE localizable and attracts tourists closer to geosites, sometimes far from the main leading tourist destinations. However, whatever the scientific interest of the sites, the success of these tourism projects must be based on a structured network of actors, gathering scientists, tourism actors supported by local policy-makers. Although until now, projects were often initiated by scientists, local stakeholders aware of their heritage can bring about change by developing innovative geotourism projects.

In France, the longest dinosaur track in the world was discovered in 2009 near the small village of Plagne. The 135 footprints left by the ichnogenre *Brontopodus plagnensis* can be followed over a length of 155 m. An ambitious development project is currently underway. Dinoplagne will open to the public in the summer of 2019. The main objective of the interpretation of the site will be to take the visitor back in time along a walkway that will take him or her from current biodiversity to that of the past.

Whatever the strategy adopted, the tourist trajectory of dinosaur sites must above all respect the "sense of the place" and allow the visitor to apprehend a fascinating world forever gone.

References

Akademie für Geowissenschaften und Geotechnologien e.V (2018) Nationale Geotope. https://www.geoakademie.de/stern_geotope.php?seite=3. Accessed 21 Dec 2018

Alcalà L (2011a) Un modelo de desarrollo regional fundamentado en recursospaleontológicos (Dinópolis-Teruel). Paleontología y dinosauriosdesdeAmérica Latina, pp 253–260

Alcalà L (2011) Territorio Dinópolis; Dinópolis Palaeontological Park. HER & MUS 8 III(3):96–106

Andrasanu A, Grigorescu D (2012) Romania in geoheritage in Europe and its conservation PROGEO Wimbledon (W.A.P. & Smith-Meyer S (eds)), pp 275–287

Benton M, Csiki Z, Grigorescu D, Redelstorff R, Sander M, Stei K, Weishampel D (2010) Dinosaurs and the island rule: the dwarfed dinosaurs from Hategisland. Palaeogeogr Palaeoclimatol Palaeoecol 438-454

Bosquet RL (2005) Un nuevoconcepto de parque tematico: origen e impactos de dinopolis. Cuadernos de Turismo 15:149–167

Brauer S et al (2016) Saurier-Erlebnispfad Georgenthal & Tambach-Dietharz. Geo-Routen Führer 6. https://www.geopark-thueringen.de/fileadmin/user_upload/user/GeoPark-Faltblaetter/fb_georoute06_saurierer lebnispfad.pdf. Accessed 21 Dec 2018

Carpenter K (2018) Rocky start of Dinosaur National Monument (USA), the world's first dinosaur geoconservation site. Geoconservation Res 1(1):1–20

Cayla N (2018) Dinosaurian geosites: successful geotouristic destinations. Geophys Res Abstr 20. EGU2018-11979, 2018 EGU General Assembly

Cayla N, Duval M (2013) Legéotourisme. Patrimoines, pratiques, acteurs et perspectives marocaines. Collect EDYTEM 14:101–116

Cobos A, Alcàlà L (2018) Palaeontological heritage as a resource for promoting geotourism in the rural setting: El Castellar (Teruel, Spain). Geoheritage 10(3):405–414

Courel L, Demathieu GR (2000) Une nouvelle ichnoespèce *Coelurosaurichnusgrancieri*du Trias supérieur de l'Ardèche, France. Geodiversitas 22(1):35–46

David B (2018) Carton plein pour le T. rex à Paris, avec 340.000 visiteurs. https://www.la-croix.com/Culture/Carton-plein-T-rex-Paris-340-000-visiteurs-2018-11-08-1300981783. Accessed 09 Janvier 2019

Dinosaurierpark Münchehagen (o.J.) Dinosaurier-Freilichtmuseum Münchehagen. Brochure

Dino World (2018) Dino-Park Weiterstadt. https://dino-world.de/. Accessed 19 Dec 2018

Fuertes-Gutierez I, Garcia-Ortiz E, Fernandez-Martinez E (2016) Anthropic threats to geological heritage: characterization and management: a case study in the dinosaur tracksites of La Rioja (Spain). Geoheritage 8(2):135–153

Grigorescu D (2005) Rediscovery of a "Forgotten land": the last three decades of research on the dinosaur-bearing deposits from the Hateg basin. Acta Paleontologicaromaniae 5:191–204

Grigorescu D (2010) The "Tustea puzzle": hadrosaurid (Dinosauria, Ornithopode) hatchlings associated with Magaloolithidae eggs in the Maastrichtian of the Hateg Basin (Romania) Rev. Asoc Paleontol Argent 47(1):89–97

Guyomar M (2013) Patrimoine géologique : conservation et valorisation in situ des fossiles. Trente années d'expérimentation en Haute-Provence*CeROArt* [Online]. HS I 2013, Online since 18 August 2013. https://journals.openedition.org/ceroart/3485. Accessed 12 January 2019

Härtling J, Meier I (2010) Economic effects of geotourism in geopark TERRA. vita, Northern Germany. George Wright Forum 27(1):29–39

Iatu C, Ibanescu BC, Stoleriu OM, Munteanu A (2018) The WHS designation—a factor of sustainable tourism growth for romanian rural areas? Sustainability 10:626

Iorio M, Corsale A (2010) Rural tourism and livelihood strategies in Romania. J Rural Stud 26:152–162

Knötschke N (2018) Dinopark Münchehagen. Mail from 18th december 2018

Knötschke N (2019) Dinopark Münchehagen. Telephone interview at the 10th of January 2019

Kruczek Z (2015) Analysis of visitor attendance at polish tourism attractions. Tourism 25(1):47–56

Landel PA, Senil N (2009) Patrimoine et territoire, les nouvelles ressources du développement, Développement durable et territoires Ed. [Online], 12. https://developpementdurable.revues.org/7563. Accessed Jan 2019
Laws E, Scott N (2003) Developing new tourism services: Dinosaurs, a new drive tourism resource for remote regions? J Vacat Mark 9(4):368–380
Lipps JH (2009) Paleoparks: our paleontological heritage protected and conserved in the field worldwide. In: Lipps JH, Granier BCR (eds) Paleoparks – the protection and conservation of fossil sites world wide
Megerle H (2008) Geotourismus, Innovative Ansätze zur touristischen Inwertsetzung und nachhaltigen Regionalentwicklung; Band 1 Geographie in Wissenschaft und Praxis, Kersting-Verlag, Rottenburg am Neckar, 2te überarbeitete und erweiterte Auflage
Megerle H, Pietsch D (2017) Consequences of overlapping territories between large scale protection areas and Geoparks in Germany: opportunities and risks for geoheritage and geotourism In: Annales de Géographie 717:598–624. Natur- und Geopark Terra Vita (2017): Saurierfährten Bad Essen/Barkhausen. Brochure
Page K (2018) Fossils, heritage and conservation: managing demands on a precious resource. In: Reynard E, Brilha J (eds) Geoheritage. Elsevier, Chennai, pp 107–128
Peyrache-Gadeau V, Perron L, Janin C (2016) La spécificité territoriale comme alternative à la généricité des produits-ressources. In Pecqueur Bernard et GlonEric (sous la direction de) Au cœur des territoires créatifs –proximités et ressources territoriales, Presses Universitaires de Rennes 226–236
Perez y Perez L, Gomez Loscos A (2011) Efectoseconomicos y sobre el empleodel parque TerritorioDinopolis de Teruel, Documentos de trabajo Fundacion Economia Aragonesa, p 56
Perini MM, Calvo JO (2008) Paleontological tourism: an alternative income to vertebrate paleontology. Arquivos do Museu Nacional, Rio de Janeiro 66(1)
Rehberg-Loccum (2018) Freizeit und Tourismus. https://rehburg-loccum.de/staticsite/staticsite.php?menuid=15&topmenu=15. Accessed 21 Dec 2018
Richter A et al (2012) Museumsführer Dinosaurier-Freilichtmuseum und Naturdenkmal Dinosaurierfährten Münchehagen. Rehburg-Loccum
Seghedi A, Andrasanu A, Radan S (2017) The Transylvanian dinosaur museum project: the contribution of Geoecomar to valorize and promote the paleontological heritage of Romania. Geo-Eco-Marina 23:145–164
Selles A, Bravo A, DelclòsX CF, MartíX O-B, ParelladaC GA (2013) Dinosaur eggs in the Upper Cretaceous of the Coll de Nargó area, Lleida Province, south-central Pyrenees, Spain: oodiversity, biostratigraphy and their implications. Cretac Res 40:10–20
Senckenberg Gesellschaft für Naturforschung (2018) Naturmuseum Frankfurt https://www.senckenberg.de/root/index.php?page_id=5247. Accessed 19 Dec 2018
Simpson M (2018) Geotourism and geoconservation on the Isle of Wight, UK: balancing science with commerce. Geoconservation Res 1(1):44–52
Stoleriu MO (2014) National approaches to geotourism and geoparks in Romania. In: 14th SGEM Geo conference on ecology, economics education and legislation 2014-06 conference-paper
Süddeutsche Zeitung (2018) Dinosaurier suchen neue Heimat. Dinopark-Gruppe Münchehagen auf Expansionskurs. Article published 07 December 2018
Taquet P (2001) Philippe Matheron et Paul Gervais: deux pionniers de la découverte et de l'étude des os et des oeufs de dinosaures de Provence (France). Geodiversitas 23(4):611–623
Van Itterbeeck J, Markevitch V, Vlad C (2005) Palynostratigraphy of the Maastrichtian dinosaur-and mammal sites of the Râul Mare and Barbat Valleys (Haţeg Basin, Romania). Geologicacarpathica 56(2):137–147
Wings O (2012) The lower cretaceous Dinosaur tracksite in Münchehagen, Germany – track record, preservation, and potential. In: Richter A, Reich M (Hrsg.) Dinotracks. An international symposium. Abstract Volumen and Field Guide to Exkusions, p 59
Wings O et al (2012) The early cretaceous Dinosaur trackways in Münchehagen (Lower Saxony, Germany) – the natural monument 'Saurierfährten Münchehagen' and the adjacent Wesling

Quarry In: Richter A, Reich M (Hrsg.) Dinotracks an international symposium. Abstract Volumen and field guide to exkusions, p 113

Wilson RJ (2018) Facing a new frontier: defining nature and culture within dinosaur parks in Europe. In: Saul H, Waterton E (eds) Affective geographies of transformation, exploration and adventure: rethinking frontiers. Routledge, London, pp 65–82

Chapter 20
The Use of Geoheritage, Geopark and Geotourism Concepts to Conserve and Sustain Tourism Development in Zambia

Kagosi Mwamulowe and Imasiku Anayawa Nyambe

Abstract This paper examines how geoheritage, geopark and geotourism concepts can be used as tools for sustainable tourism development and conservation in Zambia. It introduces the various concepts in earth heritage namely geology, geomorphology, geological heritage (geo-heritage), geosite, geoheritage conservation, geopark, geotourism and sets out Zambia from a geopolitical context in Africa in relation to its neighbouring nations and projects its geodiversity and geoheritage in terms of the various geological formations that include the Basement Complex, Katanga Supergroup, Karoo Supergroup and the Kalahari Supergroup that houses the Barotse Floodplain of cultural heritage. Furthermore, key features such as escarpments, valleys, gorges and waterfalls are also described. Zambia is being depicted as very rich in geodiversity which has given birth to a numerous geoheritage sites. The National Heritage Conservation Commission (NHCC) Act is the principal law, which is applicable in the field of heritage protection and conservation in Zambia. This also involves declaring sites of nationally geological importance, rock art sites and caves of archaeological significance amongst many heritage typologies as national monuments. However, there are other pieces of legislation such as the National Museums Board Act and National Parks and Wildlife Act, as well as National Forest Act which are also relevant in the protection and management of earth heritage in Zambia. Besides the laws on heritage conservation and management, there are institutions that have an interest in geoheritage and these include the NHCC, the Geological Survey Department, the University of Zambia School of Mines, Water Resources Management Authority and National Museums Board. At global level, Zambia is a signatory to the 1972 UNESCO World Heritage Convention which

K. Mwamulowe (✉)
Director - East Central Region, National Heritage Conservation Commission, Box 320013, Lusaka, Zambia
e-mail: mwamsprog@gmail.com

I. A. Nyambe
Professor, Geology Department, School of Mines, University of Zambia, Box 32379, Lusaka, Zambia

Board Chairperson, National Heritage Conservation Commission, Lusaka, Zambia

R. B. Singh et al. (eds.), *Global Geographical Heritage, Geoparks and Geotourism*, Advances in Geographical and Environmental Sciences,
https://doi.org/10.1007/978-981-15-4956-4_20

protects sites (including geological/geomorphological sites of global significance) such as the Victoria/Mosi-oa-Tunya Falls as World Heritage Site. Besides declaring a few geological heritage sites as National Monuments, Zambia has recorded 219 geological heritage sites, 225 geomorphological sites, 742 rock art sites, some of these are painted caves; over 139 sites as hot and mineralized springs. Amongst the geomorphological sites, there are also sinkholes and waterfalls. Zambia has also recorded type localities and the largest open-pit mines in Africa. However, despite this potential, Zambia has no geoparks compared to South Africa and Tanzania which have. Further, the discussion on geotourism has been seldom absent in the Zambian tourism marketing scenario despite the rich geological diversity of the nation as well as setting tourism as one of the priority socio-economic sectors by Zambia's Seventh National Development Plan. The government through the NHCC has put in place deliberate measures of providing basic tourism infrastructure at most of these sites to improve on visitor experience at these sites. Entry fees are charged on visits to these heritage sites. Zambia also promotes palaeontological and archaeological research through research permits. Further, the government promotes sustainable tourism development at these geoheritage sites through tourism, public–private partnerships and community involvement policies. Any major developmental projects are subjected to environment/heritage impact assessments. The concepts of geoheritage, geopark and geotourism have not been applied but will indeed do so by fostering conservation and sustain the development of tourism in Zambia for socio-economic development.

Keywords Geoheritage · Geopark · Geotourism · Geodiversity · Sustainable tourism development

20.1 Introduction

Heritage conservation is recognized as a major milestone and indicator towards building sustainable cities and healthy communities. The realization by countries of the importance of conserving and managing heritage reached its pinnacle in 1972 when the World Heritage Convention (WHC) came into force. Since then, other Conventions such as the 1995 UNIDROIT Convention on Stolen or Illegally Exported Cultural Objects and the 2001 Convention on the Protection of under Water Cultural Heritage to mention just a few were subsequently formulated. Regional groupings such as the African Union have also taken cognizance of the need to conserve the environment and heritage. African countries currently collaborate with the African World Heritage Fund (AWHF) and the Centre for Heritage Development in Africa (CHDA) in funding and supporting the conservation of heritage in Africa.

Zambia became a party to the WHC in 1984. This was a demonstration of the realization of the importance and the need to conserve the country's heritage properties. Under the Convention, Zambia saw the need through Article 5(a) to formulate the NHCC Policy 2018 in order to meet a number of factors such as new developments in the sector both on the international and national level, challenges posed by globalization and rural–urban migration, developments in the area of good governance and community participation and the need to balance conservation and development in the context of sustainable development.

This paper describes the earth heritage related conservation, geoheritage concepts and their applicability in Zambia. It further examines the potential for their use in the light of sustainable development. In order to understand their usage in Zambia, the paper also highlights the policy, legal and institutional frameworks, which support the documentation, interpretation, conservation, management and sustainable utilization.

20.1.1 Location and Physical Description of Zambia

Zambia is a land-linked butterfly-shaped country, which is located in Southern Africa and forms part of the Southern African Development Community (SADC) and shares borders with 8 countries, namely, Malawi, Mozambique, Zimbabwe, Botswana, Namibia, Angola, the Democratic Republic of Congo and Tanzania. Zambia has approximately as an area of 752,614 km^2, and measuring up to a maximum length of 1,206 km East–West and a maximum width of 815 km North–South (https://www.nationsencyclopedia.com/Africa/Zambia). Zambia has 10 provinces with an estimated population of 17 million people and is economically very much copper–cobalt dependent owing to its rich ore deposits, which it shares with the Democratic Republic of Congo. The nation is renowned for its rich emeralds and deposits of manganese and amethyst. It has also coal, uranium as well as indications of the possible occurrence of gas and petroleum reserves.

Zambia also boasts of having the source and shares the largest portion of the fourth largest river in Africa. The Zambezi River with the mean annual discharge of 4,134 m^3/s or around 130 km^3/year at the outlet forms part of the Zambezi River Basin (International Bank for Reconstruction and Development 2010) constituting more than 40 per cent surface area of the waters of the 8 riparian states. Together with the Congo River Basin which forms the northern parts of the country makes Zambia to also boast of variant lakes ranging from open water, sunken, rift valley, deep and shallow freshwater lakes rivers with clear waters and abundant magnificent and very scenic waterfalls which arose from variations in landforms supporting rich plant and animal biodiversity.

Geologically (Fig. 20.1), Zambia is largely comprised of 6 geological units, which include the Basement Complex, Muva Supergroup, Katanga Supergroup, Karoo Supergroup and Kalahari Sediments and Alluviums (Thieme and Johnson 1974/75). Zambia forms part of the Central African Plateau, which manifests in various types of landform features such as escarpments, valleys and gorges. Some of the notable features include the Muchinga, Zambezi and Lake Tanganyika escarpments; the Barotse and Bangweulu Floodplains, the Kafue Flats, the Lukanga Swamps; the Batoka Gorge and Victoria Falls Gorge systems, the Lunsemfwa

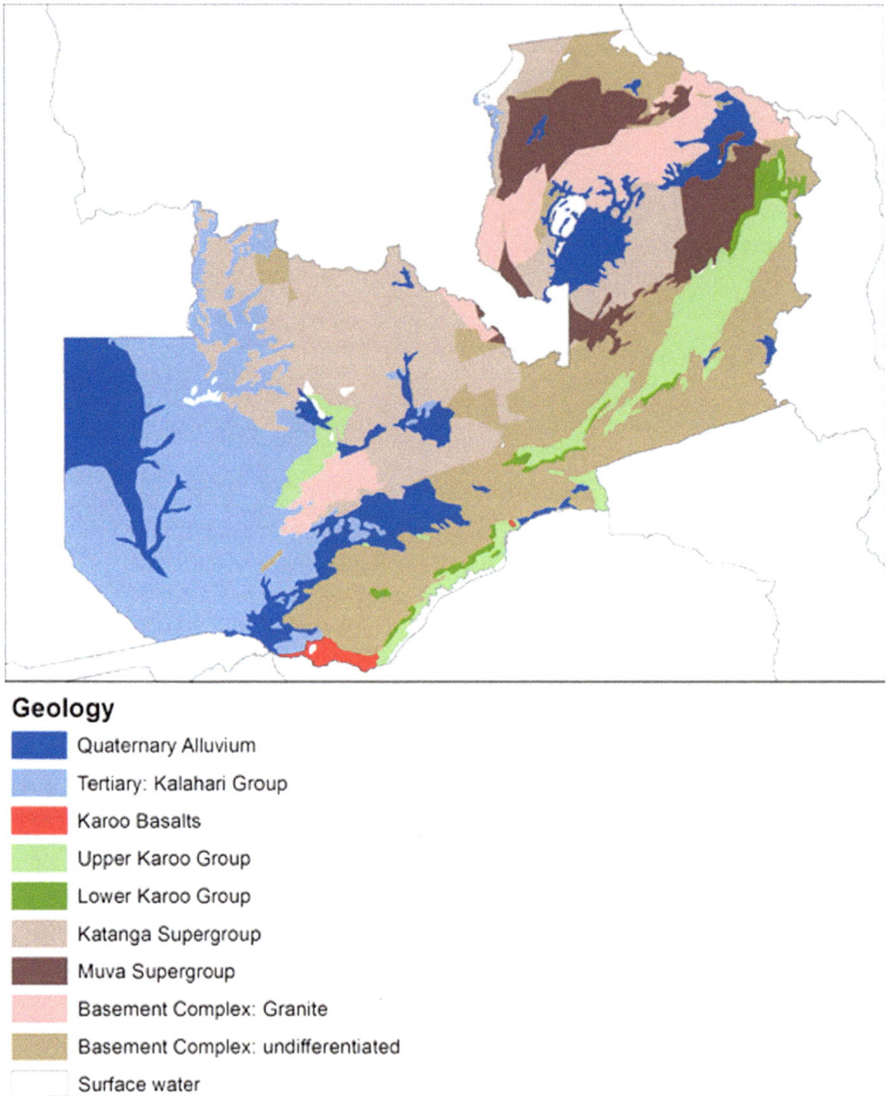

Fig. 20.1 Geology of Zambia at 1:5 million scale (Thieme and Johnson 1974/75)

Wonder Gorge, Kafue Gorge, Kabompo Gorge and the Kalambo Falls Gorge; the Luangwa, Luano, Lukusashi, Gwembe and Lower Zambezi valleys. Zambia also promotes palaeontological and archaeological research and minerals and exploration research.

20.1.2 Government Policyand NHCC Policy 2018

Taking cognizance of the fact that mining of copper is a wasting resource the Government of Zambia's policy has been to diversify its national economy by deliberately prioritizing tourism and agriculture to complement the gains in the mining industry. It is important to note that Zambia tourism despite having 36 National Parks managed under the Department of National Park, the future of its tourism industry lies in the heritage mostly earth heritage (geoheritage) related sites, which are managed by the National Heritage Conservation Commission (NHCC 2018a, b and c).

Despite Zambia prescribing to the WHC in 1984, importance attached to the conservation of heritage sites, NHCC has operated with no policy till 2018 with a vision "A prosperous Zambia that enjoys its diverse cultural and natural heritage on a sustainable basis" through the realization that

- The country's cultural and natural heritage are increasingly threatened with destruction not only by the natural causes of decay, but also by changing social and economic conditions;
- Most of the cultural and natural heritage are non-renewable and once lost cannot be regenerated, and that the deterioration or disappearance of any part of the cultural or natural heritage constitutes a destruction of the heritage of the country;
- The protection of the country's heritage at the national level often remains incomplete because of the scale of the resources required and insufficient economic, scientific and technological resources of the country;
- The existing International Conventions, recommendations and resolutions concerning the cultural and natural property to which Zambia is a party to, so as their cultural and natural heritage that are of outstanding universal value;
- The expansion and modernization of urban developments have resulted in heritage structures being altered demolished and replaced by modern structures;
- The natural sites, namely waterfalls have been greatly desired for investments in hydropower generation, which have the potential to negatively impact on the outstanding natural values of the site;
- The sustainable utilization of heritage sites demands that developments at heritage sites not only contribute to safeguarding the value of the site but also impacts positively on the livelihoods of the local communities; and
- Considering the magnitude and gravity of the new dangers threatening the heritage sites, it is incumbent on all Zambians to participate in the protection of the cultural and natural heritage of outstanding value, by the granting of collective assistance was possible.

The Policy is, therefore, pivotal in providing direction, dealing with these issues and galvanizing key stakeholders in the protection, conservation and utilization of the country's cultural and natural heritage in order to maintain or enhance the values of the sites, foster good governance, encourage stakeholders' participation for sustainable development. It is hoped that the NHCC Policy 2018 will contribute to Zambian people's vision of becoming "A Prosperous Middle Income Nation by 2030".

However, although Zambia has had a number of laws and policies designed to mitigate the negative impact of development on heritage conservation, protection of heritage resources still remains a challenge due to a number of factors such as lack of adequate financial and material resources.

20.1.3 Relevant Legal and Institutional Framework

This section describes the legal and institutional frameworks which relate to the documentation, conservation management and utilization of geoheritage (geological) resources in Zambia underpinned by relevant international conventions.

20.1.3.1 National Legal and Institutional Frameworks

The National Heritage Conservation Commission (NHCC) Act Cap 173, 23 of 1989/13 of 1994 of the laws of Zambia is mainly applicable in the field of heritage conservation in Zambia. The Act provides for the conservation of ancient, cultural and natural heritage, relics and other objects of aesthetic, historical, pre-historical, archaeological or scientific interest. It also provides for the regulation of archaeological excavations and export of relics (GRZ 1989). Thus, this piece of legislation regulates the protection, management, collection and export of Zambia's cultural and natural heritage including geoheritage sites, which are also used as geotourism sites to meet the demands of tourists. However, there are other pieces of legislation such as the National Museums Board Act and National Parks and Wildlife Act as well as National Forest Act, which are also relevant in the protection and management of earth heritage in Zambia. This geological heritage such as animal fossils as well as archaeological, anthropological and historical heritage is found in either National Parks or Game Management Areas (GMAs), which are managed under the National Parks and Wildlife Act.

In terms of institutional arrangements, the NHCC Act 1989 established and defined the functions and powers of the National Heritage Conservation Commission as an institution mandated to protect, manage and conserve the geoheritage resources of Zambia through the NHCC Act.

A number of other statutes related to Forest, Wildlife, Chiefs, War Graves and Memorials, National Archives, National Museums and National Arts have over time incorporated aspects of heritage management. Notable ones are the National Parks and Wildlife Management, which is the responsibility of the Zambia Wildlife Authority (ZAWA) established in 1998 under the Zambia Wildlife Act No. 12 of 1998, now operating as the Department of National Parks. Besides the laws on heritage conservation and management, there are other institutions that have an interest in geoheritage which include the Geological Survey Department, University of Zambia School of Mines' Geology Department, The Water Resources Management Authority, National Museums Board and Zambia Environmental Management

Authority (ZEMA). It is important to also note that museums are repositories of most of the archaeological material collected from the various heritage sites in Zambia include geoheritage sites.

20.1.3.2 Implementation of International Treaties, Protocols and Convention Versus Relevance of International Legal and Institutional Frameworks to Zambia

There are a number of International Treaties, Conventions and Protocols in the heritage sector, but Zambia has not ratified most of them. Zambia has ratified the World Heritage Convention of 1972 and the 1970 Convention on the means of Prohibiting and Preventing the Illicit Import and Export and Transfer of Ownership of Cultural Property but has not ratified the 1954 Hague Convention for the Protection of Cultural Property in the Events for Armed Conflicts, the 1995 UNIDROIT Convention on Stolen or Illegally Exported Cultural Objects and the 2001 Convention on the Protection of Under Water Cultural Heritage to mention just a few.

Some of these international instruments are a useful tool for enhancing the protection of Zambia's heritage cultural and natural heritage. The country has an obligation to implement the provisions of these International Treaties, Conventions and Protocols in the conservation of Zambia's cultural and natural heritage. However, most of these instruments have not been domesticated into national laws.

For those, Zambia has signed, which are related to heritage management, the country has an obligation to domesticate them. Hence, in the NHCC 2018 Policy, it has included an objective 10—"**To enhance the implementation of Protocols, Conventions, and other International Treaties that Zambia is party to in order to improve on compliance and domestication of these Agreements" with two policy m*easures***:

- Strengthen institutional capacity and accountability to undertake legal reviews in order to ensure compliance to, and domestication of, international treaties and agreements; and
- Strengthen collaboration with other stakeholders in order to ensure a coordinated approach in the domestication and implementation of international treaties and agreements.

For example, as a United Nations Member State, Zambia is a signatory to the United Nations **Convention concerning the Protection of World Cultural and Natural Heritage, Paris, France, 16 November 1972**. By Zambia appending its signature to this convention, it was able to meet at least one out of ten selection criteria for it to nominate the Victoria Falls (Fig. 20.2) jointly with Zimbabwe as an area of geoheritage of an outstanding universal value. Victoria Falls was designated so in 1989 by UNESCO as a World Heritage Site using the following criteria:

- *Criterion (vii)* to contain superlative natural phenomena or areas of exceptional natural beauty and aesthetic importance; and

Fig. 20.2 Mosio-a-Tunya/Victoria Falls Cataract and its zig-zag gorges at the Zambian–Zimbabwean Border

- Criterion (viii) to be outstanding examples representing major stages of earth's history, including the record of life, significant ongoing geological processes in the development of landforms, or significant geomorphic or physiographic features (UNESCO 1989).

Other areas that are being nominated because of their geoheritage features are the Barotse Flood Plain Cultural Landforms as a World Heritage Site and the Lower Zambezi National Park as a Biosphere Reserve.

20.2 Application of Policy, Legal and Institutional Framework

20.2.1 Policy and Legal Framework

In the geoheritage documentation and classification (Fig. 20.3), Zambia's Earth Heritage is strictly categorized into geological and geomorphological heritage. However, in line with the internationally accepted definition/s the Zambian case study excludes painted caves and rock shelters, rock art sites, hot and mineralized springs. Besides declaring a number of heritage sites as National Monuments, Zambia through the NHCC has recorded 219 geological heritage sites, 225 geomorphological

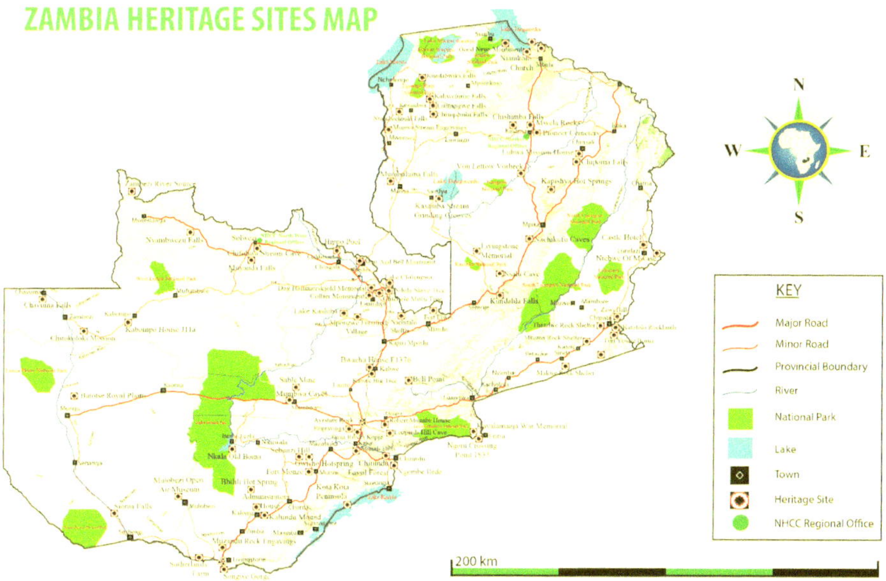

Fig. 20.3 Map of Heritage Sites in Zambia (NHCC 2018a)

sites, 742 rock art sites, some of these are overpainted caves, 200 over 139 sites as hot and mineralized springs (NHCC 2018a). Amongst the geomorphological sites, there are also sinkholes and waterfalls about 470 waterfalls (Quentine et al. 2014).

There are only a few geological sites which are declared as National Monuments. Zambia has also recorded type localities and some of the largest open pits in Africa.

The NHCC Act Cap 173 of the Laws of interprets heritage to include natural and cultural heritage. Though these concepts are not mentioned in the Act 1989, the specific mention of geological in the natural heritage is important in determining the applicability of the concepts of **Geoheritage conservation, Geopark and** Geotourism, which are covered in either or both categories of the major heritage typologies. Thus, there is a law though not explicitly that provides a platform for defining the three terminologies and guides the geoconservation processes. Further, the World Heritage Convention also provides an opportunity for Zambia to implement UNESCO Geoparks Programmes. The Tourism Masterplan of Zambia is recognizing the fact that the future of Zambia's Tourism lies in water-based heritage, which is largely dependent on the geoheritage conservation especially waterfalls all of which enjoy blanket protection by law as national heritage sites.

20.2.2 The Seventh National Development Plan (SNDP) 2017–2021 of Zambia

To attain the Vision 2030, Zambia is embarking on diversification and make economic growth inclusive (inclusive and diversified growth) (Ministry of National Development Planning 2017). This strategic objective addresses the negative consequences of Zambia's traditional adherence to copper as a dominant source of revenue and maize as the main staple food. Both policies have not sufficiently supported poverty reduction and hunger eradication in rural areas. The strategy is adopted in the remaining 15 years (2017–2030), divided into stages of diversification of agricultural products, development of agro-business and other manufacturing sectors and expansion of the service sector, ***in particular tourism*** hinged on the ***Development Outcome 3: a diversified tourism sector***.

In that chapter, Zambia itself as a country endowed with rich natural heritage and other tourism attractions such as waterfalls, lakes, rivers, gorges and wildlife, singling out Mosi-oa-Tunya Falls (Victoria Falls), one of the seven natural wonders of the world, as the largest tourist attraction site. However, it is clear from the statement from the SNDP, 2017–21 that waterfalls, gorges and lakes are non-geoheritage, which as Zambians is something we have to work on for geotourism is to succeed in Zambia. However, it is encouraging that in the SNDP, Zambia will embark on a number of strategies including

- Strategy 1: Promote tourism-related infrastructure such as develop, upgrade and rehabilitate roads, viewing loops and airstrips to and within the major national parks, namely Kafue, Lower Zambezi and the Mosi-oa-Tunya to efficiently interlink the major destinations in the southern tourism circuit and South Luangwa to prolong the tourism season from 7 to 12 months per annum through the following programmes: (a) Southern tourism circuit linkage development; (b) Northern tourism circuit linkage development; (c) South Luangwa National Park accessibility development; (d) Road and local air travel infrastructure and service development; and (e) Arts and cultural infrastructure Development;
- Strategy 2: Promote diversification of tourism products such as development and expansion of non-traditional modes of tourism, such as eco- and agro-tourism and cultural and community-based tourism under programmes: (a) Tourism products promotion; (b) Theme parks, accommodation and conference facilities development promotion; (c) Music and arts festivals development promotion; (d) Adrenaline and adventure tourism promotion; and (e) Tourism investment promotion;
- Strategy 3: Promote tourism source market segmentation focusing on growing the traditional source markets by increasing market penetration and exploiting emerging markets to expand the tourist base through a programme on "tourism promotion and marketing"; and
- Strategy 4: Promote domestic tourism to allow local citizens pay a lower rate in facilities such as accommodation during off-pick seasons through programmes:

(a) Two-tier tourism product pricing system development; and (b) Domestic tourism campaigns (Ministry of National Development Planning 2017).

This aspiration by Government should give NHCC an impetus to work extra hard so that geotourism sites are included and not left out. There is, therefore, a need for Zambia to package and classify, protect, interpret and present geoheritage sites of Zambia to the world following the works of Quentine, 2014 which provided a good starting point for Zambia. It is saddening that despite the NHCC Act being in place since 1989, Zambia has not classified earth science or heritage conservation in detail as outlined by Hose (2016). Thomas A. Hose, 2016, uses 3 major categories of Earth Science Conservation Classification. These are ***Exposure or Extensive Sites, Integrity Sites and Finite Sites all of which are divided into 8, 4 and 4,*** respectively. Zambia, on the other hand, has simply classified its earth heritage system into two broad categories, namely geological and geomorphological heritage with no coding. There is need for Zambia to come up with an appropriate detailed classification and coding system for recording and documentation purposes so that geoheritage sites could benefit from the programmes and strategies in SNDP 2017–2021.

20.2.3 Geoheritage Protection and Conservation

Out of the 90 sites declared as National Monuments, there are only 13 sites which are considered as of geoheritage. However, Table 20.1 shows an example of various national monuments, which touch on the sites that have some bearing on geoheritage though in some cases, they are far-fetched on the subject matter.

From Table 20.1 above, it is clear that the list of national monuments has included sites such as Kopje No. 2 at latitude 14°27′45″S, 28°26′27″E in Kabwe Central Province of Zambia which is the type locality of a rare mineral tarbuttite discovered in 1907 and Chinyunyu Hot Spring, Longola Hot Springs, Chipota Falls was recently declared as a national monument. However it has still excluded some sites such as type locality for Isokaite at Nkumbwa Hill Carbonatite (Nkombwa Hill Carbonatite), Isoka District, Muchinga Mountains, Muchinga Province, at latitude 10° 10′ 00″S and 32° 51′ 00°E. In addition, there is also Nchanga Open Pit, which was at some point the largest Open Pit Copper Mine in Africa.

20.2.4 Applicability Matrix of the Geoheritage Sites

Zambia's hot springs (Table 20.2 and examples in Fig. 20.5) are, for instance, found along fault lines at a number of localities in the northern and southern parts of the Kafue National Park and at sites close to the Park where Karoo Supergroup rocks meet older rocks. The most impressive of these are the Chibemba saltpan between Lufupa and Moshi and the Chipushi spring near Lubungu Pontoon. Chipushi hot springs

Table 20.1 Showing the various geoheritage-connected national monuments in Zambia (NHCC 2018a)

	Name of national monument	Legal significance	Other key accepted significance	Key attribute	Recommended typology	State of conservation presentation and interpretation
1	The Zambezi Source	Political	River Source of the fourth largest	Source of Country Name (Zambia)	Historical Geomorphic Botanical	Excellent
2	Victoria Falls	Largest Cataract in the World (WHS) Eastern cataract Archaeological	Outstanding Natural Beauty	Largest Cataract at 1,708 m Zig-zag gorge formation (Active and Static) Rainbows *Cataractagenitus* Transboundary Gorges are habitat of *Taita falcon*	Archaeological Anthropological Botanical (IBA)	Excellent
3	Chishimba Falls	Scenic beauty	Series of Waterfalls	3 Cascading and Cataract Waterfalls Combination falls	Geomorphic	Excellent
4	Chipoma Falls	Scenic beauty	A set of waterfalls and rapids	2 Cascading falls Numerous smelting Kilns	Geomorphic Archaeological Anthropological	Good
5	Kalambo Falls	Scenic beauty second highest Waterfall in Africa	Nest places of Marabou Stork	Plunge/Cascades Combination falls The first discovery of ancient fire making technology in Africa was made here Transboundary National Forest No 1	Geomorphic Archaeological Anthropological	Excellent

(continued)

Table 20.1 (continued)

	Name of national monument	Legal significance	Other key accepted significance	Key attribute	Recommended typology	State of conservation presentation and interpretation
6	Kundalila Falls	Scenic beauty owing to its variety of wild flowers	Mistry about the site name Cooing doves which have since disappeared	Inclined Meta-quartzite	Geomorphic Archaeological Ecological (Botanical) Anthropological	Good
7	Lumangwe Falls/Kabwelume Falls Chimpempe Falls Complex (Boundary Extension ongoing)	Scenic beauty	Miniature Victoria Falls	Series of Cascades	Geomorphic Ecological (Botanical) Anthropological	Excellent
8	Ntumbachushi Falls	Scenic beauty	Cultural	Parallel falls More than 4 waterfalls Escarpment and Plain Ancient Rock Art on whitish meta-quartzites	Geomorphic Archaeological Ecological (Botanical) Anthropological	Very Good
9	Barotse Flood Plains Cultural Landscape	Flood Plain Manifestation of Traditional Management System Ancient Civilization	Cultural	Interaction between nature and culture Cultural Landscape	Geomorphic Historical Anthropological Ecological	Good
10	Lunsemfwa Wonder Gorge	Scenic beauty V-shaped	Geological	Taita falcon breeding sites	Geomorphic Historical Anthropological Ecological (IBA)	Excellent
11	Lake Chilengwa	Sunken Lake	Geological	Underground limestone lake system	Geological	Very good

(continued)

Table 20.1 (continued)

	Name of national monument	Legal significance	Other key accepted significance	Key attribute	Recommended typology	State of conservation presentation and interpretation
12	Lake Kashiba	Sunken Lake	Geological	Underground limestone lake system	Geological	Very good
13	Chirundu Fossil Forest	Plant Fossils	Palaeontological	Trunks of Fossilized wood	Geological (Palaeontological)	Poor
14	Rockland Farm (Katolola) Rock Shelter	Rock Paintings	Geological	Granitic Rock Outcrop	Archaeological	Good
15	Thandwe Rock Shelter	Rock Paintings	Geological	Felsic Granitic Rock Shelter	Archaeological	Good
16	Mkoma Rock Shelter	Rock Paintings	Geological	Felsic Granitic Rock Shelter	Archaeological	Good
17	Zawi Hills	Rock Paintings	Geological	Felsic Granitic Rock Shelter	Archaeological	Poor
18	Kifubwa Rock Engravings	Rock Engravings	Geological	Rock Shelter	Archaeological	Good (Graffiti)
19	Gwisho Hot springs	Archaeological	Geological	Hot spring Ancient human occupation	Archaeological Geological	Good
20	Kalemba Rock Art	Rock Paintings	Geological	Felsic Granitic Rock Shelter	Archaeological	Good
21	Makwe Rock Shelter	Rock Paintings	Geological	Rock Shelters Miniature Caves	Archaeological	Good

(continued)

Table 20.1 (continued)

	Name of national monument	Legal significance	Other key accepted significance	Key attribute	Recommended typology	State of conservation presentation and interpretation
22	Good News (on shores of Lake Tanganyika	Historic	Geological	Extension of Great Rift Valley System High Fish Endemism	Historical Geological (Rift Valley) Ecological	Poor
23	Collier Monument Roan Antelope Copper Mine	commemorate the original copper claim by W. C Collier in 1902	Mineralogical	Mining history	Geological	Very Good
24	Moir and Bell Monument: Mufulira Copper Mines	Discover of Copper deposits	Mineralogical	Mining history	Geological	Very Good
25	Mwela Rock Art	Rock Paintings	Geological	Rock Shelters	Archaeological	Poor (Threatened)
26	Mumbwa Caves	Rich Archaeological Site		Ancient human occupation	Archaeological	Poor
27	Leopards Hill Caves	Ancient human occupation	Archaeological	Bat-life (rare bat disease discovery site)	Archaeological Speleological Geological Ecological	Poor
28	Libala Limestone	Karstic	Geological	Static Geomorphological	Geological Geomorphological	Poor

(continued)

Table 20.1 (continued)

	Name of national monument	Legal significance	Other key accepted significance	Key attribute	Recommended typology	State of conservation presentation and interpretation
29	Kota Kota	Geomorphological	Peninsula		Geomorphological	Very Good
30	Chinyunyu Hot Spring	Geological	Geological	Geophysical/Geohydrological	Geophysical/Geohydrological	Good
31	Longola Hot Springs	Geological	Geological	Geophysical/Geohydrological	Geophysical/Geohydrological	Excellent
32	Chipota Falls	Waterfalls	Geomorphological	Active Geormorphological	Geomorphological	Excellent
33	Tarbuttite Discovery	Type Locality Site	Mineralogical	Discovery of Mineral	Geological	Very poor

(a) (b)

Fig. 20.4 **a** Tarbuttite Discovery Site in Kabwe, Central Province and **b** Nchanga Open Pit in Chingola, Copperbelt Province, Zambia

Fig. 20.5 Kabilubilu Hot springs (top left) Msaope Hot Springs (top right) in Mambwe and Longola Hot Springs in Itezhi itezhi

Table 20.2 Showing some of the hot springs in Zambia which could be recorded as geoheritage sites

	Geoheritage			Geopark none	Geotourism level	Conservation status
	Typology	Level representation	No			
Luapula	Waterfalls	Very High		None	Very High	Very Good
	Shallow Lakes	Highest		None		Largest lake in Zambia
	Hot springs	Low	5	None	Low	Good
Northern	Waterfalls	Very High		None	Very High	Highest No. of Waterfalls
	World's second deepest Lake	Low		None		World Record Biodiversity Depth
	Rock Art on Quartzite	Very		None		Threatened
	Hot springs	Very high	17			Good
Muchinga	Vertebrate and Plant Fossils	Very High		None		
	Waterfalls	Low				
Eastern	Rock Art on Granitic Felsic Batholiths	High		None		Good
	Hot springs	Very high	33	None		Good
Lusaka	Karsts	Very high	8	None		Destroyed
	Caves	Very high		None		Destroyed
	Archaeology			None		Destroyed
Central	Rock Art/Shelter Gorge	Very high		None		Good
	Caves	Very high		None		Fair some with fossil record in Breccia
	Hot springs	Very High	19			
	Archaeological record					Homo Rhodensiesis discovery site
Southern	Hot springs	Very High	18	None	Very High	
	Waterfalls	Low				The largest curtain of falling waters
Western	Flood plain	Very High		None		

(continued)

Table 20.2 (continued)

	Geoheritage			Geopark none	Geotourism level	Conservation status
	Typology	Level representation	No			
North Western	Hot springs	Low	5	None		
Copperbelt	Mines	Many		None		World largest open-pit mine
	Hot springs		4			
	Sunken lakes	Many		None		

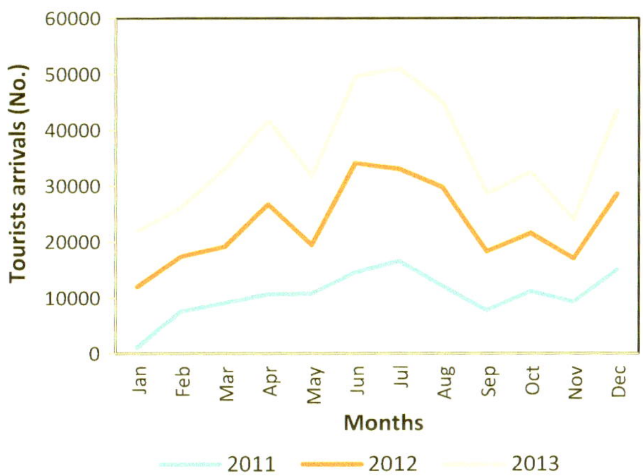

Fig. 20.6 Tourists statistics from 2011 to 2013 at the Victoria Falls World Heritage Site for both Zambia and Zimbabwe (Zambia–Zimbabwe 2014)

are believed to be the most spectacular in Zambia. Longola hot springs are only the third locality in the world with records of *Onkonemacamptum* and *Synechococcus sulphuricus*, forms of blue-green algae, after they were recorded in Java and Israel, respectively. The Longola hot springs water temperature reaches up to 87 °C, and has some thermophilic Cyanophyta (*Synechococcus lividus*) flourishing at 75 °C (Obrdlik 1988).

20.3 Geoheritage Marketing

The Zambia Tourism Agency has traditionally been marketing the Victoria Falls (Fig. 20.2) as its flagship and part of the tourism brand. This waterfall brand is also true to the situation on the ground considering the 470 waterfalls recorded and documented in Quentine et al. (2014) as well as however through the National Heritage Conservation Commission. The visitor statistics at this Geoheritage Conservation

Table 20.3 2014–2015 visitor statistics for both Zambia and Zimbabwe for the Victoria Falls World Heritage Site (State of Conservation Report Zambia–Zimbabwe)

Year	2014	2015	Up to 31 October 2016
Total	353,025	544,104	80,059 (Zambian side)

Table 20.4 2016–2018 Visitor statistics for both Zambia/Zimbabwe at the Mosi-oa-Tunya/Victoria Falls World Heritage Site (Zambia–Zimbabwe 2018)

Year	2016	2017	2018
Total	417,372	467,408	463,578

Site are illustrated here to indicate how important these sites are to the overall tourism in Zambia:

(i) **The Victoria Falls** as the World Heritage Site shows interesting statistics of tourists visiting both Zambia and Zimbabwean sides (Fig. 20.6, Tables 20.2 and 20.3).

Further, there has been an increase in tourist numbers between 2016 and 2017. A total of 417,372 and 467,408 tourists visited the site in 2016 and 2017, respectively, representing a 12% increase. The increase in the numbers could be attributed to marketing efforts and tourism product development within the destination. The captured statistics for 2018 are from January to October. NHCC has a deliberate marketing strategy to increase tourism at Victoria Falls. This marketing is not only confined to the Victoria Falls but all other heritage sites and mostly geoheritage conservation sites.

(ii) **The Northern Circuit** (Table 20.4 and corresponding graph). Table 20.4 and corresponding graph show statistics relating to sites which are depicted in the Map of Heritage Sites in the three provinces of Northern Zambia most of which are waterfalls (geohydrological) sites.

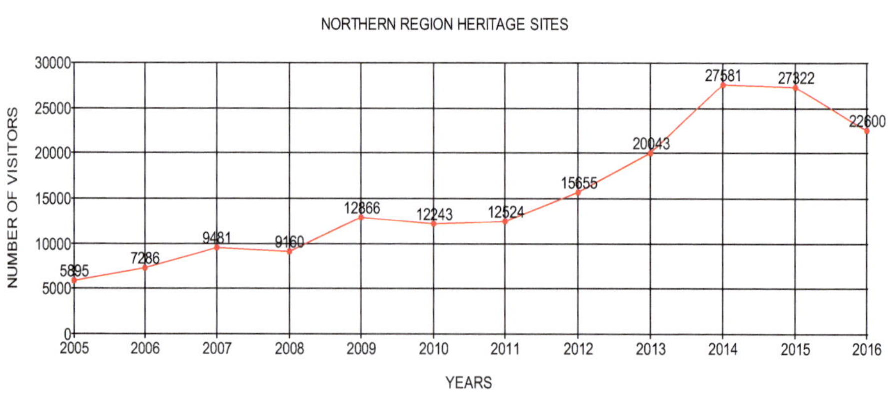

Table 20.5 A table of visitor statistics relating to Zambia's Northern Circuit (**NHCC Annual Reports 2005–2016**)

	2005	2006	2007	2008	2009	2010	2011	2012	2013	2014	2015	2016
Chilambwe Falls	00	00	00	00	00	146	134	110	156	338	250	261
Chipoma Falls*	00	00	00	00	173	220	167	173	531	1320	1053	280
Chishimba Falls	3450	4249	5490	4985	6512	6073	6591	10,273	12,985	15,552	15,866	14,757
Kalambo Falls	482	693	1191	610	741	674	593	414	620	1111	1361	916
Kundabwika Falls**	00	00	00	00	00	29	05	06	00	00	00	00
Lumangwe Falls	442	432	741	763	674	777	660	660	823	834	408	674
Lufubu Falls	00	00	00	00	00	00	00	477	468	404	863	579
Mumbuluma Falls	00	00	00	00	898	794	1067	756	945	1497	1101	834
Mwela Rock Art	255	821	634	739	1074	883	1057	907	1185	1790	1712	1367
Nachikufu Cave	329	429	416	379	399	530	382	372	372	375	239	251
Ntumbachushi Falls	937	574	945	1519	2350	2,088	1839	1477	1958	4340	4465	2673
Von Lettow-Vorbeck	00	00	64	74	45	29	29	00	00	20	04	06
Total	5895	7286	9481	9160	12,866	12,243	12,524	15,655	20,043	27,581	**27,322**	**22,600**

*Data not received for some months

**No Care Taker for the period 2005–2016. A new Care Taker has since been deplored (2017)

Fig. 20.7 The various scenic views of Lumagwe Falls, a pristine geoheritage conservation site, Northern circuit, Zambia

The statistics from Northern Region of the National Heritage Conservation Commission are indicative of an increase in visitors to various heritage sites in the Northern Circuit of Zambia. Most of these sites are waterfalls (Fig. 20.7) or rock art sites (Fig. 20.8) which fall within the category of geoheritage.

20.4 Non-implementation Geopark Concept in Zambia

Geopark is a relatively new concept in Zambia in a similar line as geoheritage. In the NHCC Act of 1989, both terms are not used and the only term that appears is "geological" under the definition of natural heritage. The concept Geopark has only surfaced in NHCC National Policy of 2018. No wonder why Zambia has no national geopark whilst other countries such as South Africa and Tanzania have despite that a Zambian Committee for this purpose was established at one point. However, Zambia has a number of opportunities to come up with a Geopark. A candidate for a good example as a Geopark is the Victoria Falls as a joint venture with Zimbabwe.

(a)

(b)

Fig. 20.8 **a** Rock art and shelter atThandwe National Monument depicting two different types of media used and **b** Mwela Rock Art of Kasama, Northern Province, Zambia

20.5 Sustainable Development Agenda

The discussion on geotourism has been seldom absent in the Zambian Tourism arena despite Zambia being richly endowed with geoheritage and that tourism is placed as one of three key priority socio-economic sectors. This is due to the fact that the geological significance which is behind the tourist sites in Zambia is not known by the people marketing these sites. What is being marketed, for example, for the Victoria Falls is the water that is falling (visible appearance) without taking into consideration the geological significance behind the water. The lack of interpretive and directional signage to places of geological significance in Zambia has resulted in little appreciation of these sites with their subsequent neglect.

With the emergence of these relatively new concepts of geoheritage, geotourism, and geoparks, Zambia has an opportunity to promote geotourism destinations owing to its geological diversity. The Government through the NHCC has put in place deliberate measures of providing basic tourism infrastructure at most of these sites to improve on visitor experience and has embarked on a programme of heritage concessions with the private sector.

The upkeep and sustainable management of these sites are dependent on entry fees by visitors to the heritage sites. There is, therefore, a need for NHCC to organize as a national geoconservation lobby group comparable to what South Africa did as the ongoing implementation of the Tourism Development Fund and the presence of the Zambia Tourism Agency alone will not offer hope for developing, managing, and effectively marketing these geosites. That lobby group or conference should identify geoheritage sites to be considered as geotourism and or geopark sites through the involvement of local communities at an early planning stage of developing the geopark proposals.

In order to succeed, geotourism should be promoted in Zambia aiming at

- having very good stories that will inspire people;
- using simple, familiar and appropriate language to a specific group and ensure easy access to information at all times;
- engaging specialists not only geologist but also people with good experience in information dissemination and requirements of the audience;
- establishing partnerships with the local community, existing organizations and networks;
- linking geoheritage with cultural heritage of the particular area; and
- using geoheritage to inspire art and encourage people to see their surroundings from a different perspective.

In summary, in order to promote sustainable development of geoheritage sites, the Government of Zambia through NHCC is encouraging public–private partnerships, community involvement partnerships and networking. Therefore, every site undergoes Environmental and Heritage Impact Assessments prior to implementation as a geoheritage site in Zambia.

20.6 Way Forward (Conclusion)

In conclusion, Zambia is very rich in geodiversity giving birth to numerous geoheritage sites. Whilst conservation of geological heritage is very practical in Zambia through the National Heritage Conservation Commission Act, the concepts of geopark and geotourism are uncommon. However, there is NHCC 2018 policy now that would enable the implementation of the concepts of geoheritage, geopark and geotourism which are severally but partially applied in Zambia.

References

Appleton JD, Bland DJ, Nancarrow PH, Styles MZ, Mambwe SH, Zambezi P (1992) The occurrence of daqingshanite-(Ce) in the Nkombwa Hill carbonatite, Zambia. Mineral Mag-Miner Mag 56:419–422. https://doi.org/10.1180/minmag.1992.056.384.16

GRZ (1989) The national heritage conservation commission act (Chap. 173). Laws of Zambia, Government Printers, Republic of Zambia, Lusaka

Hose TA (2016) Geoheritage and geotourism a European perspective (Hose TA (ed)) The International Centre for Cultural and Heritage Studies, Woodbridge UK

https://www.nationsencyclopedia.com/Africa/Zambia

https://www.thejourneytourism.com/case-studies/strategy-and-planning/zambia-tourism-master-plan/

International Bank for Reconstruction and Development (2010) The Zambezi River Basin: a multi-sector investment opportunities analysis, summary report, vol 1. Word Bank Water Resources Management Africa Region, Washington DC, pp 3–4

Mineral Data Publishing (2001–2005) Tarbuttite version 1. Handbook of mineralogy

Ministry of National Development Planning (2017) Seventh National Development Plan (SNDP) 2017–2021, Lusaka

NHCC (2005–2016) National Heritage Conservation Commission Annual Reports, Lusaka

NHCC (2018a) The National Heritage Register, Lusaka.

NHCC (2018b) The National Heritage Policy, Lusaka

NHCC (2018c) The National Heritage Conservation Commission Strategic Plan 2018–2021, Lusaka

Obrdlik P (1988) The Longola hot springs of Zambia: the need for conservation. Biol Conserv 43(2):81–86. Elsevier, Germany

Quentine A, Mwanza I, Mandandi M, Mututkwa TK (2014) Guide to little known waterfalls of Zambia, vol II. Horizon Printers, Lusaka

Thieme JG, Johnson RL (1974/75) Geological map of the Republic of Zambia. Geological Survey Department, Zambia

UNESCO (1989) Mosi-oa-tunya/Victoria Falls World Heritage Nomination - IUCN Summary (Evaluation) No. 509-Zambia-Zimbabwe. Paris

Zambia–Zimbabwe (2014) The Mosi-oa-tunya/Victoria Falls World Heritage Site 2011–2013 State of Conservation Report Livingstone Victoria Falls

Zambia–Zimbabwe (2018) The Mosi-oa-tunya/Victoria Falls World Heritage Site 2016–2018 State of Conservation Report Livingstone Victoria Falls

Chapter 21
Highlight of Geotouristic Values of a Volcanic Landform on the Mount Manengouba Eastern Slopes: Case of Djeu-Seh Basin

Ghislain Zangmo Tefogoum, Armand Kagou Dongmo, David Guimolaire Nkouathio, and Merlin Gountié Dedzo

Abstract Mount Manengouba (MM) is one of the most voluminous volcanic apparatus along the Cameroon Volcanic Line. It is made up of two nested calderas notably, Eboga and Elengoum. Numerous landforms are found on the calderas floor, rim and external slopes. Among them, there is the Djeu-Seh Basin (DSB). It is located on the Eastern slopes of MM, in the Mbouroukou Village. It is made up of volcanic rock, namely pyroclastic ejectas and basalts. DSB is funnel-like shaped with internal slopes very steeper in certain parts. This landform has been assessed through (Reynard et al., Geoheritage 8:43–60, 2016) methodology. The scientific value of the DSB is high (0.88), because it is well preserved (1), rare (1), and representative (1) of the whole geomorphology of the eastern slopes of MM. It also plays an important role in the geographical history (0.5) of MM. DSB has an average ecological value due to the fact that it is not protected. However, the ecological influence remains quite high because it is important for biological diversity. The aesthetic value is high because the relief structuring and it is lengthened in a southwest–northeast direction. In addition, the color contrast embellishes the landscape of DSB. The cultural value of the site is insignificant. The site is characterized by a quiet environment and good accessibility. This landform is unique in the MM and constitutes the natural laboratory for outdoor training and is the main attraction for visitors who are still insignificant nowadays. It appears that the geotouristic values of DSB are underexploited.

Keywords Geomorphosite · Geotourism · Djeu-Seh Basin · Mount Manengouba

G. Zangmo Tefogoum (✉) · A. Kagou Dongmo · D. G. Nkouathio · M. Gountié Dedzo
Department of Earth Sciences, University of Maroua, P.O. Box 814, Maroua, Cameroon
e-mail: zangmotefogoum@gmail.com

R. B. Singh et al. (eds.), *Global Geographical Heritage, Geoparks and Geotourism*, Advances in Geographical and Environmental Sciences,
https://doi.org/10.1007/978-981-15-4956-4_21

21.1 Introduction

21.1.1 Geological Context

Since several tens of millions of years, volcanic, plutonic and tectonic phenomena occurred in Cameroon along an N30°E axis called Cameroon Volcanic Line (CVL) gives rise to the construction of volcanoes and plutonic complexes in the oceanic and continental branches (Tchoua 1974; Lee et al. 1994; Wandji 1995; Aka et al. 2004; Kagou Dongmo et al. 2010). Mount Manengouba is one of the huge volcanoes along the continental branch of CVL (Fig. 21.1). It is set up by a volcano-tectonic horst

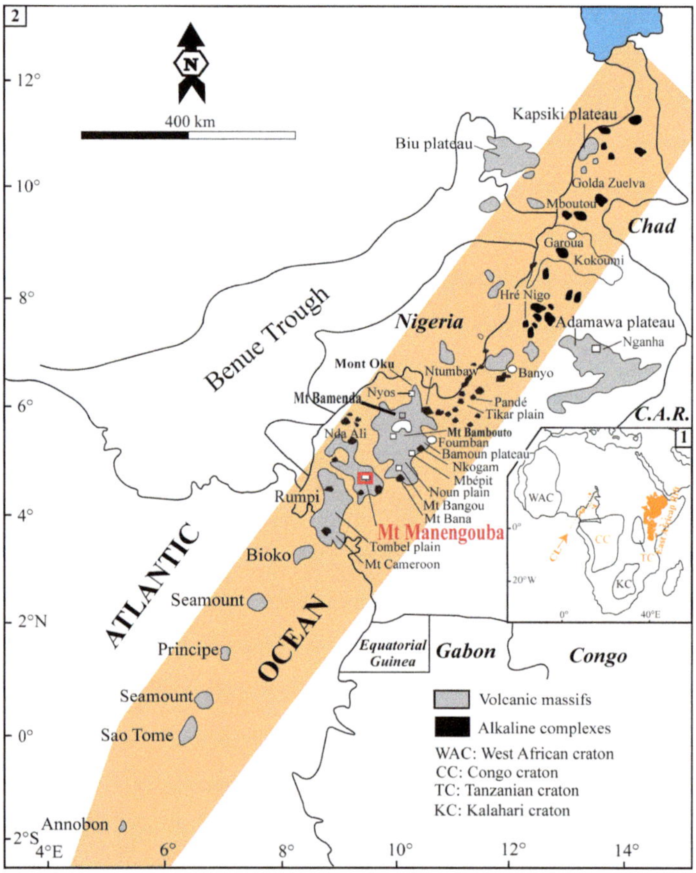

Fig. 21.1 (1) Cameroon Line (CL) in Africa; the main geologic features of Africa are indicated. (2) Map of Cameroon showing the distribution of the Cameroon Volcanic Line magmatism including Mount Manengouba. Locations of seamounts are after Burke (2001)

trending N40°E and crosscut by N40°E, N130°E and N0° faults (Kagou Dongmo et al. 1998, 2001).

The volcano was built on a 800 m uplifted granite-gneiss substratum. It resulted from successive building from 1.55 Ma to recent time (Kagou Dongmo et al. 2001; Kagou Dongmo 2006), of two volcanoes, Elengoum and Eboga. The Elengoum was made of piled basaltic to trachytic lavas and pyroclastites. Its upper part sunk, between 0.7 and 0.56 Ma, in a more than 6–7 km large caldera. The western margin of this old caldera was strongly eroded. The Eboga erected inside the former caldera as a basaltic to Hawaii lava cone that also collapsed to form the second caldera that measures 3–4 km in diameter (Kagou Dongmo 2006; Zangmo Tefogoum 2016). Numerous post-caldera manifestations occurred in the whole volcano. They gave rise to the formation of a range of volcanic landforms that include Crater Lakes, Domes, Broken Cones, and Basins. These landforms are mainly constituted by well-defined geomorphosites among which a maar called Djeu-Seh Basin (DSB) that has been chosen for the present work. DSB (Fig. 21.2) is located on the Eastern slopes of Elengoum Volcano, in the Mbouroukou county, precisely at 09°53′05′′E

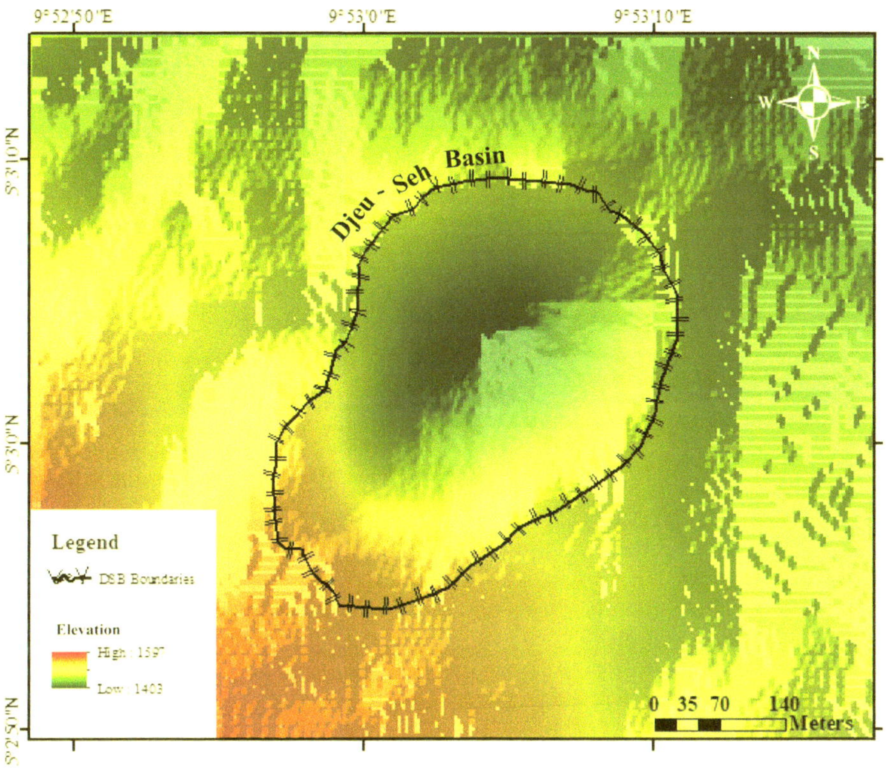

Fig. 21.2 DEM of the Djeu-Seh Basin

and 05°03′03′'N (Zangmo Tefogoum et al. 2014, 2015a). It constitutes one of the main interesting geomorphosites of Mount Manengouba.

The concept of geomorphosite results from a dozen years of study conducted by geologists and geomorphologists on the environment. In fact, during the past two decades other terms have been used and can be considered as synonymous (Coratza and Hobléa 2018): *geomorphological assets* (Panizza and Piacente 1993; Quaranta 1993), *geomorphological goods* (Carton et al. 1994), *geomorphological sites* (Hooke 1994), *geomorphological geotopes* (Grandgirard 1995), sites of geomorphological interest (Rivas et al. 1997). Finally, Panizza (2001) proposed the term "***geomorphosites***" that is the most widely used in the literature for qualifying landforms that make up geomorphological heritage. According to the author, geomorphosites are geomorphological landforms that have acquired a scientific, cultural/historical, aesthetic, and/or social/economic value due to human perception or exploitation. Thus, geomorphosites addresses specific issues concerning geomorphological heritage (Reynard et al. 2009). Geomorphosites, witnesses of the geodiversity and the work of the time, testify to the long history of the construction of the landscape (Brancucci et Gazzola 2002; Ilies and Josan 2009). They are natural forms of relief because their construction is done without the intervention of the Man (Portal 2010). This natural character is the main support of geomorphosites values. Hence, geomorphosites have three main characteristics that make them unique and distinctive types of geoheritage: the aesthetic dimension, the dynamic dimension, and the imbrication of scales (Reynard 2009).

Numerous studies focus on volcanic geomorphosites have been carried out in the past years (Erfurt-Cooper and Cooper 2010; Závada et al. 2010; Gao et al. 2013; Moufti and Nemeth 2013; Wang et al. 2014; Erfurt-Cooper 2010, 2014; Field and Newsome 2014; Różycka and Migon 2014; Moufti et al. 2015; Zangmo Tefogoum et al. 2014, 2015a, 2017; Fepuleai et al. 2017; Németh et al. 2017; Asfawossen 2018; Raška et al. 2018; Pérez-Umaña et al. 2018; Khalaf et al. 2018). The assessment of gomorphosites constituted the main issues of geological and geomorphological heritage experts. Indeed, during the past decade, many authors have developed different assessment methods for geomorphosites that have proved to be useful in several analyses: Bruschi and Cendrero (2005), Pralong and Reynard (2005), Serrano and Gonzalez Trueba (2005), Pereira et al. (2007), Zouros (2007), Feuillet and Sourp (2011), Cocean (2011), Coratza et al. (2012), Brilha (2015), Reynard et al. (2007, 2016). Some of these methods have been utilized for the assessment of volcanic geomorphosites abroad; in order to promote the geotouristic activities in a given geological field (Cedro et al. 2009; Migoń and Pijet-Migoń 2010; Costa 2011; Różycka and Migoń 2014). According to Newsome and Dowling (2010), geotourism is defined as a form of tourism practised on natural spaces which specifically focusses on landscape and geology and their promotion, interpretation, and conservation through appreciation and education.

In Cameroon, few works, have been focused on the assessment of volcanic geomorphosites for geotouristic issues (Zangmo Tefogoum et al. 2017, 2018). DSB is the volcanic geomorphosite that need some specific studies. In this work, we present some geological features of the DSB. Moreover, the DSB will be assessed in view of raising the level of awareness of the local authorities and population about his assets for geotouristic activities.

21.2 Methodology

The methodology of the present work has been defined about the main issues of this paper notably, the geological features and the assessment of the DSB.

For both aspects, a bibliographic reviewing has been done through previous geological studies carried out on the whole Manengouba Volcano in general and on the DSB in particular. In addition, bibliographic studies have also been made on the previous work on geoheritage and geomorphosites assessment in a Cameroon and abroad. This allowed to identify the deficiencies in these works and define the thematic of the present study. The bibliographic study was followed by a preliminary cartographic study in order to locate the study area and, to highlight its geomorphological features and that of the surrounding terrains (eastern slopes of the Elengoum Volcano). It has been achieved through the use of the topographic map, the aerial photos and satellite images. Moreover, several field studies have been made. They were followed by laboratory works which marked the final stage of this work.

21.2.1 Field Studies

In the field, rock outcrops, structures and samples have been described. The geographical coordinates of rock samples have been obtained via the Global Positioning System (GPS). Moreover, field works allowed to acquire a good knowledge of the geotouristic assets of the DSB and to perform its complete assessment. The method adopted for the assessment of DSB is that of Reynard et al. (2016). This method has been used for the assessment of geomorphosites in some countries (Zangmo Tefogoum et al. 2017, 2018; Mauerhofer et al. 2017; Boukhchim et al. 2017; Clivaz and Reynard 2018). It includes the documentation, the assessment of intrinsic value and the use and management characteristics of the site (Fig. 21.3).

21.2.2 Laboratory Studies

In the laboratory, the geographical coordinates of the different sampling points of rocks have been plotted on the topographic map of the study area. Through these different points and the description of the samples, a geological map and a cross section were realized. In addition, the maps and some parameters of the

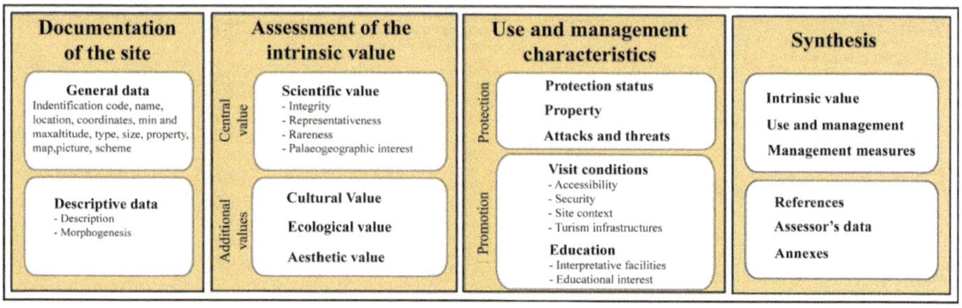

Fig. 21.3 Methodology of Assessment of geomorphosites (*From* Reynard et al. 2016)

DSB have been made/determined using ArcGIS, Google Earth Pro, Adobe Illustrator CS4, aerial photographs, satellite images coupled with data obtained from field observations.

21.3 Results

21.3.1 Geological Features of the Djeu-Seh Basin

21.3.1.1 Morphogenesis

The formation of DSB has been ruled by one of the adventive volcanic events that preceded, accompanied and followed the collapse of both calderas. In fact, during the Mount Manengouba history, some fissures were created on the eastern slopes. These features have been taken by basaltic magmas. During the ascent of the basaltic magmas, there was an encounter between them and underground phreatic layers. The temperature of that water is usually low (around 20 °C). Thus, the irruption of basaltic magmas whose temperature is regularly more or less than 1000 °C, induced the vaporization. That vaporization of waters led to the generation of significant volumes of gases that combined with magmatic gases. Since the degassing of magma governs the degree of explosivity of volcanoes, the addition of water vapour would increase this explosivity, thus modifying the normal course of a volcanic eruption. The magmas were rapidly cooled by the water and were on one hand vertically thrown and scattered (in the form of loose materials) by plume and on other hand, carried away on the ground through the lateral expansion of gases. These explosions were later followed by the subcircular collapse of the crater that gave rise to the formation of DSB. This morphogenetic process is known as the phreatomagmatic phenomenon.

Fig. 21.4 Picture displaying the marshy bottom of the Djeu-Seh Basin

21.3.1.2 Morphology

From the sky, DSB looks like a tropical avocado whose head is facing the eastern side of the Elengoum Volcano. Hence, it is sub-circular/elliptical in shape and it covers an area of 134,609 m^2. DSB is a volcano negatively erected, i.e., below the surface of the ground. A cross section shows that DSB is funnel-like shaped with internal slopes very steeper in certain parts notably the southern, eastern, and northern rims. The slopes of the western rim are very smooth. That is why the deepest point of the DSB is found on the eastern side of the floor. DSB is not occupied by the water as found in the classical crater lakes. This maar presents a marshy bottom (Fig. 21.4).

21.3.1.3 Petrography

DSB has been built through the stacking of pyroclastic ejecta and basalts.

The Pyroclastic Ejecta

The pyroclastic ejectas are made up of *basal surges* and *air fall deposits*. These rocks are layered and they constitute a lowered ring which borders the crater. *Basal surges* are characterized by lapilli that constitute more than 70% of the lithology of the DSB. In addition, there are scorias and bombs which are well observed at the basis of the maar. As scorias and bombs, lapilli vary in shape and sizes. *Aerial falls* are represented in the DSB by fine and coarse ashes and well-graded lapilli (Fig. 21.5).

Basalts

Basalts found in the DSB and vicinities constitute the main petrographic units of the Low-Manengouba shield volcano. They are called lower basalts and constitute the bottom on which the maar has been built. In the DSB, they crop out in multiform

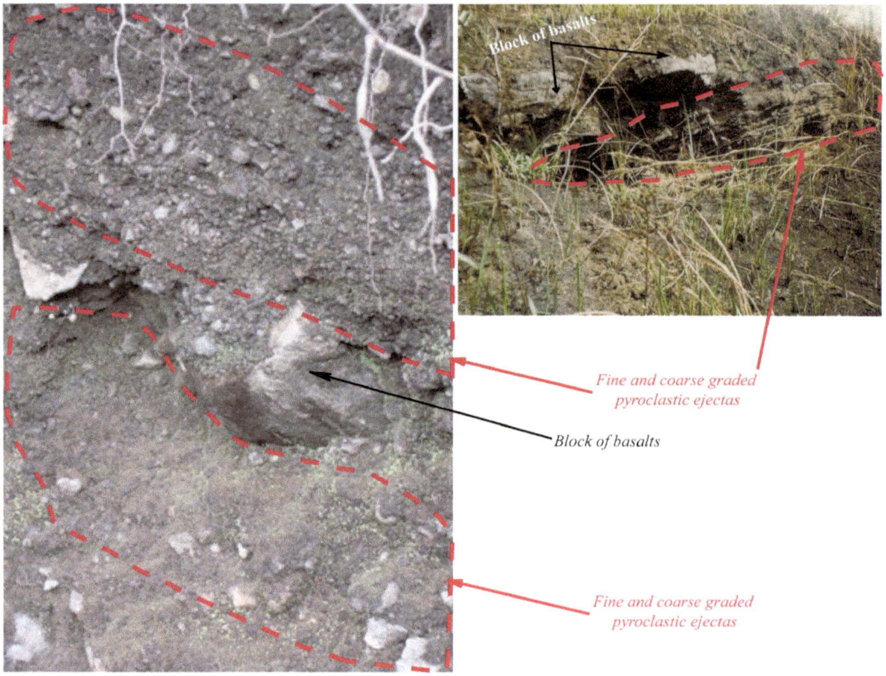

Fig. 21.5 Pictures displaying the interpolation of the pyroclastites and basalts in the Djeu-Seh Basin

and multidimensional blocks interspersed in layers of pyroclastic ejectas (Figs. 21.5 and 21.6).

21.3.2 Assessment of the Djeu-Seh Basin

21.3.2.1 Intrinsic Value

- **Central value**
  The central value of the DSB is characterized by a significant scientific value that is around 0.88. DSB is well preserved (1), rare (1), and representative (1) of the whole geomorphology of the eastern slopes of Elengoum Volcano. It also plays an important role in the geographical history (0.5) of MM.
- **Additional value**
  DSB has an average ecological value due to the fact that it is not protected. However, the ecological influence remains quite high. Moreover, although the aesthetic value of the DSB is also high, its cultural value is insignificant.

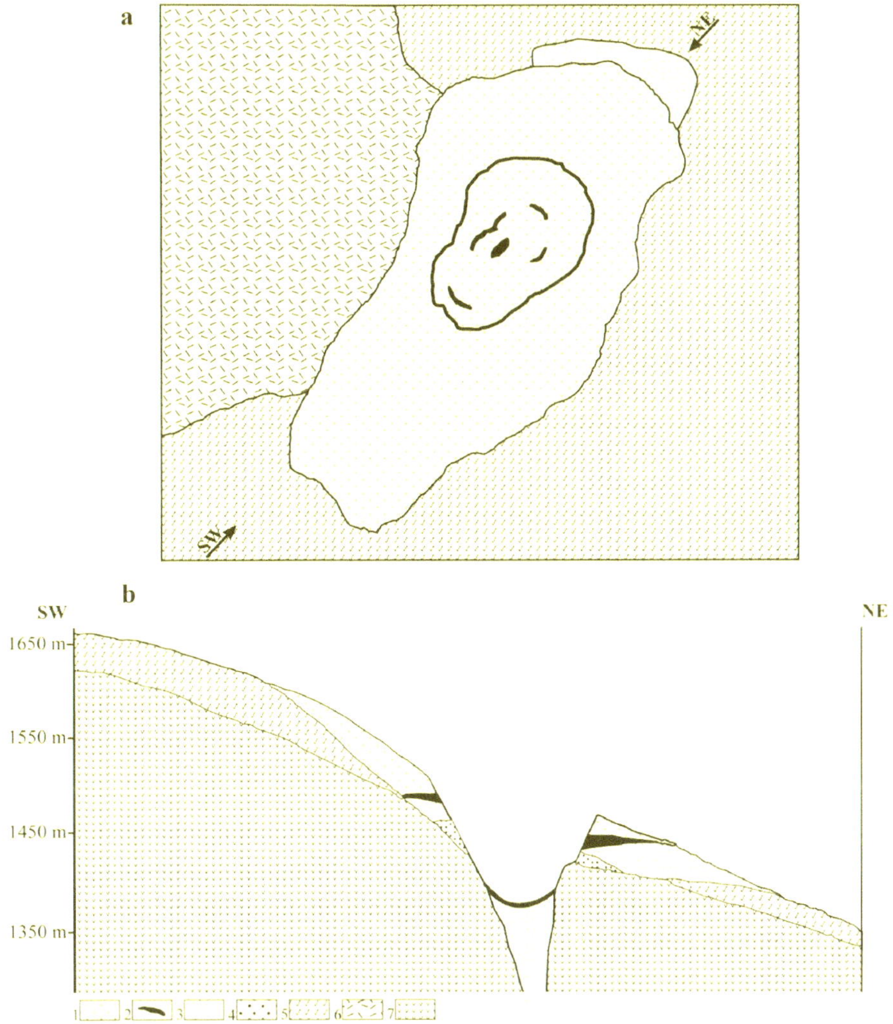

Fig. 21.6 **A**- Simplified Geological map of the Djeu-Seh Basin. **B**- Cross section 1- Pyroclastic ejectas; 2- Scorias, 3- Hydro-magmatic deposits; 4- Breccias; 5- Recent flow, 6- Elengoum flow, 7- Basalts

21.3.2.2 Use and Management Characteristics

- **Protection**
  The DSB is a public property although there is no legal protection of the DSB. The synthetic map (Fig. 21.7) gives a broad outline of the attacks and threats that disturb the maar. The main threats are attributed to human interactions. Building and farming activities constitute the principal threats for the integrity of the DSB.

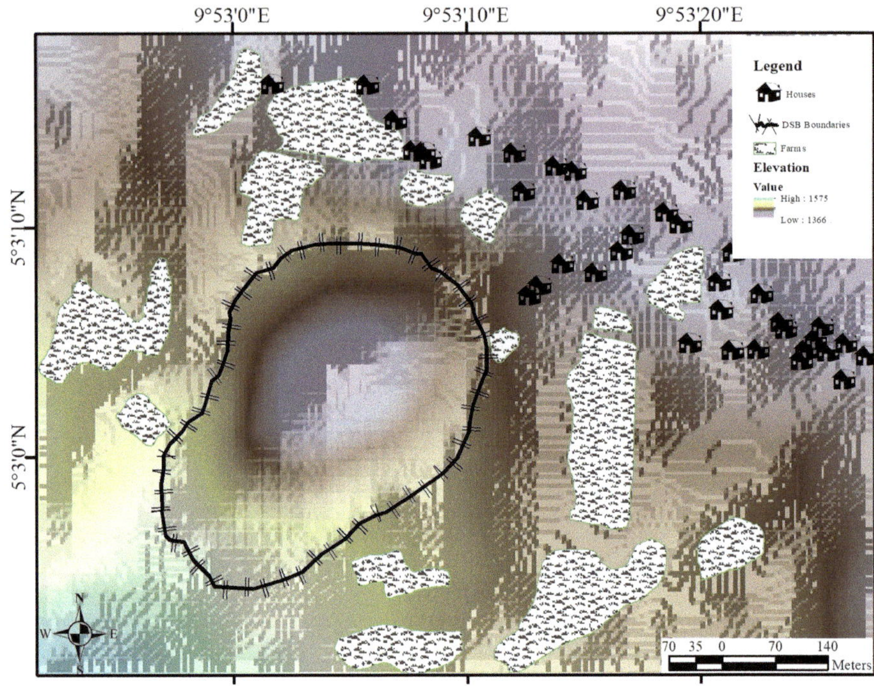

Fig. 21.7 Map displaying the threats about the Djeu-Seh Basin

- **Promotion**
  From Melong city, tourists can get to Mbouroukou village through an off-road vehicle. Since there is no road fitted for such vehicles to reach the DSB, the tourists must take a pedestrian track.

The waymarked paths do not exist in the DSB environment. Moreover, the existing paths are uneven and are made of pyroclastic ejectas and clumps of grasses. These constitute the real obstacles for tourists when walking. For some security reason, it is recommend to tourists to inform the local authorities before getting to the DSB.

At the present time, the environment of the DSB is quiet because road traffic and noise nuisance are a little bit far or absent. However, the occupation of green spaces through house building and population settlement on the north-eastern environment of the maar could influence the tranquility of the DSB in the forthcoming year.

There are no tourism infrastructures around the DSB environment. But tourists can accommodate in some hotels in the city of Melong.

The education of the DSB relies on its readability. The sub-circular form of the DSB informs us that it is a volcanic crater. DSB is a maar resulting from the explosive dynamism type called phreatomagmatism. The tokens of the explosive episodes rely on the present of pyroclastic ejectas scattered in the maar and on its outer slopes. The blocks of xenoliths found between the layers of pyroclastic ejectas are the main indicator of the nature of the material on which the maar has been built. These information are important for the aimed public during the geotouristic activity.

Nowadays, the DSB has not benefited from the educational valuing (no interpretation equipment such as educational panels, brochures, postcards, etc.) despite its good readability.

21.4 Discussion and Interpretation

Volcanic geomorphosites are one of the best touristic attractions around the Earth Globe. That's why several studies where focussed on volcanic landforms for geotouristic perspectives.

DBS is one of the three maars of the Mount Manengouba Volcano after the Female Lake and the Male Lake. The latter are occupied by water, unlike the DSB which has no water and has a swampy bottom (Kagou Dongmo et al. 2005; Zangmo Tefogoum et al. 2014). DSB has been formed through phreatomagmatism mechanism as observed on other volcanic provinces in Cameroon and beyond (Le Guern et al. 1982; Kagou Dongmo et al. 1999; Zangmo Tefogoum et al. 2015b, 2017; Strehlow et al. 2017). The eruption gave rise to the stacking of pyroclastic ejectas layers that are made up of basal surges and air fall deposits as encountered in the El Elegante maar, Pinacate volcanic field, Sonoran Desert, Mexico (Atlas recondito 2017) and in the El Muweilih maar, Bayuda Volcanic Field, Sudan (Lenhardt et al. 2018). The number of layers corresponds to the rhythmicity of the eruptive events. These layered pyroclastic ejectas are sloping like those observed in the Bayt al-nur cinder cone, Sudan (Lenhardt et al. 2018). These pyroclastic deposits are interpolated in some places by former basalts from the Low-Manengouba shield volcano testifying the high intensity of explosions that ruled the eruptive episodes.

The assessment of the DSB shows that the intrinsic value is significant due to the scientific and the additional values. In fact, the scientific value of DSB is 0.88 because—the integrity got a score equal to 1 since the maar is covered by endemic grasses that preserve its original shape—the representativeness also got a score equal to 1; this is due to the fact that DSB is representative of the whole geomorphology of the eastern slopes of Elengoum Volcano—the rareness that also got a score equal to 1 because it is the single maar without water in the mount Manengouba. Moreover, it is the single maar on the eastern slopes of Elengiuba Volcano which is dominated by flows and the broken cones—the palaeogeographic interest that obtained the lowest score (0.5) because the DSB is the witness of a long geological history of the eastern slopes of Elengoum Volcano by modifying the previous natural landscape. However, DSB has an average additional value. In fact, it has an average ecological value due to the fact that it is not protected. However, the ecological influence remains quite high because it is important for biological diversity and that sometimes harbor typical wildlife and plant species on slopes and the basin floor. The aesthetic value is high because of its general shape close to that of tropical avocado, its role in the structuration of the relief, its lengthened structure in a southwest–northeast direction (Fig. 21.8) and the presence of some viewpoints (from the upper slopes of the volcano).

In addition, the color contrast due to the alternation between the dark gray bands of chaotic scoria/basalts and the green of the vegetation embellish the landscape of DSB (Fig. 21.9).

The cultural value of the site is insignificant because of a few scientific works carried out and there is the absence in the exhibition. Moreover, there is a lack of ritual practices as observed in the Female Lake (Zangmo Tefogoum et al. 2014, 2015a). DSB does not benefit from legal protection because in Cameroon the nature

Fig. 21.8 Image of the Djeu-Seh Basin

Fig. 21.9 Picture of the band of rocks on the rim of the Djeu-Seh Basin

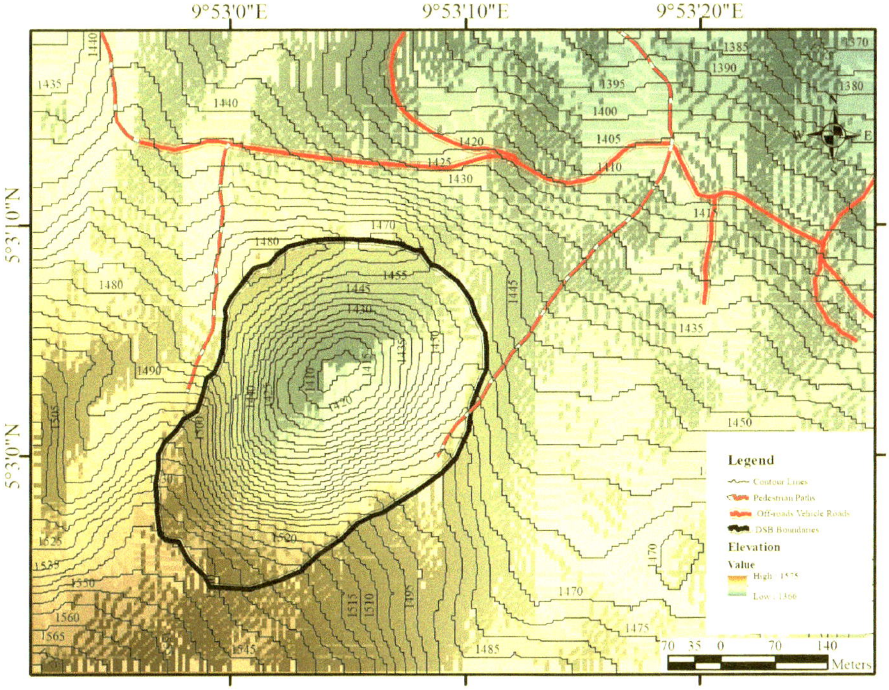

Fig. 21.10 Map showing the accessibility of the Djeu-Seh Basin

conservation policies implemented are focused on biodiversity conservation concerns (Zangmo Tefogoum et al. 2017). DSB is characterized by a quiet environment although the fact that the human activities are perceived as the main threats in the forthcoming future. Thus, according to Bussard (2014) inaccessible geomorphosites and those remote from human activities are on the whole preserved and less threatened. DSB is accessible (Fig. 21.10) although the fact that tourists must pay attention to the loose state of ejectas that makes slippery the soil path. These paths must be marked for the protection of tourists and for preventing the sporadic opening of tracks that could endanger the integrity of the site.

DSB constitutes one of the best geotouristic potentialities along the CVL. According to Pereira et al. (2007), the geomorphosites assessment has been considered a fundamental tool for supporting management decisions for its protection and geotourism implementation. Moreover, before a geosite can be developed for tourism, it first has to be assessed in relation to its geoheritage (geological value for human society) and geoconservation needs (need for protection) (Newsome and Dowling 2018). Thus, the DSB has been assessed and it emerges that its values are significant for the implementation of geotourism. The geomorphosites, that are the main components of the geological heritage in the perspective of geotourism promotion, are not well recognized in Cameroon as exemplified in many countries in Africa, notably Cape Verde (Costa 2010, 2011), Uganda (Bakka Male 2011), Ethiopia (Asrat et al.

2012), and São Tomé Island (Henriques and Neto 2015). Yet though, Africa has a "wild and unspoilt" landscape and, therefore there is no doubt that the continent has much unexploited geotourism potential Ngwira (2015). The touristic offer is insignificant in Cameroon. The main touristic attractions rely on the components of ecotourism (Reserve Forests, National Parks) and cultural tourism (traditional festivals). DSB has significant values that must be exploited. The promotion of geotourism in the DSB as in other volcanic provinces in Cameroon will diversify the touristic offer while increasing the number of tourists (Zangmo Tefogoum, 2016; Zangmo Tefogoum et al. 2017). According to Neches and Erdeli (2015), geotourism is now being considered as a sustainable alternative with manageable forms of tourism taking place alongside educational activities. Moreover, geotourism is defined by its essential educative content (Newsome and Dowling 2018). The education interest in huge; then the geotourists could learn the volcanic events that occurred several centuries ago and were responsible for the deposition of numerous pyroclastic ejectas in the eastern slopes of the Elengoum Volcano. Similar educational assets for geotouristic activity are observed in Krakatau, where there are unconsolidated volcanic ash beds whose interpretation could be a central theme in geotourism (e.g., Newsome and Dowling 2010). Since the Cameroonian general public and decision-makers are not really aware of the geological heritage, several geomorphosites are destroyed through the opening of quarries (Zangmo Tefogoum et al. 2012, 2014, 2017). Moreover, some are threatened by human settlement and pressure. Geotourism thus has a role to play in raising public awareness about geoheritage providing that we do not lose sight of what we are attempting to appreciate and that we do not fail to confer adequate protection (Newsome and Dowling 2018). DSB must be turned into a geotouristic destination in order to provide protection and preservation for future generations. Education (Fig. 21.11) still has a strong role to play for that issue since according to Stoffelen and Vanneste (2015) geotourism is a niche form of tourism having a focus on geoheritage with the goal of attaining geoconservation by education.

Fig. 21.11 Pictures of scientists around the Djeu-Seh Basin

21.5 Conclusion

DSB is a volcanic landform built on the eastern slope of the Elengoum Volcano. The phreatomagmatism is the only mechanism responsible for this formation. It gave rise to the deposition of layers of pyroclastic ejectas that include basal surges and air fall deposits interpolated by basalts. DSB has been assessed through the Reynard et al. (2016) methodology. It emerges that the scientific value of the DSB is high (0.88), because it is well preserved (1), rare (1), and representative (1) of the whole geomorphology of the eastern slopes of MM. It also plays an important role in the geographical history (0.5) of MM. DSB has an average ecological value due to the fact that it is not protected. However, the ecological influence remains quite high because it is important for biological diversity and that sometimes harbor typical wildlife and plant species on slopes and the basin floor. The aesthetic value is high because of its special geomorphological features and the color contrast due to the alternation between the dark gray bands of rocks and the green of the vegetation. The cultural value of the site is insignificant. However, DSB has a quiet environment and is accessible although some obstacles related to steep slopes and the loose state of pyroclastic. There are not infrastructures for the geotourism promotion despite the high educational interest of the DSB. However, the integrity of the latter could be threatened by human interactions in the coming years. Geological landforms take millions of years or more to build and model themselves; but man, through his activities only needs a minute or less to destroy it. The promotion of geotourism will help in protecting the site while creating job opportunities in the Mbouroukou village. To achieve this, it is recommended to the Cameroon's stakeholders:

- To prohibit the gradual occupation of slopes of DSB;
- To improve the accessibility conditions of DSB;
- To use the data provided in this work out the interpretative panels, leaflets, which must be placed on the DSB;
- To popularize the DSB through field trips, leaflets, social media, and exhibition;
- To train some Mbouroukou villagers as tourist guides.

Acknowledgments We thank Profs. Emmanuel Reynard for providing us with documents on the concepts gemorphosites assessment and geotourism. Moreover, we are grateful to Mbouroukou villagers for their hospitality during our field campaigns.

References

Aka FT, Nagao K, Kusakabe M, Sumino H, Tanyileke G, Ateba B, Hell J (2004) Symmetrical helium isotope distribution on the Cameroon volcanic line, West Africa. Chem Geol 203:205–223

Asrat A, Demissie M, Mogessie A (2012) Geoheritage conservation in Ethiopia: the case of the Simien Mountains. Quaestiones Geographicae 31(1), Bogucki Wydawnictwo Naukowe, Poznań, pp 7–23. https://doi.org/10.2478/v10117-0120001-0

Asrat A (2018) Potential geoheritage sites in Ethiopia: challenges of their promotion and conservation. In: Reynard E., Briha J. (Eds), Geoheritage: assessment, protection, and management. Elsevier, 339–353. http://dx.doi.org/10.1016/B978-0-12-809531-7.00019-8

Atlas_recondito (2017) El Elegante Crater. https://reconditosite.wordpress.com/2017/04/19/el-elegante-crater/, Accessed 14 Jan 2018

Bakka Male MK (2011) Geosites in Uganda. In: Proceeding abstracts of the 1st international conference of African and Arabian geoparks, pp 42

Boukhchim N, BenFraj T, Reynard E (2017) Lateral and B "Vertico-Lateral" cave dwellings in Haddej and Guermessa: characteristic geocultural heritage of southeast Tunisia. https://doi.org/10.1007/s12371-017-0251-2

Brancucci G, Gazzola A (2002) Geositi e percezione sociale degli elementi naturali. Geologia dell'Ambiente 10(2)

Brilha J (2015) Concept of geoconservation. In: Tiess G, Majumder T, Cameron P (eds) Encyclopedia of mineral and energy policy. Springer, Berlin, Heidelberg, p 2. https://doi.org/10.1007/978-3-642-40871-7_2-1

Bruschi VM, Cendrero A (2005) Geosite evaluation; Can we measure intangible values? Il Quaternario 18(1):293–306

Burke K (2001) Origin of the Cameroon line of volcano-capped swells: J Geol 109:349–362

Bussard J (2014) Protection et valorisation du patrimoine géomorphologique du Parc naturel régional Gruyère Pays-d'Enhaut États des lieux et perspectives. Université de Lausanne, Lausanne, Suisse, Mémoire de Master, p 120

Carton A, Cavallin A, Francavilla F, Mantovani F, Panizza M, Pellegrini GG, Tellini C (1994) Ricerche ambientali per l'individuazione e la valutazione dei beni geomorfologici – metodi ed esempi. Il Quaternario 7(1):365–372

Cedro B, Mianowicz K, Zawadzki D (2009) Ocena walorów geoturystycznych stanowisk pochodzenia wulkanicznego Gór i Pogórza Kaczawskiego. Problemy Turystyki I Rekreacji 2:25–35

Clivaz M, Reynard E (2018) How to integrate invisible geomorphosites in an inventory: a case study in the Rhone River Valley (Switzerland). Geoheritage 10:527–541. https://doi.org/10.1007/s12371-017-0222-7

Cocean G (2011) Munţii Trascău Relief Geomorfosituri. Turism. Presa Universitară Clujeană, Cluj-Napoca

Coratza P, Galve JP, Soldati M, Tonelli C (2012) Recognition and assessment of sinkholes as geosites: lessons from the Island of Gozo (Malta). Questiones Geographicae 31(1):25–35

Coratza P and Hobléa F (2018) The specificities of geomorphological heritage. In: Reynard E, Briha J (eds) Geoheritage: assessment, protection, and management, Elsevier, pp 87–106. https://doi.org/10.1016/B978-0-12-809531-7.00005-8

Costa FL (2011) Volcanic geomorphosites assessment of the last eruption, on April to May 1995, within the natural park of Fogo island, Cape Verde. GeoJ Tour Geosites 2(8):167–177

Costa F (2010) Volcanic landforms of the last eruption, on April to May 1995, within the Natural Park of Fogo island, Cape Verde. In: Proceeding abstracts of the international conference on geoheritage and geotourism. Lisbon, Portugal, pp 13–14

Erfurt-Cooper P (ed) (2014) Volcanic tourist destinations. Springer, Berlin, Heidelberg

Erfurt-Cooper P (2010) The importance of natural geothermal resources in tourism. In: Proceedings world geothermal congress 2010 Bali, Indonesia, 25–29 April 2010, p 10

Erfurt-Cooper P, Cooper M (2010) Volcano & geothermal tourism sustainable geo-resources for leisure and recreation. Earthscan Publishing 409 p

Fepuleai A, Weber E, Németh K, Muliaina T, Iese V (2017) Eruption styles of Samoan volcanoes represented in Tattooing, language and cultural activities of the indigenous people. Geoheritage 9:395–411. https://doi.org/10.1007/s12371-016-0204-1

Feuillet T, Sourp E (2011) Geomorphological heritage of the Pyrenees National Park (France): assessment, clustering, and promotion of geomorphosites. Geoheritage 3:151–162

Field R, Newsome D, (2014) Krakatau: tourism and the recovery of a volcanic rainforest. In: Erfurt-Cooper P (ed) Volcanic tourist destinations. Springer, Berlin, Heidelberg, pp 217–229

Gao W, Li J, Mao X, Li H (2013) Geological and geomorphological value of the monogenetic volcanoes in Wudalianchi National Park NE China. Geoheritage 5(2):73–85

Grandgirard V (1995) Méthode pour la réalisation d'un inventaire de géotopes géomorphologiques. Ukpik, Cahiers De l'Institut De Géographie, Fribourg 10:121–137

Le Guern F, Tazieff H, Faivre Pierret R (1982) An example of health hazard: people killed by gas during a phreatic eruption: Dieng Plateau (Java, Indonesia), February 20 1979. Bull Volcanol 45:153–156

Henriques MH, Neto K (2015) geoheritage at the equator: selected geosites of São Tomé Island (Cameron Line, Central Africa). Sustainability 7:648–667. https://doi.org/10.3390/su7010648

Hooke JM (1994) Strategies for conserving and sustaining dynamic geomorphological sites. In: O'Halloran D et al (eds) Geological and landscape conservation. Geological Society, London, pp 191–195

Ilies DC, Josan N (2009) Geosites-geomorphosites and relief. Geo J Tour Geosites 3(1):78–85

Kagou Dongmo A, Wandji P, Pouclet A, Vicat JP, Cheilletz A, Nkouathio DG, Alexandrov P, Tchoua FM (2001) Evolution volcanologique du mont Manengouba (Ligne du Cameroun); nouvelles données pétrographiques, géochimiques et géochronologiques. C R Acad Sci Terre et planets 333:155–162

Kagou Dongmo A, Nkouathio DG, Pouclet A, Bardintzeff JM, Wandji P, Nono A, Guillou H (2010) The discovery of Late Quaternary basalt on Mount Bambouto: implications for recent widespread volcanic activity in the southern Cameroon line. J Afr Earth Sci 57:87–108

Kagou Dongmo A, Wandji P, Pouclet A, Nkouathio DG, Tchoua FM (2005) Le mont Manengouba (Ligne du Cameroun), Un volcan bénéfique, mais potentiellement dangereux, typologie des aléas et évaluation des risques naturels associés. Afr Geosci Rev 12(2):97–109

Kagou Dongmo A (2006) Le mont Manengouba: Evolution volcanologique, caractères magmatologiques et risques naturels; comparaison avec les monts Bambouto et Bamenda (Ligne du Cameroun). Th Doct d'Etat, Univ Yaoundé 1, Yaoundé, 230p

Kagou Dongmo A, Wandji P, Nkouathio DG, Tchoua FM, Tchouankoue JP, Vicat JP (1999) Etude d'un environnement volcanique à aléas multiformes et en plein essor démographique : les régions du mont Manengouba (Cameroun). In: Vicat JP, Bilong P (eds) Géologie et environnement au Cameroun. Coll. Geocam 2 Press. Université Yaoundé, Cameroun, pp 61–70

Kagou Dongmo A, Wandji P, Tchoua FM, Nkouathio DG, Vicat JP (1998a) Volcanologie, pétrographie et géochimie d'un volcan polygénique: le stratovolcan Manengouba (Secteur sud continentale de la Ligne du Cameroun). Ann. Fac. Sci. Univ. Yaoundé I, série Sci. Nat. et Vie, (2) 34:169–183

Khalaf EA, Wahed MA, Maged A and Mokhtar H (2018) Volcanic geosites and their geoheritage values preserved in monogenetic neogene volcanic field, Bahariya depression, western desert, Egypt: Implication for climatic change-controlling volcanic eruption. Geoheritage. https://doi.org/10.1007/s12371-018-0336-6

Lee DC, Halliday AN, Fitton JG, Poli G (1994) Isotopic variation with distance and time in the volcanic islands of the Cameroon Line, evidence from mantle plume origin. Earth Planet Sci Lett 124:119–138

Lenhardt N, Borahn SB, Lenhardt SZ, Adam Bumby AJ, Ibinoof MA, Salih SA (2018) The monogenetic Bayuda Volcanic Field, Sudan—New insights into geology and volcanic morphology. J Volcanol Geoth Res 356:211–224

Mauerhofer L, Reynard E, Asfawossen A, Hurn H (2017) Contribution of a geomorphosite inventory to the geoheritage knowledge in developing countries: the case of the Simien Mountains National Park, Ethiopia. https://doi.org/10.1007/s12371-017-0234-3

Migoń P, Pijet-Migoń E (2010) Problemy udostępniania turystycznego obiektów wulkanicznego dziedzictwa Ziemi na przykładzie Pogórza Kaczawskiego. Zeszyty Naukowe Wyższej Szkoły Bankowej we Wrocławiu 17:129–151

Moufti MR, Nemeth K (2013) The intra-continental Al Madinah volcanic field, Western Saudi Arabia: a proposal to establish Harrat Al Madinah as the first volcanic geopark in the kingdom of Saudi Arabia. Geoheritage 5(3):185–206
Moufti MR, Nemeth K, El-Masry N, Qaddah A (2015) Volcanic geotopes and their geosites preserved in an arid climate related to landscape and climate changes since the neogene in Northern Saudi Arabia: Harrat Hutaymah (Hai'il Region). Geoheritage 7(2):103–118
Neches I-M, Erdeli G (2015) Geolandscapes and geotourism: integrating nature and culture in the Bucegi Mountains of Romania. Lands Res 40(4):486–509
Newsome D and Dowling R (2018) Geoheritage and geotourism In: Reynard E and Briha J (eds) Geoheritage: assessment, protection, and management, Elsevier, pp 305–321. https://doi.org/10.1016/B978-0-12-809531-7.00017-4
Newsome D, Dowling RK (2010) Geotourism: the tourism of geology and landscape. Good¬fellow Publishers Ltd, Oxford. 260p
Ngwira PM (2015) Geotourism and geoparks: Africa's current prospects for sustainable rural development and poverty alleviation. In: Errami E, Brocx M, Semeniuk V (eds) From geoheritage to geoparks case studies from Africa and beyond. Springer International Publishing, Switzerland, pp 25–33. https://doi.org/10.1007/978-3-319-10708-0_2
Nobuo Geshi N, Iguchi M, Shinohara H (2016) Phreatomagmatic eruptions of 2014 and 2015 in Kuchinoerabujima Volcano triggered by a shallow intrusion of magma. J Nat Disaster Sci 37(2):67–78
Németh K, Wu J, Sun C, Liu J (2017) Update on the volcanic geoheritage values of the pliocene to quaternary Arxan-Chaihe volcanic field, Inner Mongolia, China. Geoheritage 9:279–297. https://doi.org/10.1007/s12371-017-0224-5
Panizza M, Piacente S (1993) Geomorphological assets evaluation. Z Geomorph Suppl Bd 87:13–18
Panizza M (2001) Geomorphosites: concepts, methods and example of geomorphological survey. Chinese Sci Bull 46(Suppl. Bd.):4–6
Pereira P, Pereira D, Caetano AI (2007) Geomorphosite assessment in Montesinho Natural Park (Portugal). Geographica Helvetica 62(3):159–168
Pralong JP, Reynard E (2005) Proposal for a classification of geomorphological sites depending on their tourist value. Il Quaternario 18(1):315–321
Pérez-Umaña D, Quesada-Román A, Rojas JCDJ, José Juan Zamorano-Orozco J, Dóniz-Páez J, Becerra-Ramírez R (2018) Comparative analysis of geomorphosites in volcanoes of Costa Rica, Mexico, and Spain. Geoheritage. https://doi.org/10.1007/s12371-018-0313-0
Quaranta G (1993) Geomorphological assets: conceptual aspect and application in the area of Crodo da Lago (Cortina d'Ampezzo, Dolomites). In Panizza M, Soldati M, Barani D (eds) European Intensive Course on Applied Geomorphology. Proceedings, Modena – Cortina d'Ampezzo, 24 June–3 July 1992, pp 49–60
Raška P, Pokorný R, Krmíček L, Koubušková S, Mortensen L (2018) Basaltic dyke with specific volcanogenic structures and its geomorphic evolution: unique geoheritage of the Faroe Islands (North Atlantic Ocean). Geoheritage. https://doi.org/10.1007/s12371-018-0296-x
Reynard E, Fontana G, Kozlik L, Scapozza C (2007) A method for assessing « scientific » and « additional values » of geomorphosites. Geographica Helvetica 62(3):148–158
Reynard E, Perret A, Bussard J, Grangier L, Martin S (2016) Integrated approach for the inventory and management of geomorphological heritage at the regional scale. Geoheritage 8(1):43–60. https://doi.org/10.1007/s12371-015-0153-0
Reynard E (2009) Geomorphosites: definitions and characteristics. In: Reynard E, Coratza P, Regolini-Bissig G (eds) Geomorphosites. Pfeil, München, pp 9–20
Reynard E, Coratza P, Regolini-Bissig G (2009) Scientific research on geomorphosites over the last eight years: improvements and aims of the book. In: Reynard E, Coratza P, Regolini-Bissig G (eds) Geomorphosites. Pfeil Verlag, Mu¨nchen, pp 5–8.
Rivas V, Rix K, Frances E, Cendrero A, Brunsden D (1997) Geomorphological indicators for environmental impact assessment: consumable and nonconsumable geomorphological resources. Geomorphology 18:169–182

Różycka M, Migoń P (2014) Visitors' background as a factor in geosite evaluation. The case of Cenozoic volcanic sites in the Pogórze Kaczawskie region, SW Poland. Geotourism 3–4 (38–39):3–18

Różycka M, Migoń P (2018) Customer-oriented evaluation of geoheritage-on the example of volcanic geosites in the West Sudetes, SW Poland. Geoheritage 10:23–37. https://doi.org/10.1007/s12371-017-0217-4

Serrano E, González-Trueba JJ (2005) Assessment of geomorphosites in natural protected areas: the Picos de Europea National Park (Spain). Géomorphologie: Relief. Processus, Environ 3:197–209

Strehlow K, Sandri L, Gottsmann JH, Kilgour G, Rust AC, Tonini R (2017) Phreatic eruptions at crater lakes: occurrence statistics and probabilistic hazard forecast. J ApplVolcanol 6:4. https://doi.org/10.1186/s13617-016-0053-2

Stoffelen A, Vanneste D (2015) An integrative geotourism approach: bridging conflicts in tourism landscape research. Tour Geogr 17(4):544560

Tchoua MF (1974) Contribution à l'étude géologique et pétrographique de quelques volcans de la "Ligne du Cameroun" (Monts Manengouba et Bambouto), Thèse Doct. Etat, Univ. Clermont-Ferrand, France, 337p + Biblio +75 figures

Wandji P (1995) Le volcanisme récent de la plaine du Noun (Ouest- cameroun). Volcanologie, Pétrologie, Géochimie et Pouzzolanité. Thèse Doct. d'Etat Univ. Yaoundé 1, Yaoundé, Cameroun, 295p + 3 planches + 1 carte

Wang L, Tian M, Wen C, Zhao L, Song J, Sun M et al (2014) Geoconservation and geotourism in Arxan-Chaihe volcano area, Inner Mongolia, China. Quat Int 349:384–391

Zangmo Tefogoum G, Dongmo Kagou A, Nkouathio DG, Dedzo Gountié M, Kamgang P (2017) The volcanic geoheritage of the mount Bamenda Calderas (Cameroon Line): assessment for geotouristic and geoeducational purposes. Geoheritage 9(3):255–278. https://doi.org/10.1007/s12371-016-0177-0

Zangmo Tefogoum G, Kagou Dongmo A, Nkouathio DG, Wandji P, Gountié Dedzo M (2014) Geomorphological features of the Manengouba Volcano (Cameroon Line): assets for potential geopark development. Geoheritage 6(3):225–239

Zangmo Tefogoum G (2016) Caractérisation volcanologique de quelques caldeiras de la Ligne Volcanique du Cameroun. Etude du géohéritage et évaluation des risques naturels associés: Cas des caldeiras des monts Manengouba, Bambouto et Bamenda. Thèse Doctorat, Univ Dschang, 252p.

Zangmo Tefogoum G, Kagou Dongmo A, Nkouathio DG, Gountié Dedzo M (2015b) Study of Environmental and Human Impacts of Mount Manengouba Lakes (Cameroon Line). The TropiLakes Conference, Bahir Dar (Ethiopie) du 23–29 Octobre 2015, p 63

Zangmo Tefogoum G, Kagou Dongmo A, Nkouathio DG, Gountié Dedzo M (2018) The assessment of geomorphosites of Eboga and Elengoum calderas (Manengouba Volcano): a contribution for the promotion of geotourism. First International Congress on Geosciences (ICG1), Joint Congress CAAWG9 - AIC2 - ICGAME3, 20–24 March 2018 El Jadida – Morocco, pp 219–220

Zangmo Tefogoum G, Kagou Dongmo A, Nkouathio DG, Wandji P (2015a) Geomorphological features of the Manengouba volcano (Cameroon line): assets for geotourism and other anthropogenic activities. In: Errami E, Brocx M, Semeniuk V (eds) From geoheritage to geoparks. Geoheritage, Geoparks and Geotourism, pp 183–198. https://doi.org/10.1007/978-3-319-10708-0_13

Zangmo Tefogoum G, Kagou Dongmo A, Nkouathio DG, Wandji P, Gountié Dedzo, M (2014) Geomorphological features of the Manengouba volcano (Cameroon line): assets for potential geopark development. Geoheritage 6(3):225–239

Zangmo Tefogoum G, Kagou Dongmo A, Nkouathio DG, Wandji P, Gountié Dedzo, M (2012) The geodiversity of the Mount Manengouba Calderas (Cameroon Volcanic Line): values, threats and geoconservation. The 6th conference of the african association of women in geosciences, 23–27 April 2012. p 138

Zangmo Tefogoum G., Dongmo Kagou A, Nkouathio DG, Dedzo Gountié M, Kamgang P (2017) The volcanic geoheritage of the Mount Bamenda Calderas (Cameroon Line): assessment for geotouristic and geoeducational purposes. Geoheritage 9(3):255–278. https://doi.org/10.1007/s12371-016-0177-0

Zouros N (2007) Geomorphosite assessment and management in protected areas of Greece. Case study of the Lesvos Island-coastal geomorphosites. Geogr Helv 62(3):69–180

Závada P, Dědeček P, Mach K, Lexa O, Potužák M (2010) Emplacement dynamics of phonolite magma into maar-diatreme structures-Correlation of field, thermal modeling and AMS analogue modeling data. J Volcanol Geotherm Res. https://doi.org/10.1016/j.jvolgeores.2010.07.012

Chapter 22
Estimating Carrying Capacity in a High Mountainous Tourist Area: A Destination Conservation Strategy

Ravinder Jangra and S. P. Kaushik

Abstract In the sector of tourism, all issues are associated with the "magic number" of tourists that visit certain destination. The assessment of physical carrying capacity is an important component in the planning of spatial development to maintain sustainability for establishing standards in tourism sector. Study area has rich cultural and religious identity of tribal community as well as eco-system of cold desert which make it unlike any other in the world. Estimating tourist's threshold value of the destinations becomes vital considering the recent tragedies at the pilgrimage destinations in nearby hilly areas in Himalayas. Therefore, the present research assesses the physical carrying capacity of three selected tourist spots in Kinnaur based on the guidelines suggested by International Union of Conservation of Nature and Natural Resources (IUCN). As per calculations, the number of PCC, RCC, and ECC of selected tourist spots in Kinnaur are 64,835, 9595, and 5928, respectively. It is found that ECC is the most acceptable type of carrying capacity and the present status of tourism activity is highly under exploited vis-a-vis its carrying capacity in the study area.

Keywords Land use/land cover (LULC) · Physical carrying capacity (PCC) · Real carrying capacity (RCC) · Effective carrying capacity (ECC)

22.1 Introduction

The natural environment, local culture, and geographically diverse habitats are the major attractions that play an important role to attract mass tourism at any destination (Hasan et al. 2014). Thriving economy of Asia and dirt cheap transportation and

R. Jangra (✉) · S. P. Kaushik
Department of Geography, Kurukshetra University, Kurukshetra 136119, Haryana, India
e-mail: ravinderjangra13@gmail.com

S. P. Kaushik
e-mail: spk34@rediffmail.com

R. B. Singh et al. (eds.), *Global Geographical Heritage, Geoparks and Geotourism*, Advances in Geographical and Environmental Sciences,
https://doi.org/10.1007/978-981-15-4956-4_22

communication leads to flourishing tourism. According to UNWTO (2019), the international tourist arrivals reached 1.5 billion in 2019 which was 25 million in 1950 and the figure will cross the number of 1.8 billion in 2030. It nowadays provides enormous prospect to develop diverse type of tourist's products and supplement the income (Kruk et al. 2007). But, development of tourism not only creates positive impacts but also causes negative impacts, if it is not planned well, managed, and developed (Zhong et al. 2011) and all these issues are associated with the "magic number" of tourists that visit certain destination. Every destination has its carrying capacity to absorb tourist's activities, beyond that it should not be developed for long term benefit industry. Therefore, to avoid this situation, the concept of tourism carrying capacity has been discussed, thought, calculated, and also implemented (Hasan et al. 2014). This concept was used in 1945 for the first time by "Dasmann" in wildlife management for assessing the capacity of the forests for grazing by animals (Wall 1983). The first experiments were conducted in the 1960 in natural parks within the United States. Also, for the purpose of determining the ecological disturbance from use, this concept was applied in recreational research in early 1960 (Lucas 1964; Wagar 1964). In the 1970s and 1980s, emerged the first proposal that addressed the use of tourism carrying capacity (Davis and Tisdell 1995; Coccossis and Mexa 2004). A methodology for calculating tourism carrying capacity proposed by Cifuentes et al. (1990) has been applied in various protected natural areas of Latin America (Rodriguez 1992, 1997) as well as island systems, i.e., Galapagos National Park in Ecuador (Cayot et al. 1996), Fernando de Noronha National Park in Brazil (Mitraud 1998). As new evidence, tourism carrying capacity is a very effective strategy that addressing not only environmental questions but also economic and social issues (Davis and Tisdell 1995; Coccossis and Mexa 2004). The physical carrying capacity is concerned with the amount of space in undeveloped natural areas (Shelby and Heberlein 1986), or alternatively, as the maximum number of "use units" (people, vehicles, boats) which can be physically accommodated in an area (Pigram 1983). The concept is very helpful for sustainability in tourism sector for the planning of spatial development to establishing standards. In the milieu of Himalaya, study area has a lot of potentiality for tourism with its many products such as eco-tourism, rural tourism, adventure tourism, high altitude trekking, cultural tourism, tribal festivals and ethnics, medical tourism, and many more. Tourism is one of the more promising alternative livelihood options for people living in the Himalayas (Sharma 2000; East et al. 1998). The demand of mountain tourism is growing rapidly that provide opportunities as well as challenges by providing gainful employment, income, and other socioeconomic benefits. As a result of mass tourism, Shimla and Manali which are the major tourist destinations in Himachal Pradesh have also faced many problems. The unplanned development of infrastructure affects the destination's natural aesthetic attractions and creates a situation of losing recovery capacities of eco-system. Often, water crisis has been occurred in peak seasons, i.e., happened in June 2018. At high altitude, mass tourism also affects the natural phenomenon of destinations and with the passage of time it would cause climate change such as melting of glaciers (Times of India, June 13, 2018). A few months ago, the Archaeological Survey of India has issued statement to restrict the number of tourists visiting the Taj Mahal to 40,000 daily which reflect the importance of the concept. State government has recently allocated Rs 500 million for the development of unexplored tourist's destinations under the new scheme named "*Nai Raahein Nai Manzilien*" many of which are also located in the study area (Times of India, June 12, 2018). Therefore, it is necessity of the hours to regulate unmanaged tourism activities to maintain the ecological balance and facilitating tourism especially in the ecologically fragile agriculturally backward mountainous area by adopting inclusive tourism strategy. The present research seeks to estimate the specific physical carrying capacity for each selected destinations in cold desert area of Himachal Pradesh which has the potential to emerge a hub of tourist activities in near future. The estimation of carrying capacity should help establish mechanisms leading to the participation of performers connected to the high altitude and thereby contribute to a pertinent eco-friendly tourism proposal that would be important for the welfare of Kinnaur. It represents an important component of planning spatial development in tourism and is one of the mechanisms to establish the standards for sustainable tourism.

22.2 Study Area Description

Himalaya, a mighty and youngest mountain range of the world gives huge exposure to tourism from all over the world. The landscape and environment of the Himalayas provide an unforgettable experience to tourists and explorers. Therefore, people from every nook and corner come here to explore and meet their expectations on tourism. Kinnaur is situated in the north-eastern part of Himachal Pradesh between 31°05′50″ and 32°05′15″ North latitudes and 77°45′ and 79°00′35″ East longitudes nearby the Indo-China border (Fig. 22.1) with 6401 km^2. It is spread on both the banks of Sutlej and demarcates a well-defined natural international boundary between India and Tibet toward the east. Due to geographical situation of Kinnaur, the variance of climate is as varied as the area, from the heat of tropical zone to freezing temperature in winter. Kinnaur can be divided into three physical units at broad level upper, middle, and lower Kinnaur (Sanan and Swadi 2002). The upper reaches largely remains occupied with permanent snow, however, marginal change has been observed over the last few decades. Broadly, winter is from October to May (snowy season), summer from June to September, April to May is spring, and from September to October is autumn in Kinnaur (Bajpai 1981). There are number of places, namely Nichar, Kothi, Ribba, Moorang, Sangla, Rakchham, Chitkul, Namgya, Pooh, Chango, Nako, Leo, and Lippa that attracts lots of tourists. The Chini-Kalpa-Reckong-Peo area is the heartland of Kinnaur district. For the present research, major tourist destinations, i.e., Chitkul, Kalpa, and Nako have been selected. It is worthwhile to calculate the carrying capacity of these selected tourist destinations of the study area to regulate tourists for safe and sustainable tourism.

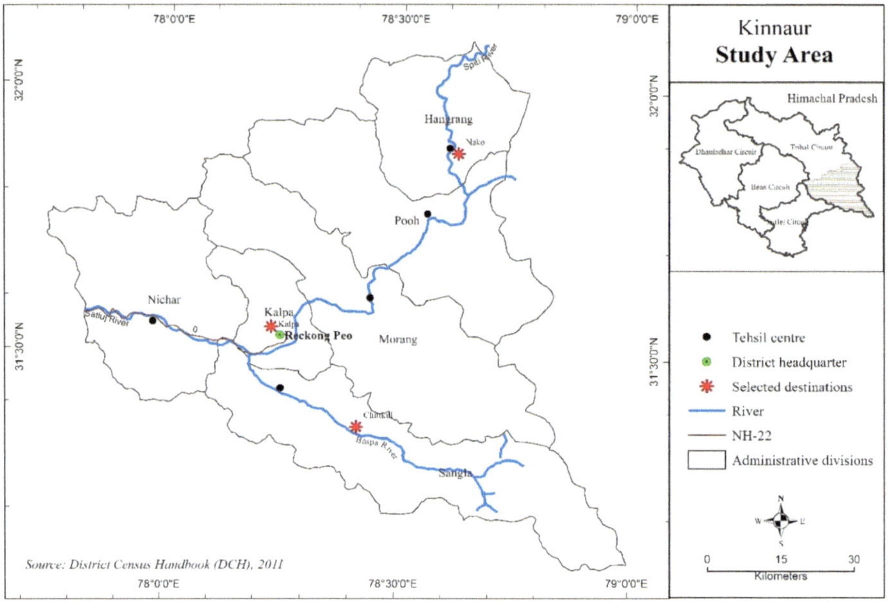

Fig. 22.1 Map of study area, Kinnaur. *Source* District Census Handbook (2011)

22.3 Database and Methodology

Study area has rich attraction for tourism both in terms of destinations and types. Main approach to achieve the intended objective is to calculate the carrying capacities of nine tourist spots in selected three tourist destinations at Kinnaur. To assess the tourist carrying capacity of these spots, adopted methods and materials is presented under the following heads.

22.3.1 Methods

The assessment of tourist carrying capacity was based on the model developed by Cifuentes (1992) and further suggested by International Union of Conservation of Nature and Natural Resources (1996). This method is based on site-specific peculiarities and characteristics of area which attempts to establish the maximum number of visits an area can tolerate. This methods is widely acceptable and often used to assess the carrying capacity of tourist destination of all types (natural, protected areas, coastal, beaches). A number of authors (Aryasa et al. 2017; Bera et al. 2015; Boullon 1985; Cifuentes et al. 1990; Gallo et al. 2003; Lawson et al. 2003; Lagmoj et al. 2013; Nghi et al. 2007; Papageorgiou and Brotherton 1999; Ruschmann et al.

2008; Zacarias et al. 2011) used same methods with some modifications. Following are the types of carrying capacity:

22.3.1.1 Physical Carrying Capacity (PCC)

It is defined as maximum number of tourists who can physically fit into a specific area over a given time. The general equation used to calculate PCC are

$$\text{PCC} = A \times V/a \times Rf$$

where

PCC = Physical carrying capacity

A = Area available for tourism (m^2)

V/a = Appropriate space for displacement of tourists/tourist density (tourists/m^2)

Rf = Rotation factor (number of visits per day)

Area available for tourism (A): A is available area for tourist/public use. In natural areas, area can be determined by natural boundary such as mountain range and river, whereas in conservation/protected area, space is used by tourist in different activities or they can do camping which is considered as an available area for tourism.

Appropriate space for displacement of tourists (V/a): The area required per tourist to do their activities comfortably. It is measured in square meters of the space.

Rotation factor (Rf): It is the number of permissible visits over a specified time (Usually calculated by daily open hours) and calculated through dividing the amount of time usable in day for tourists on the meantime of a visit.

$$Rf = \text{Open Period/average Time of Visit}$$

Duration of usability: It is the number of visits hours per day of the tourist destination. Since there are no entrance gates in the protected areas (lake and riverbank), the daily hours of sunshine were defined as a parameter. Total 13 h (5.50 AM–6.50 PM) considered as open period on the basis of average timing of sunrises and sunsets during the months May to September. Besides, on the time basis of worship (6–11 AM and 4–8 PM), total 9 h has been considered as open period for a religious spot in Kinnaur.

Visit duration: It is the average time required by tourists for touring and visiting various attractions of the region.

22.3.1.2 Real Carrying Capacity (RCC)

RCC is defined as "the maximum allowable number of tourists to the specific site within a given area, once the reductive/corrective factors (Cf) derived from the particular characteristics of the site has been applied to the PCC". The general formula where Cf is a corrective factor expressed as

$$\mathrm{RCC} = \mathrm{PCC}{-}\mathrm{Cf}_1 - \mathrm{Cf}_2 - \mathrm{Cf}_3 \ldots \mathrm{Cf}_n$$

The following formula better explains the RCC with corrective factors in percentages that is used in study:

$$\mathrm{RCC} = \mathrm{PCC} \times (100 - \mathrm{Cf}_1)/100 \times (100 - \mathrm{Cf}_2)/100 \ldots (100 - \mathrm{Cf}_n)/100$$

The Cf is based on the observation that certain factors such as environmental, ecological, and social, and management characteristics are closely linked to the specific conditions and characteristics of each site. They are expressed as follows:

$$\mathrm{Cf} = M_1/\, M_t \times 100$$

where

Cf = Corrective factors

M_l = Limiting magnitude of the variable

M_t = Total magnitude of the variable

Correction Factor (Cf)

At any natural areas, the tourism industry is highly dependent on climatic variables that have on particular destination. The climatic factor works as a two way; first it attracts tourist and second is limiting the touristic activities. Moreover, at many times the local condition, i.e., social attitude, crime rates, crowding, etc., of destination is also considered as a correction factor. Therefore, in the context of sustainability, evaluation of the potential factors that affects tourism number or limit touristic activities is essential. In this study, total five natural factors (rainfall, snowfall, frost, landslide, and temperature) and two man-made factors (accommodation and transportation) have been selected which limiting the tourist development in Kinnaur. These are as follows.

Table 22.1 Average month wise climatic limiting factor, 1961–2002

Months	Climatic variable's, 1961–2002		
	Mean monthly temperature (°C)	Frost frequency	Precipitation (mm)
January	4.3	14.8	35.4
February	5.5	12.5	44.2
March	9.8	5.3	40.5
April	14.5	1.6	32.8
May	17.5	1.5	42.2
June	19.1	0.8	120.9
July	18.4	0.1	240.0
August	17.7	0.1	221.2
September	16.5	0.8	129.0
October	13.8	1.7	19.8
November	9.9	4.7	11.2
December	6.7	9.6	14.1

Source www.waterportal.org

Natural Correction Factor

A long term climatic data during the period of 1961–2002 was used to obtain the natural correction factor which was downloaded (Table 22.1) from the government site www.waterportal.org. On the basis of these climatic variables the magnitude of limiting factor was determined.

- **Rainfall correction factor (Cf_1)** = Higher rainfall occurs during monsoon season in the months of July and August. Due to the terrain of study area, the accessibility has become low in the rainy season. Therefore, two months of rainfall have been considered as the period of limited access. The limiting magnitude for this factor was determined as 62 days per total magnitude (365 days).

$$\begin{aligned} Cf_1 &= 62/365 \times 100 \\ &= 16.98 \end{aligned}$$

- **Snowfall correction factor (Cf_2)** = Kinnaur receives snowfall from the month of November end or early December till March. Sometimes, the time of snowfall extends till April month. The study area remains isolated during this period. Besides, the higher peaks of Kinnaur has been covered with snow throughout the year (Central Ground Water Board 2013). This is a foremost factor that is limiting the tourist during these periods. Heavy snow also affects the road structure and damaged it. The limiting magnitude has been considered 135 days to available days in a year.

$$Cf_2 = 135/365 \times 100$$
$$= 36.98$$

- **Frost correction factor (Cf_3) =** Frequency of frost has increased during the period of December to February month. Moreover, the amount of snowfall has been high in that period which limits the touristic activities in Kinnaur. The limiting magnitude considered for this period was 90 days per year.

$$Cf_3 = 90/365 \times 100$$
$$= 24.65$$

- **Temperature correction factor (Cf_4) =** Temperature is pleasant throughout the year except for the time period of December to February. Human being needs a pleasant temperature condition to feel comfortable. At study area, almost 90 days per year have a temperature lower than 7° C. The mean monthly temperature varies from 5 to 23 °C.

$$Cf_4 = 90/365 \times 100$$
$$= 24.65$$

- **Landslides correction factor (Cf_5) =** The Hindustan-Tibet road named NH-22 alignment between Jeori and Khab, passing through slope wash debris, jointed gneiss and compact granites of Rakchham Granite formation which is widely affected by landslides in Sutlej valley, Kinnaur. According to Geological Survey of India, nine major landslides at Kinnaur are Poo, Mailing, Urni, Han, Akpa, Shillu, Shiasu, Dabling, and Chango. The limiting period of landslides is same as precipitation and frost time periods because rain and snow work as lubricants in increasing the quantity of landslides. During the period of 2000–09, 85 landslides events have occurred in Kinnaur that is the 2nd highest in Himachal Pradesh. In Himachal Pradesh during the period 1971–2009, 703 landslides events out of 919 have occurred during monsoon months from July to September (Kahlon et al. 2014). The total limiting period of this factor is 92 days.

$$Cf_5 = 92/365 \times 100$$
$$= 25.20$$

Man-Made Correction Factor

To determine the man-made correction factor, a field survey of 280 tourists on five-point Likert scale has been done in June 2016, to analyze their satisfaction from accommodation and transport facilities.

- **Accommodation correction factor (Cf_6) =** Out of the total sample tourists 63 tourists have not given high rating to accommodation availability at Kinnaur.

$$\begin{aligned} Cf_6 &= 63/280 \times 100 \\ &= 22.50 \end{aligned}$$

- **Transportation correction factor (Cf_7) =** The condition of road is an important element in accessibility. The study area lied at Trans-Himalayan topography where it has number of blind curves, deep valley, gorges, canyons, landslides, and many more. Therefore, the good condition is very important for a smooth transportation. During the field survey, 2016, author has found that 34.98 km length of roads was fully damaged out of total 262.1 km visited roads. The correction factor is 13.34 in the context of this factor when considered the total magnitude

$$\begin{aligned} Cf_7 &= 34.98/262.1 \times 100 \\ &= 13.34 \end{aligned}$$

22.3.1.3 Effective Carrying Capacity (ECC)

It is the maximum number of tourists that a spot can sustain, given the management capacity (Mc) that is available at a particular place. It is the results of the combination of RCC and available management capacities of the area. Management capacities (Mc) were based on the available infrastructure, facilities, parking, amenities, equipment, staff, etc (Table 22.2).

Table 22.2 Indicators of management capacity applied to assess ECC at tourist spots, Kinnaur

Indicators		Classification		
		Low (1)	Medium (2)	High (3)
Accessibility	Mc_1	Absent	Little	Adequate
Parking	Mc_2	Absent	Little	Adequate
Lighting	Mc_3	Absent	Little	Adequate
Public bathrooms	Mc_4	Absent	Little	Adequate
Drinking Water	Mc_5	Absent	Little	Adequate
Waste bins	Mc_6	Absent	Little	Adequate
Safety	Mc_7	Absent	Little	Adequate
Personnel	Mc_8	Absent	Little	Adequate
Food-joints	Mc_9	Absent	Little	Adequate
Presence of animals	Mc_{10}	Frequent	Moderate	Absent

The equation is as follows:

$$\mathrm{ECC} = \mathrm{RCC} \times \mathrm{Mc}_1 \times \mathrm{Mc}_2 \ldots \times \mathrm{Mc}_n$$

where

ECC = Effective carrying capacity

RCC = Real carrying capacity

Mc = Management capacities

22.3.2 Field Visits

A survey based on evaluation usually involves acquiring a desired or undesired result (Best and Kahn 1998). A simple random sampling method was applied to collect the required data such as comfortable space and time to visit any tourist's spots. The last section of questionnaire covered the questions concern with physical carrying capacity of destination. A total of 280 questionnaires were collected over the two-week period in June 2016 due to location of study area. To fulfill the requirement of formula given by IUCN, there is need to know the comfortable level of tourist on any destination in terms of space and time. The available area for tourism at 9 tourist spots was calculated through land use/land cover map which is based on imagery downloaded from open-source Google earth pro for the year 2016. In the context of area required per tourist, the average standard of individual comfort is depending on the decision of management and planners that may be varied from 2 to 10 m^2 depending on particulars of visited spot (Masum et al. 2013). Different studies have adopted different criteria of space requirement of a visitor, i.e., 1 m^2 per visitor (Sayan and Atik 2011; Queiroz et al. 2014), 2 m^2 per visitor (Masum et al. 2013), 2.5 m^2 per visitor (Daneshvar et al. 2017), 5 m^2 per visitor (Bera et al. 2015), and 5–10 m^2 for tourists in beaches or riverside (Zacarias et al. 2011). It may differ from person to person as well as destination to destination. Table 22.3 shows the observation of average time and area required by a tourist collected through field survey.

22.4 Results and Discussion

22.4.1 Assessment of Carrying Capacities at Selected Destinations, Kinnaur

Different types of carrying capacities have been assessed for nine destinations in Kinnaur district. To fulfill the requirement of IUCN guidelines recommended for

Table 22.3 Space and time for different zones assigned by tourists

Tourist spots	Activities/usage	Area required (m^2)	Average time		
			Nako	Chitkul	Kalpa
Temple	Movement	2	–	20 min	20 min
	Seating		–	15 min	15 min
Monastery	Movement		2 h	15 min	15 min
	Seating		15 min	15 min	15 min
Lake	Walking	5	30 min	–	–
River bank	Relaxation/enjoyment	10	–	4 h	–

Source Field survey (2016)

calculating carrying capacities, land use/land cover has been calculated by using Google Earth pro imageries.

22.4.1.1 Land Use/Land Cover of Selected Tourist Spots, Kinnaur

Religious Spots

Pilgrimage tourism is one of the major sectors in the growing Indian economy. The dominant religion is Hindu at lower Kinnaur and Buddhism in upper areas. Therefore, study area has rich cultural and religious identity with a number of Hindu temples and Buddhist monastery. The selected religious spots are as follows:

- Mathidevi temple
- Durga temple
- Chitkul monastery
- Devi Chandika temple
- Kalpa monastery
- Narayan-Nagni temple
- Nako monastery

For the estimation of physical carrying capacity, different categories were identified through land use/land cover maps which were made of selected tourist spots. Only those zones that can be used by visitors been considered for PCC analysis. Broadly, two major zones such as movement zone and seating stair have been used by tourists at religious spots in Kinnaur. The other zones, i.e., temple/monastery building, pedestal, slaughterhouse, surrounding space, storeroom are of relatively insignificance.

Natural Spots

Kinnaur has emerged as desirable tourist destination in Himachal Pradesh with its physiographic peculiarities and ensuing climatic conditions. An eco-system of cold desert as well as cultural uniqueness of tribal community make it unlike any other in the world. Moreover, there is a lot of opportunities in the sector of adventure tourism. Destination also offers a widespread range of possibilities such as unique natural attractions, hiking, skiing, rafting, kayaking, rock climbing, mountain biking, bungee jumping, and paragliding that attracts mass tourism. There two natural tourist spots that have been selected for assessing PCC are

- Baspa riverbank
- Nako lake

Baspa River offers a beautiful valley as well as numerous trekking trails which is a tributary of Sutlej River. It is also known as anglers region that have varieties of fishes found in river. River crossing is a popular activity done by tourists here. The bank of river is a soothing place that is used for sitting and relaxing. The “camp” is an adventure camp in the delightful valley that offers beauty of nature and calmness of hills along with fun-filled excitement. Besides, at the height of 11,000 feet “Nako Lake” is offering a most stunning Himalayan panorama. It is a quaint lakefront in middle of the remotest villages in Kinnaur. A foot like of “Saint Padmasambhava” has been built nearby lake. The periphery of lake has plantations of “willow” and “poplar” trees. The lake has facilities of ice skating practiced on the lake’s surface during the winter season when it gets frozen. Figure 22.2 shows the land use/land cover maps of identified categories of tourist spots in Kinnaur.

22.4.1.2 Estimation of Physical Carrying Capacity (PCC)

To investigate the PCC, there is a need to calculate and assess required data, i.e., suitable area for tourism, area available per user, rotation factor that depends on visit duration per day, etc. In the context of the present study, there had been dearth of official data on the number of tourists that visit particular tourist spots and other essential parameters. So, the carrying capacity was estimated empirically based on field observations done in June 2016. The areas of destination where tourists do camping have been considered to calculate the PCC. The daily open hours have been taken based on their timing limitations such as 13 h (sunrises to sunset’s timing) for natural protected areas (lake and riverbank) and worship timing hours (total 9 h) for religious spots to evaluate PCC for tourist spots in Kinnaur. Furthermore, formula given by IUCN was applied to calculate final PCC for each nine selected tourist spots located in three separate zones in Kinnaur (Table 22.4). The maximum PCC has been observed of Nako monastery with 20,022 number of visitors per day, whereas Durga temple has the minimum capacity, i.e., 784 visitors per day. As discussed in methodology, PCC depends on available land for tourism and duration (opening time) of visit. In Chitkul, the religious spots are situated within premises of

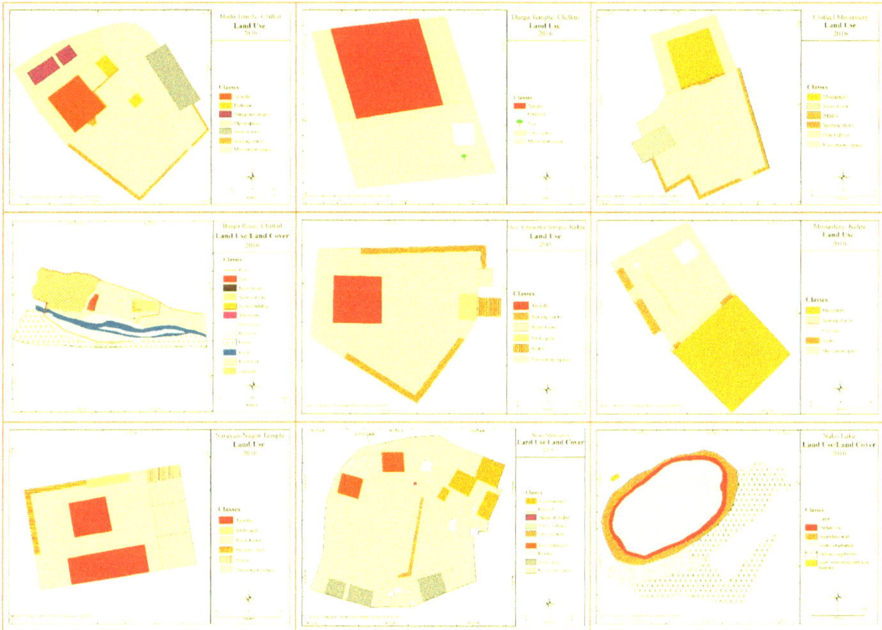

Fig. 22.2 Land use/Land cover of selected tourist destinations, Kinnaur. *Source* Google earth imagery 2016

settlement. Durga temple has less availability of open space within and periphery of temple. Similarly, Monastery has no basic infrastructure for the organization of any activities. The building of Mathi temple has adequate space for religious activities as compared to others in Chitkul. The bank of Baspa River attracts a lot of tourists, situated at the periphery of village settlements. A government school has been built in the premises of riverbank. Kalpa has number of famous religious tourist spots with its scenic natural beauty. Devi Chandika temple is located nearby the government school that offers an additional space for any occasion, whereas other spots have congested location due to surrounding settlement. Nako has famous and beautiful lake and an old monastery. In May 1998, the University of Vienna in association with the Indian National Trust for Art and Cultural Heritage (INTACH) launched a research project focusing to preserve art & culture. Similarly, in July 2002, the Nako Preservation Project (NPP) has been launched for the conservation of monastery and other cultural heritage buildings in Nako (District Census Handbook 2011). A new structure has been made nearby the heritage old monastery building with facility of parking. Therefore, maximum capacity has been noticed by Nako monastery. Besides, lake has a very congested pedestrian and the construction within buffer of lake affects the destination attraction adversely. Lower side is surrounded by village settlement and recently even the upper catchment area which is the source of water replenishment for the lake is also experiencing appearance of hotel, campsite, and guest house buildings.

Table 22.4 Physical carrying capacity estimation of selected tourist spots, Kinnaur

Destinations	Zones	Derived PCC (A × V/a × Rf)		PCC (Diurnal)
Chitkul				**26,668**
Mathidevi temple				9744
	Movement space	$666.78 \times \frac{1}{2} \times \frac{9}{20 \div 60}$	9002	
	Seating stairs	$41.24 \times \frac{1}{2} \times \frac{9}{15 \div 60}$	742	
Durga temple				784
	Movement space	$58.08 \times \frac{1}{2} \times \frac{9}{20 \div 60}$	784	
Chitkul monastery				4629
	Movement space	$236.52 \times \frac{1}{2} \times \frac{9}{15 \div 60}$	4257	
	Seating stairs	$20.67 \times \frac{1}{2} \times \frac{9}{15 \div 60}$	372	
Baspa river				11,511
	River bank	$35418.16 \times \frac{1}{10} \times \frac{13}{4}$	11,511	
Kalpa				**10,006**
Devi Chandika temple				6196
	Movement space	$396.56 \times \frac{1}{2} \times \frac{9}{20 \div 60}$	5354	
	Seating stairs	$46.8 \times \frac{1}{2} \times \frac{9}{15 \div 60}$	842	
Kalpa monastery				2467
	Movement space	$134.41 \times \frac{1}{2} \times \frac{9}{15 \div 60}$	2419	
	Seating stairs	$2.62 \times \frac{1}{2} \times \frac{9}{15 \div 60}$	47	
Narayan-Nagni temple				1343
	Movement space	$68.80 \times \frac{1}{2} \times \frac{9}{15 \div 60}$	1238	
	Seating stairs	$5.82 \times \frac{1}{2} \times \frac{9}{15 \div 60}$	105	
Nako				**28,161**
Nako monastery				20,022
	Movement space	$7998.18 \times \frac{1}{2} \times \frac{9}{2}$	17,996	
	Seating stairs	$112.54 \times \frac{1}{2} \times \frac{9}{15 \div 60}$	2026	
Nako lake				8139
	Pedestrian	$1565.28 \times \frac{1}{5} \times \frac{13}{30 \div 60}$	8139	
Total				**64,835**

Source Calculated by author as per IUCN guidelines

22.4.1.3 Estimation of Real Carrying Capacity (RCC)

To estimate the RCC, different natural and man-made correction factors were used which are specific to each tourists site. Total seven correction factors that are rainfall, snowfall, frost, landslide, temperature, accommodation, and transportation have been considered for this study (Table 22.5). In the present research, the total limiting

Table 22.5 Correction factor applied on Kinnaur

Corrective factor (Cf)	Alias	Cf	$Cf(100-Cf_1)/100$
Rainfall correction factor	Cf_1	16.98	0.83
Snowfall correction factor	Cf_2	36.98	0.63
Frost correction factor	Cf_3	24.65	0.75
Temperature correction factor	Cf_4	24.65	0.75
Landslides correction factor	Cf_5	25.20	0.75
Accommodation correction factor	Cf_6	22.50	0.77
Transportation correction factor	Cf_7	13.34	0.87

values of correction factor was about 0.157. The snowfall correction factor (Cf_2) had the greatest influence with limiting factor (0.63) on the overall RCC. The study area is situated in cold desert area of Himalaya hence in this context it led to considerable reduction in the number of tourists. The values of around 1 for the correction factor "transportation" indicate that it did not affect RCC. Instead, visitors enjoy the high terrain drive/biking that offering adventure in tourist visit. After applying the correction factors to PCC, the real carrying capacities were calculated of different selected tourist spots in Kinnaur (Table 22.6). The RCC was estimated as follows:

$$RCC = PCC \times Cf_1 \times Cf_2 \times Cf_3 \times Cf_4 \times Cf_5 \times Cf_6 \times Cf_7$$

As for the results of the correction factors, the total RCC of Kinnaur (for selected destinations) is 9595 visitors per day and it is the frequency of the maximum permissible number of people that should be allowed or support by particular tourist spots.

22.4.1.4 Estimation of Effective Carrying Capacity (ECC)

Generally, ECC is the sum of the conditions that tourism management of any destination requires if it is to carry out its objectives and functions. Therefore, in this context, optimal management capacity (Mc) is defined as the best state of conditions that have to develop its activities and reach its objectives by the administration of tourist destinations (Cifuentes et al. 1992). Correspondingly, ten management capacities were considered to calculate ECC according to the methodology of Cifuentes (1992). These variables are as follows:

Table 22.6 Correction factor applied to assess RCC at tourist spots, Kinnaur

Destinations	PCC (Diurnal)	Indicators							RCC (Diurnal)
		Cf_1	Cf_2	Cf_3	Cf_4	Cf_5	Cf_6	Cf_7	
Chittkul	**26,668**								**3941**
Mathidevi temple	9744	0.83	0.63	0.75	0.75	0.75	0.77	0.87	1440
Durga temple	784	0.83	0.63	0.75	0.75	0.75	0.77	0.87	116
Chitkul monastery	4629	0.83	0.63	0.75	0.75	0.75	0.77	0.87	684
Baspa river bank	11,511	0.83	0.63	0.75	0.75	0.75	0.77	0.87	1701
Kalpa									**1492**
Devi Chandika temple	6196	0.83	0.63	0.75	0.75	0.75	0.77	0.87	916
Kalpa monastery	2467	0.83	0.63	0.75	0.75	0.75	0.77	0.87	365
Narayan-Nagni temple	1343	0.83	0.63	0.75	0.75	0.75	0.77	0.87	211
Nako	**28,161**								**4162**
Nako monastery	20,022	0.83	0.63	0.75	0.75	0.75	0.77	0.87	2959
Nako lake	8139	0.83	0.63	0.75	0.75	0.75	0.77	0.87	1203
Total									**9595**

Source Calculated by author as per IUCN guidelines

- Accessibility
- Parking
- Lighting
- Public bathrooms
- Drinking Water
- Waste bins
- Safety
- Personnel
- Food-joints and
- Presence of animals

In addition, these selected components were evaluated on the basis of quantity and location criteria. Quantity just estimates the relationship between existing or optimal quantity which is based on the judgment of the author. Alternatively, location accounts for the appropriate spatial distribution of managing capacities at particular areas. Moreover, it was important to convert qualitative criteria into quantitative to undertake a more objective estimation of the Mc. Subsequently, all selected managing factors were evaluated at three levels, i.e., 1 (absent), 2 (little), and 3 (adequate). The Mc of each tourist spot was estimated by calculating the value of all selected variables. In case of Mathidevi temple, average score (0.53) of the management capacities (Mc) is based on the actual weighted score ($Mc_1 + Mc_2 + \ldots Mc_n = 16$) divided by the maximum possible weightage score of all indicators ($10 \times 3 = 30$). Further, according to RCC values and above mentioned managing factors, daily ECC values were estimated as per following formula:

$$ECC = RCC \times Mc$$

Table 22.7 expressed the number of visitors per day that ranged between 50 and 2367 persons per day at any tourist spots. These differences mainly depend on their facilities developed by local administration as well as due to correction factors included in the RCC. Each level constitutes a corrected capacity level of the preceding level thus PCC is always greater than the RCC and RCC is greater than the ECC. Consequently, ECC is more acceptable type of carrying capacity than the other two. Moreover, it is most useful for tourism management to develop policies. As Table 22.7 indicates the optimum number of tourists that would be allowed on the tourist spots with existing condition and management capacity, the total ECC for selected tourist spots in Kinnaur is 5928 visitors per day. Considering the total ECC as well as the yearly tourist flow in the selected tourist spots of Kinnaur, there is a risk of saturation or carrying capacity overload particularly in the peak seasons. Nevertheless, through better management practices, there is immense scope to enhance it many times. Further, technological interventions can reduce the impact of constraints such as rainfall, landslide, and poor roads to achieve the full potential of the selected destinations. Rather, some of the constraints such as snowfall can be converted into an additional attraction for the tourist like Rohtang (Himachal Pradesh) or Pahalgam (Jammu and Kashmir) by developing slopes of Chitkul and Kalpa for ski and other winter sports. The cost of developing infrastructure can be recovered in a very short period of time if world-class road infrastructure is created and landslides are controlled using the best of technical know-how at global level. It will be a boon to the area if exploited systematically.

22.5 Conclusion and Recommendations

Tourism has a diverse type of impacts and assessing carrying capacity is vital to address negativity in the system if any. They are dependent on different parameters and also varying with the changing nature of interaction. A three-level procedure to estimate physical, real, and effective carrying capacities were applied with seven correction factors (Cf) and ten management capacities (Mc). From the present study results, the PCC, RCC, and ECC value of total selected tourist spots in Kinnaur are 64,835, 9595, and 5928, respectively. The maximum PCC has been observed of Nako monastery with 20,022 number of visitors per day, whereas Durga temple has the minimum capacity, i.e., 784 visitors per day. As a cold desert location, major constraints snowfall and landslide are worked as barriers in the development of

Table 22.7 Indicators of management capacity (Mc) applied to assess ECC at tourist spots, Kinnaur

Destinations	RCC (Diurnal)	Indicators										Mc	ECC (Diurnal)
		Mc_1	Mc_2	Mc_3	Mc_4	Mc_5	Mc_6	Mc_7	Mc_8	Mc_9	Mc_{10}		
Chittkul	**3941**												**2180**
Mathidevi temple	1440	1	1	2	1	1	2	3	1	1	3	0.53	763
Durga temple	116	1	1	1	1	1	1	3	1	1	2	0.43	50
Chitkul monastery	684	1	1	1	1	1	1	3	1	1	2	0.43	295
Baspa river bank	1701	3	3	1	2	2	1	3	1	1	2	0.63	1072
Kalpa	**1492**												**780**
Devi Chandika temple	916	1	1	2	1	1	2	3	1	1	3	0.53	486
Kalpa monastery	365	1	1	2	1	1	1	3	1	1	3	0.50	182
Narayan-Nagni temple	211	1	1	2	1	1	2	3	2	1	3	0.53	112
Nako	**4162**												**2968**
Nako monastery	2959	3	3	3	2	2	2	3	2	1	3	0.80	2367
Nako lake	1203	1	1	1	1	1	3	3	1	1	2	0.50	601
Total	**9595**												**5928**

Note: Absent (1), Medium (2) and Adequate (3)
Source Calculated by author as per IUCN guidelines

tourism at Kinnaur. To minimize the impact of natural constraints, there is a need to develop good infrastructure comparable to tourist destinations in Europe and Russia that may minimize the man-made constraint and help to increase the capacity of the destinations. ECC is the most acceptable type of carrying capacity, and there have dearth of management practices/facilities at all tourist spots that would have been very helpful to increase the capacity of the particular tourist spots. A number of destinations like Museum, Akshardham temple, and many more have good management practices which have increased the number of visitors. It is found that the present status of tourism activity in the study area is highly under exploited vis-a-vis its carrying capacity. Like the daily limit fixed of the number of tourists that can visit TajMahal, authorities should fix the upper limit of visitors to each destination for managing sustainability at Kinnaur. Present research may also be very helpful to restrict the number of tourist arrivals in Kinnaur and regulate the tourism.

Acknowledgments Article is part of a Ph.D. research thesis on the topic "*Assessment of Tourism Carrying Capacity in Selected Destinations of Cold Desert Area of Kinnaur District, Himachal Pradesh*" awarded in September 2019. Authors acknowledged to research team who helped in field survey at cold desert destination and also thankful to ICSSR (Indian Council of Social Science Research), Government of India, for financial support in carrying out the said work under RPR research project scheme.

References

Acevedo-Ejzman M (1997) Determinacio´n de la capacidad de cargaturı´stica en dos sitios de visitadel Refugio de Vida Silvestre La Marta, e identificacio´n de supunto de equilibriofinanciero. Universidad Latinoamericana de Ciencia y Tecnologı´a, San Jose

Aryasa AM, Bambang AN, Muhammad F (2017) The study of environmental carrying capacity for sustainable tourism in TelagaWarnaTelagaPengilon Nature Park, DiengPlateu, Central Java. Proceedings of 3rd international conference of planning in the era of uncertainty, IOP publishing

Bajpai SC (1981) Kinnaur in the Himalayas: mythology to modernity. New Delhi, Concept Publishers

Bera S, Majumdar DD, Paul AK (2015) Estimation of tourism carrying capacity for Neil island, south Andaman India. J Coast Sci 2(2):46–53

Best JW, Kahn JV (1998) Research in education. Prentice Hall, New York

Boullon RC (1985) Planification Del EspacioTuristico. Trillas, Mexico

Cayot L, Cifuentes M, Amador E (1996) Determinacio´ndelacapacidad de cargaturı´stica en los sitios de visitadelParqueNacionalGala´pagos. ServicioParqueNacionalGala´pagos, InstitutoEcuatorianoForestal y de A ´ reasNaturales y Vida Silvestre

Ceballos-Lascurain H (1996) Tourism, ecotourism and protected areas: the state of nature- based tourism around the world and guidelines for its development. Gland and Cambridge

Cifuentes AM (1992) Determinacion de capacidad de cargeturistica en areas protegidas. CATIE, Turrialba, Costa Rica

Cifuentes M, Alpizar W, Barroso F, Courrau J, Falck ML, Jimenez R, Ortiz P, Rodriguez PV, Romero JC, Tejada J (1990). Capacidad de CargaTurística de la ReservaBiológicaCarara. Informe de Consulta, Servicio de ParquesNacionales/Programa de ManejoIntegrado de Recursos, Centro Agronómico Tropical de Investigación y Enseñanza (CATIE).Turrialba, Costa Rica

Cifuentes M, Alpizar F, Barroso F (1992) Determinación de capacidad de carga turística en áreas protegidas. CATIE Serie técnica, Turrialba

Coccossis C, Mexa A (2004) The challenge of tourism carrying capacity assessment. Ashgate Publishing, Ashgate

Daneshvar MRM, Khatami F, Zahed F (2017) Ecological carrying capacity of public green spaces as a sustainability index of urban population: a case study of Mashhad city in Iran as a sustainability index of urban population Model. Earth Syst Environ 3:1161–1170

Davis D, Tisdell C (1995) Recreational scuba-diving and carrying capacity in marine protected areas. Ocean Coast Manag 26(1):19–40

District Census Handbook (2011). Village and town directory, primary census abstract, Census of India, Kinnaur (Himachal Pradesh). Directorate of Census Operations Series (3) Part XII B. Shimla

East P, Luger K, Inmann K (eds) (1998) Sustainability in mountain tourism: perspectives for the himalayan countries. Book Faith India and StudienVerlagAudtria, Innsbruck-Vienna

Gallo F, Martı´nez A, Rı´os JI (2003) Capacidad de carga de visitantes en areas de buceo de San Andres Isla (Colombia). Universidad Tecnologica de Pereira, Pereira

Hasan SR, Hasan K, Islam S (2014) Tourist-group consideration in tourism carrying capacity assessment: a new approach for the Saint Martin's island, Bangladesh. J Econ Sustain Dev 5(19):150–158

Kahlon S, Chandel VBS, Brar KK (2014) Landslides in Himalayan mountains: a study of Himachal Pradesh, India. Int J IT, Eng Appl Sci Res (IJIEASR) 3(9):28–34

Kruk E, Hummel J, Banskota. (2007) Facilitating sustainable mountain tourism resource. Int Centre Integr Mount Dev 1:3–18

Lagmoj MA, Shokry AK, Hashemi SA, Zadegan HK (2013) Defining the ecotourism carrying capacity of Langeroud city case study: Khorma forest. Greener J Soc Sci 3(9):447–457

Lawson SR, Manning RE, Valliere WA, Wang B (2003) Proactive monitoring and adaptive management of social carrying capacity in Arches National Park: an application of computer simulation modeling. J Environ Manag 68:305–313

Lucas RC (1964) Wilderness perception and use: the example of boundary water canoe area. Nat Resour J 3(3):394–411

Masum KM, Mamun AA, Rahman ZMM, Newaz MS, Redowan M, Science F (2013) Ecotourism carrying capacity and the potentiality of the safari park of Bangladesh. J Forest Sci 29(4):292–299

Mitraud, S (1998) Projeto de Desenvolvimiento de TilhasIterpreta- tivas do ParqueNacionalMarinho de Fernando de Noronha-PE. Actividade 3: Determinac¸a˜o da Capacidade de Cargae Sistema de Monitoramento de Impacto de Visitac˜o, VelatorioTe¸a´cnico 2, Convenio IBAMA-WWF No.006.97, WWF-Brazil

Nghi T, Lan NT, Thai ND, Mai D, Thanh DX (2007) Tourism carrying capacity assessment for PhongNha -Ke Bang and Dong Hoi, QuangBinh Province. VNU J Sci Earth Sci 23:80–87

Papageorgiou K, Brotherton I (1999) A management planning framework based on ecological, perceptual and economic carrying capacity: the case study of Vikos-Aoos National Park Greece. J Environ Manag 56(4):271–284

Pigram P (1983) Outdoor recreation and resource management. St. Martin's Press, New York, NY

Queiroz RE, Ventura MA, Guerreiro JA, Tristao da Cunha R (2014) Carrying capacity of hiking trails in Natura 2000 sites: a case study from north Atlantic islands (Azores, Portugal). J Integr Coas Zone Manag 14(2):233–242

Rodriguez V (1992) Determinación de Capacidad de Carga Turística para el Parque Nacional Manuel Antonio. Dissertation, Centro Agronómico Tropical de Investigación y Enseñanza

Ruschmann D, Paolucci L, Maciel N (2008). Capacidade de carga no planejamentoturístico: estudo de caso da Praia Brava–Itajaífrente à implantação do ComplexoTurísticoHabitacional Canto daBrava. *RevistaBrasileira de PesquisaemTurismo.* 2(2):41–63 (NOTE)

Sanan D, Swadi D (2002) Exploring Kinnaur & Spiti in the Trans-Himalayas, 2nd edn. Indus Publishing, India

Sayan MS, Atik MS (2011) Recreation carrying capacity estimates for protected areas: a study of Termessos National Park. J Ecol 20(78):66–74

Sharma P (2000) Tourism as development: case studies from the Himalaya. Himal Books and StudienVerlag, Innsbruck

Shelby B, Heberlein TA (1986) Carrying capacity in recreation settings. Oregon State University Press

UNWTO (2017) UNWTO tourism highlights: 2017 edition. https://www.e-unwto.org/doi/pdf/10.18111/9789284419029

Wagar J (1964) The carrying capacity of wild lands for recreation. Forest Service Monograph, Society of American Foresters 7:23

Wall G (1983) Cycles and capacity: a contradiction in terms! Ann Tour Res 268–269

Zacarias DA, Williams AT, Newton A (2011) Recreation carrying capacity estimations to support beach management at Praia de Faro, Portugal. Appl Geogr 31:1075–1081

Zhong L, Deng J, Song Z, Ding P (2011) Research on environmental impacts of tourism in China: progress and prospect. J Environ Manage 92(11):2972–2983

Chapter 23
Collaborative Creation of Educational Geo Routes: A Strategy for Teaching and Learning Sciences and Geography, Puchuncaví, Chile

Andoni Arenas-Martija, Jael Orrego-Araya, Carolina Ortiz-Pavéz, Camila Gajardo-Tapia, Nicolás Zanetta-Colombo, Karla Inzunza-López, Nataly Cisternas-Silva, Carolina Córdova-Veas, Ana María Goehler-Marchant, Natalia Báez-Córdova, Valeska Bustamante-Bernal, Ximena Galleguillos-Bustamante, Karolyn Gavilán-Maturana, María Gómez-Miqueles, Cynthia Iturrieta-Ortiz, Jeannette Nilo-Fuentes, Carolina Báez-Donoso, Irene Constanza Chacana-Faras, and Ximena Rivera-Silva

Abstract This chapter describes the collaborative development of a teaching innovation called Educational Geo Routes (GRE, for its name in Spanish), with a group of teachers working in public schools of the city of Puchuncaví, Chile. We will first describe the theoretical–practical components that give meaning to the GREs as well as the educational emphasis we have tried to give to them. We will then describe

This chapter has been prepared in the context of the researcher "Teacher Professional Development Trajectories in the field of science education: (Re)framing the contribution of collaborative work" funded by the National Science and Technology Fund – Chile (FONDECYT N° 1,181,834, 2017–2020).

A. Arenas-Martija (✉)
Geography, Education and Social Actors of Space Laboratory. Institute of Geography, Pontificia Universidad Católica de Valparaíso, Valparaíso, Chile
e-mail: andoni.arenas@pucv.cl

J. Orrego-Araya · C. Ortiz-Pavéz · C. Gajardo-Tapia
Graduated Students Pedagogy in Primary Education, Pontificia Universidad Católica de Valparaíso, Valparaíso, Chile
e-mail: jael.nooemi@gmail.com

C. Ortiz-Pavéz
e-mail: am.carolinaortiz@gmail.com

C. Gajardo-Tapia
e-mail: camila.gajardo.t@mail.pucv.cl

N. Zanetta-Colombo
Versus Sole and Research Associate Geography, Education and Social Actors of Space Laboratory, Pontificia Universidad Católica de Valparaíso, Valparaíso, Chile
e-mail: nicolaszanetta@gmail.com

R. B. Singh et al. (eds.), *Global Geographical Heritage, Geoparks and Geotourism*, Advances in Geographical and Environmental Sciences,
https://doi.org/10.1007/978-981-15-4956-4_23

the context in which its design took place and the collaboration focus based on the interactive teaching training models for both the work between teachers and the relationship of the teachers with the support team. The designing process was developed during 2018 and was focused on the relationship between the socio-environmental context of the students and their schools with the national curriculum. We will present some of the GREs designed by the teachers and the results obtained when evaluating this experience, which led us to conclude that GREs provide a strong strategy for teachers to achieve relevant and quality learning in their students. In addition, we refer to the benefits of collaborative work for the participants without setting aside the personal, professional and institutional elements that promote or limit it.

Keywords Educational geo routes · Science teaching · Geography teaching · Continued training · Collaborative work · Teacher professional development

23.1 Introduction

23.1.1 The Sources for Educational Geo Routes

The GRE proposal draws on elements broadly described from other Scientific-educational strategies: Field Work, Pedagogical Field Trips and Tourist Routes. These characteristics are combined with the addition of educational value to the

K. Inzunza-López
Los Maquis School, Puchuncaví, Chile
e-mail: profekarin@hotmail.com

N. Cisternas-Silva · C. Córdova-Veas
General José Velásquez School, Puchuncaví, Chile
e-mail: nat.cisternas.s@gmail.com

C. Córdova-Veas
e-mail: carola.corve@gmail.com

A. M. Goehler-Marchant · J. Nilo-Fuentes
Maitencillo Elementary School, Puchuncaví, Chile
e-mail: amgoehler@gmail.com

J. Nilo-Fuentes
e-mail: monita6927@gmail.com

N. Báez-Córdova · V. Bustamante-Bernal · X. Galleguillos-Bustamante · K. Gavilán-Maturana · M. Gómez-Miqueles · C. Iturrieta-Ortiz
The Chocota School, Puchuncaví, Chile
e-mail: nakbaez@gmail.com

V. Bustamante-Bernal
e-mail: vale.anto@live.cl

X. Galleguillos-Bustamante
e-mail: xigabu@hotmail.com

socio-spatial context of the schools, students and teachers. This is conceived as a great opportunity to teach contents and achieve academic goals with the students, strengthening the spatial identity of boys, girls and teenagers for active citizenship (Arenas et al. 2018). It also evidences the difficulties teachers face in order to design more complex and inclusive learning strategies. This challenges teachers to use all their knowledge and pedagogical skills to integrate elements that are disciplinary (Natural Science–Geography), pedagogical (curricular–evaluation) and contextual (socio-environmental problems, heritage components) in order to guide, through education, the relationships between the spatial context of students with their formal learning. These are matters of the very recent development in Chile but have a long tradition in other contexts (Collet-Sabé and Subirats 2016; Copatti et al. 2018).

GREs are based on Field Work, on its long scientific tradition and especially in its educational dimension (Pedone 2000; Grindsted et al. 2013). It is valued that the world that has to be explained and understood needs to be known as directly as possible, needs to be explored in order to describe its features, explain its components and understand its spatial relationships (Godoy and Sánchez 2007). For this, it is necessary to collect information, make observations and use a different range of techniques, instruments and methodological procedures (Scoffham 2010).

For Lara de González (2011), Field Work becomes a teaching–learning group strategy in which the Teacher adapts the contents to the interests and academic level of students in order to motivate and connect learning of contents with the learning of procedures in specific situations. Following Godoy and Sanchez (2011), we can say that Field Work not only is a generic teaching strategy, but is also indispensable in Sciences as it is in Geography, since it allows to obtain meaningful experiences in order to understand phenomenon of the earth surface, its current problems and its consequences. This is an open lab, a unique opportunity to awake curiosity in students that will lead them to discover the information that appears imprinted in a natural–cultural context (Godoy and Sanchez 2011). Quintero (2005) refers to this strategy as an experience that opens a space for re-observation, analysis and creativity.

K. Gavilán-Maturana
e-mail: karolyngavilan@gmail.com

M. Gómez-Miqueles
e-mail: mceciliagomezm@hotmail.com

C. Iturrieta-Ortiz
e-mail: cviturri@gmail.com

C. Báez-Donoso · I. C. Chacana-Faras
La Laguna Elementary School, Puchuncaví, Chile
e-mail: baezcarolina166@gmail.com

I. C. Chacana-Faras
e-mail: chacanacontanza@gmail.com

X. Rivera-Silva
El Rungue Elementary School, Puchuncaví, Chile
e-mail: ximenita-isi@hotmail.com

Pedagogical Field Trips are planned academic activities that are carried outside the school. They are practical in nature, framed in learning units with the goal of promoting the development of skills, and linking content with the context of students (Mohamed et al. 2017). This didactic strategy promotes learning from experience, favoring educational, social and personal development of students, strengthening their curiosity (Romero 2010).

Pedagogical Field Trips in Chile do not have a pedagogical orientation. The Ministry of Education (2014), in its national guidelines, establishes the emphasis on administrative procedures that will protect participants' safety and integrity. Another guiding element are the Tourist Routes (Figueroa, Negrete and Moreira, 2021) that draw on the importance of geodiversity sites, land registries and inventories of attractions, available tourist services, and cultural activities and resources, which intends to determine relevant milestones and the most appropriate routes. In educational terms, in 1995 Sánchez had already mentioned the need of developing didactic itineraries that will promote geographic conceptualization, procedure dexterity, observation skills and in situ experimentation (Sánchez 1995).

23.2 GREs as Teaching–Learning Strategies

Academic tasks are "*events from the classroom that provide opportunities for the students to use their cognitive and motivational resources in favor of achieving personal and educational goals*" (Winne and Marx 1989: 242). They are varied and dynamic and are carried out considering contextual and personal aspects of whom they are directed to (Paoloni and Rinaudo 2009). So, GREs would be a group of interrelated academic tasks. On the other hand, Feo (2010) states that didactic strategies are procedures, methods, techniques and activities that teachers carry out intentionally for the development of their students. An outstanding feature of them is that they have a high level of adaptation to the particular needs of each context: student diversity and content objectives of complex learning.

For Arenas-Martija and Margalef-García (2006), Teaching–Learning Strategies are the ensemble of activities, tasks and procedures that have a common purpose. Zabala (2001) notes that this ensemble and these goals are set as an action course, a process and a sequence that has a specific order given by certain criteria. Estrada (2014) proposed that teaching strategies are all the acts, activities, processes or procedures a teacher uses in a flexible and reflective way to promote and stimulate the achievement of meaningful learning in students.

Díaz-Barriga and Hernández (2010), Arenas-Martija (2009) establish the difference between teaching strategies and learning strategies. The latter are "*a group of interrelated functions and resources capable of generating action schemes that make it possible for the student to face general and specific situations of their learning in a more effective way*" (González 2003: 3), which means that they are student-centered and focus on their academic work and learning.

Under the previous premises, GREs are teaching and learning strategies, since it is the Teacher who designs the actions and procedures in order to achieve learning in their students, at the same time her students combine reflection from experience, theoretical understanding and practical activities when they work on the geographical, cultural and social contexts to which they belong.

23.3 Continuous/Permanent Teacher Training

23.3.1 Education Models

According to Justi and Van Driel (2006), there is agreement on the importance of Teacher Professional Development. However, this agreement becomes relative when addressing matters regarding their research and promotion. Following Sprinthall et al. (1996), Justi and Van Driel indicate that there are three models to explain the professional development of teachers.

The first model states that teachers develop as the result of *becoming experienced teachers*, through their profession. Their development comes from their classroom experiences although it is not clear why some teachers only replicate the same experience many times without moving beyond it. For its part, *the expert model* is centered around teachers learning to teach through an expert in a training fashion, generally outside the classroom and with the perspective of *"learning in order to apply"*. We understand here that the experts hold the knowledge (Vezub 2013). Because of this, the relationships that this model produces are vertical and oriented toward correcting teacher shortfalls. For Ávalos (2007), the external pole is emphasized: the needs of the educational system and its reforms with the purpose of providing new predetermined guidelines and work models for teachers.

Interactive models promote active learning processes where the needs and experiences of teachers are important, as well as dialogue with the school practice. These models intend to recover practical knowledge, experiences, needs and problems emerging from the classroom, understanding that theoretical knowledge and educational research knowledge are nurtured and articulated by teacher practical knowledge (Vezub 2013). Although most of these spaces are directed and supported by teachers' non-peer experts under different training forms, experts contribute through questioning, connecting their work with social, political and cultural aspects. The teacher here views their practice, classrooms and institutions as research and reflection spaces, in which it is possible to discuss from an internal pole (Ávalos 2007) the interests and needs of teachers through horizontal and collaborative relationships.

For the GRE design process, we took some elements of the expert model, but mostly from the interactive models. This is explained in greater depth in Sect. 23.3.

23.3.2 *Difficulties in Permanent Teacher Training in Sciences and Geography*

Considering the proposals by Vezub (2013) about the "ineffectiveness" of teacher improvement and the little value that has been given to the teachers' perspective about their own professional learning, the critical exam of the training models has been limited. Therefore, the difficulties that are seen in the training of Science and Geography teachers stem not only from the general models, but also from specific issues.

According to Parga (2016), the social, political, cultural and economic changes that are currently evidenced in society demand the transformation of educational systems, and therefore, the review of teacher training. Imbernón (2011) notes that a difficulty in teacher education systems is the lack of coordination, follow-up and evaluation from the institutions and services that are part of the education plan. Miranda and Rivera (2009) claim that little progress has been made in the understanding of the role of teacher educators, not only in relation to their impact on teacher learning but on their personal and professional character as well. In this sense, it is the key to review the role they have and the training model where the teacher educator is situated.

Regarding continuous training in Sciences, according to Miranda et al. (2010) a relevant difficulty is that in the last levels of primary education, sciences are taught by teachers whose initial generalist education does not adequately train them in relation to disciplinary and didactic content for scientific literacy. This problem is similar to what happens in Geography education that, being in the same class as History, is given less importance than the latter.

Another problem is the impact of research in teacher education: There is a wide range of research related to Science Education such as Geography, but a minority of teachers read, reflect and adapt the results to their educational practices (Parga 2016). In the case of Geography, continuous training is hindered by the current teaching that is distanced from developing academic research and socio-territorial topics of current public interest (Fernández 2008).

23.4 The Experience

23.4.1 *Socio-environmental Context*

On the morning of August 21, 2018, more than 50 students had to be evacuated from their schools in the cities of Quintero and Puchuncaví due to fainting, nausea

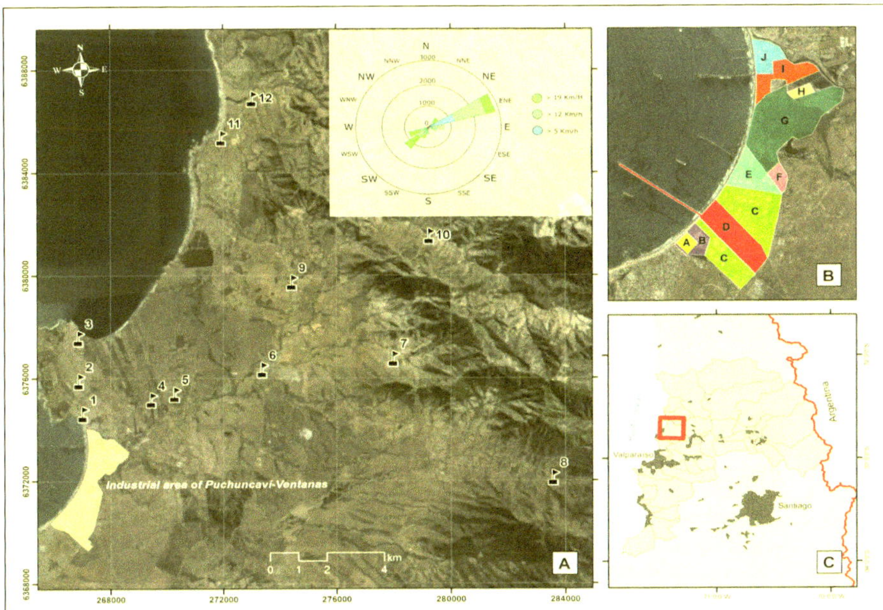

Quadrant (A) presents the location of the Schools of the Puchuncaví city and the Ventanas Industrial Complex (CIV for its name in Spanish). Quadrant (B) presents the detail of the companies that form the CIV. Quadrant (C) presents the location of the Puchuncaví city in the context of the Valparaiso-Chile region. The name of the schools, GRE and CIV companies' milestones, are presented in Table 1.

Source: Own elaboration

Fig. 23.1 General Situation of the Work Area

and vomiting.[1] At the end of the week, the number of intoxication cases admitted to medical centers of the area was greater than 350, making this one of the most serious events happening in the central area of Chile in the last years[2] (see Fig. 23.1 and Table 23.1, where you can see the industrial complex, the location of schools and the predominant direction of wind).

In November of 1968, an official document from the Ministry of Agriculture received the complaints of landowners in the cities of Puchuncaví and Quintero (north coast of the Valparaiso Region), in which they reported the loss of their crops

[1]Investigations made on toxic cloud that affected more than 30 children in Quintero. Daniela Astudillo. https://www.latercera.com/nacional/noticia/indagan-origen-nube-toxica-afecto-mas-30-ninos-quintero/290017/ Recovery date: December 10, 2018.

[2]Chronology of an emergency: The 15 days of Quintero not knowing the origin of toxic episodes. Gonzalo Núñez and Bárbara Osses. https://www.emol.com/noticias/Nacional/2018/09/04/919543/Cronologia-de-una-emergencia-Los-15-dias-de-Quintero-sin-conocer-el-origen-de-los-episodios-toxicos.html Recovery Date: December 10, 2018.

Table 23.1 Names of Schools of the Puchuncaví City and the Ventanas Industrial Complex Companies. The details of their location are presented in Fig. 23.1

Schools		Ventanas Industrial Complex	
Quadrant A	ID	Quadrant B	ID
Sarge Aldea Educational Complex	1	SHELL (Fuels)	A
The Chocota School	2	COPEC (Fuels)	B
Horcón School	3	ENAP (Petroleum)	C
La Greda School	4	ENAP (Petroleum)	C
Campiche School	5	Quintero Liquid Gas Terminal (Liquid gas)	D
General José Velásquez School	6	Oxiquim S.A. (Chemical products)	E
El Rincón School	7	GASMAR (Liquid gas)	F
Los Maquis School	8	CODELCO Ventanas (Copper Refinery)	G
El Rungue School	9	Bio Bio Cements	H
La Quebrada Elementary School	10	Ventanas Port S.A	I
Maitencillo Elementary School	11	AES Gener S.A. (Generation of Electricity)	J
La Laguna School	12		

Source Own elaboration

as a result of the settling of a copper refinery and foundry in the Ventanas division of the National Copper Corporation (CODELCO, by its name in Spanish), 4 years prior.

Today, 50 years later, in the 50 thousand people that inhabit the area, the majority of the students and teachers of the Puchuncaví schools among them have had to live with the operations of 15 companies and a series of pollution episodes: at least 3 major hydrocarbon spills in the ocean in the last 3 years, almost 300 discharges of coal in the beach and massive intoxication of people, mainly children. In 2011, the La Greda Public School had to be relocated because of the presence of heavy metals in the students' bodies.

23.4.2 The Group of Sciences and Geography Teachers

The Sciences and Geography Network is a group formed in 2017, primary and secondary teachers that work in different public schools of the Puchuncaví city, in the Valparaiso region, Chile.

The network was formed with all the teachers who participated in a program called Scientific Inquiry for the Education in Sciences (ICEC, for its name in Spanish), between October of 2015 and July of 2016. This program had the purpose of "*strengthening teaching and learning of the Natural Sciences in public primary and secondary schools of our country and contribute to teacher professional development through training on scientific inquiry*".[3] During the year 2017, they shared their experiences in teaching science: innovations, work spaces and learning; they also made exhibitions about teaching science within and outside the commune of Puchuncaví. From November of 2017 to May of 2018, some of these teachers continued in a second ICEC for more in-depth work.

At the beginning of 2018, the Department of Administration for Municipal Education (DAEM for its name in Spanish) of Puchuncaví[4] requested support for the Science Network from an interdisciplinary team performing research on collaborative work and science teaching.[5] The team proposed to perform a theoretical–practical work that would consolidate this budding network out of the need of linking the territory with the Natural Sciences and Geography curriculum, in a shared work space, through the creation of Educational Geo Routes.

During 2018, we carried out different work sessions (see Table 23.2) that contributed to the creation of one Educational Geo Route for each one of the participant teachers, and as a group, the configuration of didactic material for schools of the city and other teachers and schools in the country (Figs. 23.2, 23.3, 23.4, 23.5, 23.6 and Tables 23.3, 23.4, 23.5, 23.6, 23.7 and 23.8).

23.4.3 The Permanent Training Proposal

The proposal considers the needs of teachers, and that practical knowledge is a source of knowledge in itself, that nurtures and strengthens educational research. Therefore, the workshop planning considered the teacher as an active, reflective and critical participant, capable of generating knowledge according to their own experiences.

We planned a sequence of 8 half-day workshops, distributed monthly, starting in the month of May. Out of the 8 workshops, only 6 were implemented, due to the socio-environmental conflicts that took place in the area and that caused the suspension of classes for a month. Two of the workshops were carried out during one full day in order to make up for the two missed sessions.

---

[3]The Scientific Inquiry Program for the Education in Sciences from the Ministry of Education in Chile is carried out in the Valparaiso region by the Pontificia Universidad Católica de Valparaíso. https://icec.ucv.cl/web/ Recovered: December 16, 2018.

[4]The Department of Administration of Municipal Education (DAEM) has the goal of designing and implementing the educational project of the city (Municipality), through the administrative, financial and technical-pedagogical management of schools under its jurisdiction, with the purpose of improving their quality and equality. This includes management of human resources that work in schools.

[5]Teacher Professional Development Trajectories in the field of science education: (Re)framing the contribution of collaborative work, FONDECYT N° 1,181,834, 2017–2020.

Table 23.2 Summary of the executed workshop sequence

Workshop 1	*Date: May 23*	*Topic: support proposal to Science Group*	
General objective	*Specific objectives*	*Step-stages*	*Learning products*
To visualize the curricular and territorial context of the GREs	1. To situate the GRE as a didactic strategy 2. To diagnose the potential and problems of Scientific–Geographic Education 3. To diagnose the potential and problems of the territorial context of Puchuncaví	1. Greetings—Purpose of sessions 2. Presentation of the driving ideas: The GRE as a Teaching–Learning Strategy and its relation to Socio-Scientific Issues and Socially Relevant Issues and the National Curriculum 3. Target Shoot Technique (Vargas and Bustillos 1990) 4. Schedule and Closing	– Hierarchically organized list of difficulties and opportunities of science and geography teaching in the Puchuncavi area – Diagnosis performed by the Science Group regarding their experiences and collective purpose (see Fig. 23.2)
Workshop 2	*Date: June 22*	*Topic: participative cartography*	
General objective	*Specific objectives*	*Step-stages*	*Learning products*
To introduce general aspects of Google Earth and participative cartographies as useful tools for the construction of the GRE	1. To know the uses of Google Earth 2. To learn about participative cartographies 3. To acknowledge possible geographical milestones to visit through the participative cartography (Azócar 2016)	1. Greetings—Purpose of the session 2. Introduction 2.1 Presentation of Google Earth 2.2 Presentation of participative cartographies 3. Participative Cartography Workshop 3.1 Distribute participants in groups 3.2 Answer a territory-related SWOT evaluation 3.3 Plenary session 4. Closing and schedule of next activities	– Represent the SWOT analysis through the participative cartography of the Puchuncavi city (see Fig. 23.7) – First approach to the use of cartographic software, Google Earth
Workshops 3–4	*Date: August 31*	*Topic: design of 1 GRE*	
General objective	*Specific objectives*	*Step-stages*	*Learning products*

(continued)

Table 23.2 (continued)

To design a GRE from the given chart	1. To learn the curricular contents that could be addressed in GREs 2. Use Google Earth 3. Present chart using a completed example	1. Greeting—Purpose of Workshop 2 2. Contents that can be addressed in a GRE 3. Use of Google Earth 3.1 Practice 3.2 GRE Examples, presentation of GRE chart 3.3 Plenary session 4. Closing and schedule of next activities	– Google Earth Practice – GRE Chart after feedback and modifications by participants
Workshop 5	*Date: September 26*	*Topic: curricular mapping*	
General objective	*Specific objectives*	*Step-stages*	*Learning products*
To analyze the agreement between GRE and curricular objectives	1. To give feedback to GREs designed by teachers 2. Curricular mapping exercise (Jacobs 2004) 3. To analyze themes addressed in GREs	1. Greeting—Purpose of the workshop 2. Presentation and group feedback to GREs from the model (Anijovich 2010) 3. GRE Analysis with the curricular framework 4. Plenary session 5. Closing and Schedule of next activities	– Analysis of contents, skills and procedures set in the curriculum and those that can be addressed in the GREs – Analysis of each GRE's theme (see Fig. 23.3)
Workshop 6	*Date: November 21*	*Topic: design of GRE 2*	
General objective	*Specific objectives*	*Step-stages*	*Learning products*
To perform the GRE route in Google Earth	1. To give feedback to GREs designed by teachers 2. To learn how to trace a route on Google Earth (Robinson et al. 1995) 3. To trace GRE in software	1. Greeting—Purpose of the workshop 2. Individual feedback of GRE 3. Google Earth route practice 4. Discussion about the design of possible publishing of GREs 5. Closing and Schedule of next activities	– Google Earth Route Maps – Preliminary version of GREs
Workshop 7–8	*Date: December 21*	*Topic: submission–closing*	
General objective	*Specific objectives*	*Step-stages*	*Learning products*

(continued)

Table 23.2 (continued)

To perform GRE designed by the research team	1. To experience a GRE from the perspective of a student 2. To evaluate the GRE creation process (López de Ceballos et al. 2001) 3. To comment on the projections of the Science Network	1. Greeting—Purpose 2. Execution of GRE 2.1 Milestone 1—Ventana Lookout 2.2 Milestone 2—Playa Larga, Horcón Cove 2.3 Milestone 3—Former Melinka Detention Center 3. Evaluation of workshops carried out by the Science Network 4. Plenary session 5. Presentation of Scientific–Environmental Research School Kit 6. Projections and Closing of year	– Experiencing a GRE in the city of Puchuncaví as a participant (see Figs. 23.5 and 23.6) – Completion of the GRE work chart – Results of group evaluation (see Fig. 23.4)

23.4.4 The GRE Designing Process

Table 23.2 presents a summary of the executed workshop sequence and later some important elements are specified. It must be noted that workshops 3 and 4 were carried out during the same session, as well as workshops 7 and 8, due to adjustments that had to be made as a result of the socio-environmental conflict in Puchuncaví.

23.5 Valuation of the Experience

To evaluate the experience, we performed 3 semi-structured interviews (Kvale 2011) and an evaluation among participant teachers and the support team (Workshop 8, in Table 23.2). The evaluated aspects are the following: GREs as a didactic strategy, the methodology used in the workshops and the development of the teacher group (Science Network), to represent the voice of teachers since teacher education is a fundamental part of the professional culture (Nóvoa 2009).

Fig. 23.2 Workshop 1: target practice methodology. *Source* Andoni Arenas Martija

Fig. 23.3 Workshop 5: curriculum analysis. *Source* Camila Gajardo Tapia

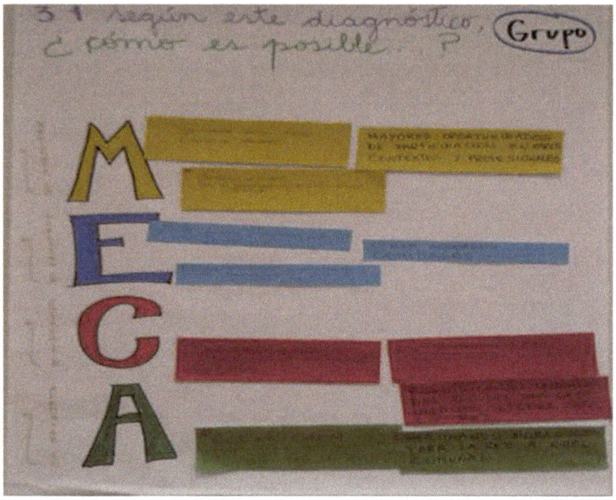

Fig. 23.4 Workshop 8: final evaluation. SODA-MECA. *Source* Jael Orrego Araya

Fig. 23.5 Workshop 7: GRE example milestone: Melinka Former Concentration Camp Site. *Source* Andoni Arenas Martija

23.5.1 The Educational Geo Routes as a Didactic Strategy

For one of the interviewed teachers (Interview 1 GC)[6] GREs are positive for both students and teachers, since they are a valid resource where their own experiences are important, and in this way, abilities are developed. In addition, GREs allow connecting the territory with the curriculum, even beyond the local area (there are students that come from other areas in Puchuncaví and even Santiago, the capital).

[6]Interviews 1 GC, 2 GC y 3 GC were carried out on November 7, 9 and 12, 2018, respectively. They were registered through audio recordings and field notes; transcribed on the 26, 14 and 27 of December 2018, respectively. The interviews were based on a flexible script as part of the Fondecyt research 1,181,834, 2017–2020 and the project DI-PUCV 39.486.

Fig. 23.6 Workshop 7: GRE example milestone: Ventanas Port. *Source* Andoni Arenas Martija

Therefore, if GREs are included in the experiential space at different levels, children will develop a local identity based on elements such as environmental and cultural heritage. This also evidences that there is a difference between the traditional field trip and the GREs: in the latter it is possible to establish clearer elements, such as the ones that result from answering the following: What are the children going to do? What will they learn, with what purpose and why?

Another Teacher states (Interview 2 GC) that GREs allow to address territorial problems, since it is impossible to educate without considering the context. For her, the texts given by the Ministry of Education are general, but it is the Teacher that has to mediate considering the group characteristics and their interests. Therefore, it is fundamental to develop a sense of belonging in students, and GRE then becomes a great strategy that can be used to mediate these learnings. In addition, she explains that one of the differences between Field Trips and GREs is that the latter combine or unify different items toward a common goal. The first ones are much simpler, and in general only one activity per trip takes place. In comparison, GREs are much more complete in several aspects, such as the evaluation, which takes place live, during the activity.

From the other perspective, another teacher (Interview 3 GC) developed, along with another colleague, a GRE based in the real history of Puchuncaví. They stated that the GRE draws from a structure that links pedagogical activities with particular objectives and contents from different disciplines. In addition, its effectiveness lies on what students experience during the activity. Therefore, it is necessary to link the curriculum with the territory because it promotes meaningful and constructive learning in which the teacher is a facilitator that puts at the students' disposal different strategies that allow them to understand the place where we live. This is enhanced

Table 23.3 GRE reusing the resources in our surroundings

Educational Geo Route chart "Reusing resources of our surroundings" Author: Karla Inzunza López—Los Maquis Elementary School Primary School (Grade 1–4)			
Educational objective: to recognize the importance of the natural surroundings and its resources, build technological instruments to reuse, reduce and recycle them at home and at school			Classes: Language, Natural Sciences, Arts and History, Social Sciences and Geography
General description	Importance of the topic: – The students value and explain the importance of using waste as a resource – They identify the different destinations of waste (reuse, recycling and reduction) – They observe different uses of natural resources in known places of the town of Puchuncaví		Route—Milestones or Stops: Milestone 1: House of actress Pia González Milestone 2: Karuna Educational Farm Milestone 3: Puchuncaví Museum
Milestone number in route: 1 "House of actress Pia González"	Location: Pucalán	City: Puchuncaví	Type of Milestone: Cultural
Pedagogical Objective: Explain the importance of adequately using resources, proposing actions and building technological instruments to reuse, reduce and recycle them at home and school	Summary Description of Milestone: The house of actress Pía González is being built with reused materials, using mud as insulation and coverage for her home. This promotes the reuse of different materials to create various objects		Student activities: – Students observe the construction of the house – Students listen to a talk about reuse of materials and the use of mud in construction – They collaboratively participate in a workshop about reutilization of materials, given by the actress
Milestone number in route: 2 "Karuna Farm"	Location: Pucalán	City: Puchuncaví	Type of Milestone: Cultural–Natural
Pedagogical Objective: To recognize the importance of natural surroundings and its resources, developing environment care and protection behaviors	Summary Description of Milestone: The Karuna farm of two hectares is located in the Pucalán area, in the town of Puchuncaví, in an agricultural circuit. The visit includes a tour through the flower, vegetable and fruit crops. Visitors learn how to harvest them and prepare natural juices with products grown in the farm		Student Activities: – Students observe the farm, guided and supported by an educational talk – They prepare homemade bread. – They share their experiences in relation to the farm visit, while eating a snack

(continued)

Table 23.3 (continued)

Milestone number in route: 3 "Puchuncaví Museum"	Location: Puchuncaví	City: Puchuncaví	Type of Milestone: Cultural–Natural
Pedagogical Objective: To recognize the importance of the natural surroundings and its resources, developing environmental care and protection behaviors	Summary Description of Milestone: The mission of the Museum of Natural History is to generate knowledge and promote the valuation and protection of the natural and cultural heritage and the environment of Puchuncavi. In addition, it seeks to be an exceptional research center of the Paleontology and Archeology Heritage		Student Activities: – Recycling Workshop in the Puchuncaví Museum: Students listen to a presentation about the benefits of recycling. Then, they create different art pieces with recycled elements

further when the different disciplines work and support each other in the construction of knowledge from their own perspectives.

Considering evaluations performed during Workshop 8 (see Table 23.2), teachers assessed as satisfactory (S) the fact that GREs can be articulated with different classes, achieving learning, an affective relationship and a sense of belonging with their surroundings. This also helps them to strengthen interpersonal relationships between students and teachers, since the development of GREs requires an active and important role of everyone. Then, in order to maintain (M) these aspects at a satisfactory level, teachers considered necessary to use this strategy at least once a year.

In regard to the discussion about opportunities (O), they identified collaborative work, the methodology per each project, research of the Historical context and the management of Google Earth as opportunities to develop GREs. Therefore, in order to exploit (E) the opportunities, teachers claim that collaborative work should be systematized, and Google Earth should be applied in the classroom as a pedagogical resource.

On the other hand, the protocols that schools have to process in order to take their kids out of the classroom, motivations and student organization to achieve the objectives and moving between the different milestones could make GREs development more difficult (D). However, in order to correct (C) some of these difficulties, the teachers agreed that if they had more planning time for the GRE, this could have been achieved.

The threats (A) that teachers visualize are related to the little time they have to carry out the activity outside the school and the limited or no access to certain places. However, to address them (A), they propose to socialize GREs among teachers prior to the implementation and to manage authorizations with more time.

Table 23.4 GRE coming back to the roots of Puchuncaví

Educational Geo Route chart “coming back to the roots of Puchuncavi” Authors: Nataly Cisternas Silva–Carolina Córdova Veas—General Velásquez Bórquez School Secondary Education			
Educational objective: To unfold the research abilities that involve identifying, processing and summarizing information from different sources; to organize relevant information regarding a topic or problem; to review different approaches in the light of new evidence and perspectives; and delay judgement in the absence of sufficient information			Classes: History, Social Sciences and Geography, Natural Sciences and Language
General description	Importance of the topic: The walking tour has as its starting point the General Velasquez School, located in the community of Puchuncaví, from where you can get to different places in the city Summary: The themes are based on the general observations of milestones that are important for the Puchuncaví city, recognizing the value of the tangible and intangible heritage of the city Social-cultural context aspects of the territory are developed, such as Monuments, historic architecture, orality, Human Rights and territorial Identity		Route—Milestones or Stops: Milestone 1: Historical context of the Puchuncaví Church Milestone 2: Puchuncaví Square and General Velásquez Monument Milestone 3: Puchuncaví Cerrillo Milestone 4: Cemetery Milestone 5: Melinka Milestone 6: Farm
Milestone number in Route: 1 “Puchuncaví Church”	Location: Puchuncaví	City: Puchuncaví	Type of Milestone: Cultural
Pedagogical Objective: To unfold the research abilities that involve identifying, processing and summarizing information from different sources; to organize relevant information regarding a topic or problem; to review different approaches in the light of new evidence and perspectives; and delay judgement in the absence of sufficient information	Summary Description of Milestone: Historical context of the Puchuncaví Church, which includes its age, how it was founded and the heritage importance that it has to our city		Student Activities: – Observation of Architecture – Ask and answer questions – Create a guide for the topic – Description questions regarding such criteria
Milestone number in route: 2 “General Velásquez Monument”	Location: Puchuncaví	City: Puchuncaví	Type of Milestone: Cultural

(continued)

Table 23.4 (continued)

Pedagogical Objective: To unfold the research abilities that involve identifying, processing and summarizing information from different sources; to organize relevant information regarding a topic or problem; to review different approaches in the light of new evidence and perspectives; and delay judgement in the absence of sufficient information	Summary Description of Milestone: This monument, built in honor of one of the Pacific War heroes, is located in the Puchuncaví Square and was built by the "Los Peucos" Hiking Association of Valparaíso. It was inaugurated on March 31 of 1934 and torn down on July 30 of 1988		Student Activities: – Observation of the Monument – Narrate and personify the story of General Velasquez – Ask and answer questions – To create a guide on the topic
Milestone number in Route: 3 "Puchuncaví Cerrillo"	Location: Puchuncaví	City: Puchuncaví	Type of Milestone: Cultural–Natural
Pedagogical Objective: To unfold the research abilities that involve identifying, processing and summarizing information from different sources; to organize relevant information regarding a topic or problem; to review different approaches in the light of new evidence and perspectives; and delay judgement in the absence of sufficient information	Summary Description of Milestone: In the nineteenth Century, at the request of the parish priest Fermín del Real, the city established a cemetery on the hill, in order to take those infested with leprosy or smallpox. According to some versions, there was a leprosarium there, brought by sailors that arrived on this land		Student Activities: – Trekking – Observation of the site – Ask and answer questions – Create a guide for the topic – Recognize the geography of the site (Medicinal Plants)
Milestone number in Route: 4 Puchuncaví Cemetery	Location: Puchuncaví	City: Puchuncaví	Type of Milestone: Cultural–Natural

(continued)

Table 23.4 (continued)

Pedagogical Objective: To unfold the research abilities that involve identifying, processing and summarizing information from different sources; to organize relevant information regarding a topic or problem; to review different approaches in the light of new evidence and perspectives; and delay judgement in the absence of sufficient information	Summary Description of Milestone: It serves the community since 1967. The Historical context is covered since it is one of the oldest cemeteries of the region, its origins dating back to colonial times		Student Activities – Tour of the site – Photo registry – Language descriptive folder
Milestone number in Route: 5 Melinka	Location: Puchuncaví	City: Puchuncaví	Type of Milestone: Cultural–Natural
Pedagogical Objective: To unfold the research abilities that involve identifying, processing and summarizing information from different sources; to organize relevant information regarding a topic or problem; to review different approaches in the light of new evidence and perspectives; and delay judgement in the absence of sufficient information	Summary Description of Milestone: Political Prisoner Camp during the Military Dictatorship that today serves as part of the Central coast memorials		Student Activities: – Trekking – Comparison and analysis with primary sources (photos and narratives)
Milestone Number in Route: 6 Los Pinos Estate	Location: Puchuncaví	City: Puchuncaví	Type of Milestone: Cultural–Natural

(continued)

Table 23.4 (continued)

Pedagogical Objective: To unfold the research abilities that involve identifying, processing and summarizing information from different sources; to organize relevant information regarding a topic or problem; to review different approaches in the light of new evidence and perspectives; and delay judgement in the absence of sufficient information	Summary Description of Milestone: Historical context is covered, recognizing the importance of agriculture and livestock farming in the Puchuncaví area. This tradition continues until today	Student Activities: Tour of the site Observation Answering Questions Guide elaboration

Table 23.5 GRE: learning about my natural–cultural heritage

Educational Geo Route chart "learning about my natural–cultural heritage" Author: Ana María Goehler Marchant—Maitencillo Primary School Primary School (Grades 5–8)			
Educational objective: To get to know places that are related to the historical, natural and cultural heritage of the Maitencillo area, in order to value its surroundings			Classes: Natural Sciences, Technology, Language and History, Social Sciences and Geography
General description	Importance of the topic: The tour will begin at the Maitencillo Elementary School, up to the Avenida del Mar, located in the Puchuncaví area, in Valparaíso region. Maitencillo is one of the oldest towns in Chile. It seems that it originally was an old Incas settlement that was later occupied by the mapuches ("the beautiful lands of the Chief Gaspar de Catapilco")		Route Milestones or Stops: Milestone 1: El abanico beach Milestone 2: Caja 18 Hotel Milestone 3: Human Treasure
Milestone Number in Route: 1 "Tourist meeting points: El Abanico Beach"	Location: Maitencillo	City: Puchuncaví	Type of Milestone: Mixed (Natural–Cultural)
Pedagogical Objective: Recognize tourist meeting points	Summary Description of Milestone: El Abanico is the center of the Maitencillo area; it holds several activities during most of the year, especially in its high season, summer. There are sports, cultural and entertainment activities		Student Activities: – Ask and answer questions about the topic – Make a photographic record – Complete a guide related to each specific milestone.
Milestone number in Route: 2 "Rescue of the main architectural and emblematic monuments of Maitencillo"	Location: Maitencillo	City: Puchuncaví	Type of Milestone: Tourist-Natural
Pedagogical Objective: Observe and appreciate the landscape, highlighting the architectural and emblematic monuments	Summary Description of Milestone: The purpose is to value and recognize sites from the past, touring them in order to rescue the history of the monuments		Student Activities: – Spatial orientation, observation of landscape – Recognize tourist meeting points such as beaches and restaurants – Ask and answer questions regarding the topic – Photo record

(continued)

Table 23.5 (continued)

Milestone number in Route: 3 "Human Treasure"	Location: Maitencillo	City: Puchuncaví	Type of Milestone: Historic–Cultural
Pedagogical Objective: To ask and answer questions to people that may express some form of local tradition (craftsmanship, music, oral literature, religious festivities, among others)	Summary Description of Milestone: To find people that express some manifestation of local tradition (craftsmanship, music, oral literature, religious festivities, among others)		Student Activities: – Explore and observe the geographical space, asking and answering questions to the people on site about topics and facts of interest (myths or legends of the place) – Photo and video records

23.5.2 Workshop Methodology

Regarding work methodology (Interview 1 GC), it is noted that never before had they addressed such relevant topics in city teacher meetings as they did this year. They observed the participation in the workshops: teachers attend with enthusiasm, offer their opinion, and everything is useful to their practices. The methodology is dynamic, participative, the team was welcoming and did not impose their knowledge; they did not present themselves from a superior position or arrogance when teaching. She argues: "*It has been a constant learning process, of collaborative work and we must take advantage of that. I hope that in every school there is a network like this since it is very meaningful to share experiences with teachers from other communities*".

Another teacher (Interview 2 GC), considering the socio-environmental conflict, stated that the workshops were very meaningful, with different group activities in the first sessions and later more individual activities in which she actively participated. For a third teacher (Interview 3 GC), their participation in the ICEC program allowed her to make sense of the contents addressed in the workshops. The methodology used was innovative and relevant since there was a theory–practice combination. The first sessions were very meaningful since they contributed with guidelines to create and develop GREs. In addition, the management of time and questions was adequate, although she mentions that teachers were still lacking in responsibility and persistence but understands that the socio-environmental conflict was a relevant cause for the suspension of a few workshops.

Upon evaluation, different impressions emerged from the group. Teachers considered as satisfactory (S) the fact that a planning matrix was proposed for the GREs, since it organizes the process and makes the curriculum transversal. In addition, they assessed the systematic support from the research team, from the proposal, construction and application of a GRE, as satisfactory. Another relevant aspect considered was that the Science Network worked, and the initially discussed purpose and activity

Table 23.6 GRE: use of water resources

Educational Geo Route chart "Use of water resources" Authors: Natalia Báez C., Valeska Bustamante B., Ximena Galleguillos B., Karolyn Gavilán M., María Gómez M., Cynthia Iturrieta O. and Jeannette Nilo F.— La Chocota Elementary School, 5th Grade			
Educational objective: To research through observation, exploration and description, the positive and negative effects caused by human activity in different water sources, giving value to this natural resource, present in some areas of Puchuncaví			Classes: Natural Sciences, Art, Language and History, Social Sciences and Geography
General description	Importance of topic: Will take place in relation to the use of water resources, fresh and salt water, with the purpose of recognizing and comparing forms of exploitation, addressing the various impacts on nature caused by human activity		Route Milestones or Stops: Milestone 1: Low Humedal (wetland) Milestone 2: Las Ventanas cove Milestone 3: Las Salinas Estate
Milestone number in Route: 1 Low Humedal (wetland)	Location: La Chocota	City: Puchuncaví	Type of Milestone: Mixed (Natural–Cultural)
Pedagogical Objective: Recognize the water resource, its use and human impact in the natural surroundings	Summary Description of Milestone: The name La Chocota comes from the Mapudungun "Eyes of water", due to the existence of a wetland in the area called "The Low". It is a land with a diversity of species that create water and land ecosystems that are also part of the supply for some of the townspeople that live close by		Student Activities: – They will observe the site, especially the area where the underground flow of water is evident – They will report as to what type of water they are observing and their reasons – Photo records of the site – Socialize with townspeople – Register observations and relevant data of the area

(continued)

Table 23.6 (continued)

Milestone number in Route: 2 "Las Ventanas cove"	Location: Las Ventanas	City: Puchuncaví	Type of Milestone: Natural–Cultural
Pedagogical Objective: To recognize the positive and negative aspects of the use of the water resource by industries	Summary Description of Milestone: Las Ventanas is a town with a long beach and bay used as a small cove. It currently is an area of tourist interest, thanks to its three beaches. Toward the south of the cove we find the industrial area, established there since 1964, and we can see the companies Puerto Ventanas S.A, CODELCO, AES Gener, among others Source: https://www.munipuchuncavi.cl/2.0/sitio10/ventanas.php		Student Activities: – Observe the water resource that is prevalent in the area – Communicate the situation of the exploited resource by the Industrial Park – Make predictions about the use of the resource during the Old Age – Observe the booth used for the exploitation of the resource in past decades – Socialize about the negative and positive effects of the use of the resource by the industry – Photo and written record of the observations
Milestone number in Route: 3 "Las Salinas Estate"	Location: Maitencillo	City: Puchuncaví	Type of Milestone: Natural–Cultural
Pedagogical Objective: Recognize the use of the water resource in older ages and compare it to its current exploitation	Summary Description of Milestone: The Las Salinas estate, right next to the Campiche estuary, is a salt producer that operates with pre-Hispanic techniques since its creation (S. XIX) Source: puchuncavi.blogspot.com/2006/01/las-salinas-de-la-greda.html		Student Activities: – Guided tour through the estate – Compare the environmental impact – Photo and written record of observations

Table 23.7 GRE: from the coastal zone to the heights of a hill

Educational Geo Route chart "From the coastal zone to the heights of a hill" Authors: M. Carolina Báez Donoso and Irene Constanza Chacana Farías La Laguna Elementary School. Primary School (Grades 5–8)			
Educational objective: To develop a relationship of the students with the natural, cultural and social surroundings of their community			Classes: History, Social Sciences and Geography, Natural Science and Physical Education
General description	Importance of the topic: The importance of the following Geo Route is related to generating a sense of belonging in the students, toward the Puchuncaví and La Laguna communities, developing in them a relationship with the natural, cultural and social surroundings of their community. We begin at the museum, enjoying the great scientific richness of the region and then continue to the La Laguna wetland where we can find flora and fauna that is characteristic of the area and that make it possible to study birds. The tour ends with a trip to the El Rincon mine		Route Milestones or Stops: Milestone 1: Puchuncaví Museum Milestone 2: La Laguna de Puchuncaví wetland Milestone 3: El Rincón mine
Milestone number in Route: 1 "Puchuncaví Museum"	Location: Puchuncaví	City: Puchuncaví	Type of Milestone: Cultural–Natural
Pedagogical Objective: Recognize the cultural and Historical heritage of the Puchuncaví area through a museum visit	Summary Description of Milestone: The Puchuncaví museum is an institution that preserves the identity of the city, so it becomes necessary to get to know its legacy and the activities they develop		Student Activities: – Listening to explanations regarding the theme of each classroom – Excavate in order to find remains of fossils – Visit the Quirilluca cliffs
Milestone number in Route: 2 "Humedal"	Location: La Laguna	City: Puchuncaví	Type of Milestone: Natural–Cultural

(continued)

Table 23.7 (continued)

Pedagogical Objective: To explain the consequences of erosion on the surface of the earth, identify the agents that cause it, such as wind, water and human intervention	Summary Description of Milestone: The wetland is connected to the Catapilco estuary which runs into the Pacific Ocean. There we can find birds such as the small Tagua (red-gartered coot), the small and big heron, ducks and cormorants		Student Activities: – Through observation, students generate hypothesis about the erosion of some sectors of the wetland – Photo and written record of the site – Research which were the causes for erosion in the previously observed areas – Ask questions to local inhabitants – Come to conclusions about the type of erosion existing in the area, why did erosion happen, attaching their photos to a written report
Milestone number in Route: 3 "El Rincón mine[a]"	Location: El Rincón	City: Puchuncaví	Type of Milestone: Natural–Cultural
Pedagogical Objective: Explain the consequences of erosion on the Surface of the Earth, identifying the causing agents such as wind, water and human activity. 6th grade. Perform physical activity in contact with nature, promoting the care of our surroundings	Summary Description of Milestone: Entrance to the mine is prohibited due to safety reasons. In order to arrive there, a road built by the old mining workers must be traveled. This is located around 400 m above sea level where the mine shaft is. This is an abandoned copper mine		Student Activities: – We proposed a pedagogical trekking activity for the students

[a]The authors thank maestro Martín Pérez for his help in developing this milestone

schedule was fulfilled. Lastly, they assessed as satisfactory the fact that leadership was distributed. In order to maintain (M) these satisfactory aspects, teachers decided to continue with their Science Network meetings and establish as their work methodology the participative work they had performed, maintaining the distributed leadership based on co-responsibility.

Regarding opportunities (O), teachers noted how innovative the Methodology was and that it had allowed them to approach different places, strategies and experiences, such as their participation in seminars. They considered the application of new technologies related to the GRE an opportunity. To exploit (E) the opportunities, teachers expressed their need for continuous improvement and constantly enriched the methodology, participating in seminars at the city, regional and national levels.

Table 23.8 Name of Schools of the Puchuncaví City and Educational Geo Routes. The detail of their location is featured in Fig. 23.8

Schools			
Quadrant A	ID		ID
Sarge Aldea Educational Complex	1	El Rincón School	7
The Chocota School	2	Los Maquis School	8
Horcón School	3	El Rungue School	9
La Greda School	4	La Quebrada Elementary School	10
Campiche School	5	Maitencillo Elementary School	11
General José Velásquez School	6	La Laguna School	12
Milestones–geo routes			
GEO ROUTES 1	ID	GEO ROUTES 3	ID
El Abanico Beach	H1	Humedal Bajo	H1
Caja 18 de September Hotel	H2	Las Ventanas Cove	H2
Don Ruperto Salazar Bernal	H3	Las Salinas Estate	H3
GEO ROUTES 2	ID	GEO ROUTES 4	ID
Puchuncaví Museum	H1	Puchuncaví Church	H1
Humedal	H2	General Velásquez Monument	H2
Mine El Rincón	H3	Puchuncaví Cerrillo	H3
		Puchuncaví Cemetery	H4
		Melinka (Former Prisoner Camp)	H5

Source Own elaboration

The main difficulties (D) were the Yellow Alert originated by the socio-environmental conflict that caused the suspension of school classes for more than a month and the irregular attendance of the whole group that prevented the development of a systematic process of the workshops. However, in order to correct (C) some of these difficulties, the group suggested selecting a day to meet at the city level and have a "plan b" to address emergencies. Also, to have an effective coordination between the Science Network and the leaders at the schools with the purpose of informing what they are working on, the school can provide the necessary spaces and times to develop collaborative work.

Finally, threats (A) that the teachers observe in the workshop methodologies correspond to the time for creation and coordination, low attendance and the Yellow Alert that interrupted the systematization of workshops. To address the mentioned threats, the teachers suggested incentives to improve attendance and the demand for new environmental policies to be applied in the city.

Fig. 23.7 Workshop 2: participative cartography *Source* Andoni Arenas Martija

23.5.3 About the Group of Teachers

For one of the teachers (Interview 1 GC), the group that was formed this year has been a very good group: teachers are friendly, welcoming, hard workers and willing to work in teams. Together with some teachers (from 3 different schools), they agreed on applying the GREs with their students, but due to the socio-environmental conflict it was impossible to make the necessary coordination. They have worked collaboratively within the network. They have all worked with the best disposition. However, in order for collaborative work to be effective, workshops should take place more frequently, establishing places and schedules for the whole year. She thinks the group could continue, with or without the support of the university team to carry out the workshops, since they should not depend on what comes from outside the network. If the Direction of Administration of Municipal Education (DAEM) does not continue to support this Network, the need to meet and maintain the group should emerge from the teachers.

Another opinion (Interview 2 GC) states that the collaboration that has taken place has not been much, although if there are questions they would consult with the members they feel closer to, the ones they think can work with others and not those who tend to work by themselves. They can count on them, but they don't have the closeness or the confidence to ask for help. This teacher also proposes ideas about organization and sense of the group: the DAEM should allocate the work hours at

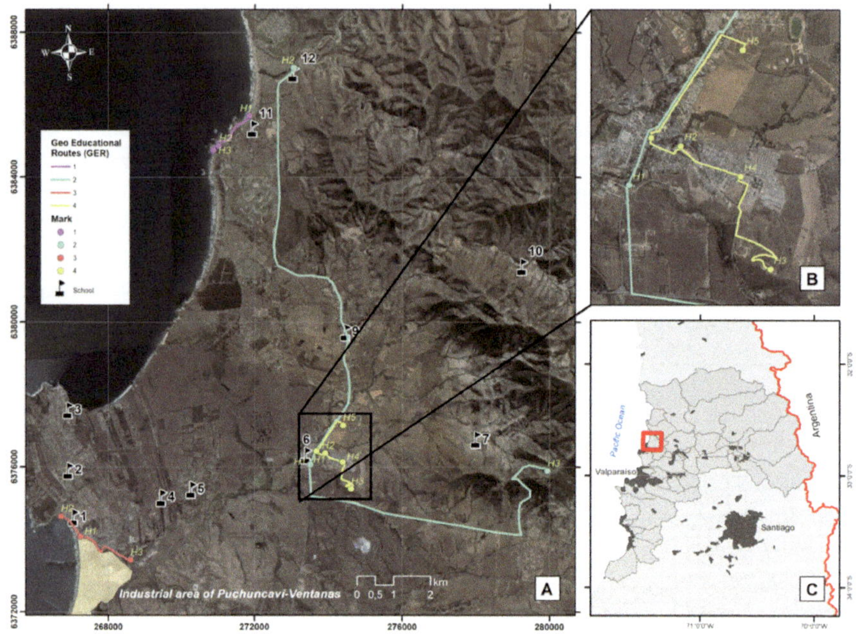

Source: Own elaboration

Fig. 23.8 Location of educational geo routes.

least once every two months, that would make the initiative more constant, and give meaning to the activities. For her, if these two elements existed as a group, great goals could be achieved, considering that when working with others great work can be done. She indicated that belonging is important too since it changes the disposition, the group feels more of their own and not imposed.

Another teacher (Interview 3 GC) addressed the internal nuances of the group indicating that they could have gotten to know each other better, since they didn't have the time–space to exchange ideas and/or plan in a more in-depth way. They think there was a subgroup that had more effective feedback among them for the GREs, because they knew each other better so there was a trusting environment which led to collaborative work. For her, teachers should generate collaborative work opportunities, in which the necessary accuracy exists in order to protect the time and disposition of each member.

In the general evaluation, the formation of the group has had some satisfactory elements (S): good relationships, development of communication, the exchange of meaningful experiences between the teachers and with the research team. Something important is the development of an effective collaborative work that has allowed to grow in knowledge and make it possible for some teachers to participate in seminars. To maintain (M) these satisfactory aspects, they mention their desire to continue in 2019 with the same work team, regularly attend meetings and continue with the support of the research team.

Regarding opportunities (O) acknowledged by the group, teachers consider as relevant to have the possibility of reflecting around a collective purpose and exchange experiences and realities considering that they represent various schools, as well as the opportunity of working around the same disciplines. They expressed that to exploit (E) these opportunities, they could continue the interaction with other teachers and guides, make city-level agreements, in addition to sharing non-successful experiences to promote feedback within the group.

The difficulties (D) the group observes are related to the lack of basic technical vocabulary for scientific instruments, intermittent attendance and meeting times. In order to correct (C) these difficulties, we proposed a scientific instruments literacy workshop and an annual plan with an internal organization for the Network. This organization, according to some members of the group, is not the responsibility of the group but of the DAEM coordination, which shows that there is disagreement regarding the leadership in the development of the meetings.

The threats (A) that were identified correspond to the time each teacher had to meet, which was not always supported by school leadership, causing rotations among participant teachers. These obstacles can be addressed (A) by DAEM management by assigning set hours for the group to meet and develop their work.

23.6 Conclusions

Considering the presented evidence through the designed GREs, interviews, final evaluation and reflections of the authors, the following questions were asked in order to systematize the conclusions as answers to them:

Do GREs *have the required characteristics to be considered Teaching and Learning Strategies?*

The GREs designed by the teachers have most of the characteristics of Teaching and Learning Strategies. From more instrumental aspects, such as the systematization of processes through the reformulated planning chart, to those features related to the opportunity provided to students to develop their abilities.

We observe that the GRE activities promote opportunities for knowledge, interaction, observation and reflection, fostered and generated by teachers in a reflective and flexible way (Estrada 2014). It is a strategy to build not only knowledge of a specific topic, but also activities that teachers carry out intentionally for the learning of their students (Feo 2010).

They potentiate knowledge about science, geographic, social and cultural situations by considering contextual and personal factors of the target group, developing topics that are familiar and pertinent to a sequence, an order given by certain criteria (Zabala 2001). The real value of the GREs as learning strategies will be evidenced when they are implemented during the 2019 school year.

Are they present in the attributes of the three sources of GRE, Field Word, Heritage Tourist Routes and Pedagogical Field Trips?

Shared features with Field Work are the *in-situ* collection of information, connecting academic content with everyday situations that at the same time promote relationship bonds and a territorial sense of belonging. GREs also project the need to assign meaning to curricular contents, taking the classroom to other places.

In relation to the Geo-Tourist Routes, we take the idea of connecting different milestones that together as a group conform a unit, that is toured and experienced directly, since they are representative elements of the said unit. Of the Pedagogical Field Trips, we value that the world to be known has to be explored in order to describe its characteristics, explain its components and understand its spatial interrelations (Godoy and Sánchez 2007). We must then observe, use a different array of techniques, instruments and methodological procedures with educational purposes, according to Scoffham (2010).

How and to what extent does the territory relate to the school curriculum?

The territory offers a pedagogical potential that can be related to the curriculum at different levels, from using some elements of the territory to understanding one or several curricular contents, to having one or several curricular contents to understand the territory and its identity, as can be seen in the GREs. This becomes easier when we understand that teaching and learning are not isolated from the context, and therefore designing and developing activities for some classes allow a smooth approach between school and the student context, as defined by teachers in the interviews and final evaluation (Workshop 8).

GREs draw on the need for developing abilities, competences and learning in the space one inhabits. They teach/learn, from experience, where space not only is an instrument for illustration. It allows to problematize the territory since teachers selected the milestones they acknowledged as essential, even controversial, promoting reflection regarding facts that happen in their own context, addressing criticism received to textbooks, activity guides, standardized tests, in relation to the lack of context of the contents.

How, and to what extent, did collaborative work take place?

From the teacher perspective, collaborative work happened every time they performed group activities during the workshops. For example, during Workshop 1 (target practice) collaboration was evidenced when teachers agreed and ranked their points of view regarding learning and the difficulties in the scientific-geographic education (see Fig. 23.2). During the development of the participative cartography in Workshop 2 (see Fig. 23.3), teachers worked collaboratively analyzing the potential and problems of the territorial context in Puchuncaví and decided as a group on representative sites for the participant educational community to analyze.

Collaborative work was evidenced when teachers shared their GRE progress during Workshop 3 (see Fig. 23.4), giving feedback through comments and suggestions in order to strengthen the design. It is important to highlight their willingness to listen, understand and contribute to the work of others, centering their opinions around the ideas and not the personal aspects of who was presenting those ideas.

The third element in collaboration emerged between the support team and participant teachers, like the diagnosis about the difficulties of teaching Sciences and Geography, and the learning they have acquired by teaching them (Workshop 1), in the process of construction of the GREs, with the reformulation of the work chart, in which teachers proposed improvements and changes to the document (Workshop 3 and 4). This is in addition to the different decisions that were made as a group (Workshop 6).

What can we say about teaching and learning of Sciences and Geography from this experience?

The participative construction of the Educational Geo Routes has allowed, in the case of Natural Science teaching, the connection between some elements of the Scientific method and the development of the activities at the milestones: formulation of hypothesis and exploration. For the teaching of Geography, it has strengthened in teachers the acknowledgement of the territorial Identity that contextualizes the Puchuncaví schools, supporting the educational task from the students' own reality.

The need to generate new didactic strategies and apply new teaching methods led teachers to think about new ways of interest for students with the purposes of scientific education, overcoming their uneasiness regarding traditional ways of improvement. The designed GREs favor the development of inquiry abilities and competences (Scientific, Geographical and Historical), with the possibility of connecting different classes with others from practical and meaningful learning.

What can we say about Teacher Professional Development from this experience?

Within the context of the development of the GREs, teachers discussed the potential of networking within the educational community and toward other schools in a relationship that emerged from the signified milestones and created routes. In this way, collaboration with other teachers is promoted, both from the same School and other schools. This collaboration develops better professional capabilities in participant teachers.

It is possible to evidence in this process, as stated by Vezub (2013), that expert knowledge and educational research knowledge is nurtured and articulated with the practical knowledge of teachers, which becomes a specific way of supporting Teacher Professional Development. When considering the proposals of Sprinthall et al. (1996), Justi and Van Driel (2006) to explain models associated with professional training and development, we can say that the work dynamic that took place between the research team and the group of teachers became a strategy that intends to approach interactive models by centering in the recovery of practical knowledge, experience, and teacher needs and problems (Vezub 2013).

Acknowledgments The authors would like to thank The National Fund for Science and Technology for funding the research project: "Teacher Professional Development Trajectories in the field of science education: (re)framing the contribution of collaborative work" (FONDECYT N° 1181834, 2017–2020) in which context this text has been elaborated.

To the Research Department of the Pontificia Universidad Católica de Valparaíso, which through the funding of the project: "Learn how to research by researching: Development of Research Abilities in Science Teaching. 39.486" of the Undergraduate Interdisciplinary Research Fund 2018 allowed the participation of Camila Gajardo, Carolina Ortiz and Jael Orrego in the Fondecyt research project mentioned above.

To the Department of Administration of Municipal Education (DAEM) of Puchuncaví and especially to Elizabeth Franjola, for the creation and support of the Science Network.

References

Anijovich R (2010) La retroalimentación en la evaluación. In: Anijovich R (ed) La evaluación significativa. Paidós, Buenos Aires, pp 129–149

Arenas-Martija A (2009) Las estrategias, actividades y tareas de enseñanza en el aprendizaje comprensivo de la Historia, la Geografía y las Ciencias Sociales Escolares. Tesis Doctoral, Facultad de Educación, Universidad de Alcalá, Madrid, España. Sin Publicar. https://hdl.handle.net/10017/9041. Accessed 10 Dec 2018

Arenas-Martija A, Margalef-García L (2006) ¿Qué entendemos por innovación educativa? A propósito del Desarrollo Curricular. Perspect Educ 47:13-31

Arenas-Martija A, Figueroa-Sterquel R, Salinas-Silva V (2018) Educación geográfica y formación ciudadana. Rev Interdiscip do Direitos Humanos 6:17–66

Ávalos B (2007) Formación Docente Continua y factores asociados a la política educativa en América latina y el Caribe. Informe preparado para el Diálogo Regional de Política, Banco Interamericano de Desarrollo, pp 1–43

Azócar P (2016) Nuevas prácticas cartográficas: democratización de la cartografía mediante las geotecnologías y su impacto en el desarrollo local. Rev de Estudios Políticas Estratégicas 4:54–71

Collet-Sabé J, Subirats J (2016) Educación y territorio: 15 años de Proyectos Educativos de Ciudad (PEC) en Cataluña (España). Análisis, evaluación y perspectivas. Scr Nova XX:1–23. https://www.ub.edu/geocrit/sn/sn-532.pdf. Accessed 10 Dec 2018

Copatti C, Ziech M, Copetti H (2018) Identidad y pertenencia en Escuelas del Campo: vivencias cotidianas en relación con el lugar. Anekumene 11:38–47. https://revistas.pedagogica.edu.co/index.php/anekumene/article/view/8043. Accessed 10 Dec 2018

Díaz-Barriga F, Hernández G (2010) Estrategias docentes para un aprendizaje significativo. Una interpretación constructivista. McGraw-Hill, México

Estrada E (2014) Factores que contribuyen y dificultan el desarrollo de la enseñanza aprendizaje mediada por dispositivos móviles en instituciones de educación superior en Colombia. Maestría en informática educativa, centro de tecnologías para la academia, Universidad de la Sabana, Chía, Colombia. https://intellectum.unisabana.edu.co/handle/10818/11596. Accessed 10 Dec 2018

Feo R (2010) Estrategias instruccionales para promover el aprendizaje estratégico en estudiantes del Instituto Pedagógico José Manuel Siso Martínez Sapiens. Rev Univ de Investig 11(2):90–112

Fernández M (2008) Formación docente continua y procesos de transformación curricular. Aportes para pensar el cambio y la innovación en la enseñanza de la geografía. Enseñanza de las Ciencias Soc 7:99–107

Figueroa-Sterquel R, Negrete-Sepúlveda J, Moreira-Muñoz A (2021) New routes of Geotourism for the La Campana-Peñuelas Biosphere Reserve, Chile. In Singh RB, Anand WS, Majumdar X (eds) Global Geographical Heritage, Geoparks and Geotourism. Series Title: Advances in Geographical and Environmental Sciences. Springer Nature Singapore Pte Ltd. ISBN:978-981-15-4955-7. London

Godoy I, Sánchez A (2007) El trabajo de campo en la enseñanza de la Geografía. Rev Univ de Investig Año 8 2:137–146
González V (2003) Estrategias de enseñanza y aprendizaje. Editorial Pax México, México
Grindsted T, Madsen L, Nielsen T (2013) One just better understands.... when standing out there': fieldwork as a learning methodology in University Education of Danish Geographers. Rev Int Geogr Educ Online 3. https://rigeo.org/vol3no1/RIGEO-V3-N1-1.pdf. Accessed 10 Dec 2018
Imbernón F (2011) Un nuevo desarrollo profesional del profesorado para una nueva educación. Rev de Ciencias Humanas 12:75–86. Agosto, Chile. https://www.revistas.fw.uri.br/index.php/revistadech/article/view/343/622. Accessed 10 Dec 2018
Jacobs H (2004) Getting results with curriculum mapping. Association for Supervision and Curriculum Development. Alexandria: Virginia, USA
Justia R, Van Driel J (2006) The use of the interconnected model of teacher professional growth for understanding the development of science teachers' knowledge on models and modelling. Teach Teach Educ 22:437–450
Kvale S (2011) Las entrevistas en investigación cualitativa. Ediciones Morata S. L, Madrid
Lara de González S (2011) El trabajo de campo desde la perspectiva del docente. Rev Univ de Investig 1:76–93
López de Ceballos P, García H, Merlo J (2001) Un método de evaluación formativa en el campo social. Madrid: Popular
Miranda C, Rivera P (2009) Formación permanente de profesores: ¿Quién es el formador de formadores? Estudios Pedagógicos XXXV(1):155–169
Miranda C, Rivera P, Salinas R, Muñoz E (2010) ¿Que hace a la formación permanente de profesores eficaz?: Factores que inciden en su impacto. Estudios Pedagógicos XXXVI(2):135–151
Mohamed M, Pérez M, Montero M (2017) Salidas pedagógicas como metodología de refuerzo en la Enseñanza Secundaria. ReiDoCrea 6:194–210. https://www.ugr.es/~reidocrea/6-16.pdf. Accessed 10 Dec 2018
Nóvoa A (2009) Para una formación de profesores construida dentro de la profesión. Rev de Educ 359:203–218
Parga D (2016) El continuo de la formación del profesorado de ciencias. Tecné, Episteme y Didaxis: TED (40):1–12. https://www.scielo.org.co/scielo.php?script=sci_arttext&pid=S0121-38142016000200001&lng=en&tlng=es. Accessed 10 Dec 2018
Pedone C (2000) El trabajo de campo y los métodos cualitativos. Necesidad de nuevas reflexiones desde las geografías latinoamericanas. Scripta Nova 4:55–78. https://revistes.ub.edu/index.php/ScriptaNova/issue/view/16. Accessed 10 Dec 2018
Robinson A, Morrison J, Muehrcke P, Kimerling A, Guptill S (1995) Elements of cartography, 6th edn. John Wiley and Sons Inc., New York
Romero M (2010) El aprendizaje experiencial y las nuevas demandas formativas. Rev de Antropol Exp 10:89–101
Sánchez A (1995) El trabajo de campo y las excursiones. In: Moreno A (ed) Enseñar Geografía: de la teoría a la práctica. Síntesis, Madrid, pp 159–184
Scoffham S (ed) (2010) Primary geography handbook. Geographical Association, Sheffield
Sprinthall NA, Reiman AJ, Thies-Sprinthall L (1996) Teacher professional development. In: Sikula J, Buttery TJ, Guyton E (eds) Handbook of research on teacher education. Macmillan Library Reference, New York, pp 666–703
Vargas L, Bustillos G (1990) Técnicas participativas para la educación popular. Centro de Investigaciones y Desarrollo de la Educación, CIDE, Santiago
Vezub L (2013) Hacia una pedagogía del desarrollo profesional docente. Modelos de formación continua y necesidades formativas de los profesores. Páginas de Educación 6(1):1–31
Winne P, Ronald M (1989) A cognitive-processing analysis of motivation within classroom tasks. In: Ames G, Ames R (eds) Research on Motivation in education. Academic Press, Orlando, pp 223–257
Zabala A (2001) Cómo trabajar los contenidos procedimentales en el aula. Graó ICE, Barcelona